Lecture Notes in Artificial Intelligence 8272

Subseries of Lecture Notes in Computer Science

Stephen Cranefield   Abhaya Nayak (Eds.)

AI 2013: Advances in Artificial Intelligence

26th Australasian Joint Conference
Dunedin, New Zealand, December 1-6, 2013
Proceedings

Volume Editors

Stephen Cranefield
University of Otago
Department of Information Science
Dunedin, New Zealand
E-mail: stephen.cranefield@otago.ac.nz

Abhaya Nayak
Macquarie University
Department of Computing
Sydney, NSW, Australia
E-mail: abhaya.nayak@mq.edu.au

ISSN 0302-9743 e-ISSN 1611-3349
ISBN 978-3-319-03679-3 e-ISBN 978-3-319-03680-9
DOI 10.1007/978-3-319-03680-9
Springer Cham Heidelberg New York Dordrecht London

Library of Congress Control Number: Applied for

CR Subject Classification (1998): I.2, H.3-4, F.1, H.2.8, I.4-5, J.3

LNCS Sublibrary: SL 7 – Artificial Intelligence

Typesetting: Camera-ready by author, data conversion by Scientific Publishing Services, Chennai, India

Printed on acid-free paper

Springer is part of Springer Science+Business Media (www.springer.com)

Preface

This volume contains the papers presented at the 26th Australasian Joint Conference on Artificial Intelligence (AI 2013). The conference was held from 3-6 December 2013 in Dunedin, home to the University of Otago in the South Island of New Zealand. This annual conference remains the premier event for artificial intelligence researchers in the Australasian region, and it is only the second time in its 26-year history that it was held outside Australia. It was co-located with the 16th International Conference on Principles and Practice of Multi-Agent Systems (PRIMA 2013).

AI 2013 received 120 submissions with authors from 34 countries. Each submission was reviewed by at least three Program Committee members or external referees. Subsequent to a thorough discussion and rigorous scrutiny by the reviewers and the dedicated members of the Senior Program Committee, 54 submissions were accepted for publication: 35 as full papers and 19 as short papers. The acceptance rate was 29% for full papers and 45% overall (including short papers).

AI 2013 featured keynote speeches by two eminent scientists. Fangzhen Lin (Hong Kong University of Science and Technology) talked about the connection between satisfiability and linear algebra. Pascal Van Hentenryck (NICTA), on the other hand, spoke of the role that optimisation has to play in effective disaster management.

Four workshops with their own proceedings were held on the first day of the conference:

- The Third Australasian Workshop on Artificial Intelligence in Health (AIH 2013)
- The Workshop on Machine Learning for Sensory Data Analysis (MLSDA'13)
- The 4th International Workshop on Collaborative Agents — Research and Development (CARE 2013)
- The 16th International Workshop on Coordination, Organisations, Institutions and Norms in Agent Systems (COIN@PRIMA2013)

These workshops were complemented by a tutorial on "Theory and Applications of State Space Models for Time Series Data", presented by Peter Tino (University of Birmingham).

AI 2013 would not have been successful without the support of authors, reviewers, and organisers. We thank the many authors for submitting their research papers to the conference, and are grateful to the successful authors whose papers are published in this volume for their collaboration during the preparation of final submissions. We thank the members of the Program Committee and the external referees for their expertise and timeliness in assessing the papers. We also thank the organisers of the workshops and the tutorial for their commitment and dedication. We are very grateful to the members of the Organising

Committee for their efforts in the preparation, promotion, and organisation of the conference. We acknowledge the assistance provided by EasyChair for conference management, and we appreciate the professional service provided by the Springer LNCS editorial and publishing teams.

September 2013

Stephen Cranefield
Abhaya Nayak

Organisation

General Chairs

Michael Winikoff	University of Otago, New Zealand
Abdul Sattar	Griffith University, Australia

Program Chairs

Stephen Cranefield	University of Otago, New Zealand
Abhaya Nayak	Macquarie University, Australia

Finance Chair

Stephen Hall-Jones	University of Otago, New Zealand

Workshop Chair

Alistair Knott	University of Otago, New Zealand

Tutorial Chair

Lubica Benuskova	University of Otago, New Zealand

Doctoral Consortium Chair

John Thangarajah	RMIT University, Australia

Publicity Chair

Brendon Woodford	University of Otago, New Zealand

Local Arrangements Chair

Bastin Tony Roy Savarimuthu	University of Otago, New Zealand

Website

Mariusz Nowostawski	Gjøvik University College, Norway
Heather Cooper	University of Otago, New Zealand

Sponsorship Chair

Mike Barley	University of Auckland, New Zealand

Senior Advisors

Nikola Kasabov	Auckland University of Technology, New Zealand
Mengjie Zhang	Victoria University of Wellington, New Zealand
Hans Guesgen	Massey University, New Zealand
Wai Kiang Yeap	Auckland University of Technology, New Zealand

Senior Program Committee

Hussein Abbass	UNSW Canberra @ ADFA, Australia
James Bailey	University of Melbourne, Australia
Longbing Cao	University of Technology, Sydney, Australia
Jeremiah D. Deng	University of Otago, New Zealand
Byeong-Ho Kang	University of Tasmania, Australia
Jimmy Lee	Chinese University of Hong Kong, SAR China
Michael Maher	UNSW Canberra @ ADFA, Australia
Thomas Meyer	CAIR, UKZN and CSIR Meraka, South Africa
Mehmet Orgun	Macquarie University, Australia
Mikhail Prokopenko	CSIRO, Australia
Claude Sammut	University of New South Wales, Australia
Abdul Sattar	Griffith University, Australia
Sylvie Thiebaux	Australian National University and NICTA, Australia
Michael Thielscher	University of New South Wales, Australia
Dianhui Wang	La Trobe University, Australia
Chengqi Zhang	University of Technology, Sydney, Australia
Dongmo Zhang	University of Western Sydney, Australia
Mengjie Zhang	Victoria University of Wellington, New Zealand

Program Committee

Shafiq Alam	University of Auckland, New Zealand
Grastien Alban	NICTA, Australia
Quan Bai	Auckland University of Technology, New Zealand
Yun Bai	University of Western Sydney, Australia
Lubica Benuskova	University of Otago, New Zealand
Wei Bian	University of Technology, Sydney, Australia
Ivan Bindoff	University of Tasmania, Australia
Blai Bonet	Universidad Simón Bolívar, Venezuela

Wray Lindsay Buntine	NICTA, Australia
Lawrence Cavedon	NICTA and RMIT University, Australia
Jeffrey Chan	University of Melbourne, Australia
Ling Chen	University of Technology, Sydney, Australia
Songcan Chen	Nanjing University of Aeronautics and Astronautics, China
Geoffrey Chu	University of Melbourne, Australia
Vic Ciesielski	RMIT University, Australia
Dan Corbett	Optimodal Technologies, USA
Xuan-Hong Dang	Aarhus University, Denmark
Hepu Deng	RMIT University, Australia
Grant Dick	University of Otago, New Zealand
Clare Dixon	University of Liverpool, UK
David Dowe	Monash University, Australia
Atilla Elçi	Aksaray University, Turkey
Cesar Ferri	Universitat Politècnica de València, Spain
Eibe Frank	University of Waikato, New Zealand
Tim French	University of Western Australia, Australia
Marcus Gallagher	University of Queensland, Australia
Junbin Gao	Charles Sturt University, Australia
Xiaoying Gao	Victoria University of Wellington, New Zealand
Yang Gao	Nanjing University, China
Edel Garcia	CENATAV, Cuba
Enrico Gerding	University of Southampton, UK
Manolis Gergatsoulis	Ionian University, Greece
Guido Governatori	NICTA, Australia
Christian Guttmann	IBM Research, Australia
Patrik Haslum	Australian National University
Bernhard Hengst	University of New South Wales, Australia
José Hernández-Orallo	Universitat Politècnica de València, Spain
Geoffrey Holmes	University of Waikato, New Zealand
Xiaodi Huang	Charles Sturt University, Australia
Mark Johnston	Victoria University of Wellington, New Zealand
George Katsirelos	INRA, Toulouse, France
C. Maria Keet	University of KwaZulu-Natal, South Africa
Yang Sok Kim	University of New South Wales, Australia
Michael Kirley	University of Melbourne, Australia
Frank Klawonn	Ostfalia University of Applied Sciences, Germany
Thomas Kleinbauer	Monash University, Australia
Reinhard Klette	University of Auckland, New Zealand
Alistair Knott	University of Otago, New Zealand
Mario Koeppen	Kyushu Institute of Technology, Japan
Kevin Korb	Monash University, Australia
Norbert Krueger	University of Southern Denmark
Rudolf Kruse	University of Magdeburg, Germany

Gerhard Lakemeyer	RWTH Aachen University, Germany
Jérôme Lang	Université Paris-Dauphine, France
Nung Kion Lee	Universiti Malaysia Sarawak, Malaysia
Li Li	Southwest University of China
Yuefeng Li	Queensland University of Technology, Australia
Wei Liu	University of Melbourne, Australia
Wei Liu	University of Western Australia, Australia
Xudong Luo	Sun Yat-sen University, China
Eric Martin	University of New South Wales, Australia
Michael Mayo	University of Waikato, New Zealand
Brendan McCane	University of Otago, New Zealand
Nina Narodytska	University of Toronto and University of New South Wales, Australia
Richi Nayak	Queensland University of Technology, Australia
Kourosh Neshatian	University of Canterbury, New Zealand
Vinh Nguyen	University of Melbourne, Australia
Kouzou Ohara	Aoyama Gakuin University, Japan
Lionel Ott	University of Sydney, Australia
Laurence Park	University of Western Sydney, Australia
Adrian Pearce	University of Melbourne, Australia
Laurent Perrussel	Université de Toulouse, France
Bernhard Pfahringer	University of Waikato, New Zealand
Duc-Nghia Pham	Griffith University, Australia
Sarvapali Ramchurn	University of Southampton, UK
Mark Reynolds	University of Western Australia
Deborah Richards	Macquarie University, Australia
Juan A. Rodriguez-Aguilar	IIIA, Spanish National Research Council, Spain
Ji Ruan	Auckland University of Technology, New Zealand
Torsten Schaub	University of Potsdam, Germany
Daniel Schmidt	University of Melbourne, Australia
Rolf Schwitter	Macquarie University, Australia
Yanir Seroussi	Monash University, Australia
Lin Shang	Nanjing University, China
Hannes Straß	Leipzig University, Germany
Maolin Tang	Queensland University of Technology, Australia
John Thornton	Griffith University, Australia
James Underwood	University of Sydney, Australia
Ben Upcroft	Queensland University of Technology, Australia
Jaime Valls Miro	University of Technology, Sydney, Australia
Pascal Van Hentenryck	NICTA and University of Melbourne, Australia
Pradeep Varakantham	Singapore Management University
Ivan Varzinczak	CAIR, UKZN and CSIR Meraka, South Africa
Wamberto Vasconcelos	University of Aberdeen, UK
Karin Verspoor	NICTA and University of Melbourne, Australia
Meritxell Vinyals	University of Southampton, UK

Kewen Wang	Griffith University, Australia
Peter Whigham	University of Otago, New Zealand
Wayne Wobcke	University of New South Wales, Australia
Brendon J. Woodford	University of Otago, New Zealand
Bing Xue	Victoria University of Wellington, New Zealand
Roland Yap	National University of Singapore
Rui Zhang	University of Melbourne, Australia
Zili Zhang	Southwest University and Deakin University, China/Australia
Yi Zhou	University of Western Sydney, Australia
Zhi-Hua Zhou	Nanjing University, China

Additional Reviewers

Benjamin Andres
Alex Bewley
Christian Braune
Giovanni Casini
Raphael Cobe
Matthew Damigos
Meng Fang
James Hales
Nader Hanna
Daniel Harabor
Yujing Hu
Jin Huang
Jing Huo
Jing Jiang
Holger Jost
Manolya Kavaklı
Phil Kilby
Szymon Klarman
Carlo Kopp
Nan Li
Martin Liebenberg
Kar Wai Lim
Mufeng Lin
Chunyang Liu
Dongwei Liu
Mingxia Liu
Linda Main
Christian Moewes
Christos Nomikos
Yuki Osada
Jianzhong Qi
Chao Qian
Gavin Rens
Mahdi Rezaei
Valentin Robu
Javier Romero
Orkunt Sabuncu
Christoph Schwering
Yinghuan Shi
Bok-Suk Shin
Yu Shyang Tan
Junli Tao
Huihui Wang
Liping Wang
Zhe Wang
Zeyi Wen
Andy Yuan Xue
Wanqi Yang
Xiangfu Zhao
Zhiqiang Zhuang

Sponsoring Institutions

Asian Office of Aerospace Research and Development, US Air Force Office of Scientific Research
University of Otago
University of Auckland

Abstracts of Keynote Talks

From Satisfiability to Linear Algebra

Fangzhen Lin

Hong Kong University of Science and Technology

Satisfiability of boolean formulas (SAT) is an interesting problem for many reasons. It was the first problem proved to be NP-complete by Cook. Efficient SAT solvers have many applications. In fact, there is a huge literature on SAT, and its connections with other optimisation problems have been explored. In this talk, I discuss a way to map clauses to linear combinations, and sets of clauses to matrices. Through this mapping, satisfiability is related to linear programming, and resolution to matrix operations.

Computational Disaster Management

Pascal Van Hentenryck

NICTA

The frequency and intensity of natural disasters have significantly increased over the past decades and this trend is predicted to continue. Natural disasters have dramatic impacts on human lives and on the socio-economic welfare of entire regions; they are identified as one of the major risks of the East Asia and Pacific region. Dramatic events such as Hurricane Katrina and the Tohoku tsunami have also highlighted the need for decision-support tools in preparing, mitigating, responding, and recovering from disasters.

In this talk, I will present an overview of some recent progress in using optimisation for disaster management and, in particular, in relief distribution, power system restoration, and evacuation planning and scheduling. I will argue that optimisation has a significant role to play in all aspects of disaster management, from policy formulation to mitigation, operational response, and recovery, using examples of systems deployed duting hurricanes Irene and Sandy. Moreover, I will indicate that disaster management raises significant computational challenges for AI technologies, which must optimize over complex infrastructures in uncertain environments. Finally, I will conclude by identifying a number of fundamental research issues for AI in this space.

Table of Contents

Agents

AI Applications

Cognitive Modelling

Computer Vision

Constraint Satisfaction, Search and Optimisation

Evolutionary Computation

Game Playing

Knowledge Representation and Reasoning

Machine Learning and Data Mining

Natural Language Processing and Information Retrieval

Planning and Scheduling

Natural Language Processing and Information Retrieval

Planning and Scheduling

A Logical Framework of Bargaining with Integrity Constraints*

Xiaoxin Jing[1], Dongmo Zhang[2,**], and Xudong Luo[1,**]

[1] Institute of Logic and Cognition, Sun Yat-sen University, China
[2] Intelligent System Lab, University of Western Sydney, Australia
d.zhang@uws.edu.au, luoxd3@mail.sysu.edu.cn

Abstract. This paper proposes a logical framework for bargaining with integrity constraints (IC) in multi-agent and multi-issue bargaining environments. We construct a simultaneous concession solution for bargaining games under IC, and show that the solution is uniquely characterised by a set of logical properties. In addition, we prove that the solution also satisfies the most fundamental game theoretic properties such as symmetry and Pareto optimality. Finally, we discuss the relationship between merging operators and bargaining solutions under integrity constraints.

1 Introduction

Bargaining is a process to settle disputes and reach mutually agreements. It has been investigated from many perspectives, including economics, social science, political science and computer science [1–7]. Different from other disciplines where quantitative approaches dominate bargaining analysis, studies of bargaining in computer science, especially in artificial intelligence, pay more attention to logical reasoning behind bargaining processes. Thus, a number of logical frameworks were proposed for specifying reasoning procedures of bargaining [7–12]. In particular, similar to Nash's axiomatic model of bargaining, in [7] Zhang proposed an axiomatic model of bargaining in propositional logic. With his model, bargainers' demands are represented in propositional formulae and the outcome of bargaining is viewed as a mutual acceptance of the demands after necessary concessions from each bargainer.

Although Zhang's model provides a purely qualitative approach for bargaining analysis, there is a difficulty to apply his approach to the real-life bargaining. As mentioned in [7], the demands of a player are not necessarily the player's real demands but could be the player's beliefs, goals, desired constraints or commonsense. For example, a couple bargains over where to go for dinner: either a French restaurant (denoted by f) or an Italian restaurant (denoted by i). The husband prefers Italian food to French food but his wife likes more the romantic environment in French restaurants than Italian ones, even though they

* This paper is partly supported by the International Exchange Program Fund of 985 Project and Bairen Plan of Sun Yat-sen University, China and the Australian Research Council through Discovery Project DP0988750.

** Corresponding author.

S. Cranefield and A. Nayak (Eds.): AI 2013, LNAI 8272, pp. 1–13, 2013.

have some favourite dishes in common, which may or may not be offered in both restaurants. Obviously, each player can express his/her demands in propositional language by writing down their favourite restaurant and dishes, say $\{i, pizza\}$. However, if we use Zhang's model, all the domain constraints, such as $\neg f \vee \neg i$, $pizza \rightarrow i$, have to be included in the demand set of each player, which does not seem intuitive. Thus, this paper is devoted to providing a solution to this issue.

Similar to belief merging [13], specifying domain constraints (or called integrity constraints) in a bargaining model gives a number of challenges to the modelling of bargaining reasoning. First, simply assuming logical consistency of individual demand sets is not enough because new constraints may be generated after combining all constraints from individual bargainers as logical consequences. Second, preference ordering relies on constraints, so a logical requirement for the rationality of preference ordering has to be applied. Finally, constraints and demands from individuals may be described in different forms. It is crucial that a bargaining solution does not rely on the syntax of description, which is actually the case for Zhang's system. As we will see, our model of bargaining is syntax-irrelevant, which in fact reshapes the whole axiomatic system.

The rest of the paper is organised as follows. Section 2 defines our bargaining model. Sections 3 and 4 introduce its solution concept and some of its properties. Finally Section 5 discusses the related work and concludes the paper.

2 Bargaining Model with Integrity Constraints

This section presents our bargaining model. We consider a propositional language $\mathcal{L}$ built from a finite set $\mathcal{P}$ of propositional letters and the standard propositional connectives $\{\neg, \vee, \wedge, \rightarrow, \leftrightarrow\}$. Propositional sentences are denoted by $\phi, \psi, \cdots$. We use $\vdash$ to denote the logical deduction relation in classical propositional logic. Cn represents the corresponded local consequence closure. Furthermore, we say that a set Φ of formulae in $\mathcal{L}$ is *consistent* if there is no formula ϕ such that $\Phi \vdash \phi$ and $\Phi \vdash \neg\phi$. A set K of sentences in $\mathcal{L}$ is *logically closed* if and only if $K = Cn(K)$, where $Cn(K) = \{\phi \in \mathcal{L}, K \vdash \phi\}$. Let Φ be a finite set of propositional formulae. A binary relation $\geq$ over Φ is a *pre-order* if and only if it is a reflexive and transitive relation over Φ. A pre-order is *total* if for all $\phi, \psi \in \Phi$, $\phi \geq \psi$ or $\psi \geq \phi$. Given a pre-order $\geq$ over Φ, we define $\phi > \psi$ as $\phi \geq \psi$ and $\psi \not\geq \phi$, and $\phi \simeq \psi$ as $\phi \geq \psi$ and $\psi \geq \phi$. Moreover, if $\phi \geq \psi$ then $\psi \leq \phi$ and if $\phi > \psi$ then $\psi < \phi$.

2.1 Bargaining Games

Following [7], we assume that each bargainer has a set of demands and a preference order over the demand set. As we will show later, the domain constraints, common sense knowledge, and other integrity constraints will be specified separately and so need not to be included in the individual demand set.

Definition 1. *A* **demand structure** *D is a pair $(X, \geq)$, where X is a finite, logically consistent set of demands that are represented by a set of sentences in $\mathcal{L}$, and $\geq$ is a total pre-order on X, which satisfies:*

(LC) *If* $\phi_1, ..., \phi_n \vdash \psi$, *then there exists at least one* $k \in \{1, \cdots, n\}$ *such that* $\psi \geq \phi_k$.

Intuitively, a demand structure represents the statements a bargainer wants to put into the agreement, and the total pre-order over the demands is the description for bargainer's preference over the demands, i.e., $\phi \geq \psi$ means the bargainer holds demand ϕ more firmly than demand ψ. In addition, the logical constraint LC, introduced by Zhang and Foo in [14], places a rationality requirement on the preference ordering, which says that if a demand of ψ is a logic consequence of demands $\phi_1, ..., \phi_n$ then the firmness to keep ψ should not be less than at least one formula in $\phi_1, ..., \phi_n$.

In a bargaining scenario, an integrity constraint means a rule that all participants in the bargaining must follow. Such a rule could be something like domain restrictions, generic settings, commonsense knowledge and so on. As we will see below, we assume that any integrity constraint can be represented by a propositional formula and all integrity constraints for each bargaining situation are logically consistent. The following definition extends [7]'s bargaining model to allow integrity constraints.

Definition 2. *A* **bargaining game** *is a tuple of* $\langle (X_i, \geq_i)_{i \in N}, IC \rangle$, *where*
(i) $N = \{1, 2, ..., n\}$ *is a set of bargainers;*
(ii) each $(X_i, \geq_i)$ *is the demand structures of a bargainer; and*
(iii) IC *is a consistent set of sentences (i.e., integrity constraints).*
The set of all bargaining games in language $\mathcal{L}$ *is denoted by* $G^{IC}_{n,\mathcal{L}}$.

A bargaining game specifies a snapshot of a bargaining procedure. As we will demonstrate, we model a bargaining procedure as a sequence of bargaining games. Normally a bargaining starts with a situation in which the demands of the bargainers conflict each other. With the proceeding of negotiation, bargainers may make concessions in order to reach an agreement. Eventually, the bargaining terminates with either an agreement or a disagreement. The terminal situations can be specified in the following two specific games.

Definition 3. *A bargaining game* $\langle (X_i, \geq_i)_{i \in N}, IC \rangle$ *is* **non-conflictive** *if* $\bigcup_{i=1}^n X_i \cup IC$ *is logically consistent. It is a* **disagreement** *if there is* $k \in N$ *such that* $X_k = \emptyset$.

Note that a disagreement means that there is a bargainer who has nothing to give up.[1]

2.2 Demand Hierarchy and Comprehensiveness

In order to develop a solution concept for our bargaining model, we need to introduce a set of concepts based on single player's demand structure.

[1] In the real-life bargaining, a bargainer may declare a disagreement when he finds that an agreement would not be reached without giving up all his reservation demands.

Definition 4. *Give a demand structure $D = (X, \geq)$ where $X \neq \emptyset$, $P = (X^1, \cdots, X^L)$ is the* **partition** *of D if it satisfies:*
(i) $X = \bigcup_{l=1}^{L} X^l$;
(ii) $X^l \subseteq X$ *and* $X^l \neq \emptyset$ *for all* $l\,(1 \leq l \leq L)$;
(iii) $X^k \cap X^l = \emptyset$ *for any* $k \neq l$; *and*
(iv) $\forall \phi \in X^k\ \psi \in X^l$, $\phi > \psi$ *if and only if* $k < l$.

We define the demand hierarchy under IC using the partition above.

Definition 5. *Given a demand structure $D = (X, \geq)$ and a set of integrity constraints IC, let $P = (X^1, .., X^L)$ be a partition of D. Then the* **hierarchy** *of D under IC is defined as follows:*
(i) $H^1 = Cn(X^1 \cup IC)$, *and*
(ii) $H^{k+1} = Cn(\bigcup_{i=1}^{k+1} X^i \cup IC) \setminus \bigcup_{i=1}^{k} H^i$.

$\forall \phi \in Cn(X \cup IC)$, we define $h(\phi) = k$ if and only if $\phi \in H^k$, where k is ϕ's hierarchy level in D. And we write $h_D = \max\{h(\phi) \mid \phi \in Cn(X \cup IC)\}$ as the height of D. In addition, $\forall \phi, \psi \in X$, suppose $\phi \in H^k$ and $\psi \in H^j$, we write

$$\phi \geq^{IC} \psi \ \text{ iff } \ k \leq j.$$

For simplicity, we assume that $H^i \neq \emptyset$ for all i. In fact, if there is $k \in N^+$ such that H^k is $\emptyset$, we can remove all empty levels and let the remaining hierarchy as the hierarchy of D. Since $\geq$ is a total pre-order on X, it is easy to see that $\geq^{IC}$ is also a total pre-order on $Cn(X \cup IC)$.

Definition 6. *Given a demand structure $D = (X, \geq)$ and a set of integrity constraints IC, Ω is an* **IC-comprehensive set** *of D if:*
(i) $\Omega \subseteq Cn(X \cup IC)$;
(ii) $\Omega = Cn(\Omega)$; *and*
(iii) For any $\phi \in \Omega$ *and* $\psi \in Cn(X \cup IC)$, $\psi \geq^{IC} \phi$ *implies* $\psi \in \Omega$.

In other words, a subset of $Cn(X \cup IC)$ is IC-comprehensive if it is logically and ordinally closed under $\geq^{IC}$. We denote the set of all IC-comprehensive sets of D by $\Gamma^{IC}(D)$, or $\Gamma(D)$ if IC is obvious from the context.

The following theorem is important to our bargaining solution.

Theorem 1. *Given a demand structure $D = (X, \geq)$ and a set of integrity constraints IC, a set Ω is an IC-comprehensive set of D if and only if there exists $k \in \{1, \cdots, h_D\}$ such that $\Omega = \bigcup_{i=1}^{k} H^i$.*

Proof. ($\Rightarrow$) We first prove that if $\Omega \in \Gamma(D)$ then there exists $k \in \{1, \cdots, h_D\}$ such that $\Omega = T^k$, where $T^k = \bigcup_{i=1}^{k} H^i$. Let $k^0 = \min\{k \mid \Omega \subseteq T^k\}$. Obviously, $1 \leq k^0 \leq h_D$. We aim to prove $\Omega = T^{k^0}$. By the definition of k^0, $\Omega \subseteq T^{k^0}$. So, we just need to prove $T^{k^0} \subseteq \Omega$. Suppose it is not the case. Then there must exist ψ such that $\psi \in T^{k^0}$ but $\psi \notin \Omega$. Since $\Omega \in \Gamma(D)$, $\forall \phi \in \Omega$, we have $\psi \in Cn(X \cup IC)$ and $\psi \notin \Omega$, and then $\psi <^{IC} \phi$. So, $h(\psi) > h(\phi)$. $\forall \phi \in \Omega$, we have $1 \leq h(\phi) \leq k^0$. In addition, $k^0 = \min\{k \mid \Omega \subseteq T^k\}$. Therefore, $h(\psi) > k^0$. However, $\psi \in T^{k^0}$, and then $1 \leq h(\psi) \leq k^0$, which is contradicting. Therefore, the assumption is false, i.e., we have $T^{k^0} \subseteq \Omega$. As a result, if $\Omega \in \Gamma(D)$, we can find $k = k^0$ such that $\Omega = T^{k^0}$.

($\Leftarrow$) $\forall k \in \{1, \cdots, h_D\}$, we need to prove $T^k \in \Gamma(D)$. Because $T^k = \bigcup_{i=1}^k H^i = Cn(\bigcup_{i=1}^k X^i \cup IC)$, then T^k is closed. In addition, $\forall k \in \{1, \cdots, h_D\}$, $\bigcup_{i=1}^k X^i \subseteq X$, and then $T^k = Cn(\bigcup_{i=1}^k X^i \cup IC) \subseteq Cn(X \cup IC)$. $\forall k \in \{1, \cdots, h_D\}$, $\forall \phi \in T^k$, and $\psi \in Cn(X \cup IC)$, if $\psi \geq^{IC} \phi$, we need to prove $\psi \in T^k$. Suppose it is not this case, i.e., $\psi \notin T^k$. Because $T^k = Cn(X \cup IC) \setminus \bigcup_{i=k+1}^{h_D} H^i$, then $\psi \in \bigcup_{i=k+1}^{h_D} H^i$, and so $k+1 \leq h(\psi) \leq h_D$. In addition, since $\phi \in T^k$, $T^k = \bigcup_{i=1}^k H^i$, $1 \leq h(\phi) \leq k$. So, we can get $h(\psi) > h(\phi)$, and then $\phi >^{IC} \psi$, which contradicts premise $\psi \geq^{IC} \phi$. Therefore, the assumption is false. □

In the following, we define the equivalence of demand structures under an integrity constraints, which plays an important role for describing syntax independency:

Definition 7. *Let $D = (X, \geq)$ and $D' = (X', \geq')$ be two demand structures, where $X \neq \emptyset$ and $X' \neq \emptyset$, IC is a set of integrity constraints. We say D and D' are* **equivalent** *under IC, denoted as $D \Leftrightarrow^{IC} D'$, if and only if there is $\Gamma(D) = \Gamma'(D')$.*

3 Bargaining Solution

In this section, we will develop our solution concept for the bargaining model we introduced in the previous section.

Definition 8. *A* **bargaining solution** *s is a function from $G^{IC}_{n,\mathcal{L}}$ to $\prod_{i=1}^n \Gamma(D_i)$, i.e., $\forall G \in G^{IC}_{n,\mathcal{L}}$, $s(G) = (s_1(G), \cdots, s_n(G))$, where $s_i(G) \in \Gamma(D_i)$ for all i. $s_i(G)$ denotes the i-th component of $s(G)$. $Cn(\bigcup_{i=1}^n s_i(G))$ is called the agreement of the bargaining game, denoted by $A(G)$.*

Intuitively, the agreement of a bargaining is a set of demands mutually accepted by all the bargainers. A bargaining solution is then to specify which demands from each bargainer should be put into the finial agreement.

In the following, we will construct a concrete bargaining solution that satisfies a set of desirable properties. The intuition behind the construction can be stated as follows: assume a bargaining situation where all bargainers agree on a set of integrity constraints IC. Firstly, all the bargainers submit their demands to an arbitrator who also knows IC. The arbitrator then judges if the current bargaining situation forms a non-conflictive game or a disagreement game. If so, the bargaining ends with either an agreement, which is the collection of all the demands, or a disagreement, which is an empty set. Otherwise, the arbitrator requests each bargainer to make a concession by withdrawing their least preferred demands. We call such a solution simultaneous a concession solution. Formally, we have:

Definition 9. *Given a bargaining game $G = \langle (X_i, \preceq_i)_{i \in N}, IC \rangle$,* **simultaneous concession solution** *$S(G)$ is constructed as follows:*

$$S(G) = \begin{cases} (H_1^{\leq h_{D_1}-\rho}, \cdots, H_n^{\leq h_{D_n}-\rho}) & \textit{if } \rho < L; \\ (\emptyset, \cdots, \emptyset) & \textit{otherwise.} \end{cases}$$

where $\forall i \in N$, $H_i^{\leq j} = \bigcup_{k=1}^{j} H_i^k$ *(*H^k *is defined in Definition 5),* h_{D_i} *is the height of* D_i*,* $\rho = \min\{k \mid \bigcup_{i=1}^{n} H_i^{\leq h_{D_i}-k}$ *is consistent},* and $L = \min\{h_{D_i} \mid i \in N\}$.

For a better understanding of the above definition, let us consider the restaurant example we show in the introduction section again.

Example 1. *A couple bargains over which restaurant to go to celebrate their wedding anniversary: either Italian (i) or French (f). The husband (h) likes to eat pizza (p). Alternatively, he is also fine with beefsteak (b) and vegetable salads (v). In fact, he does not mind to go to the French restaurant but cannot stand people eating snails (s) . The wife (w) leans towards the romantic French restaurant and particular likes the vegetable salads. She would like to try snails once as all her friends recommend it. Both know that pizza is only offered in the Italian restaurant (*$p \to i$*) and snails only offered in the French restaurant (*$s \to f$*). Obviously they can only choose one restaurant for the dinner (*$\neg i \vee \neg f$*).*

Putting all the information together, the husband's demands can be written as $X_h = \{\neg s, p, v, b\}$ with the preference: $\neg s \geq_h p \geq_h v \geq_h b$; the wife's demands are $X_w = \{v, f, s\}$ with the preference: $v \geq_w f \geq_w s$; and the integrity constraints can be represented by $IC = \{\neg i \vee \neg f, p \to i, s \to f\}$. Thus, we can model the game as $G = \langle (X_h, \geq_h), (X_w, \geq_w), IC \rangle$.

Table 1. Player's hierarchies from high (top) to low (bottom)

Husband	**Wife**
$\neg s$, $\neg i \vee \neg f$, $p \to i$, $s \to f$	
p, i, $\neg f$	v, $\neg i \vee \neg f$, $p \to i$, $s \to f$
v	f, $\neg i$, $\neg p$
b	s

To solve the problem, we first calculate the normalised hierarchy for each player according to Definition 5 as shown in Table 1. According to the table, it is easy to see that $h(D_h) = 4$, $h(D_w) = 3$, $\rho = 2$ and $L = 3$. Then the solution of the bargaining game is:

$$s_h(G) = H_h^{\leq 2} = \{\neg s, \neg i \vee \neg f, p \to i, s \to f, p, i, \neg f\},$$
$$s_w(G) = H_w^{\leq 1} = \{v, \neg i \vee \neg f, p \to i, s \to f\}.$$

As a result, the agreement of the bargaining is:

$$A(G) = Cn(H_h^{\leq 2} \cup H_w^{\leq 1}) = \{\neg s, \neg i \vee \neg f, p \to i, s \to f, p, i, \neg f, v\}.$$

4 Properties of the Solution

In this section, we investigate the properties of the solution that we construct in the previous section. To this end, we need to introduce a few concepts first.

Definition 10. *Given two bargaining games* $G = \langle (D_i)_{i\in N}, IC\rangle$ *and* $G' = \langle (D_i')_{i\in N}, IC'\rangle$, *we say* G *and* G' *are* **equivalent**, *denoted by* $G \equiv G'$, *if and only if*

(i) Both G *and* G' *are disagreement games; or*
(ii) None of G *and* G' *is a disagreement game,* $\vdash IC \leftrightarrow IC'$ *and* $D_i \Leftrightarrow^{IC} D_i'$ $\forall i \in N$.

Definition 11. *Given a bargaining game* $G = \langle (D_i)_{i\in N}, IC\rangle$, *a bargaining game* $G' = \langle (D_i')_{i\in N}, IC'\rangle$, *where* $D_i'{=}(X_i', \geq_i)$, *is a* **subgame** *of* G, *denoted by* $G' \subseteq G$, *if and only if for all* $i \in N$,

(i) $IC \vdash IC'$ *and* $IC' \vdash IC$, *and we write it simply as* $\vdash IC \leftrightarrow IC'$;
(ii) $Cn(X_i' \cup IC')$ *is an IC-comprehensive set of* D_i; *and*
(iii) $\geq'^{IC'}_i = \geq^{IC}_i \cap (Cn(X_i' \cup IC') \times Cn(X_i' \cup IC'))$.

Furthermore, G' is a proper subgame of G, denoted by $G' \subset G$, if $Cn(X_i' \cup IC') \subset Cn(X_i \cup IC)$ for all $i \in N$. Specially, given a bargaining game $G = \langle (D_i)_{i\in N}, IC\rangle$, and $h_{D_i} = 1$ for any D_i in G, then G does not have any proper subgame. Moreover, the following concept follows Zhang's idea in [7].

Definition 12. *A proper subgame* G' *of* G *is a* **maximal proper subgame** *of* G, *denoted by* $G' \subset_{max} G$, *if for any* $G'' \subset G$, $G'' \subseteq G'$.

4.1 Logical Characterisation

We first consider the logical properties of our bargaining solution. In general, we expect any bargaining solution (refer to Definition 8) satisfies:

(i) *Consistency*: If IC is consistent, then $A(G)$ is consistent.
(ii) *Non-conflictive*: If G is non-conflictive, then $s_i(G) = Cn(X_i \cup IC)$ for all i.
(iii) *Disagreement*: If G is a disagreement, then $A(G) = \emptyset$.
(iv) *Equivalence*: If $G \equiv G'$, then $s_i(G) = s_i(G')$ for all i.
(v) *Contraction independence*: If $G' \subset_{max} G$ then $s_i(G) = s_i(G')$ for all i unless G is non-conflictive.

Intuitively the above properties are basic requirements for bargaining solutions. Property 1 states that if the integrity constraints are consistent, then the outcome of the bargaining, i.e., the agreement should also be consistent. Property 2 says that if there are no conflict among all the bargainers' demands and the integrity constraints, then nobody has to make any concession to reach an agreement. Property 3 indicates that a disagreement means that no agreements are reached. Property 4 is the principle of irrelevancy of syntax, i.e., if two bargaining games are equivalent, then the solutions of bargaining are the same. This property is crucial for the bargaining, while as we can see, it is not satisfied in [7]. The last property requires that a bargaining solution should be independent of any minimal simultaneous concession of the bargaining game.

In the following, we will show that our simultaneous concession solution satisfies all the five properties. To this end, we need the following lemma first:

Lemma 1 *Given two bargaining games G($h(D_i) > 1$ for any D_i in G) and G', G' is a maximal proper subgame of G if and only if* $\forall i$,

(i) $\vdash IC \leftrightarrow IC'$;

(ii) $Cn(X'_i \cup IC') = H_i^{\leq h_{D_i}-1}$; *and*

(iii) $\geq'^{IC'}_i = \geq^{IC}_i \cap (Cn(X'_i \cup IC') \times Cn(X'_i \cup IC'))$.

Proof. ($\Leftarrow$) We will prove that if G' satisfies properties (i)-(iii), G' is a maximal proper subgame of G. Because property (ii) is satisfied, G is not a disagreement. Thus, we need to prove $G' \subset G$ first. Since we find properties (i) and (iii) are the same as (i) and (iii) in Definition 11, we just need to prove (ii), i.e., $Cn(X'_i \cup IC') = H_i^{\leq h_{D_i}-1}$ is an IC comprehensive set of D_i and $Cn(X'_i \cup IC') \subset Cn(X_i \cup IC)$. Because $h_{D_i} > 1$ for any D_i in G, by Theorem 1, $H_i^{\leq h_{D_i}-1} = \bigcup_{j=1}^{h_{D_i}-1} H_i^j$ is an IC comprehensive set of D_i. In addition, since $H_i^{h(D_i)} \neq \emptyset$, $T_i^{h_{D_i}-1} = Cn(X_i \cup IC) \backslash H_i^{h(D_i)} \subset Cn(X_i \cup IC)$.

Next, for $G'' = \langle D''_{i \in N}, \geq'' \rangle$, if $G'' \subset G$, we need to prove $G'' \subseteq G'$. Because $G'' \subset G$ and $G' \subset G$, (i) $IC'' \leftrightarrow IC \leftrightarrow IC'$; (ii) $Cn(X''_i \cup IC'')$ and $Cn(X'_i \cup IC')$ are IC comprehensive sets of D_i for all i; and (iii) $\geq''^{IC}_i = \geq^{IC}_i \cap (Cn(X''_i \cup IC'') \times Cn(X''_i \cup IC'')) = \geq'^{IC}_i \cap (Cn(X''_i \cup IC'') \times Cn(X''_i \cup IC''))$. In addition, $Cn(X''_i \cup IC'') \subset Cn(X_i \cup IC)$ and $Cn(X'_i \cup IC'') \subset Cn(X_i \cup IC)$ for all i. So, we just need to prove $Cn(X''_i \cup IC'')$ is an IC comprehensive set of D'_i for all i.

We prove $Cn(X''_i \cup IC'') \subseteq Cn(X'_i \cup IC')$ first. Suppose not. Then there is $\phi \in Cn(X''_i \cup IC'')$ but $\phi \notin Cn(X'_i \cup IC')$. From $Cn(X'_i \cup IC') = H_i^{\leq h_{D_i}-1} = Cn(X_i \cup IC) \backslash H_i^{h_{D_i}}$ and $Cn(X''_i \cup IC'') \subset Cn(X_i \cup IC)$, we derive that $\phi \in H_i^{h_{D_i}}$. Therefore, $h(\phi) = h_{D_i}$. Because $\forall \psi \in Cn(X_i \cup IC)$, $h(\psi) \leq h_{D_i}$, $\psi \geq^{IC} \phi$, and $Cn(X''_i \cup IC'')$ is an IC-comprehensive set of D_i, $\psi \in Cn(X''_i \cup IC'')$, which implies $Cn(X_i \cup IC) \subseteq Cn(X''_i \cup IC'')$. However, this contradicts $Cn(X''_i \cup IC'') \subset Cn(X_i \cup IC)$. So, the assumption cannot hold. Thus, $Cn(X''_i \cup IC'') \subseteq Cn(X'_i \cup IC')$.

Because $G' \subset G$ and $G'' \subset G$, for all i, $Cn(X'_i \cup IC')$ and $Cn(X''_i \cup IC'')$ are both IC-comprehensive sets of D_i. Thus, $\forall \phi \in Cn(X''_i \cup IC'')$, $\forall \psi \in Cn(X'_i \cup IC')$ (thus we have $\psi \in Cn(X_i \cup IC)$), if $\psi \geq^{IC} \phi$ then $\psi \in Cn(X''_i \cup IC'')$. Therefore, $Cn(X''_i \cup IC'')$ is an IC comprehensive set of D'_i. Furthermore, $G'' \subseteq G'$.

($\Rightarrow$) If G' is a maximal proper subgame of G, then properties (i), (ii), and (iii) in this lemma are satisfied.

Firstly, because G has at least one proper subgame G', $\forall i \in N$, $X_i \neq \emptyset$ and $Cn(X'_i \cup IC') \subset Cn(X_i \cup IC)$. If G' is a maximal proper subgame of G, by Definitions 11 and 12, properties (i) and (iii) of this lemma are satisfied, and so we just need to prove its property (ii). Suppose not, i.e., $Cn(X'_i \cup IC') \neq H_i^{\leq h_{D_i}-1}$. Because G' is a proper subgame of G, $Cn(X'_i \cup IC')$ is an IC comprehensive set of D_i. Thus, by Theorem 1, there is a $k \in [1, h(D_i)]$, such that $Cn(X'_i \cup IC') = \bigcup_{j=1}^{k} H_i^j$. So, because $Cn(X'_i \cup IC') \subset Cn(X_i \cup IC)$ and $Cn(X'_i \cup IC') \neq H_i^{\leq h_{D_i}-1}$, we have $k < h_{D_i} - 1$.

Here we can find a subgame of G'' as proved above such that: (i) $\vdash IC \leftrightarrow IC''$; (ii) $Cn(X''_i \cup IC'') = H_i^{\leq h_{D_i}-1}$; and (iii) $\geq''^{IC''}_i = \geq^{IC}_i \cap (Cn(X''_i \cup IC'') \times Cn(X''_i \cup IC''))$. Then $G'' \subset G$. Then, because $k < h_{D_i} - 1$, $\bigcup_{j=1}^{k} X_i^j \subset \bigcup_{j=1}^{h_{D_i}-1} X_i^j$, $\bigcup_{j=1}^{k} H_i^j = Cn(\bigcup_{j=1}^{k} X_i^j \cup IC) \subset Cn(\bigcup_{i=1}^{h_{D_i}-1} X_i^j \cup IC) = H_i^{\leq h_{D_i}-1}$ and thus $G'' \supset G'$. This contradicts that G' is a maximal proper subgame of G. So, the assumption is false. Thus, $Cn(X'_i \cup IC') = H_i^{\leq h_{D_i}-1}$. □

Now we are ready to prove the following theorem:

Theorem 2. *The simultaneous concession solution satisfies all the five properties listed in the beginning of this subsection.*

Proof. (i) Suppose IC is consistent. Given an IC bargaining game G, if $\rho \geq L$ then $S(G) = \{\emptyset, ..., \emptyset\}$. So, $A(G) = \emptyset$. Obviously, it is consistent. If $\rho < L$, then $S(G) = (H_1^{\leq h_{D_1}-\rho}, ..., H_n^{\leq h_{D_n}-\rho})$, because $\rho = \min\{k \mid \bigcup_{i=1}^{n} H_i^{\leq h_{D_i}-k}$ is consistent$\}$ as defined in Definition 9. So, $A(G) = Cn(\bigcup_{i=1}^{n} H_i^{\leq h_{D_i}-\rho})$ is consistent. That is, the consistency property holds.

(ii) If G is non-conflictive, by Definition 3, $\bigcup_{i=1}^{n} X_i \cup IC$ is consistent. Then we can easily get $\rho = 0$ and $L \geq 1$. Thus, $\forall i \in N$, $S^i(G) = H_i^{\leq h_{D_i}-0} = H_i^{\leq h_{D_i}}$. Then, noticing $h_{D_i} = \max\{h(\phi) \mid \forall \phi \in Cn(X_i \cup IC)\}$, We have $H_i^{\leq h_{D_i}} = Cn(\bigcup_{j=1}^{h_{D_i}} X_i^j \cup IC) = Cn(X_i \cup IC)$. So, $S^i(G) = Cn(X_i \cup IC)$.

(iii) If G is a disagreement, by Definition 3, there exists k such that $X_k = \emptyset$, and then $L = 0$; but $\rho \geq 0$. So $\rho \geq L$, and thus $S^i(G) = \emptyset$ for any i. Furthermore, $A(G) = Cn(\bigcup_{i=1}^{n} S^i(G)) = \emptyset$.

(iv) Given two bargaining games $G = \langle (X_i, \geq_i)_{i\in N}, IC\rangle$ and $G' = \langle (X'_i, \geq'_i)_{i\in N}, IC'\rangle$ such that $G \equiv G'$. By Definitions 10 and 7, if G is a disagreement, so is G'; otherwise, $IC \leftrightarrow IC$ and $\Gamma(D) = \Gamma'(D')$. Therefore, if G is non-conflictive, it is easy to see that G' is non-conflictive. So, $S^i(G) = S^i(G')$ for all i. Otherwise, we can easily have $h_{D_i} = h_{D'_i}$, $\rho = \rho'$ and $L = L'$. In addition, noticing $H_i^{\leq h_{D_i}-\rho} = {H'}_i^{\leq h_{D'_i}-\rho'}$, we have $S^i(G) = S^i(G')$ for all i. That is, the equivalence property holds.

(v) Consider a bargaining game $G = \langle (X_i, \geq_i)_{i\in N}, IC\rangle$. (a) If $L = 0$, then $\exists k, X_i^k = \emptyset$, which means G has no proper subgames. Then $S(G)$ satisfies the contraction independence property trivially. (b) If $L > 0$ then because G is not non-conflictive, $\rho > 0$. Assume $G' = \langle (X'_i, \geq'_i)_{i\in N}, IC'\rangle$ is a maximal proper subgame of G. Let $\rho' = \min\{k' \mid \bigcup_{i=1}^{n} {H'}_i^{\leq h_{D'_i}-k'}$ is consistent$\}$ and $L' = \min\{h'_{Di} \mid i \in N\}$. By Lemma 1, $\forall i$, $\vdash IC \leftrightarrow IC'$; $Cn(X'_i \cup IC') = H_i^{\leq h_{D_i}-1}$; and $\geq'^{IC'}_i = \geq^{IC}_i \cap (Cn(X'_i \cup IC') \times Cn(X'_i \cup IC'))$. Obviously, $L' = L - 1$ and $\rho' = \rho - 1$, and $h_{D'_i} = h_{D_i} - 1$. Therefore, $\rho' < L'$ if and only if $\rho < L$. If $\rho \geq L$, $S(G) = S(G') = \{\emptyset, ..., \emptyset\}$. Otherwise, when $\rho < L$, $\forall i \in N$, we have

$$
\begin{aligned}
S_i(G) = H_i^{\leq h_{D_i}-\rho} &= Cn(X_i \cup IC)\backslash \cup_{j=h_{D_i}-\rho+1}^{h_{D_i}} H_i^j \\
&= (Cn(X_i \cup IC)\backslash H_i^{h_{D_i}})\backslash \cup_{j=h_{D_i}-\rho+1}^{h_{D_i}-1} H_i^j \\
&= H_i^{\leq h_{D_i}-1}\backslash \cup_{j=h_{D'_i}-\rho'+1}^{h_{D'_i}} H_i^j \\
&= Cn(X'_i \cup IC')\backslash \cup_{j=h_{D'_i}-\rho'+1}^{h_{D'_i}} H_i^j \\
&= H_i^{\leq h_{D'_i}-\rho'} \\
&= S_i(G').
\end{aligned} \quad (1)
$$

Therefore, the contraction independence property holds. □

The following theorem shows that the five properties exactly characterise the simultaneous concession solution (therefore putting these two theorems together forms a representation theorem of our solution):

Theorem 3. *If a bargaining solution s satisfies the properties of consistency, non-conflictive, disagreement, equivalence, and contraction independence, it is the simultaneous concession solution.*

Proof. We prove that if a bargaining solution satisfies the five properties, $s(G) = S(G)$ for any G by induction on ρ.

For the base case that $\rho = 0$, there are two situations. (i) If G is non-conflictive, according to the non-conflictive property, $\forall i$, $s_i(G) = Cn(X_i \cup IC)$. Because $\rho = 0$ and $L \geq 1$, $\rho < L$. Thus, by Definition 9, $S_i(G) = H_i^{\leq h_{D_i}-0} = H_i^{\leq h_{D_i}} = Cn(\bigcup_{j=1}^{h_{D_i}} X_i^j \cup IC) = Cn(X_i \cup IC)$. So, $s_i(G) = S_i(G)$ for any i. (ii) If G is a disagreement, by the disagreement property, $\forall i$, $s^i(G) = \emptyset$, and there must exist a k such that $X_k = \emptyset$ and so $L = 0$. Since $\rho = 0$, $\rho = L$. Thus, by Definition 9, $S_i(G) = \emptyset$. So, $s_i(G) = S_i(G)$.

Now we assume that for any game G' such that $\rho' = k$, $s(G') = S(G')$. Now for a game G in which $\rho = k+1$, we aim to prove $s^i(G) = S^i(G)$ for all i. Because $\rho = k+1 \geq 1$ in G, G is not a disagreement game nor a non-conflictive game. Let $G' = \langle (X'_i, \succeq'_i)_{i \in N}, IC' \rangle$, where: (a) $\vdash IC \leftrightarrow IC'$; (b) $Cn(X'_i \cup IC') = H_i^{\leq h_{D_i}-1}$; and (c) $\succeq'^{IC'}_i = \succeq^{IC}_i \cap (Cn(X'_i \cup IC') \times Cn(X'_i \cup IC'))$ for any i. So, G' is a maximal proper game of G. Here $\rho' = \rho - 1 = k$, so by inductive assumption, we have $s^i(G') = S^i(G') = H_i^{\leq h_{D'_i}-\rho'}$. In addition, by the contraction independence property, $s_i(G') = s_i(G)$ for any i. So, we just need to prove $S_i(G) = H_i^{\leq h_{D_i}-\rho} = H_i^{\leq h_{D'_i}-\rho'}$, which is similar to formula (1). □

4.2 Game-Theoretic Properties

In this subsection, we show that our solution satisfies two fundamental game theoretical properties: Pareto efficiency and symmetry. Because game-theoretical bargaining model is based on utility functions but our bargaining model is defined on bargainers' demands, we need to restate Pareto efficiency and symmetry for our model. Firstly we need two relevant definitions.

Definition 13. *Given bargaining game* $G=\langle (D_i)_{i\in N}, IC\rangle$, *where* $D_i=(X_i,\geq_i)$, *an* **outcome** *of* G *is a tuple of* $O=(o_1,...,o_n)$, *where* $\forall o_i\in\Gamma(D_i)$, $\bigcup_{i=1}^n o_i$ *is consistent.*

Definition 14. *Given two bargaining games* $G=\langle D, IC\rangle$ *and* $G'=\langle D', IC'\rangle$, *where* $D=(X_i,\geq_i)_{i\in N}$ *and* $D'=(X'_i,\geq'_i)_{i\in N}$. *We say* G *and* G' *are* **symmetric** *if and only if there is a bijection* g *from* D *to* D' *such that* $\forall i\in N, g(D_i)\Leftrightarrow^{IC} D_i$, *and* $\vdash IC\leftrightarrow IC'$.

Theorem 4. *The simultaneous concession solution satisfies:*

(i) Pareto Efficiency: Given bargaining game $G=\langle (X_i,\geq_i)_{i\in N}, IC\rangle$ *satisfying* $s(G)\neq(\emptyset,...,\emptyset)$, *let* O *and* O' *be two possible outcomes of* G. *If* $o'_i\supset o_i$ *for all* $i\in N$, *then* $s(G)\neq O$.

(ii) Symmetry: Suppose that two bargaining games G *and* G' *are symmetric with bijection* g. *Then* $A(G)=A(G')$. *Moreover, for any* $i,j\in N$, *if* $g(D_i)=D'_j$, *then* $s_i(G)=s_j(G')$.

Proof. Firstly, we prove our simultaneous concession solution S satisfies the Pareto efficiency by the contradiction proof method. Suppose $S(G)=O$, then by Definition 9, $O=(H_1^{\leq h_{D_1}-\rho},\cdots,H_n^{\leq h_{D_n}-\rho})$, where $\rho=\min\{k\mid \bigcup_{i=1}^n H_i^{\leq h_{D_i}-k}$ is consistent$\}$. Because $o'_i\supset o_i$ for all $i\in N$, $o'_i=H_i^{\leq h_{D_i}-\rho'}$, where $h_{D_n}-\rho'>h_{D_n}-\rho$. Then $\rho'<\rho$. By the definition of ρ, $\bigcup_{i=1}^n H_i^{\leq h_{D_i}-\rho'}$ is inconsistent, which means that $\bigcup_{i=1}^n o'_i$ is inconsistent, and then O' is not an outcome of G. This conclusion is conflict with the premise. So, $S(G)\neq O$.

Then we prove the simultaneous concession solution satisfies the symmetry.

(i) If G is a disagreement, there exist at least $X_k (k\in N)$ in D_k such that $X_k=\emptyset$, and so $L=0$ for bargaining game G. Because G and G' are symmetric, there must be a $D'_{k'}$ in G' such that $D'_k\Leftrightarrow^{IC} D'_{k'}$, $X'_{k'}(k'\in N)=\emptyset$, and so $L'=0$ for bargaining game G'. Thus, $\rho\geq L$ and $\rho'\geq L'$ in G and G', respectively. Therefore, by Definition 9, $S(G)=S(G')=\emptyset$.

(ii) In the case that G and G' are not disagreements, since G and G' are symmetric, and $g(D_i)=D'_j$, by Definition 14, $D_i\Leftrightarrow^{IC} D'_j$ for any $i,j\in N$, and $\vdash IC\leftrightarrow IC'$. Then for any D_i in G and D'_j in G', $\forall\Omega\in\Gamma(D_i)$, $\exists\Omega'\in\Gamma'(D'_j)$ such that $\Omega=\Omega'$; and vice visa. It is easy to see that $\rho=\rho'$, $h_{D_i}=h_{D'_j}$ and $L=L'$ for G and G', respectively. Then $\forall i,j\in N$, $S^i(G)=H_i^{\leq h_{D_i}-\rho}=H'^{\leq h_{D'_i}-\rho'}_j=s^j(G')$. In this case, $A(G)=Cn(\bigcup_{i=1}^n s^i(G))=Cn(\bigcup_{j=1}^n s^j(G'))=A(G')$. □

5 Conclusion and Related Work

This paper proposes a logical framework for bargaining with integrity constraints. More specifically, we construct a simultaneous concession solution to the bargaining game, which satisfies five logical and two game theoretical properties. Moreover, we prove that our solution can be characterised uniquely by the five logical properties.

This work is built up on Zhang's framework on the logical axiomatic model of bargaining in [7] but extends his work in several aspects. Firstly, our model solves the problem of mixing integrity constraints with players' demands. Secondly, we add a logical requirement to ensure the rationality of the preference ordering over the bargainers' demands. Most importantly, our solution is syntax independent, which makes more sense for constraint integration.

This work has also related to belief merging or database merging with constraints [13, 15]. Following Lin and Mendelzon's work [15] on database merging, Konieczny and Pérez extended the framework of belief merging [16] by adding integrity constraints, which leaded to a new framework of belief merging [13]. Although both bargaining and belief merging require to incorporate information from different sources (thus share some similar properties, for instance, the integrity constraints should be consistent with the outcomes of bargaining or merging), their ways of treating information sources are totally different. In belief merging, sometimes, an item is included in the merging outcome relies on how many sources contains this item, say the majority rule, while in bargaining, the outcome of a bargaining relies on how firmly the bargainers insist on their demands. In addition, with belief merging, each data source normally does not have a preference over the items in the belief base, while in a bargaining, players' preferences over their demands are essential. All these differences have been reflected in the frameworks of belief merging and logical based bargaining.

References

1. John, F., Nash, J.: The bargaining problem. Econometrica 18(2), 155–162 (1950)
2. Binmore, K., Rubinstein, A., Wolinsky, A.: The Nash bargaining solution in economic modelling. The RAND Journal of Economics, 176–188 (1986)
3. Shubik, M.: Game theory in the social sciences: Concepts and solutions. MIT Press (2006)
4. Raiffa, H.: The Art and Science of Negotiation. Harvard University Press (1982)
5. Jennings, N.R., Faratin, P., Lomuscio, A.R., Parsons, S., Wooldridge, M.J., Sierra, C.: Automated negotiation: prospects, methods and challenges. Group Decision and Negotiation 10(2), 199–215 (2001)
6. Fatima, S.S., Wooldridge, M., Jennings, N.R.: An agenda-based framework for multi-issue negotiation. Artificial Intelligence 152(1), 1–45 (2004)
7. Zhang, D.: A logic-based axiomatic model of bargaining. Artificial Intelligence 174(16-17), 1307–1322 (2010)
8. Kraus, S., Sycara, K., Evenchik, A.: Reaching agreements through argumentation: a logical model and implementation. Artificial Intelligence 104, 1–69 (1998)
9. Parsons, S., Sierra, C., Jennings, N.R.: Agents that reason and negotiate by arguing. Journal of Logic and Computation 8(3), 261–292 (1998)
10. Luo, X., Jennings, N.R., Shadbolt, N., Leung, H.F., Lee, J.H.M.: A fuzzy constraint based model for bilateral, multi-issue negotiations in semi-competitive environments. Artificial Intelligence 148(1), 53–102 (2003)

11. Zhang, D., Zhang, Y.: An ordinal bargaining solution with fixed-point property. Journal of Artificial Intelligence Research 33(1), 433–464 (2008)
12. Zhan, J., Luo, X., Sim, K.M., Feng, C., Zhang, Y.: A fuzzy logic based model of a bargaining game. In: Wang, M. (ed.) KSEM 2013. LNCS, vol. 8041, pp. 387–403. Springer, Heidelberg (2013)
13. Konieczny, S., Pérez, R.P.: Merging information under constraints: a logical framework. Journal of Logic and Computation 12(5), 773–808 (2002)
14. Zhang, D., Foo, N.: Infinitary belief revision. Journal of Philosophical Logic 30(6), 525–570 (2001)
15. Lin, J., Mendelzon, A.O.: Merging databases under constraints. International Journal of Cooperative Information Systems 7(01), 55–76 (1998)
16. Konieczny, S., Pino-Pérez, R.: On the logic of merging. In: KR 1998: Principles of Knowledge Representation and Reasoning, pp. 488–498. Morgan Kaufmann (1998)

Security Games with Ambiguous Information about Attacker Types

Youzhi Zhang[1], Xudong Luo[1,*], and Wenjun Ma[2]

[1] Institute of Logic and Cognition,
Sun Yat-sen University, Guangzhou, China
[2] School of Electronics, Electrical Engineering and Computer Science,
Queen's University Belfast, Belfast, UK
luoxd3@mail.sysu.edu.cn

Abstract. There has been significant recent interest in security games, which are used to solve the problems of limited security resource allocation. In particular, the research focus is on the Bayesian Stackelberg game model with incomplete information about players' types. However, in real life, the information in such a game is often not only incomplete but also ambiguous for lack of sufficient evidence, *i.e.*, the defender could not precisely have the probability of each type of the attacker. To address this issue, we define a new kind of security games with ambiguous information about the attacker's types. In this paper, we also propose an algorithm to find the optimal mixed strategy for the defender and analyse the computational complexity of the algorithm. Finally, we do lots of experiments to evaluate that our model.

1 Introduction

Nowadays the study of security games is a very active topic [1–6]. For example, the security game solving algorithm of DOBSS [1] is the heart of the ARMOR system [2, 4], which has been successfully utilised in security patrol schedule at the Los Angeles International Airport [2, 4]. In a security game, the defender has to protect the people and critical infrastructure from the attacker. However, security resources that can be allocated are usually limited. And the attacker can observe the defender, for a period of time, to understand his security strategy and then attack a target accordingly. So, the defender has to take a random strategy to provide a security cover for everything. That is, the defender should find an optimal mixed strategy to maximise his expected utility.

Lots of work (*e.g.*, [1, 3, 5]) on the topic of security games is to find the *strong Stackelberg equilibrium* [4]. That is, the attacker has perfect knowledge of the defender's optimal strategy and accordingly chooses an optimal response strategy to maximise his expected utility. The defender and the attacker have no incentive to change their strategies of the optimal response. However, in real life, there are different types of terrorist, *e.g.*, ideological, ethno-separatist,

* Corresponding author.

S. Cranefield and A. Nayak (Eds.): AI 2013, LNAI 8272, pp. 14–25, 2013.

religious-cultural, and single issue terrorist [7]. Attackers with diverse types use different ways (*e.g.*, kidnapping, bombing, and so on) to attack targets and so gain different payoffs [8]. If the defender just knows they will be attacked by the attacker with several possible types and the probability distribution over these types, we can use Bayesian Stackelberg game models [1–4] to solve the problem. Nevertheless, sometimes they are unclear about the probability of each type for lack of sufficient evidence. For example, the intelligence shows that an attacker, who wants to attack the airport, often attends the meeting with ideological and religious-cultural property or the meeting with religious-cultural and ideological property. However, the available evidence only shows the probability that this attacker is one of the types *ideological terrorist* and *religious-cultural terrorist* is 0.4, and the probability that this man is one of the types *religious-cultural terrorist* and *ideological terrorist* is 0.6. So, in this case, the problem cannot be solved using the Bayesian game model because it assumes every player knows clearly the probability of each type of other players.

On the other hand, the evidence theory of Dempster and Shafer (D-S theory) [9, 10] is a powerful tool to handle the ambiguous information (*i.e.*, imprecise probabilities) in terms of mass function. Based on D-S theory, Strat [11] introduces the concept of interval-valued expected utility and gives the method for calculating it from a mass function [9, 10]. Ma *et al.* [6] extend the model of Strat [11] to form a framework of security games with ambiguous payoffs. Its key point is that a point-valued preference degree for a choice of strategy can be calculated from an interval-valued expected utility that Strat [11] defines. However, Ma *et al.* [6] do not deal with ambiguous types of the attacker.

Based on the ambiguity decision framework [11–13], this paper proposes our ambiguous games model to handle ambiguous information in security games. More specifically, in our model, the belief about the attacker' types is ambiguous and modeled by mass functions [9, 10]. That is, the probability could be of not only one type but also many types together for lack of evidence for the defender. Thus, the expected utility of a strategy becomes an interval value by Strat's method [11]. Then, in our model, the player can get a point-valued preference degree over a strategy from its interval-valued utility according to the model of Ma *et al.* [6, 12]. Finally, the defender can find the optimal strategies according to the preference ordering.

The main contributions of this paper are as follows. (i) We deal with the defender's ambiguous belief about the attacker's types in security games. (ii) We propose an algorithm for solving security games with such ambiguous attacker types and analyse its computing complexity. And (iii) we evaluate our model by lots of experiments and find that our model is efficient and safe for handling security games with ambiguous attacker types.

The rest of this paper is organised as follows. Section 2 recaps the ambiguity decision framework. Section 3 proposes our security games with ambiguous attacker types and algorithm for solving our security games. Section 4 does lots of experiments to evaluate our game model. Section 5 discusses the related work. Finally, Section 6 concludes our paper with future work.

2 Preliminaries

This section will recap an ambiguity decision framework based on D-S theory [9–12, 14]. D-S theory [9, 10] can model the ambiguous belief when the evidence is insufficient (even missing). Formally, we have:

Definition 1. *Let Θ be a finite set of mutually disjoint atomic elements, called a* **frame of discernment** *and 2^Θ be the set of all the subsets of Θ. Then a* **basic probability assignment**, *or called a* **mass function**, *is a mapping of $m : 2^\Theta \to [0,1]$, which satisfies $m(\emptyset) = 0$ and $\sum_{A\subseteq\Theta} m(A) = 1$. Subset A ($\subset \Theta$) satisfying $m(A) > 0$ is called a* **focal element** *of mass function m.*

The more elements in focal elements of a mass function and the bigger the mass function values of focal elements, the more ambiguous the belief is. This can be captured by the ambiguity degree of the mass function, which is defined as follows [12–15]:

Definition 2. *The* **ambiguity degree** *of a mass function m over discernment frame Θ, denoted as δ, is given by*

$$\delta(m) = \frac{\sum_{A\subseteq\Theta} m(A)\log_2|A|}{\log_2|\Theta|}, \tag{1}$$

where $|A|$ and $|\Theta|$ are the cardinality of sets A and Θ, respectively.

Formula (1) reflects well that the more ambiguous the belief, the higher the ambiguity degree. In particular, $\forall A \subseteq \Theta$, if $|A| = 1$ when $m(A) > 0$, then $\delta(m) = 0$, which represents the precise belief; and if $m(\Theta) = 1$, then $\delta(m) = 1$, which represents the most ambiguous belief.

Strat [11] defines the expected utility interval as follows:

Definition 3. *Given a choice of c corresponding to mass function m over the possible consequence set, denoted as Θ, of all the choices, let $u(a_i, c)$ be the utility of choice c corresponding to element a_i in the focal element A. Then the* **expected utility interval** *of choice c is $EUI(c) = [\underline{E}(c), \overline{E}(c)]$, where*

$$\underline{E}(c) = \sum_{A\subseteq\Theta} \min\{u(a_i, c) \mid a_i \in A\} m(A), \tag{2}$$

$$\overline{E}(c) = \sum_{A\subseteq\Theta} \max\{u(a_i, c) \mid a_i \in A\} m(A). \tag{3}$$

Here $\underline{E}(c)$ and $\overline{E}(c)$ are called the lower and high boundaries of choice c, respectively.

If two choices' expected utility intervals do not overlap, people can make a choice easily (i.e., choose the one which lower boundary is higher than the other's upper boundary); otherwise, the choice is unclear [11, 12]. In this case, more evidence is required in order to make the expected utility intervals no longer overlap. However, what should we do if we cannot have more evidence? To handle this issue, the work of [12] presents a reasonable way to get a point-valued preference degree of a choice with the expected utility interval as follows:

Definition 4. *Given mass function m over discernment frame Θ corresponding to choice c, the* **preference degree** *of choice c is defined as:*

$$\rho(c)=\frac{2\underline{E}(c)+(1-\delta(m))(\overline{E}(c)-\underline{E}(c))}{2}, \tag{4}$$

where $[\underline{E}(c),\overline{E}(c)]$ is the expected utility interval of choice c and $\delta(m)$ is the ambiguity degree of mass function m.

In the ambiguous environment, first we can get the expected utility interval by formulas (2) and (3), and then we can get the point-valued preference degree of each choice by formula (4).

3 Ambiguous Security Games

This section presents our security game model with ambiguous types of the attacker.

Definition 5. *An* **ambiguous security game** *is a tuple of $(N, T_{\boldsymbol{a}}, m_{\boldsymbol{d}}, S, X, U)$, where:*

(i) $N=\{\mathbf{d},\boldsymbol{a}\}$ *is the set of players, where* $\mathbf{d}$ *stands for the defender and* $\boldsymbol{a}$ *stands for the attacker;*
(ii) $T_{\boldsymbol{a}}=\{t_1,\cdots,t_m\}$ *is the attacker's disjoint type set.*
(iii) $m_{\boldsymbol{d}}$ *is the defender's mass function over the frame of discernment* $T_{\boldsymbol{a}}$*;*
(iv) $S=S_{\boldsymbol{d}}\times S_{\boldsymbol{a}}$*, where* $S_{\boldsymbol{d}}=S_{\boldsymbol{a}}=\{s_1,\ \cdots,\ s_n\}$ *is the pure strategy set of the attacker and the defender, representing attacking or defending targets, and n is the target number;*
(v) $X=\{X_i \mid i=1,\cdots,n\}$*, where* $X_i=\{p_{i,1},\cdots,p_{i,n}\}$ *is one mixed strategy of the defender,* $p_{i,j}$ *is the probability distribution for the pure strategy* s_i *(satisfying* $p_{i,j}\in(0,1]$ *and* $\sum_{p_{i,j}\in X_i}p_{i,j}=1$*), and given the defender's mixed strategy* X_i*,* $s_{\boldsymbol{a}}(t,X_i)(\in S_{\boldsymbol{a}})$ *is the strategy taken by each type t of the attacker and* $s_{\boldsymbol{a}}(T_{\boldsymbol{a}},X_i)=\prod_{t\in T_{\boldsymbol{a}}}s_{\boldsymbol{a}}(t,X_i)$ *is the strategy profile chosen by all the types of the attacker; and*
(vi) $U=\{u_{\boldsymbol{d}}(s,t),u_{\boldsymbol{a}}(s,t)\mid s\in S,t\in T_{\boldsymbol{a}}\}$*, where* $u_{\boldsymbol{d}}(s)$ *is the defender's payoff function from strategy profile s to* $\mathbb{R}$ *given the attacker of type t,* $u_{\boldsymbol{a}}(s,t)$ *is the payoff function of the attacker of type t over strategy profile s to* $\mathbb{R}$*.*

From the above definition, we know that in security games with ambiguous attacker types first the defender commits an optimal mixed strategy to the attacker, and then the attacker tries to find the optimal strategy for himself. However, the information about the attacker's types for the defender is ambiguous, *i.e.*, the defender is unclear about the probability of each type of the attacker, but only knows a probability over a set of types for lack of sufficient evidence. Moreover, our security games with ambiguous attacker types are different from the Bayesian Stackelberg games because Bayesian Stackelberg games only consider the precise probability of the attacker's types.

Given a target of s_i, the attacker of type t receives reward $U_{\boldsymbol{a}}^{+}(s_i,t)$ if the attacker attacks target s_i, which is not covered by the defender; otherwise, the attacker receives penalty $U_{\boldsymbol{a}}^{-}(s_i,t)$. Correspondingly, the defender receives penalty

$U_{\boldsymbol{d}}^{-}(s_i,t)$ in the former case and receives reward $U_{\boldsymbol{d}}^{+}(s_i,t)$ in the latter case. The outcomes of a security game is only relevant to whether or not the attack is successful [3, 5]. So, the payoffs of a pure strategy profile with respect to the attacker and the defender can be calculated, respectively, by:

$$u_{\boldsymbol{d}}(s,t) = x \times U_{\boldsymbol{d}}^{-}(s_i,t) + (1-x) \times U_{\boldsymbol{d}}^{+}(s_i,t); \tag{5}$$

$$u_{\boldsymbol{a}}(s,t) = x \times U_{\boldsymbol{a}}^{+}(s_i,t) + (1-x) \times U_{\boldsymbol{a}}^{-}(s_i,t), \tag{6}$$

where $x \in \{0,1\}$ and $x=0$ means *fail* while $x=1$ means *success*. In a security game, the payoff could be non-zero-sum [3–5] and so we can suppose $U_{\boldsymbol{a}}^{+}(s_i,t) > U_{\boldsymbol{a}}^{-}(s_i,t)$ and $U_{\boldsymbol{d}}^{+}(s_i,t) > U_{\boldsymbol{d}}^{-}(s_i,t)$ without loss of generality.

Now we will design an algorithm to find an optimal mixed strategy for the defender. In our model, we deal with the security games with ambiguous attacker types, which are only for the defender. That is, the attacker knows the defender's strategy with perfect information.

To select an optimal mixed strategy from all the mixed strategies, the defender should consider the optimal strategy of the attacker against every mixed strategy of the defender because the attacker can observe the defender's strategy and thus knows the probability of each target defended by the defender. The attacker of different types may have different preference degrees on the same strategy. So, the defender should consider all the optimal strategies of all the attacker's types. Formally, we have:

Definition 6. *Given defender's mixed strategy* $X_i = \{p_{i,1}, \cdots, p_{i,n}\}$, $s_{\boldsymbol{a}}^*(t, X_i) \in S_{\boldsymbol{a}}$ *is an* **optimal response strategy** *taken by every type* t *of the attacker if*

$$\forall s_{\boldsymbol{a}}(t,X_i) \in S_{\boldsymbol{a}}, U_{\boldsymbol{a}}(X_i, s_{\boldsymbol{a}}^*(t,X_i), t) \geq U_{\boldsymbol{a}}(X_i, s_{\boldsymbol{a}}(t,X_i), t), \tag{7}$$

and satisfies that if there is some strategy $s_{\boldsymbol{a}}(t, X_i)$ *such that*

$$U_{\boldsymbol{a}}(X_i, s_{\boldsymbol{a}}^*(t,X_i), t) = U_{\boldsymbol{a}}(X_i, s_{\boldsymbol{a}}(t,X_i), t), \tag{8}$$

then

$$\textstyle\sum_{l=1}^{n} p_{i,l} u_{\boldsymbol{d}}(s_l, s_{\boldsymbol{a}}^*(t,X_i), t) > \sum_{l=1}^{n} p_{i,l} u_{\boldsymbol{d}}(s_l, s_{\boldsymbol{a}}(t,X_i), t), \tag{9}$$

where $u_{\boldsymbol{d}}(s_l, s_{\boldsymbol{a}}(t,X_i), t)$ *is calculated by formula (5) and*

$$U_{\boldsymbol{a}}(X_i, s_{\boldsymbol{a}}(t,X_i), t) = \textstyle\sum_{l=1}^{n} p_{i,l} u_{\boldsymbol{a}}(s_l, s_{\boldsymbol{a}}(t,X_i), t), \tag{10}$$

where $u_{\boldsymbol{a}}(s_l, s_{\boldsymbol{a}}(t,X_i), t)$ *is calculated by formula (6). All types' optimal response strategies form an optimal response strategy profile, i.e.,*

$$\prod_{t \in T_{\boldsymbol{a}}} s_{\boldsymbol{a}}^*(t, X_i) = s_{\boldsymbol{a}}^*(T_{\boldsymbol{a}}, X_i) \tag{11}$$

That is, the attacker of every type will take the strategy that can maximise not only the attacker's payoff but also the defender's payoff in order to implement his payoff as the assumption in Stackelberg games [1, 16]. Actually, the attacker's optimal strategy is corresponding to the mixed strategy of the defender. So, to implement his maximised payoff, he should ensure that the defender selects that strategy by maximising the defender's payoff.

Similarly, given the attacker's strategy, if the defender has two indifferent mixed strategies, the defender should consider the attacker's preference ordering for the strategies based on his belief about the attacker's types.

Definition 7. *In ambiguous security game (N, T_a, m_d, S, X, U), the* **attacker's total preference degree** *over mixed strategy X_i is defined as:*

$$\rho_a(X_i) = \frac{2\underline{E}_a(X_i)+(1-\delta(m_d))(\overline{E}_a(X_i)-\underline{E}_a(X_i))}{2}, \tag{12}$$

where $\delta(m_d)$ is the ambiguity degree of mass function m_d and

$$\underline{E}_a(X_i) = \sum_{\tau\subseteq T_a} \min\{U_a(X_i, s_a^*(t, X_i), t) \mid t \in \tau\} m_d(\tau), \tag{13}$$

$$\overline{E}_a(X_i) = \sum_{\tau\subseteq T_a} \max\{U_a(X_i, s_a^*(t, X_i), t) \mid t \in \tau\} m_d(\tau), \tag{14}$$

where $U_a(X_i, s_a^*(t), t)$ is calculated by formula (10).

In the above definition, formula (12) is the variant of formula (4). And formulas (13) and (14) are the variants of formulas (2) and (3), respectively.

After the attacker observes the defender's mixed strategy, he can have an optimal response strategy against the defender's mixed strategy. The attacker of every type may have different optimal strategies, which should be considered by the defender. For the ambiguous information about the attacker's types, we should use the ambiguous game model to handle. Then, the defender can find his optimal strategy from his all mixed strategies. Formally, we have:

Definition 8. *In ambiguous security game (N, T_a, m_d, S, X, U), given the defender's mixed strategy $X_i = \{p_{i,1}, \cdots, p_{i,n}\}$, suppose optimal response strategies of all the types of the attacker form an optimal response strategy profile $s_a^*(T_a, X_i^*)$. Then X_i^* is the* **defender's optimal mixed strategy** *if:*

$$\forall X_i \in X, \sum_{p_{i,l}\in X_i^*} p_{i,l}\rho_d(s_l, s_a^*(T_a, X_i^*)) \geq \sum_{p_{i,l}\in X_i} p_{i,l}\rho_d(s_l, s_a^*(T_a, X_i)), \tag{15}$$

and satisfies if there is some strategy X_i such that

$$\sum_{p_{i,l}\in X_i^*} p_{i,l}\rho_d(s_l, s_a^*(T_a, X_i^*)) = \sum_{p_{i,l}\in X_i} p_{i,l}\rho_d(s_l, s_a^*(T_a, X_i)), \tag{16}$$

then for the attacker's total payoff:

$$\rho_a(X_i^*) < \rho_a(X_i), \tag{17}$$

where $\rho_a(X_i)$ is calculated by formula (12) and

$$\rho_d(s_l,s_a^*(T_a, X_i))=\frac{2\underline{E}_d(s_l,s_a^*(T_a,X_i))+(1-\delta(m))(\overline{E}_d(s_l,s_a^*(T_a,X_i))-\underline{E}_d(s_l,s_a^*(T_a,X_i)))}{2}, \tag{18}$$

where $l \in \{1, \cdots, n\}$, $\overline{E}_d(s_l, s_a^*(T_a, X_i))$ and $\underline{E}_d(s_l, s_a^*(T_a, X_i))$ are calculated by formulas (13) and (14), respectively, by replacing $U_a(X_i, s_a^*(t, X_i), t)$ with $u_d(s_l, s_a^*(t, X_i), t)$ (which is calculated by formula (5)).

Algorithm 1. D-S Theory Based Solution Algorithm of SGAAT

Input: Strategy: *The strategy set*;
Type: *The type set of the attacker*;
X: *The mixed strategy set*;
U_d^+, U_d^-, U_a^+, U_a^-: *The reward and penalty for the defender and the attacker over each pure strategy profile*;
m_d: *Defender's mass function over the set Type* ;
Output: X_i^*;

for all $t \in T_a$ **do**
  for all $s \in S$ **do**
    Use formulas (5) and (6) to get both players' payoffs of every pure strategy profile s when the attacker is the type of t;
  end for
end for
for all $X_i \in X$ **do**
  for all $t \in T_a$ **do**
    Use formulas (7)–(9) to find the optimal strategy, $s_a^*(t, X_i)$, of the attacker with type t;
  end for
  Let $s_a^*(T_a, X_i)$ (i.e., $\prod_{t \in T_a} s_a^*(t, X_i)$) be the optimal strategy profile of the attacker.
  for all $s_l \in S_d$ **do**
    Use formula (18) to get the defender's payoff of the optimal strategy profile $s_a^*(T_a, X_i)$ of the attacker;
  end for
end for
Use formulas (15)–(17) to find the defender's optimal mixed strategy X_i^*;

In the above definition, formula (18) is the variant of formula (4). The defender should also take the strategy that can maximise his payoff when minimising the attacker's total payoff. This is because the defender commits a mixed strategy to the attacker first and thus determines the attacker's payoff.

Then we have our D-S theory based solution algorithm for security games with ambiguous attacker types (SGAAT) as shown in Algorithm 1. In our algorithm, the attacker's type and the reward and the penalty of the defender and the attacker are determined by domain experts [4]. For every mixed strategy of the defender, the attacker will have an optimal strategy profile formed by every type's optimal strategy. Then, the defender has a preference degree for every mixed strategy. Finally, the defender can find his optimal mixed strategy.

Clearly, our ambiguous security game can cover a Bayesian Stackelberg game [4] as a special case when the defender has a precise belief about the attacker types and then commits a mixed strategy to the attacker. Conitzer *et al.* [16] show that finding an optimal mixed strategy to commit in a Bayesian Stackelberg game is NP-hard. The computing complexity of our algorithm is similar as proved in the following theorem:

Theorem 1. *In security games with ambiguous attacker types, it is NP-hard to find an optimal mixed strategy to commit to.*

Proof. In the ambiguous security game, the defender commits a mixed strategy to the attacker, which means that the ambiguous security game is a Stackelberg game. Now, given $\tau \subseteq T_{\boldsymbol{a}}$, if $m_{\boldsymbol{d}}(\tau) > 0$ then $|\tau| = 1$. So, the probabilities of the attacker's types are precise. So, Bayesian Stackelberg games are special cases of our security games with ambiguous attacker types. In security games with ambiguous attacker types, we have to handle the ambiguous information about the attacker type using the ambiguity decision framework. So, finding an optimal mixed strategy to commit to in security games with ambiguous attacker types is not easier than finding an optimal mixed strategy to commit to in Bayesian Stackelberg games. However, by [16], it is NP-hard to find an optimal mixed strategy to commit to in Bayesian Stackelberg games. Then, all NP-complete problems, which can be reduced to Bayesian Stackelberg games with two players to find the optimal mixed strategy, can also be reduced to security games with ambiguous attacker types to find the optimal mixed strategy. So, in security games with ambiguous attacker types, it is NP-hard to find an optimal mixed strategy to commit to. □

4 Evaluation

In this section, we are going to evaluate our model by lots of experiments.

In the field of security games [1, 4, 5], most investigations could not handle the ambiguous information. In their models, they need the precise probability of the attacker' types. So, if we use their models to handle the ambiguous information, we have to transfer imprecise probabilities to precise probabilities using, for example, the transferable belief method [17] as follows:

Definition 9. *In a security game* $(N, T_{\boldsymbol{a}}, m_{\boldsymbol{d}}, S, X, U)$, *let* $m_{\boldsymbol{d}}$ *be mass function over* $T_{\boldsymbol{a}}$ *and* $\tau \subset T_{\boldsymbol{a}}$. *Then the* **uniform random probability (URP)** *of every element* $t \in \tau$ *is* $P(t) = \sum_{t \in \tau} \frac{m_d(\tau)}{|\tau|}$.

In a security game, the defender must find the optimal mixed strategy to avoid the attacking [4]. If there is an attack, when we cannot prevent from it, we hope that the less the loss the better because nobody wants to see the sad event. So, to be safe, we should make the loss as little as possible.

Observation 1. *Our model can guarantee more safety than the model based on uniform random probability in the ambiguous environment.*

Now we use experiments to show that, the worst case in our model is better than the one in the model based on uniform random probability (URP). In our experiment, the defender and attacker's rewards range from 1 to 10 and the penalties range from -1 to -10 randomly. There are two subsets of the attacker's type set for the defender: one contains only one element, denoted as A, the other one contains the rest elements, denoted as B, of type set $T_{\boldsymbol{a}}$. In the D-S theory based solution algorithm of SGAAT, the belief for only one element set is $\frac{1}{|T_a|}$, called $m_{\boldsymbol{d}}(A)$, and the belief for the other part of $T_{\boldsymbol{a}}$ is $1 - \frac{1}{|T_a|}$, called $m_{\boldsymbol{d}}(B)$. Then, by formula (1), the more types the attacker has, the higher the ambiguity

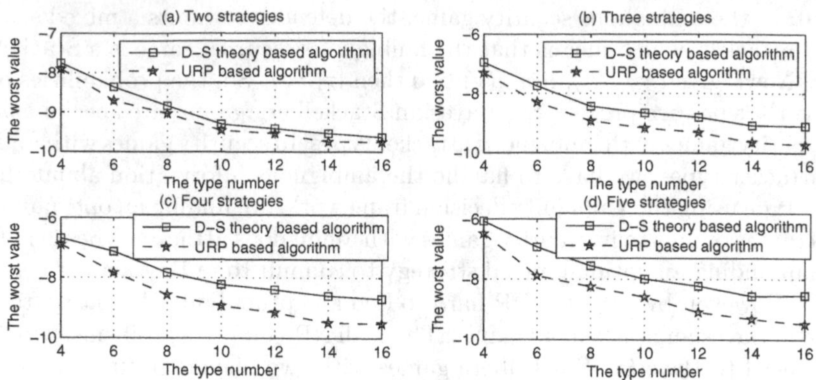

Fig. 1. The defender's worst penalty in security games with ambiguous attacker types for the defender

degree of the mass function. In the URP based solution algorithm of SGAAT, the belief for only one element set is $m_{\boldsymbol{d}}(A)$ and the belief for every element in the other part of $T_{\boldsymbol{a}}$ is $\frac{m_d(B)}{|T_a|-1}$. In our experiments, we find the optimal strategies for two models, and continue to find the worst penalty that the defender may get when the defender takes an optimal strategy. That is the worst penalty for the defender when the attacker attacks the target successfully. For every type and every strategy, we run 100 experiments and get the average value about the worst penalties that the defender may get in all experiments. For example, in every experiment, by Algorithm 1, X_i^* is the defender's optimal mixed strategy and $s_{\boldsymbol{a}}^*(t, X_i^*)$ is the optimal strategy of every type t of the attacker, then the worst penalty at the j-th experiment is $U_j^- = \min\{U_{\boldsymbol{d}}^-(s_{\boldsymbol{a}}^*(t, X_i^*), t) \mid t \in T_{\boldsymbol{a}}\}$, and thus the average worst penalty of 100 experiments is:

$$\widetilde{U}^- = \frac{\sum_{1 \le j \le 100} U_j^-}{100}. \tag{19}$$

The results of our experiments are shown in Figure 1. The cases of two, three, four, and five strategies (targets) are shown in Figure 1 (a), (b), (c), and (d), respectively. When there are four or more types of the attacker in the D-S theory based solution algorithm of SGAAT, the probability $m_{\boldsymbol{d}}(B)$ contains more than one type, which makes the belief ambiguous by formula (1), (*i.e.*, $\delta(m_{\boldsymbol{d}}) > 0$). However, in the URP based solution algorithm of SGAAT, the probability is still precise by $\frac{m_d(B)}{|T_a|-1}$. Then, when there are three or more types, the games with the URP based solution algorithm of SGAAT are still Bayesian games, but the games with the D-S theory based solution algorithm of SGAAT are not. In these cases, as shown in Figure 1, the worst penalties obtained by the D-S theory based solution algorithm of SGAAT are strictly bigger than the worst penalties obtained by the URP based solution algorithm of SGAAT.[1] For example, by formula (19), when

[1] In the D-S theory based solution algorithm of SGAAT, the types have imprecise probability except the first type. However, in the URP based solution algorithm of SGAAT, the types have precise probability.

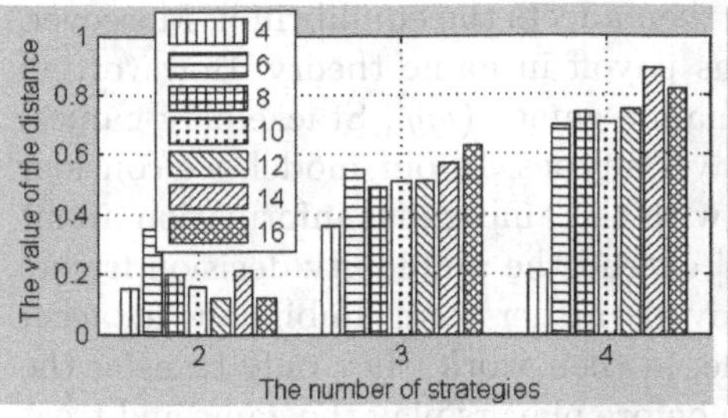

(a) The distance of worst penalties when the strategy number changes over each type number (from 4 to 16)

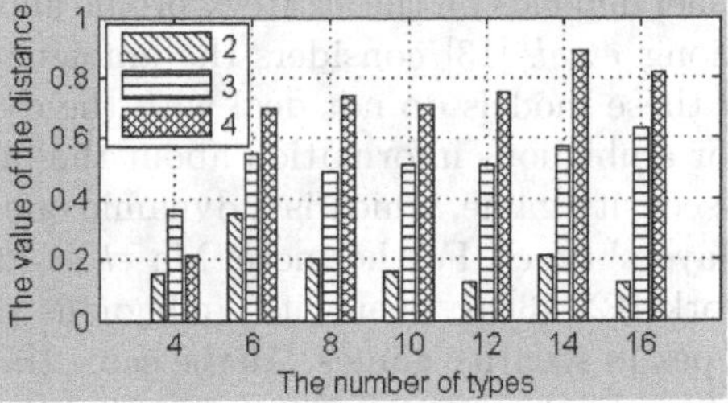

(b) The distance of worst penalties when the type number changes over each strategy number (from 2 to 4)

Fig. 2. Comparing the worst penalties of D-S theory based solution with URP based one

there are six types of the attacker, the worst penalties obtained by the D-S theory based solution algorithm of SGAAT are -7.75, -7.13, and -6.93 with respect to three, four, and five strategies respectively, whereas the worst penalties obtained by the URP based solution algorithm of SGAAT are -8.3, -7.83, and -7.89 with respect to three, four, and five strategies, respectively.

Figure 2(a) shows that the more strategies, the larger distance of the worst penalties between the D-S theory based solution algorithm of SGAAT and the URP based solution algorithm of SGAAT. For example, when the attacker has 16 types, the distances are 0.12, 0.63, and 0.82 for two, three, and four strategies, respectively. In the cases of three strategies and four strategies, as shown in Figure 2(b), the more types the attacker has, the larger distance of the worst penalties on the whole. For example, when there are three strategies, the distances are 0.36, 0.49, and 0.63 for four, eight, and sixteen types, respectively. So, the higher the ambiguity degree of the mass function is, the safer our model. Thus, we have:

Observation 2. *The more ambiguous the available information about the types of the attacker, the safer our model than the uniform random probability model.*

5 Related Work

Recently, many researchers have tried to improve the algorithms for solving security games [1, 4, 5], but most of them can only deal with the defender's precise probabilities of the attacker's types when they try to handle the uncertainty in the real world [3, 5]. For the ambiguous information, to use their model, the imprecise probabilities have to be transferred to precise ones. However, we have shown that our model can guarantee the safer result than the model based on uniform random probability in ambiguous environment.

In the field of games under ambiguity, Eichberger and Kelsey [18] define a notion of equilibrium under ambiguity to explain the hypothesis that the result, from changing an apparently irrelevant parameter, contradicts Nash equilibrium. In their two-person games, they view their opponents' behavior as ambiguous based on non-additive beliefs. And Marco and Romaniello [19] try to use an ambiguity model to remedy some defects of Nash equilibrium. In their model, the belief depends on the strategy profile and then affects the equilibrium. Moreover, Xiong *et al.* [13] considers the ambiguous payoff in game theory. However, all of these models do not deal with the dynamic games (*e.g.*, Stackelberg games) nor ambiguous information about the player's types. In our model, we consider a security game, which is a dynamic one with the ambiguous information about players' types. Furthermore, Ma *et al.* [6] extend the ambiguity decision framework [12, 13] to deal with ambiguous payoffs rather than ambiguous attacker types in security games. At the same time, in their work, they only transfer the interval-valued payoff to point-valued one before players play the game and treat security games as static games. That is, the ambiguous information is not reflected in the procedure of security games in their model. However, we consider the defender's optimal strategy influenced by the imprecise probability of the attacker' types. We also propose an algorithm to find the optimal strategy for the defender in security games, but they do not.

In addition, Wang *et al.* [20] deal with fuzzy types of players, but this model only considers static Bayesian games, while our model is dynamic. Meanwhile, this model cannot model the types with imprecise probability, but ours does.

6 Conclusion and Future work

This paper proposes a model to handle the ambiguous information about the attacker's types in security games. Moreover, we develop a D-S theory based algorithm to find the defender's optimal mixed strategy and discuss their computing complexity. Furthermore, we evaluate our model by lots of experiments. From the experiments, we find: (i) our model can guarantee more safety than the model based on uniform random probability in the ambiguous environment; and (ii) the more ambiguous the available information about the types of the attacker, the safer our model is than the model based on uniform random probability. So, our model can well handle security games in the ambiguous environment, In the future, we will develop more efficient algorithms to solve our ambiguous security game and apply it to real life to protect critical public targets.

Acknowledgments. The authors appreciate the anonymous referees for their insightful comments. Moreover, this paper is partially supported by Bairen Plan of Sun Yat-sen University, Raising Program of Major Project of Sun Yat-sen University (No. 1309089), National Natural Science Foundation of China (No. 61173019), MOE Project of Key Research Institute of Humanities and Social Sciences at Universities (No. 13JJD720017) China, and Major Projects of the Ministry of Education China (No. 10JZD0006).

References

1. Paruchuri, P., Pearce, J.P., Marecki, J., Tambe, M., Ordonez, F., Kraus, S.: Playing games for security: An efficient exact algorithm for solving Bayesian Stackelberg games. In: Proceedings of the 7th International Joint Conference on Autonomous Agents and Multiagent Systems, vol. 2, pp. 895–902 (2008)

2. Pita, J., Jain, M., Marecki, J., Ordóñez, F., Portway, C., Tambe, M., Western, C., Paruchuri, P., Kraus, S.: Deployed ARMOR protection: The application of a game theoretic model for security at the Los Angeles International Airport. In: Proceedings of the 7th International Joint Conference on Autonomous Agents and Multiagent Systems: Industrial Track, pp. 125–132 (2008)
3. Korzhyk, D., Yin, Z., Kiekintveld, C., Conitzer, V., Tambe, M.: Stackelberg vs. Nash in security games: An extended investigation of interchangeability, equivalence, and uniqueness. Journal of Artificial Intelligence Research 41(2), 297–327 (2011)
4. Tambe, M.: Security and game theory: Algorithms, deployed systems, lessons learned. Cambridge University Press, New York (2011)
5. Yang, R., Kiekintveld, C., OrdóñEz, F., Tambe, M., John, R.: Improving resource allocation strategies against human adversaries in security games: An extended study. Artificial Intelligence 195, 440–469 (2013)
6. Ma, W., Luo, X., Liu, W.: An ambiguity aversion framework of security games under ambiguities. In: Proceedings of the 23rd International Joint Conference on Artificial Intelligence, pp. 271–278 (2013)
7. Zimmermann, E.: Globalization and terrorism. European Journal of Political Economy 27(suppl. 1), S152–S161 (2011)
8. Cronin, A.K.: Behind the curve: Globalization and international terrorism. International Security 27(3), 30–58 (2003)
9. Shafer, G.: A mathematical theory of evidence. Princeton University Press, Princeton (1976)
10. Liu, L., Yager, R.R.: Classic works of the Dempster-Shafer theory of belief functions: An introduction. In: Yager, R.R., Liu, L. (eds.) Classic Works of the Dempster-Shafer Theory of Belief Functions. Studies in Fuzziness and Soft Computing, vol. 219, pp. 1–34. Springer, Heidelberg (2008)
11. Strat, T.M.: Decision analysis using belief functions. International Journal of Approximate Reasoning 4(5), 391–417 (1990)
12. Ma, W., Xiong, W., Luo, X.: A model for decision making with missing, imprecise, and uncertain evaluations of multiple criteria. International Journal of Intelligent Systems 28(2), 152–184 (2013)
13. Xiong, W., Luo, X., Ma, W.: Games with ambiguous payoffs and played by ambiguity and regret minimising players. In: Thielscher, M., Zhang, D. (eds.) AI 2012. LNCS, vol. 7691, pp. 409–420. Springer, Heidelberg (2012)
14. Dubois, D., Prade, H.: A note on measures of specificity for fuzzy sets. International Journal of General System 10(4), 279–283 (1985)
15. Jousselme, A.L., Liu, C., Grenier, D., Bosse, E.: Measuring ambiguity in the evidence theory. IEEE Transactions on Systems, Man and Cybernetics, Part A: Systems and Humans 36(5), 890–903 (2006)
16. Conitzer, V., Sandholm, T.: Computing the optimal strategy to commit to. In: Proceedings of the 7th ACM Conference on Electronic Commerce, pp. 82–90 (2006)
17. Smets, P., Kennes, R.: The transferable belief model. Artificial Intelligence 66(2), 191–234 (1994)
18. Eichberger, J., Kelsey, D.: Are the treasures of game theory ambiguous? Economic Theory 48(2-3), 313–339 (2011)
19. Marco, G.D., Romaniello, M.: Beliefs correspondences and equilibria in ambiguous games. International Journal of Intelligent Systems 27(2), 86–107 (2012)
20. Wang, C., Tang, W., Zhao, R.: Static Bayesian games with finite fuzzy types and the existence of equilibrium. Information Sciences 178(24), 4688–4698 (2008)

A Mechanism to Improve Efficiency for Negotiations with Incomplete Information

Quoc Bao Vo

Faculty of Information & Communication Technologies
Swinburne University of Technology,
Melbourne, Australia
bvo@swin.edu.au

Abstract. Classic results in bargaining theory state that private information necessarily prevents the bargainers from reaping all possible gains from trade. In this paper we propose a mechanism for improving efficiency of negotiation outcome for multilateral negotiations with incomplete information. This objective is achieved by introducing biased distribution of resulting gains from trade to prevent bargainers from misrepresenting their valuations of the negotiation outcomes. Our mechanism is based on rewarding concession-making agents with larger shares of the obtainable surplus. We show that the likelihood for the negotiators to reach agreement is accordingly increased and the negotiation efficiency is improved.

1 Introduction

Conducted experiments have shown that more often than not negotiations reach inefficient compromises [1, 2]. In relation to this phenomenon, a central question in research in economics and political science is to understand the difficulties the parties have in reaching mutually beneficial agreements. The classic result discovered by Myerson and Satterthwaite ([3]) indicates that uncertainty about whether gains from trade are possible necessarily prevents full efficiency. More precisely, their result states that, given two parties with independent private valuations, *ex post* efficiency is attainable if and only if it is common knowledge that gains from trade exist. Inefficiencies are a consequence of the incentives to misrepresent a bargainers' valuations between those with private information. The mechanism propose in this paper aims to remove such incentives by devising ways to distribute the resulting gains from trade in such a way that the bargainer who can still make a concession becomes more willing to actually make that concession.

Most games with incomplete information are modeled using some particular information structures and strategic devices to allow agents with private information to perform some action to send out a signal indicating their types.[1]

[1] The "type" of a player embodies any private information that is relevant to the player's decision making.

S. Cranefield and A. Nayak (Eds.): AI 2013, LNAI 8272, pp. 26–31, 2013.

Upon observing the action by the agent with private information, other agents can decide their own best course of actions. In various models of the bargaining problem, several mechanisms have been used to allow negotiators to communicate their private valuations to other parties.[2] These mechanisms include the use of costly delays (i.e., time delays when there are discount factors) [4, 5], transaction costs [6], or bargaining deadlines [7, 8]. Our approach, on the other hand, applies a mechanism of biased distribution of the observable gains from trade to encourage the parties with private information to truthfully reveal their types. To facilitate this mechanism we employ a negotiation protocol to allow the bargainers to concurrently submit their proposals.

The paper is organised as follows. Section 2 gives an overview of the negotiation model, including the negotiation protocol. In Section 3, we describe the use of biased surplus division as a strategic device for negotiation with incomplete information, focussing on the case of bilateral negotiation. Our results are extended for the case of multilateral negotiation in Section 4 before we conclude the paper with a discussion.

2 A Multilateral Negotiation Model

Consider the multilateral negotiation as an allocation problem with n agents. Given the set of all possible allocations A, agent i has a valuation $v_i(a, t_i)$ for the allocation $a \in A$ when its type is t_i. Assume that the status quo allocation $\tilde{a} \in A$ defines the agents' reservation utilities. We will normalise each valuation function v_i such that $v_i(\tilde{a}, t_i) = 0$. Assume also that the maximum amount of resource available for this allocation is R. Thus, an allocation $(a_1, \ldots, a_n)$ is ***feasible*** iff $\sum_{i=1}^{n} a_i \leq R$.

If the status quo allocation $\tilde{a} = (\tilde{a}_1, \ldots, \tilde{a}_n)$ is feasible then gains from trade are possible: $\tilde{G} = R - \sum_{i=1}^{n} \tilde{a}_i$. Because each agent's status quo allocation $\tilde{a}_i$ is her private information, whether or not gains from trade are possible is not common knowledge. According to Myerson and Satterthwaite's ([3]) result, this source of uncertainty is the cause for negotiation inefficiency. Throughout this paper, we assume that each agent's utility is independent of the allocations received by other agents, and that $\tilde{a}_i < R$; otherwise, agent i would not participate in the negotiation in the first place.

2.1 The Negotiation Protocol

The negotiation protocol used in our model is similar to the *Monotonic Concession Protocol* [9, 10] which proceeds in rounds. In each round, all agents make simultaneous allocation claims for themselves, i.e. they each claims an allocation a_i $(0 \leq a_i \leq R)$. The combination of all claims makes up a potential allocation $a = (a_1, \ldots, a_n)$. If a is a feasible allocation, i.e. $\sum_{i=1}^{n} a_i \leq R$, then an ***agreement*** is reached with each agent being allocated what it claims during this

[2] The literature of automated negotiation usually uses the agents' reserve prices to indicate their valuations.

round and the ***observable surplus*** $\sigma = R - \sum_{i=1}^{n} a_i$ will be divided between the agents.

REMARK: The surplus to be distributed once an agreement is reached is the *observable gains from trade* based on the agreement which could be smaller than the actual gains from trade $\tilde{G}$.

Definition 1. *Let a tuple* $\Delta = (\delta_1, \ldots, \delta_n)$ *be such that* $\forall i = 1 \ldots n. \delta_i \geq 0$ *and* $\sum_{i=0}^{n} \delta_i = 1$. *Given the agreed allocation* $a = (a_1, \ldots, a_n)$ *and the resulting surplus* $\sigma \geq 0$, *the **Δ-surplus division** provides agent i the allocation* $a_i + \delta_i \sigma$.

The **negotiation protocol** can now be described as follows:

(1) In the first round, each agent i makes an initial claim a_i^0;
(2) In each subsequent round $t > 0$, an agent i either makes a *concession* $a_i^t < a_i^{t-1}$, or stays with its previous claim $a_i^t = a_i^{t-1}$;
(3) Step (2) is iterated until either an agreement is reached or a conflict situation arises in which no agent makes a concession. When a conflict situation concludes the negotiation, all agents leave the negotiation with the status quo allocation $\tilde{a}$.

It's straightforward to see that the above negotiation protocol terminates after a finite number of rounds. In particular, the well-known seller-buyer bargaining problem can be straightforwardly rendered in our negotiation model as shown in the next section.

3 Surplus Division as a Strategic Device

Research from various disciplines as well as experimental and empirical studies has shown that cooperation is the key to achieving optimality in most interactions between a group of individuals, particularly in negotiations [11–13]. However, given the competitive nature of negotiation, not all negotiators behave cooperatively. The main purpose of our approach is to devise mechanisms that encourage the negotiators to behave cooperatively rather than competitively. We implement this via a mechanism of biased distribution of the observable surplus.

3.1 The Bilateral Bargaining Problem

Consider a bargaining between a buyer and a seller of an indivisible good. Each agent has two possible valuations high (h) and low (l) of the good that defines her types:

	Valuation	
Type	Seller	Buyer
weak	a_1^l	a_2^h
strong	a_1^h	a_2^l

where $a_1^l \leq a_2^l < a_1^h \leq a_2^h$.

Notice that gains from trade are possible unless both agents are of type strong. That is, when the buyer has a low valuation of the item a_2^l while the seller values the item high a_1^h and because $a_2^l < a_1^h$, mutually beneficial trade is not possible. We let π_1 (resp. π_2) denote the initial probability that the seller (resp. the buyer) of unknown type is strong. These probabilities are exogenously given, and are common knowledge.

Note that the above bargaining problem has the same information setting as the problem considered by Chatterjee and Samuelson ([14]). Chatterjee and Samuelson establish a (unique) Nash equilibrium for their negotiation setting by using time as a strategic variable and imposing costly delays during negotiation. While the use of impatience, via costly delays, as the strategic information device has been a common practice in the literature of bargaining, this is certainly not applicable in every negotiation. Some researchers even claim that in some negotiations some parties may prefer a later agreement to an early one [15]. Also, a necessary consequence of costly delays is that they prevent full efficiency. Our proposed approach employs biased distribution of the resulting surplus which is costless, and thus, improve the efficiency of the negotiation and allow full efficiency to be achieved under certain conditions.

In this problem setting, an agent $i \in \{1, 2\}$ can either "play tough" (denoted by the action PT) by claiming that she is of type strong, or "play soft" (denoted by the action PS) by stating that she is of type weak. Since a strong-type agent has only one strategy of playing tough all the way, most of the analysis below will be on the weak-type agents. Without loss of generality, we assume that the default surplus division scheme Δ^d distributes the surplus equally among the agents, i.e., $\delta_i^d = \frac{1}{n}$ where n is the number of negotiators. This assumption does not affect the results of this paper.

To motivate our approach, we'll consider a simple bargaining example.

Example 1. Consider a buyer B and a seller S who negotiate over the price of an item. Assume that it's common knowledge that S's cost for the item can be either \$50 or \$53 with equal probability 0.5, and it's common knowledge that the item is available for anyone at the price \$55 but B might know someone who is willing to sell him the same item for \$52. It's common knowledge that B knows the other person with probability 0.5.

Under the default surplus division which awards half of the obtainable surplus to each of the agents, this bargaining has a Nash equilibrium. In the first round of the negotiation, both agents play tough with probability 1, regardless of their types. In the second round, while a strong-type agent continues to play PT, a weak-type agent will play PT with probability 0.2 and play PS with probability 0.8. This equilibrium gives the weak-type agents an expected utility of 1.6. Note that, as the weak-type agents do not play PS with probability 1, full efficiency is not achieved.

On the other hand, consider a biased surplus division scheme in which the agent who makes a concession in the second round of the negotiation will be rewarded by getting more than half of the obtainable surplus. In particular, by giving the agent who makes a concession in the second round of the negotiation

62.5% of the surplus, we could get the equilibrium strategy of the weak-type agent to play soft (PS) with probability 1 in the second round. In other words, through biased surplus division, the full efficiency of the negotiation in this example can be achieved.

We now provide a number of formal results for the simple bargaining problem described at the beginning of this section. We first introduce the some notations:

- The default surplus division, denoted by Δ^d, is applicable when the two agents make the same number of concessions. When the seller, agent 1, makes one concession more than the buyer, the surplus division is $(\delta_1', 1-\delta_1')$ and when the buyer makes one concession more than the seller, the surplus division is $(1-\delta_2', \delta_2')$.
- $\sigma_1 = a_2^l - a_1^l$ is the surplus obtainable when the seller plays soft (PS) and the buyer plays tough (PT).
- $\sigma_2 = a_2^h - a_1^h$ is the surplus obtainable when the seller plays tough (PT) and the buyer plays soft (PS).
- $\sigma_3 = a_2^h - a_2^l$ is the surplus obtainable when both agents play PS.
- $\gamma = a_1^h - a_2^l$ is the gap between the valuations of the strong-type agents. Note that, because of the assumption that $a_2^l < a_1^h$, $\gamma > 0$.

The following theorem introduces the main result of the paper. The first part states that by increasing the share of the observable surplus for the concession making agents, we increase the chance that, in equilibrium, the weak-type agents will play soft in the second stage of the negotiation, and thus improve the efficiency of the negotiation outcome. Note that it suffices to state the result for the buyer (i.e. agent 2) since the same result and analysis apply to agent 1.

Theorem 1. (i) *By increasing the surplus share to the concession-making agent (i.e., δ_1' and δ_2'), the probability of a weak-type agent playing PS in the second stage is increased.*

(ii) *Agent 2 plays PS with probability 1 in the second stage if*

$$\frac{\sigma_1 + \gamma - \delta_1^d\sigma_3 + (1-\delta_1')\sigma_2}{\sigma_1 + \gamma - \delta_1^d\sigma_3 + (1-\delta_1')\sigma_2 + \delta_1'\sigma_1} \leq \pi_2.$$

4 Discussion and Future Work

Bargaining with private information has been a topic of much interest for many years (see e.g., [16]). It has been established that uncertainty about whether gains from trade in a negotiation are possible necessarily prevents full efficiency, resulting in negotiation breakdowns (e.g., [2]) or costly delays (e.g., [4]). In particular, Chatterjee and Samuelson ([14]) establish a sequential equilibrium for the bilateral negotiation, albeit full efficiency is not achieved due to costly delays. In our proposed approach, we introduce a mechanism to allow efficiency of the negotiation to be improved without having to use costly signal devices. That is, our mechanism is costless.

However, when the negotiation is over issues with continuous values in which agents can make as small a concession as they wish, a manipulative agent could avoid the effects of biased surplus division by making many small concessions.

References

1. Alemi, F., Fos, P., Lacorte, W.: A demonstration of methods for studying negotiations between physicians and health care managers. Decision Science 21, 633–641 (1990)
2. Weingart, L.: Knowledge matters–the effect of tactical descriptions on negotiation behavior and outcome. Journal of Personality & Social Psychology 70(6), 1205–1217 (1996)
3. Myerson, R.B., Satterthwaite, M.A.: Efficient mechanisms for bilateral trade. Journal of Economic Theory 29, 265–281 (1983)
4. Cramton, P.: Strategic delay in bargaining with two-sided uncertainty. Review of Economic Studies 59, 205–225 (1992)
5. Jarque, X., Ponsati, C., Sakovics, J.: Mediation: incomplete information bargaining with filtered communication. Journal of Mathematical Economics 39(7), 803–830 (2003)
6. Cramton, P.C.: Dynamic bargaining with transaction costs. Manage. Sci. 37(10), 1221–1233 (1991)
7. Sandholm, T., Vulkan, N.: Bargaining with deadlines. In: National Conference on Artificial Intelligence, pp. 44–51 (1999)
8. Ma, C.T.A., Manove, M.: Bargaining with deadlines and imperfect player control. Econometrica 61(6), 1313–1339 (1993)
9. Rosenschein, J.S., Zlotkin, G.: Rules of Encounter: Designing Conventions for Automated Negotiation Among Computers. MIT Press, Cambridge (1994)
10. Endriss, U.: Monotonic concession protocols for multilateral negotiation. In: AAMAS 2006: Proceedings of the Fifth International Joint Conference on Autonomous Agents and Multiagent Systems, pp. 392–399. ACM Press, New York (2006)
11. Raiffa, H.: The Art and Science of Negotiation. Harvard University Press, Cambridge (1982)
12. Fisher, R., Ury, W.: Getting to YES: Negotiating an agreement without giving in. Random House Business Books, New York (1981)
13. Gal, Y., Pfeffer, A.: Modeling reciprocity in human bilateral negotiation. In: National Conference on Artificial Intelligence, Vancouver, British Columbia, USA (2007)
14. Chatterjee, K., Samuelson, L.: Bargaining with two-sided incomplete information: An infinite horizon model with alternating offers. Review of Economic Studies 54(2), 175–192 (1987)
15. Faratin, P., Sierra, C., Jennings, N.R.: Negotiation decision functions for autonomous agents. Int. Journal of Robotics and Autonomous Systems 24(3-4), 159–182 (1998)
16. Ausubel, L.M., Cramton, P., Deneckere, R.J.: Bargaining with incomplete information. In: Aumann, R.J., Hart, S. (eds.) Handbook of Game Theory, vol. 3, pp. 1897–1945. Elsevier Science B.V. (2002)

Protein Fold Recognition Using an Overlapping Segmentation Approach and a Mixture of Feature Extraction Models

Abdollah Dehzangi[1,2], Kuldip Paliwal[1], Alok Sharma[1,3], James Lyons[1], and Abdul Sattar[1,2]

[1] Institute for Integrated and Intelligent Systems (IIIS), Griffith University, Brisbane, Australia
[2] National ICT Australia (NICTA), Brisbane, Australia
[3] University of the South Pacific, Fiji
{a.dehzangi,k.paliwal,a.sattar}@griffith.edu.au, sharma_al@usp.ac.fj, james.lyons@griffithuni.edu.au

Abstract. Protein Fold Recognition (PFR) is considered as a critical step towards the protein structure prediction problem. PFR has also a profound impact on protein function determination and drug design. Despite all the enhancements achieved by using pattern recognition-based approaches in the protein fold recognition, it still remains unsolved and its prediction accuracy remains limited. In this study, we propose a new model based on the concept of mixture of physicochemical and evolutionary features. We then design and develop two novel overlapping segmented-based feature extraction methods. Our proposed methods capture more local and global discriminatory information than previously proposed approaches for this task. We investigate the impact of our novel approaches using the most promising attributes selected from a wide range of physicochemical-based attributes (117 attributes) which is also explored experimentally in this study. By using Support Vector Machine (SVM) our experimental results demonstrate a significant improvement (up to 5.7%) in the protein fold prediction accuracy compared to previously reported results found in the literature.

Keywords: Mixture of Feature Extraction Model, Overlapping Segmented distribution, Overlapping Segmented Auto Covariance, Support Vector Machine.

1 Introduction

Prediction of the three dimensional structure (tertiary structure) of a protein from its amino acid sequence (primary structure) still remains as an unsolved issue for bioinformatics and biological science. *Protein Fold Recognition (PFR)* is considered as an important step towards protein structure prediction problem. PFR is defined as classifying a given protein to its appropriate fold (among finite number of folds). It also provides critical information about the functionality of proteins and how they are evolutionarily related to each other. Recent

S. Cranefield and A. Nayak (Eds.): AI 2013, LNAI 8272, pp. 32–43, 2013.

advancement in the pattern recognition field stimulates enormous interest in this problem.

During the last two decades, a wide range of classifiers such as, Bayesian-based learners [1], *Artificial Neural Network (ANN)* [2], *Hidden Markov Model (HMM)* [3], Meta classifiers [4,5], *Support Vector Machine (SVM)* [6–8] and ensemble methods [1, 9, 10] have been implemented and applied to this problem. Despite the crucial impact of the classification techniques used in solving this problem, the most important enhancements achieved were due to the attributes being selected and feature extraction methods being used [2, 6, 11–15]. Generally, features have been extracted to attack this problem can be categorized into three groups namely, *sequential* (extracted from the alphabetic sequence of the proteins (e.g. composition of the amino acids)), *physicochemical* (extracted based on different physical, chemical, and structural attributes of the amino acids and proteins (e.g. hydrophobicity)), and *evolutionary* (extracted from the scoring matrices generated based on evolutionary information (e.g. *Position Specific Scoring Matrix (PSSM)* [16])) feature groups.

The study of [8] and followup works explored the impact of physicochemical-based features in conjunction with sequential-based features for the PFR and attained promising results [17]. The main advantage of using physicochemical-based features is that these features do not rely on sequential similarities. Hence, they maintain their discriminatory information even when the sequential similarity rate is low. Furthermore, they are able to provide important information about the impact of physicochemical-based attributes on the folding process. However, they are unable to provide sufficient information to solve this problem individually. On the other hand, sequential-based features have the merit that they are able to provide critical information about the interaction of the amino acids in proteins based on the sequence similarity. However, they fail to maintain this information when the sequential similarity rate is low. Thus, relying solely on these two categories of features did not lead to better results.

More recent studies shifted the focus to evolutionary-based features which have significantly enhanced the performance of the PFR [6, 12]. Relying on the PSSM, evolutionary-based features are able to provide important information about the dynamic substitution score of the amino acids with each other. However, similar to the sequential-based features, they do not provide any information about the impact of different physicochemical-based attributes on the folding process. Furthermore, they lose their discriminatory information dramatically when the sequential similarity rate is low.

In this study, we propose a novel approach to enhance the protein fold prediction accuracy and at the same time to provide more information about the impact of the physicochemical-based attributes on the folding process. In our proposed approach, first we transform the protein sequence using evolutionary-based information. Then, physicochemical-based features are extracted from the transformed sequence of the proteins using segmentation, density, distribution, and autocorrelation-based methods in an overlapping style. We explore our proposed feature extraction methods for 15 most promising attributes which are selected from 117 experimentally explored physicochemical-based attributes. Finally, by applying SVM on the combinations of the extracted features, we enhance the protein fold prediction accuracy for 5.7% over previously reported results found in the literature.

2 Benchmarks

To evaluate the performance of our proposed method against previous studies found in the literature, the EDD (extended version of DD data set introduced by Ding and Dubchak [8]) and the TG (introduced by Taguchi and Gromiha [18]) benchmarks are used. In earlier studies, DD was considered as the most popular benchmark for the PFR. However, it is no longer used [12, 13] due to its inconsistency with the latest version of *Structural Classification of Proteins (SCOP)* [19]. Extracted from the latest version of the SCOP, the EDD has been widely used as a replacement for the original DD [6, 3, 11, 12]. In this study, we extract the EDD benchmark from the SCOP 1.75 consisting of 3418 proteins belonging to the 27 folds that was originally used in the DD with less than 40% sequential similarities. We also use the TG benchmark [18] consisting of 1612 proteins belonging to 30 folds with less than 25% sequential similarities.

3 Physicochemical-Based Attributes

In this study, we investigate the impact of our proposed approaches using 15 physicochemical-based attributes. These 15 attributes have been selected from 117 physicochemical-based attributes (which are taken from the AAindex [20], the APDbase [21], and previous studies found in the literature [22]) in the following manner. For a given attribute, we extracted six feature groups based on the overlapped segmented distribution and overlapped segmented autocorrelation approaches which are the subjects of this study. Then we applied five classifiers namely, Adaboost.M1, Random Forest, Naive Bayes, *K-Nearest Neighbor (KNN)*, and SVM to each feature group separately. Therefore, 30 prediction accuracies were achieved for each physicochemical-based attribute for each benchmark (five classifiers applied to six feature groups separately ($5 \times 6 = 30$)). Considering this experiment for EDD and TG benchmarks, 60 prediction accuracies ($2 \times 30 = 60$) are achieved for each individual attribute ($60 \times 117 = 7020$ prediction accuracies in total for all 117 attributes)[1]. Then we compared these results for all 117 attributes and selected 15 attributes that attained the best results in average for all 60 prediction accuracies[2]. The feature selection process was conducted manually. This process was also explored in our previous studies for the PFR and protein structural class prediction problem [15, 23]. The

[1] The experimental results achieved in this step for all five classifiers for EDD and TG benchmarks are available upon request.

[2] Details about the attribute selection process as well as the list and references of all 117 physicochemical-based attributes are available upon request.

selected attributes are: (1) structure derived hydrophobicity value, (2) polarity , (3) average long range contact energy, (4) average medium range contact energy, (5) mean *Root Mean Square (RMS)* fluctuational displacement, (6) total non-bounded contact energy, (7) amino acids partition energy, (8) normalized frequency of alpha-helix, (9) normalized frequency of turns, (10) hydrophobicity scale derived from 3D data, (11) *High Performance Liquid Chromatography (HPLC)* retention coefficient to predict hydrophobicity and antigenicity, (12) average gain ratio of surrounding hydrophobicity, (13) mean fractional area loss, (14) flexibility, and (15) bulkiness. Note that to the best of our knowledge, most of the selected attributes (attributes number 3, 4, 5, 6, 7, 10, 11, 12, 13, and 14) have not been adequately (or not at all) explored for the PFR. However, in our conducted comprehensive experimental study, they have outperformed many popular attributes that have been widely used for PFR [2,8,9,13].

4 Feature Extraction Method

In the continuation, we first use PSIBLAST for the EDD and TG benchmarks (using NCBI's non redundant (NR) database with three iterations and cut off E-value of 0.001) and extract the PSSM [12]. The PSSM consists of two $L \times 20$ matrices (where L is the length of a protein sequence) namely, PSSM_cons and PSSM_prob. PSSM_cons contains the log-odds while PSSM_prob contains the normalized probability of the substitution score of an amino acid with other amino acids depending on their positions along a protein sequence. Then four main sets of features are extracted (two sets from the transformed protein sequences and two sets directly from the PSSM). In continuation, each feature extraction approach will be explained in detail (overlapped segmented distribution, overlapped segmented autocorrelation, semi-composition, and evolutionary-based auto-covariance).

4.1 Physicochemical-Based Feature Extraction

In this study, a new mixture of physicochemical and evolutionary-based feature extraction method is proposed based on the concepts of overlapped segmented distribution and autocorrelation. The main idea of our proposed method is to extract physicochemical-based features from the transformed sequences (so called consensus sequence) using evolutionary-based information to get benefit of discriminatory information embedded in both of these groups of features, simultaneously. In our proposed method, we first extract the consensus sequence and then, two feature groups namely overlapped segmented distribution and overlapped segmented autocorrelation are extracted from it.

Consensus Sequence Extraction Procedure: An amino acid sequence when transformed using evolutionary-based information embedded in the PSSM is called a consensus sequence [12]. Previously, to extract this sequence the PSSM_cons have been popularly used [12]. In this method, each amino acid based on its position along the protein sequence $(O_1, O_2, ..., O_L)$ is replaced with the amino acid that has the highest (maximum) substitution score according to the PSSM_cons $(C_1, C_2, ..., C_L)$. Consensus sequence was also effectively used to extract sequential-based features and attained promising results for the PFR [12]. However, it fails to address an important issue. For the case of unknown proteins the PSSM_cons does not provide any information and simply returns all equal substitution scores with the other amino acids (equal to -1).

To address this limitation, we use a modified method which relies on the PSSM_prob for feature extraction. In the PSSM_prob if a sequence similarity is found in NR, it returns a substitution probability score for even unknown amino acids. Using PSSM_prob dramatically reduces the number of unknown amino acids in the consensus sequence while previous approaches. Using PSSM_prob, we have successfully replaced over 360 unknown amino acids (out of 362 unknown amino acids) for the EDD benchmark while for the TG benchmark, we have successfully replaced all of the unknown amino acids.

Overlapped Segmented Distribution (OSD): Global density of an specific attribute is considered as a popular feature for the PFR. However, it does not properly explore the local discriminatory information available in the sequence [22]. To address this issue, the distribution of different segments of a specific density is extracted and added to this feature. In this method, we first replace the amino acids in the consensus sequence $(C_1, C_2, ..., C_L)$ with the values assigned to each of them based on a given physicochemical-based (e.g. hydrophobicity) attribute $(S_1, S_2, ..., S_L)$. Then we calculate the global density $T_{gd} = \frac{\sum_{i=1}^{L} S_i}{L}$. Next, beginning from each side of the sequence, the given attribute summation again is calculated until reaching to the first $K_s\%$ of the T_{gd} as follows:

$$I_k = (T_{gd} \times L \times K_s)/100. \tag{1}$$

Finally, the number of summed amino acids divided by the length of the protein is returned as the distribution of the first $K_s\%$ of global density. For example, if the summation of the hydrophobicity of m amino acids is equal to I_k, then the output for $K_s\%$ distribution factor is m/L. In this study, K is set to 5 based on the experimental study conducted by the authors due to similar performance of using $K_s = 5$ compared to the larger distribution factors (10 or 25) and trade of between the number extracted features and achieved prediction accuracy. This process is repeated until reaching to $K_s = 75$ (5%, 10%, 15%, ..., 75%) of the global density from each side (Figure 1).

The distribution index is calculated from both sides of the proteins due to the fact that there is no rear or front for proteins. Furthermore, an approach of using one side calculation produces accumulative distribution in the other side. We also use the overlapping approach to explore distribution of the amino acids better with consideration of an specific attribute. 75% overlapping factor is selected experimentally based on the trade off between the number of features added and the discriminatory information provided. Therefore, using $K_s = 5$ distribution and 75% overlapping factors (in addition to the global density feature in each group), 31 features are extracted ($75/5 = 15$ features from each side).

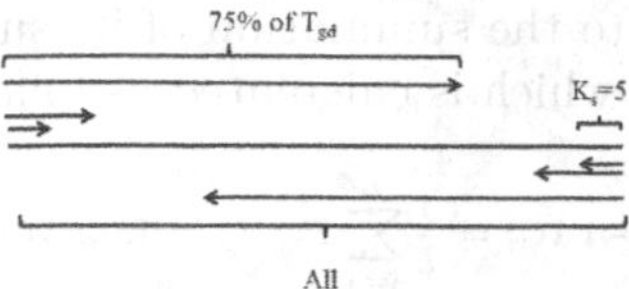

Fig. 1. Segmented distribution-based feature extraction method

Overlapped Segmented Autocorrelation (OSA): Similar to the density, autocorrelation-based features have been widely used for the PFR and attained promising results [17,2]. However, even the most sophisticated approaches failed to provide adequate local discriminatory information (e.g. pseudo amino acid composition [9]). Therefore, segmented-base approach is used in this study. In the proposed approach, we segment the protein sequence using a segmented distribution approach explained in previous subsection and then calculate the autocorrelation in each segment accumulatively (in this case, K_s is set to 10, distance factor (F) is set to 10 and overlapping factor is set to 70%). The autocorrelation in each segment is equal to:

$$\textit{Seg-Auto}_{i,a} = \frac{1}{(L(a/100) - i)} \sum_{j=m}^{n} S_j S_{j+i}\,, \quad (i = 1, \ldots, F \;\&\; a = 10, .., 70), \tag{2}$$

where L is the length of sequence, a is the segmentation factor, m and n are respectively the begin and the end of a segment, and S_j is the value of an attribute (normalized) for each amino acid. We also add the global autocorrelation (where F set to 10) which is calculated as follows:

$$\textit{Global-Auto}_{i,a} = \frac{1}{L} \sum_{j=1}^{L-i} S_j S_{j+i} \quad (i = 1, \ldots, F). \tag{3}$$

Therefore, based on each attribute the autocorrelation of the 10%, 20%, 30%, ... , 70% from each end (14 segments in total) are accumulatively calculated (*Seg-Auto* + *Global-Auto* = *OSA*). *F=10* is adopted in this study because it was showed in [6] as the most effective distance factor for the PFR. The overlapping and the segmentation factors are also adjusted based on the experimental study conducted by the authors. In results, a feature group based on this approach is extracted consisting of 150 features *(70 + 70 + 10)*.

4.2 Evolutionary-Based Feature Extraction

We also extract two sequenced-based feature groups namely, semi-composition and evolutionary-based auto-covariance directly from the PSSM.

Semi-composition (Semi-AAC): This feature group is extracted to provide more information about the occurrence of each amino acid along a protein sequence. However, instead of being extracted from the original protein sequence, we directly extract that from the PSSM. In this feature group, the composition of each amino acid is equal to the summation of its substitution scores divided by the length of the protein which is calculated as follows:

$$Semi\text{-}AAC_i = \frac{1}{L}\sum_{i=1}^{L} P_{ij}, (j = 1, ..., 20), \tag{4}$$

where P_{ij} is the substitution score for the amino acid at position i with the *j-th* amino acid in the PSSM. It was shown in [24] that Semi-AAC is able to provide more discriminatory information compared to the conventional composition feature group.

Evolutionary-Based Auto Covariance (PSSM-AC): This feature group provides crucial information about the local interaction of the amino acids from the PSSM and attained promising results for the PFR [6,24]. In the PSSM-AC the auto covariance of the substitution score of each amino acid with another amino acids with the distance factor of 10 (the distance factor is set to 10 as the most effective value as the distance factor investigated in [6]) is calculated (from the PSSM_cons). The PSSM-AC can be calculated as follows:

$$PSSM\text{-}AC_{j,f} = \frac{1}{(L-f)}\sum_{i=1}^{L-f}(P_{i,j} - P_{ave,j})(P_{i+f,j} - P_{ave,j}), \quad (j = 1, ...20 \;\&\; f = 1, ..., 10), \tag{5}$$

where $P_{ave,j}$ is the average of substitution score for the *j-th* column of PSSM. Therefore, $20 \times F$ features calculated in this feature group ($20 \times 10 = 200$).

5 Support Vector Machine (SVM)

SVM introduced by [25] aims at finding the *Maximal Marginal Hyperplane (MMH)* based on the concept of the support vector theory to minimize the error. The classification of some known points in input space $\mathbf{x}_i$ is y_i which is defined to be either -1 or +1. If x' is a point in input space with unknown classification then:

$$y' = sign\left(\sum_{i=1}^{n} a_i y_i K(\mathbf{x}_i, \mathbf{x}') + b\right), \tag{6}$$

where y' is the predicted class of point $\mathbf{x}'$. The function *K()* is the kernel function; n is the number of support vectors and a_i are adjustable weights and b is the bias. This classifier is considered as the state-of-the-art classification techniques in the pattern recognition and attained the best results for the PFR [6,12,11]. Therefore, we will only use SVM to investigate the effectiveness of our proposed methods here rather than the five classifiers that used earlier in Section 3 for the feature selection process. In this study, three different SVM-based classifiers are used to reproduce previous results as well as evaluating our proposed approaches. We use the SVM classifier implemented in the SVMLIB toolbox using *Radial Base Function (RBF)* as its kernel [26] (using grid algorithm implemented in SVMLIb to optimize its parameters (width (γ) and regularization

(C) parameters)). We also use SVM using *Sequential Minimal Optimization (SMO)* as a *polynomial kernel* which its polynomial degree is set to one (which is called linear kernel) and three (implemented in WEKA with using its default parameters [27]).

6 Results and Discussion

In the first step, the performance of the modified consensus sequence extraction method is explored by extracting occurrence (occurrence of each amino acid in a protein sequence (20 features)) and composition (percentage of the occurrence of each amino acid along a protein sequence (20 features)) feature groups. We extract these feature groups from the original sequence, the consensus sequence extracted using conventional approach and the modified consensus sequence extraction method used in this study, and applied SVM (with linear kernel). In this study, 10-fold cross validation is used as the evaluation method as it has been mainly used for this purpose in the literature [6, 8].

Table 1. Comparison of the achieved results (%) using SVM (linear kernel) to evaluate the proposed consensus sequence extraction method compared to use of original sequence as well as previously used methods for the EDD and the TG benchmarks

Methods	Composition		Occurrence	
	EDD	TG	EDD	TG
Original Sequence	32.4	31.6	41.2	33.6
Current consensus sequence extraction method	42.2	34.7	48.2	38.6
Proposed Method in this study	**44.4**	**36.3**	**48.9**	**38.8**

As shown in Table 1, the modified consensus sequence extraction method used in this study enhances the PFR performance considering composition and occurrence of the amino acids feature groups. Next, we extract features introduced in the previous stage and combine them to build the input feature vector to feed the employed SVM classifier. The input feature vector is built by combining Semi-AAC (20 features), segmented distribution (31 features), segmented autocorrelation (150 features), and PSSM-AC (200 features) feature groups in addition to the length of protein sequence feature (as used in [2, 1]). Therefore, for each attribute, a feature vector consists of 402 features is created and named *Comb_ph1* to *Comb_ph15*. The overall architecture of our proposed method is shown in Figure 2.

We then apply the SVM classifier to our extracted features. We also duplicate the study of Dong and his co-workers [6] which to the best of our knowledge attained the best results to tackle PFR. Furthermore, the 49D feature group extracted by [22] is also extracted from both of the employed benchmarks and is added to the extracted sequential-based features (Semi-AAC + PSSM-AC + length (221 features)). This feature vector consists of global density of 49 different physicochemical-based attributes (49 dimensional feature group) that has been extracted to provide sufficient physicochemical-based information for

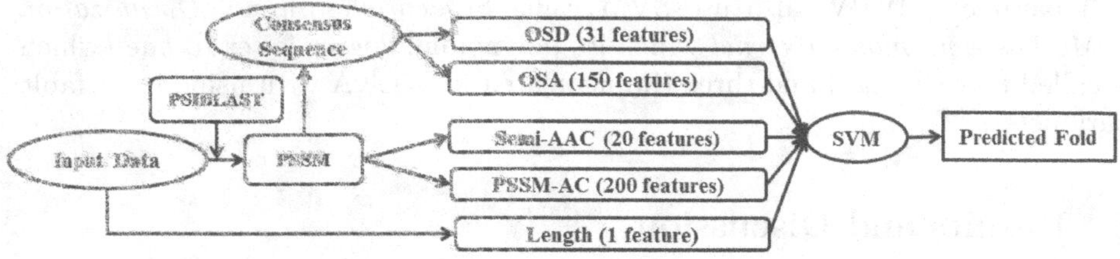

Fig. 2. The overall architecture of the proposed approach. The number of features extracted in each feature group is shown in the brackets.

the PFR [18]. In this part, we aim at comparing the impact of using a wide range of features with exploring the impact of a single attribute considering our proposed feature extraction method. Note that using SVM classifier with linear kernel attains similar results to the other two version of SVM classifier investigated in this study (SVM classifier using SMO kernel function with $p = 3$, and SVM classifier using RBF kernel function) which emphasizes on the effectiveness of the employed features rather than the kernel function used for SVM. The best results for the EDD and the TG benchmarks compared to the state-of-the-art results found in the literature are shown in Table 2.

As it is shown in Table 2, we achieve up to 82.9% and 64.6% prediction accuracies for the EDD and the TG benchmarks which are 4.8% and 5.7% better than the best results reported in the literature for the employed benchmarks respectively. Considering the small enhancement achieved in previous studies (using DD benchmark), having over 4% enhancement is considered as a significant number for the PFR [1, 4, 14]. We also achieve to over 82% and 63% prediction accuracies respectively for the EDD and the TG benchmarks using extracted features from the attributes that have not been adequately explored (attribute 14), or (to the best of our knowledge) have not been explored at all for the PFR (attributes number 1, 5, 7, and 13). Also, the significant enhancement achieved for all of the explored attributes (over 80% and 61% prediction accuracies respectively for the EDD and the TG benchmarks) emphasizes on the importance of the proposed feature extraction methods in this study. We also achieve to 41.0% and 22.7% better prediction accuracies for the EDD and the TG benchmarks respectively compared the best results achieved without using evolutionary information for feature extraction (relying solely on the original protein sequence to extract physicochemical-based features [13]). It also emphasizes on the impact of our mixture of physicochemical-based and evolutionary-based feature extraction method to enhance the protein fold prediction accuracy.

We also achieve up to 23.2% and 18.7% better prediction performance for the EDD and TG benchmarks respectively compared to use of 49D (which is extracted from the consensus sequence and combined with the Semi-AAC, PSSM-AC and the length of the amino acid sequence (221 features)). In other word, by extracting features based on a single attribute using our proposed feature extraction method, we significantly enhance the PFR performance compared to use of a wide range of physicochemical-based attributes using global density as

Table 2. The best results (in percentage) achieved in this study compared to the best results found in the literature for the EDD and the TG benchmarks

Study	Attributes (No. of features)	Method	EDD	TG
[18]	AAO original sequence (20)	LDA	46.9	36.3
[18]	AAC original sequence (20)	LDA	40.9	32.0
[13]	Physicochemical(125)	Adaboost.M1	47.2	39.1
[8]	Physicochemical(125)	SVM	50.1	39.5
[13]	Physicochemical(220)	SVM(SMO)	52.8	41.9
[22]	Threading	Naive Bayes	70.3	55.3
[2]	Bi-gram (400)	SVM	75.2	52.7
[2]	Tri-gram (8000)	SVM	71.0	49.4
[11]	Combination of bi-gram features (2400)	SVM	69.9	55.0
[3]	PSIPRED and PSSM-based features (242)	SVM	77.5	57.1
[6]	ACCFold-AAC(200)	SVM(RBF)	76.2	56.4
[6]	ACCFold-AC(4000)	SVM(RBF)	**78.1**	**58.9**
This study	Comb_ph1 (402)	SVM(SMO)	82.3	63.3
This study	Comb_ph5 (402)	SVM(SMO)	82.8	**64.6**
This study	Comb_ph7 (402)	SVM(SMO)	**82.9**	64.0
This study	Comb_ph13 (402)	SVM(SMO)	82.5	63.7
This study	Comb_ph14 (402)	SVM(SMO)	82.4	63.8
This study	Original sequence (49+221)	SVM(SMO)	44.7	35.7
This study	Consensus sequence (49+221)	SVM(SMO)	59.7	45.9

the main feature. These results emphasize on the effectiveness of the overlapped segmented-based feature extraction method to explore more discriminatory information. It is important to highlight that these results are achieved using 402 attributes, which is 10 times less than the number of attributes that was used in the ACCFold-AC model (4000 features). Besides enhancing the protein fold prediction accuracy, by proposing a mixture of physicochemical and evolutionary-based information, we introduce a new direction to obtain benefit from discriminatory power of these two groups of features simultaneously. Furthermore, by exploring physicochemical based features, the proposed method is able to provide crucial information about the impact of these attributes on the PFR. Note that our proposed features in this study (overlapped segmented-based distribution and overlapped segmented-based autocorrelation) have been investigated for the protein structural class prediction problem (in a different experiment) and obtained promising results as well [23] which highlights the generality of these approaches for similar studies.

7 Conclusion

In this study, we proposed a model to enhance the protein fold prediction accuracy as well as providing better understanding about the impact of the physicochemical-based attributes on the PFR in the following five steps. In the first step, a modified consensus sequence extraction method was proposed. It addressed the issue of unknown proteins using evolutionary-based information. Proposed method also improved the protein fold prediction accuracy over the previous methods that extracted consensus sequence. In the second step, a comprehensive study on a wide range of physicochemical-based attributes (117 attributes) were conducted and 15 most promising attributes were selected.

The selected attributes outperformed other attributes based on the density, distribution, and autocorrelation feature extraction methods. This comprehensive experimental study provided important information about the performance of these 117 physicochemical-based attributes on the PFR. In the third step, we proposed two novel feature extraction methods based on the concepts of segmented distribution and autocorrelation to provide more local and global discriminatory information for the PFR. In the next step, effective sequentially-based features that were directly extracted from the PSSM were combined with the proposed physicochemical-based features. In the final step, by using the SVM classifier (with linear kernel) to our extracted features, we achieved 82.9% and 64.6% prediction accuracies for the EDD and the TG benchmarks respectively which are 4.8% and 5.7% over the previously reported results found in the literature.

References

1. Dehzangi, A., Phon-Amnuaisuk, S., Dehzangi, O.: Enhancing protein fold prediction accuracy by using ensemble of different classifiers. Australian Journal of Intelligent Information Processing Systems 26(4), 32–40 (2010)
2. Ghanty, P., Pal, N.R.: Prediction of protein folds: Extraction of new features, dimensionality reduction, and fusion of heterogeneous classifiers. IEEE Transactions on NanoBioscience 8(1), 100–110 (2009)
3. Deschavanne, P., Tuffery, P.: Enhanced protein fold recognition using a structural alphabet. Proteins: Structure, Function, and Bioinformatics 76(1), 129–137 (2009)
4. Dehzangi, A., Phon-Amnuaisuk, S., Manafi, M., Safa, S.: Using rotation forest for protein fold prediction problem: An empirical study. In: Pizzuti, C., Ritchie, M.D., Giacobini, M. (eds.) EvoBIO 2010. LNCS, vol. 6023, pp. 217–227. Springer, Heidelberg (2010)
5. Dehzangi, A., Karamizadeh, S.: Solving protein fold prediction problem using fusion of heterogeneous classifiers. INFORMATION, An International Interdisciplinary Journal 14(11), 3611–3622 (2011)
6. Dong, Q., Zhou, S., Guan, G.: A new taxonomy-based protein fold recognition approach based on autocross-covariance transformation. Bioinformatics 25(20), 2655–2662 (2009)
7. Chmielnicki, W., Stapor, K.: A hybrid discriminative-generative approach to protein fold recognition. Neurocomputing 75(1), 194–198 (2012)
8. Ding, C., Dubchak, I.: Multi-class protein fold recognition using support vector machines and neural networks. Bioinformatics 17, 349–358 (2001)
9. Yang, T., Kecman, V., Cao, L., Zhang, C., Huang, J.Z.: Margin-based ensemble classifier for protein fold recognition. Expert Systems with Applications 38, 12348–12355 (2011)
10. Kavousi, K., Sadeghi, M., Moshiri, B., Araabi, B.N., Moosavi-Movahedi, A.A.: Evidence theoretic protein fold classification based on the concept of hyperfold. Mathematical Biosciences 240(2), 148–160 (2012)
11. Shamim, M.T.A., Anwaruddin, M., Nagarajaram, H.A.: Support vector machine-based classification of protein folds using the structural properties of amino acid residues and amino acid residue pairs. Bioinformatics 23(24), 3320–3327 (2007)

12. Yang, J.Y., Chen, X.: Improving taxonomy-based protein fold recognition by using global and local features. Proteins: Structure, Function, and Bioinformatics 79(7), 2053–2064 (2011)
13. Dehzangi, A., Phon-Amnuaisuk, S.: Fold prediction problem: The application of new physical and physicochemical- based features. Protein and Peptide Letters 18(2), 174–185 (2011)
14. Sharma, A., Lyons, J., Dehzangi, A., Paliwal, K.K.: A feature extraction technique using bi-gram probabilities of position specific scoring matrix for protein fold recognition. Journal of Theoretical Biology 320, 41–46 (2013)
15. Dehzangi, A., Sattar, A.: Protein fold recognition using segmentation-based feature extraction model. In: Selamat, A., Nguyen, N.T., Haron, H. (eds.) ACIIDS 2013, Part I. LNCS, vol. 7802, pp. 345–354. Springer, Heidelberg (2013)
16. Altschul, S.F., Madden, T.L., Schaffer, A.A., Zhang, J.H., Zhang, Z., Miller, W., Lipman, D.J.: Gapped blast and psi-blast: a new generation of protein database search programs. Nucleic Acids Research 17, 3389–3402 (1997)
17. Shen, H.B., Chou, K.C.: Ensemble classifier for protein fold pattern recognition. Bioinformatics 22, 1717–1722 (2006)
18. Taguchi, Y.H., Gromiha, M.M.: Application of amino acid occurrence for discriminating different folding types of globular proteins. BMC Bioinformatics 8(1), 404 (2007)
19. Murzin, A.G., Brenner, S.E., Hubbard, T., Chothia, C.: Scop: A structural classification of proteins database for the investigation of sequences and structures. Journal of Molecular Biology 247(4), 536–540 (1995)
20. Kawashima, S., Pokarowska, P.P.M., Kolinski, A., Katayama, T., Kanehisa, M.: Aaindex: Amino acid index database, progress report. Neucleic Acids 36, D202–D205 (2008)
21. Mathura, V.S., Kolippakkam, D.: Apdbase: Amino acid physico-chemical properties database. Bioinformation 12(1), 2–4 (2005)
22. Gromiha, M.M.: A statistical model for predicting protein folding rates from amino acid sequence with structural class information. Journal of Chemical Information and Modeling 45(2), 494–501 (2005)
23. Dehzangi, A., Paliwal, K.K., Sharma, A., Dehzangi, O., Sattar, A.: A combination of feature extraction methods with an ensemble of different classifiers for protein structural class prediction problem. IEEE Transaction on Computational Biology and Bioinformatics (TCBB) (in press, 2013)
24. Liu, T., Geng, X., Zheng, X., Li, R., Wang, J.: Accurate prediction of protein structural class using auto covariance transformation of psi-blast profiles. Amino Acids 42, 2243–2249 (2012)
25. Vapnik, V.N.: The Nature of Statistical Learning Theory. Springer (1999)
26. Chang, C.C., Lin, C.J.: LIBSVM: A library for support vector machines. ACM Transactions on Intelligent Systems and Technology 2, 1–27 (2011)
27. Witten, I., Frank, E.: Data Mining: Practical Machine Learning Tools and Techniques, 2nd edn. Morgan Kaufmann, San Francisco (2005)

Neighborhood Selection in Constraint-Based Local Search for Protein Structure Prediction

Swakkhar Shatabda[1,2], M.A. Hakim Newton[1], and Abdul Sattar[1,2]

[1] Institute for Intelligent and Integrated Systems (IIIS), Griffith University
[2] Queensland Research laboratory, National ICT Australia(NICTA)

Abstract. Protein structure prediction (PSP) is a very challenging constraint optimization problem. Constraint-based local search approaches have obtained promising results in solving constraint models for PSP. However, the neighborhood exploration policies adopted in these approaches either remain exhaustive or are based on random decisions. In this paper, we propose heuristics to intelligently explore only the promising areas of the search neighborhood. On face centered cubic lattice using a realistic 20×20 energy model and standard benchmark proteins, we obtain structures with significantly lower energy and RMSD values than those obtained by the state-of-the-art algorithms.

1 Introduction

Ab initio methods for protein structure prediction (PSP), without using any templates or structures of known similar proteins, starts searching from the scratch for the native structure that has the minimum free energy. Due to the complexity of all-atomic detailed models and unknown factors of the energy function, the general paradigm of *ab initio* PSP has been to begin with the sampling of a large set of candidate or *decoy* structures guided by a scoring function. In the final stage, the refinements [22] are done to achieve the realistic structure. Given a primary amino acid sequence of a protein, PSP can be defined as: find a *self-avoiding* walk on a discrete lattice that minimizes a contact-based energy function. However, the conformational search space still remains huge and the problem itself remains a very challenging constraint optimization problem.

Given a current partial or complete solution, selection of the neighboring solutions for further exploration is a key factor in the performance of constraint programming (CP) and local search approaches. Neighborhood selections involves selection of variables and values. Unguided random selection [23,28], costly exhaustive generation [5] and filtering or ordering techniques for enumeration [28,19] are not much effective for neighborhood selection in PSP. In other domains, such as propositional satisfiability *promising* variables are selected using different variable selection strategies [20,1].

In this paper, we propose several novel component fitness functions for neighborhood selection in PSP. These heuristics are an energy contribution function, a core-based distance function and a free lattice-neighbor count based function.

S. Cranefield and A. Nayak (Eds.): AI 2013, LNAI 8272, pp. 44–55, 2013.

These heuristics are derived from domain knowledge and are used along with a hint based variable selection strategy. The aggregate of these heuristics are used in selecting candidate structures. To the best of our knowledge, this is the first application of intelligent variable selection strategy in PSP. We also propose a new chain growth initialization for the energy model used. Experimental results show that our method significantly improves over the state-of-the-art algorithms and produces structures with lower energy and RMSD values for standard benchmark proteins on face centered cubic lattices.

2 Related Work

Simplified models such as Hydrophobic-Polar (HP) energy model [9] and discrete lattices have been studied extensively by many researchers within various frameworks such as constraint programming [7], genetic algorithms [21], and memory-based local search approaches [24].

In contrast to HP models, elaborate energy functions derived by using statistical methods [14,2] take into consideration all 20×20 amino-acid interactions. Using secondary information and constraint programming techniques, Dal Palu *et al.* [19] developed a method to predict tertiary structures of real proteins. They also proposed several generalized and problem specific heuristics [17]. Later, they also developed a highly optimized constraint solver named COLA [18].

A two-stage optimization method was proposed in [27] by Ullah *et al.* It uses CPSP tool by Backofen *et al.* [12] to provide initial structure for local search procedure on FCC lattice and an elaborate energy function. The two-stage optimization approach was reported to outperform simulated annealing-based local search procedure [25]. Ullah *et al.* also used large neighborhood search techniques on top of the COLA solver [28]. A fragment assembly method was proposed in [16] to produce low energy structures. Later, a filtering techniques for loop modeling [4] was proposed using CP techniques. In a recent work, several effective heuristics were used in a mixed fashion in [23] which produced state-of-the-art results on real proteins from standard benchmark set for contact-based energy models. Our work in this paper uses a hint-based variable selection strategy to generate the neighborhoods to be explored. Our work also uses a new chain growth initialization method.

Among other approaches in PSP are population based methods [8] and genetic algorithms [26] and CP techniques for side-chain models [13].

3 The Problem Model and CP Formulation

Proteins are polymers of amino-acid monomers. There are 20 different amino acids. In a simplified model, all monomers have an equal size and all bonds are of equal length. In a CP formulation, each monomer is modeled by a point in a three dimensional lattice (*lattice* constraint). The given amino acid sequence fits into the lattice: every pair of consecutive amino acids in the sequence are also neighbors in the lattice (*chain* constraint) and two monomers can not occupy the same point in the lattice (*self avoiding* constraint). A simplified energy function is used in calculating the energy of a structure.

Two lattice points $p, q \in \mathbb{L}$ are said to be in *contact* or *neighbors* of each other, if $q = p + \boldsymbol{v_i}$ for some vector $\boldsymbol{v}_i$ in the basis of L. The Face Centered Cubic (FCC) lattice is preferred since it provides the densest packing [6] for spheres of equal size and the highest degree of freedom for placing an amino acid. The points in FCC lattice are generated by the following basis vectors: $\boldsymbol{v_1} = (1,1,0)$, $\boldsymbol{v_2} = (-1,-1,0)$, $\boldsymbol{v_3} = (-1,1,0)$, $\boldsymbol{v_4} = (1,-1,0)$, $\boldsymbol{v_5} = (0,1,1)$, $\boldsymbol{v_6} = (0,1,-1)$, $\boldsymbol{v_7} = (0,-1,-1)$, $\boldsymbol{v_8} = (0,-1,1)$, $\boldsymbol{v_9} = (1,0,1)$, $\boldsymbol{v_{10}} = (-1,0,1)$, $\boldsymbol{v_{11}} = (-1,0,-1)$, $\boldsymbol{v_{12}} = (1,0,-1)$. In FCC lattice, each point has 12 neighbors and distance between two neighbors is $\sqrt{2}$.

In our CP model, we are given a sequence S, where each element $s_i \in S$ is an amino-acid type. Each amino acid i is associated with a point $p_i = (x_i, y_i, z_i) \in \mathbb{Z}^3$. The decision variables are the x, y and z co-ordinates of a point. For a sequence of length n, the domain of the variables is the range $[-n, n]$. Formally, $\forall_i x_i \in [-n, n]$, $\forall_i y_i \in [-n, n]$ and $\forall_i z_i \in [-n, n]$. The first point is assigned as $(0, 0, 0)$, which is a valid point in the FCC lattice. The rest of the points follows the constraint, $\forall_{i<n}(\boldsymbol{a_i}) \in \{\boldsymbol{v_1}, \cdots, \boldsymbol{v_{12}}\}$. Here, $\boldsymbol{a_i}$ is the absolute vector between points $(x_{i+1}, y_{i+1}, z_{i+1})$ and (x_i, y_i, z_i), and $\{\boldsymbol{v_1}, \cdots, \boldsymbol{v_{12}}\}$ are the basis vectors for FCC lattice. Thus all points satisfy the *lattice* constraint and *chain* constraint. The *self-avoiding* constraint is defined using the all-different constraint all-different($\forall_i p_i$). We define sqrdist(i, j) as the square of Euclidean distances between two points p_i and p_j. Now, contact$(i, j) = 1$, if sqrdist$(i, j) = 2$; and contact$(i, j) = 0$, if sqrdist$(i, j) \neq 2$. For any given protein sequence S, the energy of a structure c is defined as:

$$E(c) = \sum_{j \geq i+1}^{n} \mathsf{contact}(i, j).\mathsf{energy}(s_i, s_j) \tag{1}$$

where energy(s_i, s_j) is the empirical energy value between two amino-acids of type s_i and s_j obtained from the energy matrix given in [2]. Given this model, PSP can be defined as follows: given a sequence S of length n, find a self-avoiding walk $p_1, \cdots, p_n$ on the lattice that minimizes the energy *i.e.* obj $= E(c)$.

4 Our Approach

Our approach is based on component heuristic functions that are used with a hint based variable selection strategy and aggregate functions which are used for candidate selection. Rest of the section describes necessary details.

4.1 Search Procedure

The search starts with the greedy chain growth initialization procedure that produces a compact low energy structure. Based on a walk probability *wp* (initially set to 5%), a variable is selected randomly or from a hint based priority queue with tabu on recently selected variables and the corresponding amino-acid position is determined. Neighborhood is generated only for the position selected using a set of operators. Then the generated candidate moves are simulated. Simulation of a move temporarily calculates the changes in the heuristic functions without committing the move. At each iteration, one of the heuristics are selected randomly by a uniform random distribution. After simulation, the best candidate move is selected and executed. The execution updates all cost functions, constraints and propagate hints. Ties are broken using a uniform random distribution. The search keep tracks of the global minimum found and restarts from the last found global minimum whenever it gets stuck. Stagnation is determined by a number of non-improving moves from the last found global minimum. At stagnation, the stagnation parameter *sp* (initially set to 500) and the walk probability *wp* are multiplied by a factor (set 1.2). Parameters *wp* and *sp* are set to initial values, whenever a new global minimum is found. Pseudo-code of our algorithm is given in Algorithm 1.

Algorithm 1. localSearch(conformation C)

```
1  C ← CGInitialize()
2  nonImp ← 0
3  while nonImp ≤ stagnation do
4      evaluateHints()
5      i ← selectPosition()
6      o ← selectOperator()
7      list ← generateMoves(C, i, o)
8      h ← selectHeuristic()
9      simulateMoves(list, h)
10     m ← selectBestMove(list)
11     executeMove(C, m)
12     updateTabulist()
13     if not improving then
14         nonImp++
15     else
16         nonImp ← 0
```

For a clear outline of our contribution, it is worthwhile to note the differences of our algorithm from the algorithm in [23]. In Line 1 of Algorithm 1, we call our new initialization function while the initialization in [23] uses a structure fully contained within a sphere. In Line 4-5, we compute the hint heuristics and based on them, we select only one point, where the operators are applied. In contrast, in [23], points are selected randomly and then operators are applied in all selected points.

4.2 Heuristic Functions

The empirical energy model gives an elaborate interaction energy contribution for the amino-acid types. The points with lower contribution to the total energy functions are preferred for selection. First, we define the first component fitness function that calculates the contribution of a point to the total energy of the structure. Formally,

$$\mathsf{contr}(i) = \sum_{0 \leq k \leq n, |i-k|>1} \mathsf{contact}(i,k) \times \mathsf{energy}(s_i, s_k)$$

The intuition is to select the variables with the maximum $\mathsf{contr}(i)$ so that the corresponding operators can lower the energy contribution. Note that, the lowest energy contributions are generally negative. However, heuristics designed from domain knowledge often provide interesting insights. One such properties of protein folding is due to the solvent type water. This property lets the hydrophobic residues buried inside the structure and helps form a compact core. Based on this fact, several methods and heuristics have been developed for HP model [12,7]. However, interactions between two hydrophobic residues in contact in the core may result in repulsion (positive empirical energy) rather than attraction (negative empirical energy) and form a non-stable structure. For this reason, we don't use those heuristics directly into our model. We divide the 20 different amino-acids into two groups according to their similarity in interaction energy within each group. We run a simple k-means clustering algorithm on the empirical energy matrix to obtain two such groups (Group I: *Ala, Phe, Gly, Ile, Leu, Met, Pro, Val, Trp, Tyr*; Group II: *Cys, Asp, Glu, His, Lys, Asn, Gln, Arg, Ser, Thr*) such that interaction within each group minimizes the total energy contribution. We call these residues, *affine* residues and define an affine core, $a_c = (x_c, y_c, z_c)$, such that, $x_c = \frac{1}{|A|}\sum_{k \in A} x_k$, $y_c = \frac{1}{|A|}\sum_{k \in A} y_k$, $z_c = \frac{1}{|A|}\sum_{k \in A} z_k$, A is the set of affine amino acid positions and $|A|$ denotes the total number of affine positions. Now we expect to move the affine amino acids towards the core. We define, our next component fitness function:

$$\mathsf{sqrdist\text{-}acore}(i) = \mathsf{sqrdist}(i, a_c)$$

Naturally, we wish to move the distant affine positions nearer to the affine core, a_c, i.e. we wish to select the variables with maximum $\mathsf{sqrdist\text{-}acore}(i)$ value. However, these component fitness functions to select the variable do not work well if there is not enough free positions in the lattice neighborhood of the point assigned to the position. Therefore, we define another component fitness function, $\mathsf{free}(i)$. It counts the total number of free neighbors of a point i in a lattice. Formally,

$$\mathsf{free\text{-}count}(i) = \sum_{k \in N(i)} \mathsf{free}(k)$$

Here, $\mathsf{free}(k) = 1$, if k is free; and $\mathsf{free}(k) = 0$, if k is occupied and $N(i)$ is set of neighbors of i in the lattice. We wish to select the variables with the maximum

$\mathsf{free}(i)$ so that the number of possible moves become higher. It is interesting to note that we wish to select variables that maximizes all three component functions. Now, we aggregate the component fitness functions for all amino acid position i and derive heuristic functions for the selection of candidate structures. We denote the heuristics as follows: the energy heuristic $h_E = \sum_{i \leq n} \mathsf{contr}(i)$, the affine core heuristic $h_A = \sum_{i \leq n} \mathsf{sqrdist\text{-}acore}(i)$ and the compactness heuristic $h_F = \sum_{i \leq n} \mathsf{free\text{-}count}(i)$. We wish to guide our search using these aggregate heuristic functions. The idea of component heuristics are previously used along with extremal optimization [11]. However, the heuristics used in this paper for the given energy model are novel themselves.

4.3 Hint Based Variable Selection

Now, we define two important terms: *metric* and *hint*. Each function $f(p_1, \cdots, p_n)$ has the parameters p_js that are either variables or other functions. A function f depends on a variable x, denoted by $f \to x$, if x is itself a parameter of f or f has a parameter $p \to x$. Each function f has a non-negative metric f_m denoting its evaluation. For each $x \leftarrow f$, it also has a non-negative hint $f_h(x)$ denoting the preference of changing x's value to improve f_m. A constraint f is satisfied when $f^m = 0$ and in that case $f_h(x) = 0$ for any x, which means a constraint's metric improves when it is minimized.

In our model, all the component fitness functions are defined as constraints over the variables. For example, $\mathsf{contact}(i, j)$ is defined as a constraint, $\mathsf{sqrdist}(i, j) = 2$, which is satisfied only when the square of Euclidean distance between i and j is equal to 2. This function depends on the variables, x_i,y_i,z_i and x_j,y_j,z_j. The metric of the function is simply the evaluation of the constraint that tests equality with 2. If this constraint is not satisfied then the violation is added as the hint of these variables. Thus, variable violations for all the functions which are dependent on a particular variable is added as hint for that variable. Looking at these hint values, we decide which variable is to be selected in order to minimize the violation of the constraints. In our CP model, we take aggregate of all the component functions and take a summation of those aggregate functions to define another function on top as the hint provider. The variable violation for that top function is distributed as hints among the variables corresponding to amino acid positions. A simple heap or priority queue data structure with hint values is sufficient for us to decide which variable to select. We also maintain a tabu list to prevent recent variables to be selected. We use three different heuristics h_E, h_A and h_F, and sum them to form a top function that provides the hints for all the variables.

4.4 Chain Growth Initialization

The procedure is inspired from chain growth algorithms previously applied to HP models [3]. The initialization starts by assigning $(0, 0, 0)$ to the first amino acid position.The rest of the variables are assigned following a greedy strategy. One of the free neighbors of the last assigned amino acid position $i-1$ is assigned to position i, such that the the assignment minimizes the partial objective function, $\mathsf{obj}_i = \sum_{k<i} \mathsf{contact}(i,k) \times \mathsf{energy}(s_i, s_k) + \mathsf{free\text{-}count}(i) \times E_U[i]$. Here, k is iterated over already assigned amino acid positions and $E_U[i]$ is the per-contact expected or average energy contribution of i with possible unassigned amino acid positions in the chain. This partial objective function, obj_i is equal to the partial energy contribution, $\mathsf{partial\text{-}E}(p_1, \cdots, p_i)$. The tie-breaking is done according to a pre-defined order. Pseudo-code for variable selection for each step is given in Algorithm 3. The initialization procedure backtracks whenever, it fails to assign valid points to an amino acid position. This method guarantees to produce valid structure with low energy value. Pseudo-code of the chain growth initialization procedure is given in Algorithm 2.

Algorithm 2. CGInitialize()

```
p1 = (0,0,0)
for i ← 2 to n do
    dir=selectDirection(i)
    if dir = null then
        backtrack()
    else
        pi = pi−1 + dir
return p1, ···, pn
```

Algorithm 3. selectDirection(i)

```
MinHeap Q = {}
for all vk ∈ basis do
    pk = pi−1 + vk
    if notOccupied(pk) then
        Ek=partial-E(p0, ···, pi−1, pk)
        Q.add(vk, Ei)
if Q.isEmpty() then
    return null
else
    return Q.top()
```

4.5 Operators

After a variable is selected, we can decide which amino acid position it corresponds to and apply the operators to that position. We make use of four types of operators (see Fig. 1). First two are jump move [23] and pull move [10] which are also used in the literature. We propose a single point pull move (Fig. 1(c)) similar to that of two point pull move and a single point push move which reverses the action of a single point pull move (Fig. 1(d)).

4.6 Implementation

We implemented our algorithm using C++ on top of the constraint based local search (CBLS) system, Kangaroo [15]. The functions and the constraints are defined using invariants in Kangaroo. Invariants are special constructs that are defined by using mathematical operators over the variables. Propagation of hints, simulation of moves, execution and related calculations are performed incrementally by Kangaroo.

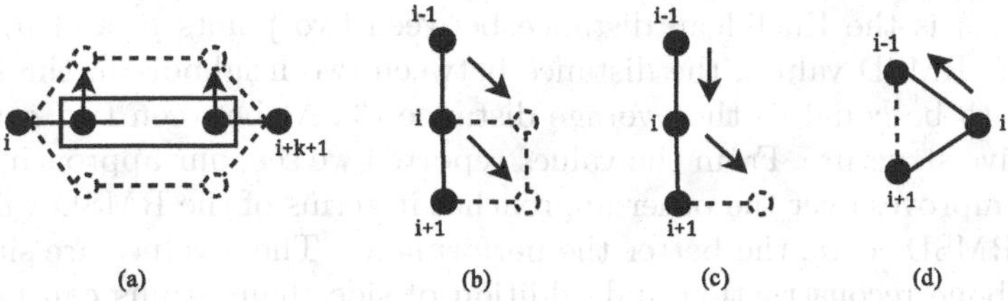

Fig. 1. Different Operators used: (a) jump moves (k=1,2,3) (b) two point pull move (c) single point pull move and (d) push move

Table 1. Results obtained by different algorithms for 12 proteins

seq no	pdb id	seq len	Our Approach energy level best	avg	Our Approach avg rmsd	Mixed [23] energy level best	avg	Mixed [23] avg rmsd	Hybrid [28] energy level best	avg	Hybrid [28] avg rmsd
1	4rxn	54	**-168.78**	**-164.28**	**5.56**	-165.21	-156.32	6.29	-157.70	-140.13	9.99
2	1enh	54	**-158.74**	**-152.43**	**5.33**	-158.75	-146.69	6.61	-154.24	-141.99	10.04
3	4pti	58	**-221.59**	**-205.42**	**5.86**	-219.52	-198.42	7.07	-213.70	-196.23	11.92
4	2igd	61	**-187.96**	**-179.36**	**6.34**	-187.20	-174.19	9.33	-184.29	-157.20	13.30
5	1ypa	64	**-257.02**	**-248.47**	**6.12**	-249.90	-239.98	7.53	-221.11	-208.10	13.42
6	1r69	69	**-223.12**	**-210.15**	**5.78**	-213.04	-204.17	6.47	-180.62	-165.11	14.78
7	1ctf	74	**-230.86**	**-220.04**	**6.14**	-224.29	-213.81	7.23	-204.88	-195.23	12.65
8	3mx7	90	**-332.79**	**-321.58**	**6.58**	-328.12	-311.56	8.18	-	-	-
9	3nbm	108	**-431.90**	**-415.02**	**6.28**	-418.60	-401.99	8.58	-	-	-
10	3mqo	120	**-476.06**	**-464.57**	**6.46**	-465.74	-455.27	8.86	-	-	-
11	3mr0	142	**-446.31**	**-435.69**	**7.32**	-445.33	-430.28	10.02	-	-	-
12	3pnx	160	**-603.78**	**-585.97**	**6.84**	-601.23	-571.13	9.38	-	-	-

5 Experiments

We ran our experiments on a cluster of computers with nodes equipped with Intel Xeon CPU X5650 processors @2.67GHz, QDR 4 x InfiniBand Interconnect. We compare our results with the mixed heuristic approach in [23] and the hybrid approach in [28]. All the algorithms are given 1 hour to finish each run, and the best and average energy levels of 50 runs are reported in Table 1 for 12 benchmark proteins. These proteins are also used in [23]. The blank values in the table are the cases where hybrid approach failed to produce any valid structure within the time limit. PDB ids, sequence length and average RMSD values are also reported in the table. The best values are indicated in bold faced font. For all the 12 proteins, our approach achieves lower energy levels. The significance of these values are confirmed from the values reported in the *rmsd* colmun. For any given structure produced by an algorithm,

$$\text{RMSD} = \sqrt{\frac{\sum_{i=1}^{n-1}\sum_{j=i+1}^{n}(\mathsf{dist}(i,j)^{\text{given}} - \mathsf{dist}(i,j)^{\text{native}})^2}{n*(n-1)/2}}$$

Here $\mathsf{dist}(i, j)$ is the Euclidean distance between two points p_i and p_j. In calculating the RMSD values, the distance between two neighbors in the lattice is considered to be equal to the average distance (3.8Å) between two α-Carbons on the native structure. From the values reported we see, our approach also significantly improves over the other approaches in terms of the RMSD values; the lower the RMSD score, the better the performance. These values are significant since backbone reconstruction and addition of side-chain atoms can guarantee to produce real protein structures within small (1-2Å) deviation [22]. However, since lattice configurations can only approximate the positions of the amino acids in the real space, lower RMSD values produced by our algorithm are satisfactory.

Table 2. Average energy level achieved by different variants of our algorithm

seq no	¬hint	+hint(h_A)	+hint(h_E)	¬hint(h_F)	¬h_{select}	init$_r$	all
1	-146.61	-149.61	-151.25	-158	-156.08	-136.23	**-164.28**
2	-136.35	-137.62	-128.15	-147.03	-142.44	-137.08	**-152.43**
3	-185.52	-192.15	-189.45	-194.62	-200.6	-197.88	**-205.42**
4	-165.5	-158.16	-157.64	-171.24	-164.12	-149.74	**-179.36**
5	-235.75	-235.45	-236.88	-244.44	-244.71	-219.91	**-248.47**
6	-193.87	-187.64	-180.61	-183.20	-182.85	-178.25	**-210.15**
7	-197.83	-204.48	-198.45	-210.27	-206.84	-182.46	**-220.04**

In order to test the effectiveness of different components of our approach, we ran different variants of our approach on first 7 proteins and report average energy level of 20 runs for each of them in Table 2. First, we report the performance of a variant without using the hint based variable selection in column $\neg hint$. It shows how the hint based system can improve on this variant to achieve the final performance shown in '*all*' column. The '*all*' column, for convenience of the reader, again shows the average energy values obtained by our approach (as shown in Table 1). Then we ran two variants with hints for h_A only (column $+hint(h_A)$) and with hints for h_E only (column $+hint(h_E)$). These two variants show the effectiveness of using these two heuristics as hint provider individually. We see that these two heuristics are showing better performance than the hintless variant for most of the proteins, and $+hint(h_A)$ is performing better than $+hint(h_E)$. It reveals that the energy function itself is not enough for providing hints for the search and heuristic approximations can actually improve the performance. However, both of them shows effectiveness of hints over the no-hint variant. Then, we ran another variant with hints for both h_E and h_A, but not using hints for h_F (column $\neg hint(h_F)$). This particular variant shows the combined performance of two heuristics as hint provider and also shows the relative strength of the other hint heuristic h_F that is absent. We get a clear idea of the strength of h_F by comparing it with the final results in column '*all*' and also strength of the combined variant by comparing the results with the individual columns of $+hint(h_A)$ and $+hint(h_E)$. We see that the combination of h_E and h_A which is $\neg hint(h_F)$ works better than the individual variants. However, its only after adding the hints for h_F, its possible to achieve the final performance. Another variant uses the hint based system but the selection of candidate structures is guided by h_E only (column $\neg h_{select}$). This indicates the effectiveness of the heuristics to select candidate structures. The penultimate column ($init_r$) shows the results achieved by replacing our chain growth algorithm by a random initialization. From the reported values we see that chain growth initialization has a greater impact on most of the proteins with respect to the random initialization methods.

To show how the search makes progress, we plot log of average energy levels achieved, (for convenience of display in the chart) added by a threshold of 250 for the protein 4pti against iteration count in Fig. 2. We see that the variant without hint ($\neg hint$) works worst. It improves only at the beginning and then gets stagnant. Variants with hints for h_E and h_A works better than this. If we add hints for all the heuristic functions but guide the search with h_E only ($\neg h_{select}$), it can further improve. However, the best performance is only after we use all the heuristics to guide the search. In case of random initialization, we find that it can gradually improve, but most of the time is spent to achieve the initial level that is achieved by the chain growth initialization for other variants.

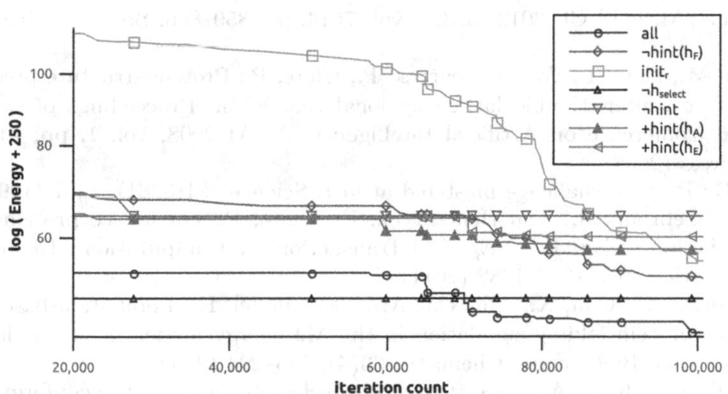

Fig. 2. Search progress of different variants for the protein 4pti

6 Conclusion

In this paper, we have proposed an intelligent variable selection strategy for candidate generation in local search methods for protein structure prediction problem using 20×20 energy model on face centered cubic lattice. In addition to this, we also have proposed a new chain growth initialization procedure and heuristic functions to select variables and candidates at each iteration. Our method significantly improves over the state-of-the-art algorithms. In future, we wish to explore the strength of our proposed scheme on other domains and on other models of protein structure prediction as well.

Acknowledgments. We gratefully acknowledge the support of the Griffith University eResearch Services Team and the use of the High Performance Computing Cluster "Gowonda" to complete this research and NICTA, which is funded by the Australian Government as represented by the Department of Broadband, Communications and the Digital Economy and the Australian Research Council through the ICT Centre of Excellence program.

References

1. Balint, A., Fröhlich, A.: Improving stochastic local search for SAT with a new probability distribution. In: Strichman, O., Szeider, S. (eds.) SAT 2010. LNCS, vol. 6175, pp. 10–15. Springer, Heidelberg (2010)
2. Berrera, M., Molinari, H., Fogolari, F.: Amino acid empirical contact energy definitions for fold recognition in the space of contact maps. BMC Bioinformatics 4, 8 (2003)
3. Bornberg-Bauer, E.: Chain growth algorithms for hp-type lattice proteins. In: Proceedings of the First Annual International Conference on Computational Molecular Biology, RECOMB 1997, pp. 47–55. ACM, New York (1997)
4. Campeotto, F., Dal Palù, A., Dovier, A., Fioretto, F., Pontelli, E.: A filtering technique for fragment assembly- based proteins loop modeling with constraints. In: Milano, M. (ed.) CP 2012. LNCS, vol. 7514, pp. 850–866. Springer, Heidelberg (2012)
5. Cebrián, M., Dotú, I., Van Hentenryck, P., Clote, P.: Protein structure prediction on the face centered cubic lattice by local search. In: Proceedings of the 23rd National Conference on Artificial Intelligence, AAAI 2008, vol. 1, pp. 241–246. AAAI Press (2008)
6. Cipra, B.: Packing challenge mastered at last. Science 281(5381), 1267 (1998)
7. Dotu, I., Cebrian, M., Van Hentenryck, P., Clote, P.: On lattice protein structure prediction revisited. IEEE/ACM Transactions on Computational Biology and Bioinformatics 8(6), 1620–1632 (2011)
8. Kapsokalivas, L., Gan, X., Albrecht, A.A., Steinhöfel, K.: Population-based local search for protein folding simulation in the MJ energy model and cubic lattices. Computational Biology and Chemistry 33(4), 283–294 (2009)
9. Lau, K.F., Dill, K.A.: A lattice statistical mechanics model of the conformational and sequence spaces of proteins. Macromolecules 22(10), 3986–3997 (1989)
10. Lesh, N., Mitzenmacher, M., Whitesides, S.: A complete and effective move set for simplified protein folding. In: Proceedings of the Seventh Annual International Conference on Research in Computational Molecular Biology, pp. 188–195. ACM, New York (2003)
11. Lu, H., Yang, G.: Extremal optimization for protein folding simulations on the lattice. Comput. Math. Appl. 57, 1855–1861 (2009)
12. Mann, M., Will, S., Backofen, R.: CPSP-tools – Exact and complete algorithms for high-throughput 3 D lattice protein studies. Bmc Bioinformatics 9(1), 230 (2008)
13. Mann, M., Hamra, M.A., Steinhöfel, K., Backofen, R.: Constraint-based local move definitions for lattice protein models including side chains. In: Proceedings of the Fifth Workshop on Constraint Based Methods for Bioinformatics, WCB 2009 (2009)
14. Miyazawa, S., Jernigan, R.L.: Estimation of effective interresidue contact energies from protein crystal structures: quasi-chemical approximation. Macromolecules 18(3), 534–552 (1985)

15. Newton, M.A.H., Pham, D.N., Sattar, A., Maher, M.: Kangaroo: An efficient constraint-based local search system using lazy propagation. In: Lee, J. (ed.) CP 2011. LNCS, vol. 6876, pp. 645–659. Springer, Heidelberg (2011)
16. Palù, A.D., Dovier, A., Fogolari, F., Pontelli, E.: Exploring protein fragment assembly using CLP. In: IJCAI, pp. 2590–2595 (2011)
17. Palù, A.D., Dovier, A., Pontelli, E.: Heuristics, optimizations, and parallelism for protein structure prediction in CLP(FD). In: PPDP, pp. 230–241 (2005)
18. Palù, A.D., Dovier, A., Pontelli, E.: A constraint solver for discrete lattices, its parallelization, and application to protein structure prediction. Softw. Pract. Exper. 37, 1405–1449 (2007)
19. Palù, A.D., Will, S., Backofen, R., Dovier, A.: Constraint based protein structure prediction exploiting secondary structure information. In: Proceedings of Italian Conference on Computational Logic, CLIC 2004 (2004)
20. Pham, D.N., Thornton, J., Gretton, C., Sattar, A.: Advances in local search for satisfiability. In: Orgun, M.A., Thornton, J. (eds.) AI 2007. LNCS (LNAI), vol. 4830, pp. 213–222. Springer, Heidelberg (2007)
21. Rashid, M.A., Hoque, M. T., Newton, M.A.H., Pham, D.N., Sattar, A.: A new genetic algorithm for simplified protein structure prediction. In: Thielscher, M., Zhang, D. (eds.) AI 2012. LNCS, vol. 7691, pp. 107–119. Springer, Heidelberg (2012)
22. Rotkiewicz, P., Skolnick, J.: Fast procedure for reconstruction of full-atom protein models from reduced representations. Journal of Computational Chemistry 29(9), 1460–1465 (2008)
23. Shatabda, S., Newton, M.A.H., Sattar, A.: Mixed heuristic local search for protein structure prediction. In: Proceedings of the Twenty-Seventh AAAI Conference on Artificial Intelligence, Bellevue, Washington, USA, July 14-18. AAAI Press (2013)
24. Shatabda, S., Newton, M., Rashid, M.A., Pham, D.N., Sattar, A.: The road not taken: retreat and diverge in local search for simplified protein structure prediction. BMC Bioinformatics 14(2), 1–9 (2013)
25. Steinhofel, K., Skaliotis, A., Albrecht, A.: Relating time complexity of protein folding simulation to approximations of folding time. Computer Physics Communications 176(7), 465–470 (2007)
26. Torres, S.R.D., Romero, D.C.B., Vasquez, L.F.N., Ardila, Y.J.P.: A novel *ab-initio* genetic-based approach for protein folding prediction. In: Proceedings of the 9th Annual Conference on Genetic and Evolutionary Computation, GECCO 2007, pp. 393–400. ACM, New York (2007)
27. Ullah, A.D., Kapsokalivas, L., Mann, M., Steinhöfel, K.: Protein folding simulation by two-stage optimization. In: Cai, Z., Li, Z., Kang, Z., Liu, Y. (eds.) ISICA 2009. CCIS, vol. 51, pp. 138–145. Springer, Heidelberg (2009)
28. Ullah, A.D., Steinhöfel, K.: A hybrid approach to protein folding problem integrating constraint programming with local search. BMC Bioinformatics 11(S-1), 39 (2010)

On Caching for Local Graph Clustering Algorithms

René Speck and Axel-Cyrille Ngonga Ngomo

Universität Leipzig, Institut für Informatik, AKSW,
Postfach 100920, D-04009 Leipzig, Germany,
{speck,ngonga}@informatik.uni-leipzig.de

Abstract. In recent years, local graph clustering techniques have been utilized as devices to unveil the structured hidden of large networks. With the ever growing size of the data sets generated in domains of applications as diverse as biomedicine and natural language processing, time-efficiency has become a problem of growing importance. We address the improvement of the runtime of local graph clustering algorithms by presenting the novel caching approach SGD*. This strategy combines the Segmented Least Recently Used and Greedy Dual strategies. By applying different caching strategies to the unprotected and protected segments of a cache, SGD* displays a superior hitrate and can therewith significantly reduce the runtime of clustering algorithms. We evaluate our approach on four real protein-protein-interaction graphs. Our evaluation shows that SGD* achieves a considerably higher hitrate than state-of-the-art approaches. In addition, we show how by combining caching strategies with a simple data reordering approach, we can significantly improves the hitrate of state-of-the-art caching strategies.

Keywords: caching, local graph clustering, large networks.

1 Introduction

Graphs are a natural representation for a large number of real-world problems and datasets ranging from protein-protein-interaction networks [1] to external memory data [2]. Over the last years, a large number of approaches have been developed to achieve the goal of clustering graphs with high accuracy [3, 4]. While the accuracy of these approaches is being studied continuously, improving their performance remains a major challenge [4, 5]. Current approaches to graph clustering can be subdivided into two main categories: global approaches, which require knowledge about the whole graph for clustering and local approaches, which find a solution vertex-wise without necessitating knowledge of the whole graph [2]. Local graph clustering algorithms were originally conceived to allow the detection of clusters around a small set $\mathcal{N}$ of nodes of interest, especially when dealing with very large graphs. However, local clustering approaches are nowadays often used to cluster whole graphs [6, 7]. One problem that then arises is the scalability of these approaches [5]. In this paper, we address the problem of improving the runtime of *local graph clustering algorithms* that allow *overlapping clusters*, especially when the magnitude of the set $\mathcal{N}$ of input vertices to process is close to the magnitude of the set of vertices. We present the novel caching strategy SGD* (Segmented Greedy Dual). SGD* combines the Segmented Least Recently Used

S. Cranefield and A. Nayak (Eds.): AI 2013, LNAI 8272, pp. 56–67, 2013.

(SLRU) [8] and Greedy Dual (GD⋆) strategies to improve the hitrate during the clustering process so as to further reduce its runtime. In addition, we show how a simple node reordering strategy can further improve the hitrate of caching algorithms. We evaluate our approaches by using the BorderFlow algorithm[1] [9] on protein-protein-interaction (PPI) networks [1]. We chose BorderFlow because of its superior accuracy on PPI networks [6] and because it has already been applied in several domains including concept location in software development [7] and query clustering for benchmarking [10]. Our experiments show that SGD⋆ outperforms state-of-the-art approaches with respect to its hit-rate and space requirements. In addition, we can more than quadruple the hitrate of common caching strategies and of SGD⋆ by combining them with node reordering. By these means, we can reduce the runtime of BorderFlow to less than 25% of its original.

The rest of this paper is structured as follows: In the next section, we present some work related to this paper. Then, we present necessary preliminaries. Thereafter, we present our approaches, SGD⋆ and RP. In the evaluation section, we compare our approaches with seven state-of-the-art caching approaches. Finally, we present relevant related work on caching for local graph clustering and conclude.

2 Related Work

A vast amount of literature has been produced to elucidate the problem of graph clustering [3, 4]. Still, with the growth of the size of the dataset at hand, improving the runtime of graph clustering becomes an increasingly urgent problem. Several approaches have been developed with the goal of improving the performance of graph clustering approaches. Overall, most of these approaches fall into one of the following two categories: sampling (also called graph sparsification) [11, 5] and caching [12]. Sampling is a generic solution to reducing the runtime of algorithms [13]. The idea here is to reduce the runtime of clustering approaches by computing a smaller representative subset of the data at hand and running the computation on this data set. While this approach can get rid of noise in the data, the alteration of the data set at hand might lead to undesired side-effects when combined with certain clustering strategies.

Caching follows a different idea and tries to store and reuse as much intermediary knowledge as possible to improve the runtime of the given algorithm. One of the most commonly used approaches is the Least Recently Used algorithm [14]. The idea behind this approach is simply to evict the entry that led to the oldest hit when the cache gets full. One of the main drawbacks of this approach is that the cache is not scan-resistant. Meanwhile, a large number of scan-resistant extensions of this approach have been created. For example, SLRU [8] extends LRU by splitting the cache into a protected and an unprotected area. The Least Frequently Used (LFU) [15] approach relies on a different intuition. Here, a count of the number of accesses to entries in the cache is kept. The cache evicts the entries with the smallest frequency count when necessary. This approach is scan-resistant but does not make use of the locality of reference. Consequently, it was extended by window-based LFU [16], sliding window-based approaches [17] and dynamic aging (LFUDA) [18] amongst others. Another commonly

[1] We used the free version of the algorithm whose code is available at `http://borderflow.sf.net`.

used caching strategy is based on the idea of first-in-first-out (FIFO) lists [19]. When the cache is full, this approach evicts the entry that have been longest in the cache. The main drawback of this approach is that it does not make use of locality. Thus, it was extended in several ways, for example by the "FIFO second chance" approach [19]. Other strategies such as Greedy Dual (GD*) [20] use a cost model to determine which entries to evict.

3 Preliminaries and Notation

3.1 Caching

The aim of caching is to reduce the runtime of algorithms by storing intermediate results of expensive computations. Formally, let $\mathcal{O} = \{o_1 ... o_n\}$ be a set of results that can be cached. Let $cost : O \rightarrow \mathbb{R}^+$ be a function that maps each object o with the cost of its computation. Furthermore, let $size : O \rightarrow \mathbb{R}^+$ be a function that maps each object o to its size. A cache $\mathcal{C}$ of maximal size $\mathcal{C}_{max}$ (with $\mathcal{C}_{max} \geq \max\limits_{o \in \mathcal{O}} size(o)$) is a subset of $\mathcal{O}$ such that $\sum\limits_{x \in \mathcal{C}} size(x) \leq \mathcal{C}_{max}$. An algorithm $\mathcal{A}$ that relies on caching issues a query sequence $\sigma : T \rightarrow \mathcal{O}$ ($T \subseteq \mathbb{N}$) to the cache $\mathcal{C}$. At each time $t \in T$, the query $\sigma(t)$ for an object $o \in O$ is sent to the cache $\mathcal{C}$. If the cache contains the object o, it simply returns the corresponding solution to the clustering problem (this is usually called a *cache hit*). Else, $\mathcal{C}$ returns $\emptyset$ (*cache miss*). In case of a hit, the cost for $cost(\sigma(t))$ is a constant c called the cache latency. In case of a miss, $\mathcal{A}$ must compute o with the cost $cost(o)$, leading to $cost(\sigma(t)) = cost(o)$. The result of the computation is then forwarded to $\mathcal{C}$. As $cost(o)$ is usually vastly superior to c, we will assume $c = 0$ in the remainder of this paper. Caching algorithms aim to minimize the total cost $\sum\limits_{t \in T} cost(\sigma(t))$ of the sequence σ by generating a sequence of cache states $\mathcal{C}^t$ for each time t that abide by $|\mathcal{C}^t|$.

3.2 Local Graph Clustering with Overlapping Clusters

Let $G = (V, E, w)$ be a graph, where V is the set of edges, $E \subseteq V \times V$ is the set of edges and $w : E \rightarrow \mathbb{R}^+$ the weight function that assign a weight to each edge of the graph G. A graph clustering algorithm aims to determine a set $\mathcal{V} = \{V_1, V_2, ..., V_n\}$ of subsets of V that maximize a certain fitness function [3]. Some local graph clustering algorithms allow for clusters to share nodes, i.e., $|V_i \cap V_j| > 0$ with $i \neq j$. Such algorithms are called *non-partitioning approaches* [21]. For the purpose of clustering, local graph clustering algorithms rely on a set of nodes $\mathcal{N} \subseteq V$ as input. For each node of interest $n \in \mathcal{N}$, they aim to discover a nearby cluster[2]. This is carried out by running iterative approaches of which most rely on local search [22, 23, 9] and random walks [24–26]. The idea behind these iterative procedures is to carry a simple operation repeatedly until the fitness function is maximized. For example, search algorithms begin with an initial solution $S^0(n)$ for $n \in \mathcal{N}$. At each step t they compute the current solution $S^t(n)$ by

[2] In most cases, n must be an element of this cluster.

altering the previous solution $S^{t-1}(n)$. This is carried out by adding a subset of the adjacent nodes of $S^{t-1}(n)$ to the solution and simultaneously removing a subset of the nodes of $S^{t-1}(n)$ from it until the fitness function is maximized. They then return the final solution $S(n)$. The insight that makes caching utilizable to reduce the runtime of local graph clustering algorithms is that if an algorithm generates the same intermediate solution for two different nodes, then the final solution for these nodes will be the same, i.e., $\forall n, n' \in \mathcal{N}\ S_1^t(n) = S_2^t(n') \rightarrow S(n) = S(n')$. Thus, by storing some elements of the sequence of solutions computed for previous nodes n and the solution $S(n)$ to which they led, it becomes possible to return the right solution of a node n' without having to compute the whole sequence of solutions. However, it is impossible to store all elements of the sequence of solutions generated by local graph clustering algorithms for large input graphs and large $\mathcal{N}$. The first innovation of this paper is a novel caching approach for local graph clustering dubbed SGD* that outperforms the state-of-the-art w.r.t. its hitrate. Note that $\mathcal{N}$ is a set, thus the order in which its nodes are processed does not affect the final solution of the clustering. Consequently, by finding an ordering of nodes that ensures that the sequence of solutions generated for subsequent nodes share a common intermediate as early as possible in the computation, we can improve the hitrate of caching algorithms and therewith also the total runtime of algorithms. This is the goal of the second innovation of this paper, our node reordering strategy.

4 Segmented Greedy Dual

SGD* combines the ideas of two caching strategies: SLRU and GD*. SLRU is a scan-resistant extension of LRU [14], one of the most commonly used caching strategies. Like LRU, it does not take the cost of computing an object into consideration and thus tends to evict very expensive objects for the sake of less expensive one. GD* on the other hand is an extension of the Landlord algorithm [27] which takes the costs and the number $hit(o)$ of cache hit that return o into consideration.

The idea behind SGD* is to combine these strategies to a scan-resistant and cost-aware caching approach. To achieve this goal, SGD* splits the cache into two parts: a protected segment and an unprotected segment. The unprotected segment stores all the $S^t(n) \subseteq V$ that are generated while computing a solution for the input node $n \in \mathcal{N}$. The protected segment on the other hand stores all the results that have been accessed at least once and contain at least two nodes. While SLRU uses LRU on both the protected and unprotected area, SGD* uses the GD* strategy on the unprotected area and the LRU approach on the protected area. An overview of the resulting caching approach is given in Algorithm 1. For each node n, we begin by computing the first intermediary result for n. Then we iterate the following approach. We ask the cache for the head (i.e., the first element) of the list S. If this element is not in the cache, we compute the next intermediary solution and add it to the head of the list. The iteration is terminated out in one of the following two cases. In the first case, the iteration terminates for n, returning $\perp$. Then, the result is added to the list S and S is cached. In the second case, a solution is found in the cache. Then this solution is cached. Note that this approach works for every caching mechanism. The main difference between caching approaches is how they implement the storage method $cachePut$ and the data fetching method $cacheGet$.

Algorithm 1. Caching for local graph clustering

```
Require: Set of nodes 𝒩
  List S
  Buffer B, id
  Result R = ∅
  Protected segment P = ∅
  Unprotected segment U = ∅
  for all n ∈ 𝒩 do
    S =compute(n)
    id =cacheGet(S)
    while id == −1 do
      B =compute(S)
      if B == ⊥ then
        cachePut(S)
        R = R ∪ (n,cacheGet(S))
        break
      end if
      S = append(B, S)
      id =cacheGet(S)
    end while
    R = R ∪ (n, id)
  end for
  return R
```

The fetching data algorithm of the SGD* cache has two functions and is summarized in Algorithm 2. First, it allows checking whether the data that is being required is in the cache. Concurrently, it reorganizes the data in the cache in case of a cachehit. The SGD* data fetching approach and works as follows: In case there is no cachehit, the cache simply returns -1. A cache hit can occur in two ways: First, the current solution can be contained in the protected segment of the cache. In this case, SGD* simply updates the credit of the entry o, i.e., of the cached entry that led to finding the cached solution to the clustering task for the current node. It then computes the id of the answer to the current caching and returns it. Note that there is then no need to evict data, as no new data is added. If the cache hit occurs within the unprotected segment U of the cache, the algorithm moves the entry o that led to the hit from U to the protected segment P of the cache. Should P exceeds its maximal size, then the elements with the smallest credit score are evicted to the unprotected segment U of the cache until there is enough space for o in P. o then gets inserted into P and its credit score is computed. The final step in case of a cachehit consists of assigning all the steps that led to the solution mapped to s. For this purpose (see Algorithm 3), each single component o_i of the solution of S is inserted into U. In case the cache would exceed its maximal size when accommodating o_i, the elements of U are evicted in ascending order of credit until enough space is available for o_i.[3]

[3] We implemented the approach and made it freely available at `http://sourceforge.net/projects/cugar-framework`

Algorithm 2. SGD*'s *cacheGet*

Require: Solution S
  $s = head(S)$
  $id = -1$
  if $s \in P$ **then**
    $credit(s) = min + \frac{(hit(o)cost(o))^{\frac{1}{b}}}{size(o)}$
    $id = id(s)$
  else
    if $s \in U$ **then**
      $id = id(s)$
      $U = U \backslash s$
      $P = P \cup s$
      while $|P| > C_{max}/2$ **do**
        $o = \arg\min_{o' \in U} credit(o')$
        $U = U \cup \{o\}$
        $P = P \backslash \{o\}$
      end while
    end if
  end if
  if $id \neq -1$ **then**
    for all $s_i \in S$ **do**
      cachePut(s_i, id)
    end for
  end if
  return id

5 Node Reordering

While SGD* outperforms the state of the art as shown in our experiments, the general behavior of caching algorithms can be further improved when assuming that the set $\mathcal{N}$ is known at the beginning of the clustering. Note that this condition is not always given, as many practical clustering approaches process the results for known nodes of interest to generate novel nodes of interest. Yet, when this condition is given and when in addition the computation of a cluster for a node n does not affect the set $\mathcal{N}$ or the computation of a cluster for another node n', the order in which the nodes are drawn from $\mathcal{N}$ does not alter the result of the clustering and can be dynamically changed during the computation. By choosing the order in which this selection is carried out, we can drastically improve the locality of caching algorithms. We propose a simple and time-efficient approach to achieve this goal: the use of a FIFO list. For this purpose, we extend the cachePut method as shown in Algorithm 4.

The FIFO list L simply stores the elements of $\mathcal{N}$ that were part of a solution (note that the elements of a solution must not all belong to $\mathcal{N}$). Instead of drawing n from $\mathcal{N}$ as described in Algorithm 1, we draw the node n by taking the first element of L if is not empty (in which case we draw one at random from $\mathcal{N}$). The rationale behind using a FIFO list is that by processing nodes n' that are closest to the input node n first, we can reduce the number of iterations necessary for a cache hit to occur. While this

Algorithm 3. SGD*'s *cachePut*

```
Require: Solution S_i
Require: Set of input nodes N
Require: ID id
  min = 0 // minimal credit
  while |P| + |U| + size(S_i) > C_max do
    o = arg min_{o' ∈ U} credit(o')
    min = credit(o)
    U = U \ {o}
  end while
  credit(S_i) = min + (hit(o)cost(o))^(1/b) / size(o)
  id(S_i) = id
  U = U ∪ {S_i}
```

Algorithm 4. SGD*'s *cachePut* with node reordering

```
Require: Solution S_i
Require: Set of input nodes N
Require: ID id
  min = 0 // minimal credit
  for all x ∈ S_i do
    if x ∉ L ∧ x ∈ N then
      L = append(L, x)
    end if
  end for
  while |P| + |U| + size(S_i) > C_max do
    o = arg min_{o' ∈ U} credit(o')
    min = credit(o)
    U = U \ {o}
  end while
  credit(S_i) = min + (hit(o)cost(o))^(1/b) / size(o)
  id(S_i) = id
  U = U ∪ {S_i}
```

assumption might appear simplistic, our evaluation shows that it suffices to reduce the space requirement of caches by a factor up to 40.

6 Evaluation

6.1 Experimental Setup

As experimental data, we used four graphs resulting from high-throughput experiments utilized in [1]. The high-throughput graphs were computed out of the datasets published in [28] (Gavin06), [29] (Ho02), [30] (Ito01) and [31] (Krogan06).[4] The graphs were

[4] All data sets used for this evaluation can be found at `http://rsat.bigre.ulb.ac.be/~sylvain/clustering_evaluation/`

undirected and unweighted. We used the BorderFlow algorithm as clustering algorithm because it has been shown to perform best on these data sets [6]. We compared our approach against the standard strategies FIFO, FIFO second chance, LRU, LFU, LFU-DA and SLRU strategies. In addition, we developed the COST strategy, which evicts the entries with the highest costs. The idea here is that solutions S^t with high costs are usually generated after a large number t of iterations. Thus, it is more sensible to store the entries $S^{t'}$ that are less costly than S^t, as a corresponding cache hit is more probable and would reduce the total runtime of the algorithm. We compare these caching approaches in two series of experiments. In the first series of experiments, we compared the hitrate of the different caching approaches without node reordering. In the second series, we clustered exactly the same data with node reordering. In each series

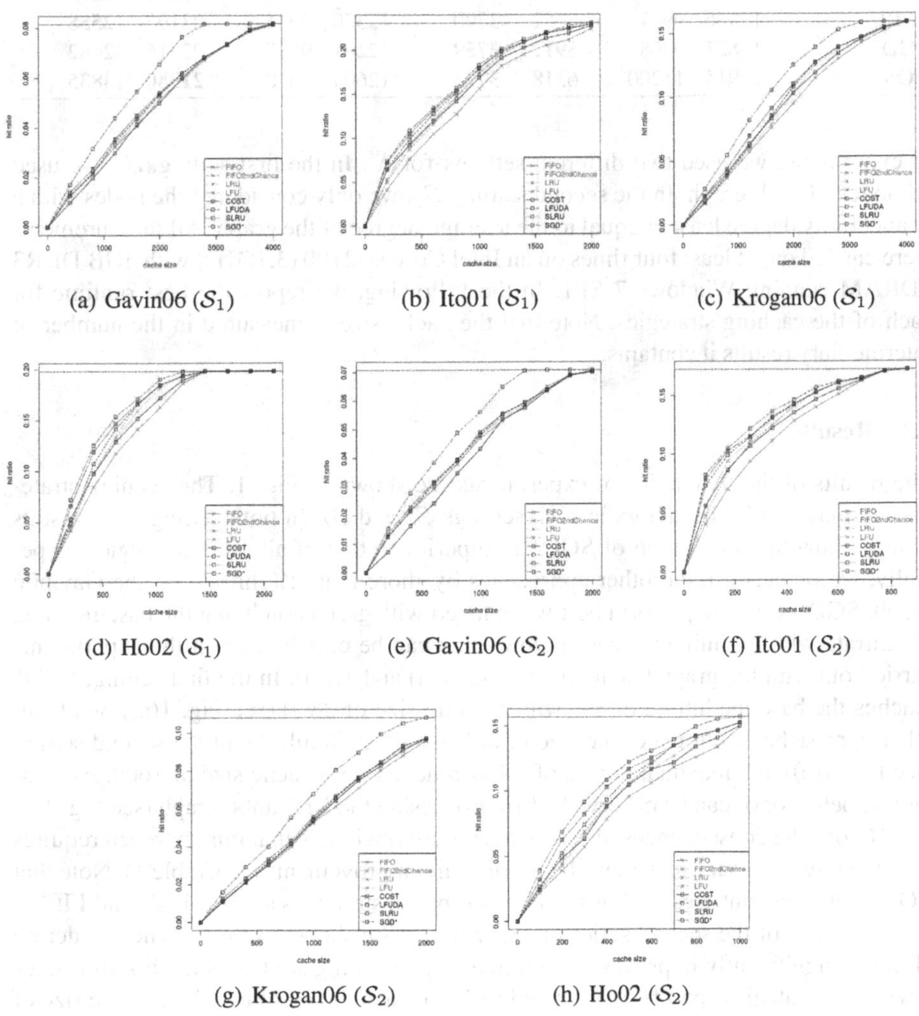

(a) Gavin06 ($\mathcal{S}_1$) (b) Ito01 ($\mathcal{S}_1$) (c) Krogan06 ($\mathcal{S}_1$)

(d) Ho02 ($\mathcal{S}_1$) (e) Gavin06 ($\mathcal{S}_2$) (f) Ito01 ($\mathcal{S}_2$)

(g) Krogan06 ($\mathcal{S}_2$) (h) Ho02 ($\mathcal{S}_2$)

Fig. 1. Comparison of the hitrate of SGD* against seven other approaches

Table 1. Comparison of runtimes with cache size 300 in setting $\mathcal{S}_1$. The best runtimes are in bold font. All runtimes are in ms. The columns labeled "Default" contain the runtime of our approaches without node reordering. The columns labeled "Reordered" contain the runtimes after the reordering has been applied.

	Gavin06		Ho02		Ito01		Krogan06	
	Default	Reordered	Default	Reordered	Default	Reordered	Default	Reordered
Baseline	14944	14944	8938	8938	42775	42775	25630	25630
FIFO	14196	**9921**	6848	**3759**	17316	9578	23758	**12558**
FIFO2ndChance	14071	9937	6692	3790	15303	9594	23868	12604
LRU	14164	9968	6708	3790	13525	9578	23899	12667
LFU	14008	12214	6099	5038	12324	11029	23244	19156
LFU-DA	14102	9984	6645	3790	12745	9609	23821	12698
SLRU	13946	9937	6052	3790	12370	**9531**	23197	**12558**
SGD*	**13821**	9968	**5912**	**3759**	**12261**	9687	22916	12682
COST	13915	11200	6318	3775	12604	9578	**21980**	14835

of experiment, we used two different settings for $\mathcal{N}$. In the first setting, $\mathcal{S}_1$, we used all nodes of each graph. In the second setting, $\mathcal{S}_2$, we only considered the nodes with a connectivity degree least or equal to the average degree of the graph. All measurements were carried out at least four times on an Intel Core i3-2100 (3.1GHz) with 4GB DDR3 SDRAM running Windows 7 SP1. In the following, we report the best runtime for each of the caching strategies. Note that the cache size is measured in the number of intermediary results it contains.

6.2 Results

The results of the first series of experiments are shown in Fig. 1. The caching strategies display similar behaviors in both settings $\mathcal{S}_1$ and $\mathcal{S}_2$. In both settings, our results clearly show that the hitrate of SGD* is superior to that of all other strategies. Especially, we outperform the other approaches by more than 2% hitrate on the Gavin06 graph. SGD* seems to perform best when faced with graphs such that the baseline (i.e., the hitrate with an infinite cache) is low. This can be clearly seen in the experiments carried out with the graph Gavin06 (see Fig. 1(a) and 1(e))). In the first setting, SGD* reaches the baseline hitrate of 8% with a cache size of 2800 (see Fig. 1(a)), while all other approaches require a cache size of at least 4000. Similarly, in the second setting (see Fig. 1(e)), the maximal hitrate of 7% is reached for a cache size of 1600. An analogical behavior of can be observed when processing the Krogan06 graph (see Fig. 1(c) and 1(g))). The consequences of this behavior are obviously that our approach requires significantly less space to achieve better runtime improvements (see Table 1). Note that COST achieves runtimes similar to that of common strategies such as LRU and FIFO.

The results of the second series of experiments are shown in Fig. 2. The reordering of nodes significantly improves the runtime of all caching strategies in all settings, allowing all strategies apart from Cost and LFU to reach the baseline with a cache size of 100 on the full Krogan06 and Ho02 graphs (i.e., in setting $\mathcal{S}_1$, see Fig 2(c) and 2(d)). In setting $\mathcal{S}_2$, the baseline hitrate is reached with a cache size of 50 on the same graphs (see Fig. 2(g) and 2(h)). Therewith, node reordering can make most caching strategies more than 40 times more space-efficient (compare Fig. 1(c) and 2(c)). The COST and LFU approaches not profiting maximally from the node reordering is simply due to cache pollution. The cost-based approach deletes those elements, which required a long processing time, the idea being that they are unlikely that they appear again. Yet, this approach leads to the content of the cache remaining static early in the computation. Consequently, reordering the nodes does not improve the hitrate of such caches as significantly as that of FIFO, SLRU and other strategies, especially when the cache is small. LFU behaves similarly with respect to the hitrate score of the elements in the cache. Overall, by combining SGD* and node reordering, we can improve the runtime of BorderFlow to less than 25% of its original runtime on the Ito01 graph.

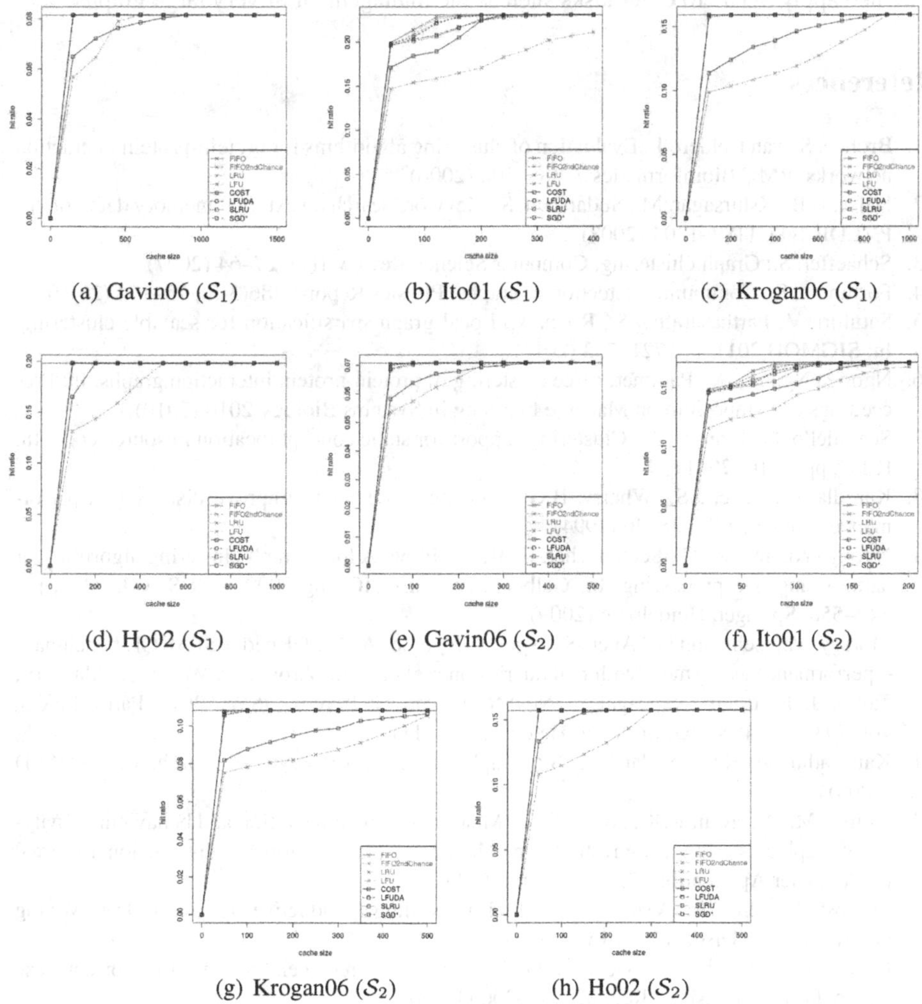

Fig. 2. Comparison of the hitrate of SGD* against seven other approaches with node reordering

7 Conclusion and Future Work

SGD* addresses the cost-blindness of SLRU by combining it with the GD* caching strategy. We showed that this combined strategy outperforms state-of-the-art approaches with respect to its hitrate. We also presented an approach to improve the locality of caching when dealing with clustering approaches where the order of the input nodes does not alter the result of the clustering. One interesting results was that once we apply reordering to the input data (therewith improving the locality of the clustering), we could boost the results of the FIFO caching approach and make it outperform all others in most cases. In future work, we will combine our caching approach with other clustering algorithms. Note that our caching approach is not limited to graph clustering and can be easily applied to any other problem that necessitates caching. Consequently, we will also apply SGD* to other tasks such as the management of very large graphs.

References

1. Brohee, S., van Helden, J.: Evaluation of clustering algorithms for protein-protein interaction networks. BMC Bioinformatics 7, 488–506 (2006)
2. Dalvi, B.B., Kshirsagar, M., Sudarshan, S.: Keyword search on external memory data graphs. PVLDB 1(1), 1189–1204 (2008)
3. Schaeffer, S.: Graph clustering. Computer Science Review 1(1), 27–64 (2007)
4. Fortunato, S.: Community detection in graphs. Physics Reports 486(3-5), 75–174 (2010)
5. Satuluri, V., Parthasarathy, S., Ruan, Y.: Local graph sparsification for scalable clustering. In: SIGMOD 2011, pp. 721–732 (2011)
6. Ngonga Ngomo, A.: Parameter-free clustering of protein-protein interaction graphs. In: Proceedings of Symposium on Machine Learning in Systems Biology 2010 (2010)
7. Scanniello, G., Marcus, A.: Clustering support for static concept location in source code. In: ICPC, pp. 1–10 (2011)
8. Karedla, R., Love, J.S., Wherry, B.G.: Caching strategies to improve disk system performance. Computer 27, 38–46 (1994)
9. Ngonga Ngomo, A.-C., Schumacher, F.: BorderFlow: A local graph clustering algorithm for natural language processing. In: Gelbukh, A. (ed.) CICLing 2009. LNCS, vol. 5449, pp. 547–558. Springer, Heidelberg (2009)
10. Morsey, M., Lehmann, J., Auer, S., Ngonga Ngomo, A.-C.: DBpedia SPARQL benchmark – performance assessment with real queries on real data. In: Aroyo, L., Welty, C., Alani, H., Taylor, J., Bernstein, A., Kagal, L., Noy, N., Blomqvist, E. (eds.) ISWC 2011, Part I. LNCS, vol. 7031, pp. 454–469. Springer, Heidelberg (2011)
11. Kanjirathinkal, R.C., Sudarshan, S.: Graph clustering for keyword search. In: COMAD (2009)
12. Kumar, M., Agrawal, K.K., Arora, D.D., Mishra, R.: Implementation and behavioural analysis of graph clustering using restricted neighborhood search algorithm. International Journal of Computer Applications 22(5), 15–20 (2011)
13. Provost, F., Kolluri, V.: A survey of methods for scaling up inductive algorithms. Data Mining and Knowledge Discovery 3, 131–169 (1999)
14. O'Neil, E.J., O'Neil, P.E., Weikum, G.: The lru-k page replacement algorithm for database disk buffering. SIGMOD Rec. 22, 297–306 (1993)
15. Breslau, L., Cao, P., Fan, L., Phillips, G., Shenker, S.: Web caching and zipf-like distributions: Evidence and implications. In: INFOCOM, pp. 126–134 (1999)

16. Karakostas, G., Serpanos, D.N.: Exploitation of different types of locality for web caches. In: Proceedings of the Seventh International Symposium on Computers and Communications, pp. 207–2012 (2002)
17. Hou, W.-C., Wang, S.: Size-adjusted sliding window LFU - A new web caching scheme. In: Mayr, H.C., Lazanský, J., Quirchmayr, G., Vogel, P. (eds.) DEXA 2001. LNCS, vol. 2113, pp. 567–576. Springer, Heidelberg (2001)
18. Arlitt, M., Cherkasova, L., Dilley, J., Friedrich, R., Jin, T.: Evaluating content management techniques for web proxy caches. SIGMETRICS Performance Evaluation Review 27(4), 3–11 (2000)
19. Tanenbaum, A.S., Woodhull, A.S.: Operating systems - design and implementation, 3rd edn. Pearson Education (2006)
20. Jin, S., Bestavros, A.: Greedydual* web caching algorithm – exploiting the two sources of temporal locality in web request streams. In: 5th International Web Caching and Content Delivery Workshop, pp. 174–183 (2000)
21. Schlitter, N., Falkowski, T., Lässig, J.: Dengraph-ho: Density-based hierarchical community detection for explorative visual network analysis. In: Springer (ed.) Proceedings of the 31st SGAI International Conference on Artificial Intelligence (2011)
22. Schaeffer, S.: Stochastic local clustering for massive graphs. In: Ho, T.-B., Cheung, D., Liu, H. (eds.) PAKDD 2005. LNCS (LNAI), vol. 3518, pp. 354–360. Springer, Heidelberg (2005)
23. Felner, A.: Finding optimal solutions to the graph partitioning problem with heuristic search. Ann. Math. Artif. Intell. 45(3-4), 293–322 (2005)
24. Alamgir, M., von Luxburg, U.: Multi-agent random walks for local clustering on graphs. In: ICDM, pp. 18–27 (2010)
25. Spielman, D.A., Teng, S.H.: A local clustering algorithm for massive graphs and its application to nearly-linear time graph partitioning. CoRR abs/0809.3232 (2008)
26. Biemann, C., Teresniak, S.: Disentangling from babylonian confusion – unsupervised language identification. In: Gelbukh, A. (ed.) CICLing 2005. LNCS, vol. 3406, pp. 773–784. Springer, Heidelberg (2005)
27. Young, N.E.: On-line file caching. In: Proceedings of the Ninth Annual ACM-SIAM Symposium on Discrete Algorithms, pp. 82–86 (1998)
28. Gavin, A.C., et al.: Proteome survey reveals modularity of the yeast cell machinery. Nature (January 2006)
29. Ho, Y., et al.: Systematic identification of protein complexes in saccharomyces cerevisiae by mass spectrometry. Nature 415(6868), 180–183 (2002)
30. Ito, T., et al.: A comprehensive two-hybrid analysis to explore the yeast protein interactome. Proc. Natl. Acad. Sci. U.S.A 98(8), 4569–4574 (2001)
31. Krogan, N., et al.: Global landscape of protein complexes in the yeast saccharomyces cerevisiae. Nature (March 2006)

Provenance-Based Trust Estimation for Service Composition

Jing Jiang and Quan Bai

School of Computing and Mathematical Sciences,
Auckland University of Technology, Auckland, New Zealand
{Jing.Jiang,Quan.Bai}@aut.ac.nz

Abstract. Provenance information can greatly enhance transparency and accountability of shared services. In this paper, we introduce a trust estimation approach which can derive trust information based on the analysis of provenance data. This approach can utilize the value of provenance data, and enhance trust estimation in open dynamic environments.

Keywords: Trust estimation, provenance, service recommendation.

1 Introduction

Nowadays, with the development of open distributed systems, increasing number of services and information are shared on open platforms. For many open distributed systems, trust is a crucial factor that reflects the quality of service (QoS) and helps manage correlation among interactive service components. Provenance data, which describes the origins and processes that relate to the generation of composite services, can provide rich context for trust estimation [1]. Especially in service-oriented computing, provenance identifies what data is passed between services, what process involved in the generation of results, who contributed to the service generation, etc. [4].

In this paper, a provenance-based trust estimation model is proposed. In this model, provenance information of a composite service is represented as a provenance graph. The similarities of different provenance graphs are analysed according to their Same Edge Contributions (SEC). Based on graph similarities and correlation to trust values, the performance of a future composite service can be predicted.

The rest of this paper is organized as follows. Section 2 describes the problems definition and some assumptions in this research. Section 3 presents the framework of the provenance-based trust evaluation model, and how to derive trust support values from provenance graph. In Section 4, we setup experiments and demonstrate the performance of the SEC model. Finally, the conclusion and future works are presented in Section 5.

2 Problem Definition

When a service consumer submits a service request, workflows which can satisfy the request will be proposed by different providers. The system will estimate each

S. Cranefield and A. Nayak (Eds.): AI 2013, LNAI 8272, pp. 68–73, 2013.

proposed workflow based on the analysis of historical provenance data (graphs). We suppose that there is a universe of n service components $S = \{S_1, S_2, ..., S_n\}$ which are loosely coupled in a SOC system. $E_x(S_i, S_j)$ represents a path leads from S_i to S_j. Firstly, we give the definition for provenance graph in knowledge base.

Definition 1: A *provenance graph* is a 2-tuple $PVG = (V_{PVG}, E_{PVG})$, where V_{PVG} is a finite set of nodes, and E_{PVG} is the finite set of edges. Furthermore, $\mid G_{PVG} \mid = \mid V_{PVG} \mid + \mid E_{PVG} \mid$ denote the size of G_{PVG}

The requests from service consumers include basic functional requirements, and then system will receive proposal graph from different providers as following definition.

Definition 2: A *proposal graph* PRG is defined as 2-tuple, i.e., $PRG =< ID, PRG = (V_{PRG}, E_{PRG}) >$. ID is the unique identifier for each service request. PRG is the proposal graph from providers that describes a finite set of service components $V_{PRG} = \{S_1, S_2, S_3, ..., S_n\}$ and a finite set of edges $E_{PRG} = \{E_1(S_1, S_2), E_2(S_1, S_3), ..., E_n(S_{n-1}, S_n)\}$.

The service components in V_{PRG} are required to achieve the functional requirement of the request, and E_{PRG} indicate the process of composite service. After the completion of the composite service, the system will generate service feedback RF which contains the proposal graph PRG and quality of composite service.

Definition 3: A *service feedback* RF is defined as a 2-tuple, $RF =< R, Q >$. R is the *service request* generated by the system which contains both unique transaction ID and provenance graph PRG. PRG describes the required service components and process in detail. Q represents the quality of composite service.

Definition 4: A sub-service graph $g = (V_g, E_g)$ is a subgraph of a graph PRG or PVG, denoted by $g \subseteq PRG/PVG$, where $V_g \subseteq V_{PRG}/V_{PVG}$ and $E_g \subseteq E_{PRG}/E_{PVG}$.

3 The Provenance-Based Trust Estimation Approach

3.1 Trust Estimation Protocol

In our approach, trust prediction is conducted by the protocol shown in Fig.1. Firstly, after the system receives a request, proposal graphs PRG based on the functional requirements from the service consumer will be generated. Then the proposal graphs will be sent to the Prediction Retrieval Module. The Prediction Retrieval Module will search Knowledge Base for all possible provenance graphs PVG which are similar to proposal graph PRG. Then, based on the previous provenance graphs PVG in the Knowledge Base, the Edge Contribution Module will update the edge contribution value for total available edges.

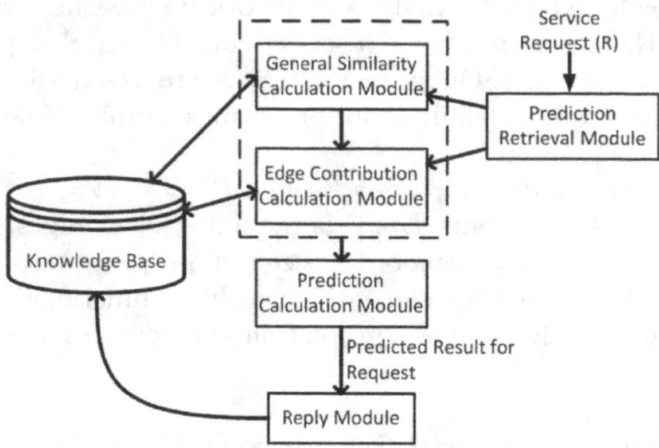

Fig. 1. Trust Estimation Protocol

At the beginning all edge are given the same weight within in the request. General Similarity Calculation Module calculates the similarity between proposal graphs PRG and provenance graph PVG based on the same service components and edges in graphs and then passed the most similar provenance graphs PVG to Prediction Calculation Module. Comparing the same edges in proposal graph PRG and provenance graphs PVG, the Prediction Calculation Module will use edge contribution value to give each provenance graph PVG support value. The system will return the class value of the provenance graph PVG which obtained the highest support value to the Reply Module.

3.2 General Similarity Calculation

The general similarity between proposal graph PRG and candidate provenance graph PVG in knowledge base is decided by an upper bound on the size of the Maximum Common Edge Subgraph (MCES) [5] [2]. First, according to service service components S in each graph, the set of vertices is partitioned into l partitions. Let g_i^{PRG} and g_i^{PVG} denote the sub-graph in i^{th} partition in graph PRG and PVG, respectively. An upper-bound on the similarity between provenance graph PRG and PVG can be calculated as follows:

$$V(PRG, PVG) = \sum_{i=1}^{l} min\{|g_i^{PRG}|, |g_i^{PVG}|\} \tag{1}$$

$$E(PRG, PVG) = \lfloor \sum_{i=1}^{l} \sum_{j=1}^{max\{|g_i^{PRG}|, |g_i^{PVG}|\}} \frac{min\{d(S_j^{PRG}), d(S_j^{PVG})\}}{2} \rfloor \tag{2}$$

$$sim(PRG, PVG) = \frac{[V(PRG, PVG) + E(PRG, PVG)]^2}{[|V(PRG)| + |E(PRG)|] \times [|V(PVG)| + |E(PVG)|]} \tag{3}$$

where $d(S_j^{PRG/PVG})$ denotes the number of adjacent service components of S_j in provenance graph PRG/PVG. Fig.2(a) and Fig.2(b) illustrate two workflow graphs for composite services. The higher $sim(PRG, PVG)$, the more same edge and nodes are share between the proposal graph PRG and candidate provenance graph PVG in knowledge base. It is necessary to specify a minimum acceptable value $sim^{threshold}$ for the general similarity measure. If $sim(PVG, PRG) \leq sim^{threshold}$, the candidate provenance graph PVG will be ignored in following edit operation cost calculation procedure.

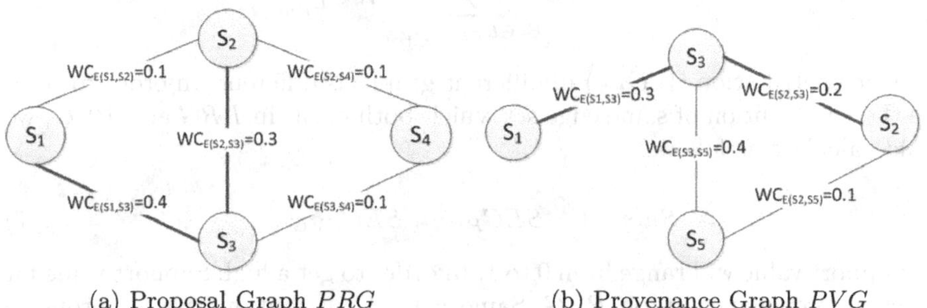

(a) Proposal Graph PRG (b) Provenance Graph PVG

Fig. 2.

3.3 Edge Contribution Calculation

In our approach, we intend to adopt the *Edge Contribution* which each edge $E(S_i, S_j)$ makes to quantify the edit operation cost. The Quality of Service (QoS) of a composite service is assumed as a random behavior. The uncertainty of such a random behavior is related with the required edges $E_G^x(S_i, S_j)$ in the process of service request, and can be reduced with the existence of a particular edge. Therefore, we firstly calculate the Quality Entropy ($H(Q)$) to measure average uncertainty of the QoS value of composite services [3]. Then, mutual information (i.e., $I(Q; E_G^x)$) [6] is used to measure how much reduction can a particular edge E_G^x make to the uncertainty of the QoS value.

$$C_{E_G^x(S_i,S_j)} = \frac{I(Q; E_G^x(S_i, S_j))}{H(Q)} \tag{4}$$

$$WC_{E_G^x} = \frac{C_{E_G^x}}{\sum\limits_{E_x \in E_G} C_{E_G^x}} \tag{5}$$

where G represents as PRG or PVG in different situations and where $WC_{E_G^x}$ is the contribution of edge E_G^x for PRG or PVG. The larger $WC_{E_G^x}$ is, the most contribution the edge E_G^x makes in the process.

Comparing proposal graph PRG and candidate provenance graph PVG passed from General Similarity Calculation step, we can get the particular same edge set between PRG and each PVG, i.e., $\{E^i_{sameSet}\} = Same(E_{PRG}, E_{PVG}) =$

$\{E_i, E_j, E_k, ...\}$, where all edges $\{E_i, E_j, E_k, ...\}$ in $\{E^i_{sameSet}\}$ both occur in PRG and PVG. For example, according to Fig. 2(a) and 2(b), $\{E_{sameSet}\} = \{E(S_1, S_3), E(S_2, S_3)\}$. Then, we should separately calculate the Same Edge Contribution rate (SEC) on proposal graph PRG and provenance graph PVG as follow:

$$SEC_{PRG/PVG} = \frac{\sum\limits_{E_x \in E^i_{sameSet}} WC_{E^x_{PRG/PVG}}}{\sum\limits_{E_x \in E_{PRG/PVG}} WC_{E^x_{PRG/PVG}}} \quad (6)$$

The edge contribution ($WC_{E^x_G}$) in different graphs is different. In order to compare the contribution of same edge set which both occur in PRG and PVG, we should calculate as follow:

$$Support = SEC_{PRG} * SEC_{PVG} \quad (7)$$

The Support value will range from 0 to 1. In order to get a high Support value for particular provenance graph PVG, Same Edge Contribution rate for proposal graph (SEC_{PVG}) and provenance graph (SEC_{PVG}) should not only as high as possible, but also as close as possible. The class which the proposal graph PRG should be classified into is dependent on the support value of each provenance graph PVG. Finally, the Reply Module generate a feed back RF for the proposal PRG after the execution, and store the information into the Knowledge Base.

4 Experiments and Analysis

Some experiments are conducted in this research. In the experiments, we included 10 service components S_i, and 45 kinds of edges $E(S_i, S_j)$. There are 2 kinds of class (i.e., Successful and Unsuccessful) are adopted for representing QoS. We except for two classification metrics: Accuracy and Precision for Successful class. We design three different scenarios for the experiment. Firstly, all training dataset and test dataset share the same set of service components and edges. Secondly, there appear new service components and edges in test, but they cannot been found in knowledge base. Thirdly, provenance graphs PVG with new service components and edges are added into the knowledge base.

Finally, following characteristics of the SEC model can be demonstrated. Firstly, even if there appear new service components and edges in *Request* provenance graph without in original knowledge base, the SEC model still can work and perform better in Precision for predicting Successful composite service. Secondly, according to the result from experiment, the Precision for the SEC model in three experiments seems to be similar, because they shared the same high contribution edge set. Thirdly, once an new edge is included into knowledge base, if it highly contribute to the class value, its $WC^x_{E_G}$ will immediately reflect it and influence the prediction ability of the SEC model.

5 Conclusion and Future Work

In this paper, we investigated the possibility of using provenance graphs in trust estimation, and proposed a trust estimation model, named the SEC model, for predicting the trustworthiness of a composite service based on related provenance information. The proposed approach can work effectively to facilitate users to analyze huge amount of provenance data, and derive trust information from them automatically for service composition in open systems. The future work of this research will mainly focus on two aspects. Firstly, we are going to investigate more advance methods to improve the accuracy of trust estimation. Secondly, we will investigate more effectively approach for estimating multi-class trust values.

References

1. Bai, Q., Su, X., Liu, Q., Terhorst, A., Zhang, M., Mu, Y.: Case-based trust evaluation from provenance information. In: 2011 IEEE 10th International Conference on Trust, Security and Privacy in Computing and Communications (TrustCom), pp. 336–343. IEEE (2011)
2. Bunke, H., Kandel, A.: Mean and maximum common subgraph of two graphs. Pattern Recognition Letters 21(2), 163–168 (2000)
3. Cover, T.M., Thomas, J.A.: Elements of information theory. Wiley, New York (1991)
4. Freire, J., Koop, D., Santos, E., Silva, C.T.: Provenance for computational tasks: A survey. Computing in Science & Engineering 10(3), 11–21 (2008)
5. Raymond, J.W., Gardiner, E.J., Willett, P.: Rascal: Calculation of graph similarity using maximum common edge subgraphs. The Computer Journal 45(6), 631–644 (2002)
6. Renner, R., Maurer, U.: About the mutual (conditional) information. In: Designs, Codes, and Cryptography (2000)

3D EEG Source Localisation: A Preliminary Investigation Using MML

Thi H. Kyaw and David L. Dowe

Computer Science, Clayton School of I.T.,
Monash University, Clayton, Vic 3168, Australia
thkya1@student.monash.edu, david.dowe@monash.edu

Abstract. Electroencephalography (EEG) source localisation (a.k.a. the inverse problem) is a widely researched topic with a large compendium of methods available. It combines the classic EEG signal processing techniques with modern methods to estimate the precise location of the sources of these signals inside the brain. Myriad factors define the differences in each of these techniques. We present here a previously untried application of the Minimum Message Length (MML) principle to the inverse problem with strictly preliminary findings. We first discuss the problem formulation of EEG source localisation and then attempt a preliminary inclusion of MML in the analysis. In this early stage, tests were conducted based on a simple head model using only artificial data.

Keywords: Electroencephalography, EEG, source localisation, inverse problem, Minimum Message Length.

1 Introduction

An Electroencephalography (EEG) is "a record of the oscillations of brain electric potential recorded from electrodes on the human scalp" [7]. EEG is a non-invasive routine which is commonly used by clinical neurophysiologists and researchers to study epilepsy, stroke and other forms of brain abnormalities [1,5,9]. Of interest to researchers is finding out the exact location of these sources of electrical activity in the human brain based on recorded electric potentials registered at the electrodes placed at specified positions on the scalp. This is known as the Inverse EEG problem – or source localisation. Despite the good temporal resolution that EEG source localisation provides [1], the latter falls short in terms of producing images that match the spatial resolutions of other functional imaging techniques. Active research to find a solution to the inverse EEG problem has consequently spawned an array of proposed methods – each with respective strengths and drawbacks.

It is well established that the inverse EEG problem is ill-posed [5,10] because of the infinite number of acceptable solutions that "fit" the data equally well, making the solution non-unique. To overcome this problem, domain knowledge and *a priori* information can be used to constrain the solution set [10]. Different assumptions may include (but are not limited to): the number of sources,

S. Cranefield and A. Nayak (Eds.): AI 2013, LNAI 8272, pp. 74–79, 2013.

biological constraints, prior probability, norms, correlation, sparsity, etc. The underlying assumption(s) that constitute the models used in various techniques define the differences in each. Some of the well-known methods include those based on the minimum-norm constraint such as Low Resolution Electromagnetic Tomography (LORETA) and Weighted Minimum Norm Estimate (WMNE) [5]. There are also other approaches such as beamforming, which are oriented more towards signal processing.

We make a preliminary attempt here to investigate source localisation using Minimum Message Length (MML) [14,13,3], see sec. 2.2. Many source localisation techniques have been documented in the literature, but not MML. We attempt to use MML as part of the EEG models – specifically in regularization.

2 Problem Definition

2.1 The Inverse Problem

The objective of EEG source localisation is to locate the active areas of the brain responsible for the registered EEG data. This comprises two problems: the forward problem and the inverse problem. The forward problem is solved by using a given electrical configuration and determining the electrode potentials at each of the electrodes on the human scalp. Both problems are integral to source localisation. This paper follows the commonly used data model for EEG source localisation; the cortical volume or region is uniformly divided into homogeneous 3-dimensional cuboids or voxels [17]. In this model, current dipoles are assumed at the active locations in the cortex. The dipole concept is introduced to aid the mathematical representation of the problem; dipoles are, in fact, hypothetical entities. The model involves a set of m electrodes (usually tens to hundreds) that receives signals at a certain time instance as a result of source k at voxel r_k. The signal received at the electrodes can be modelled as $G(r_k)\mathbf{m}_k$ where $\mathbf{m}_k \in \mathbb{R}^{3\times 1}$ and $G(r_k) = [g_x(r_k) \quad g_y(r_k) \quad g_z(r_k)] \in \mathbb{R}^{m\times 3}$ represent the dipole moment and lead field matrix, $\mathbf{G}$, respectively. $g_x(r_k)$, $g_y(r_k)$, $g_z(r_k)$ denote the $m \times 1$ lead field vectors that represent the unit amplitude sources at voxel r_k oriented in the respective Cartesian planes. These vectors that form the components of the lead field matrix are easily derived from various head models in the forward problem calculations [5,17]. The lead field matrix is a vital component of the system and will be described in later sections. For an in-depth coverage of the forward problem, see [6]. $\mathbf{m}_k$ is a product of two components: the dipole moment magnitude and the orientation for any arbitrary source k. The magnitude is a real-valued scalar and the orientation for each dipole is a 3×1 vector whose members correspond to the planes in 3-D space. The cortex is assumed to comprise n voxels, hence giving rise to a "linear superposition of the contributions" [17] of these voxels:

$$\mathbf{x} = \mathbf{Gm} + \mathbf{N} \tag{1}$$

In (1), $\mathbf{N}$ is a noise term and G now becomes $[G(r_1) \quad ... \quad G(r_n)] \in \mathbb{R}^{m\times 3n}$, with $\mathbf{m}$ being $[m_1^T \quad ... \quad m_n^T]^T \in \mathbb{R}^{3n\times 1}$. $\mathbf{x}$ refers to the vector of measured electric potentials at the 14 electrodes.

The general form described in (1) will be the basis of the inverse problem. There are also other types of formulations besides (1). The objective, in this case, would be to find a reasonable estimate $\hat{\mathbf{m}}$ to localise the sources. Having said this, the problem remains hugely underdetermined due to the number of voxels (n) being far greater than the number of observation points [or electrodes] (m). This issue can be addressed to some extent by placing certain regularisation constraints. Some of the constraints that are assumed will be discussed in the following sections. It is necessary to understand that the description of the inverse problem in this section broadly sets the context for the proceeding sections; a comprehensive description of the problem would warrant a separate piece of literature.

2.2 Minimum Message Length

Recalling sec. 1, Minimum Message Length is a Bayesian inference method which can be regarded as a quantitative version of Ockham's razor (see [3, sec. 4]). It is a useful criterion in the the acceptance, rejection and comparison of hypotheses based on some given body of data [13,14]. MML relies on the principle that the best Hypothesis [or explanation] (H) is one that produces the shortest (optimal) two-part encoding of the observed Data (D). For various desiderata of MML such as its universality, statistical invariance and general statistical consistency, see, e.g., [13,3].

3 Data Models and Assumptions

3.1 Head Model and Reducing the Number of Unknowns

The number of voxels required to represent the cortical region is usually 6000 or more [8,9]. This preliminary study focused on reducing the total number of unknowns to make the problem less underdetermined. Ideally, the more voxels there are, the better the representation of the underlying EEG sources. In this work we use a brain model composed of 4 large voxels with 4 x 3 = 12 dipole components estimated from 14 electrodes, $\mathbf{x}$, hence making the problem overdetermined (as $14 > 12$).

3.2 Initial Experiments

Tests on artificial data were conducted with the aim of finding the estimate, $\hat{\mathbf{m}}$ (of the dipole moment vector), which minimises the following expression:

$$min\{\|\mathbf{Gm} - \mathbf{x}\|_2 + \|R\mathbf{m}\|_2\} \qquad (2)$$

The second term, $\|R\mathbf{m}\|_2$, represents the regularization term, which we attempted to model using MML. Instead of using common regularizers such as Tikhonov regularization, a single latent factor model [13, sec. 6.9] was combined with MML mixture modelling [15,16][13, sec. 6.8] (similarly to [4]) and hierarchical clustering (cf. [2]) to hierarchically cluster the 4 x 3 = 12 dipole moments. Unfortunately, this elaborate regularisation term just described varied only slightly (by about 0.5 bit) across all the data which we considered in this preliminary investigation - and hence had little impact. An MML linear regression model [13, sec. 6.7] was used as the first term in (2) (this should be a minor replacement) with the regularisation term just described as having little effect.

The tests were conducted using artificially generated data for the measured electric potential vector $\mathbf{x}$ (see sec. 2.1). Pseudorandom Gaussian noise, $N(\mu, \sigma^2)$, was also introduced to $\mathbf{x}$ with $\mu = 0$ and $\sigma = 0.1$. The lead field matrix, $\mathbf{G}$ was generated under realistic settings for the 4-voxel head model and provides the link between the physical electrical potentials recorded on the scalp and the dipole components in each voxel inside the brain. An initial vector, $\mathbf{m}_0$, was generated using the Moore-Penrose pseudoinverse of the lead matrix, $\mathbf{G}$, through multiplication with the vector of observed electrode readings, $\mathbf{x}$. Doing so should approximately minimize the first term in (2) (because of our slight modification). $\mathbf{m}_0$ was put through a multi-dimensional search using simulated annealing to find the optimal estimate, $\hat{\mathbf{m}}$, which effectively minimises the objective function from (2). Each entry in the dipole component vector $\mathbf{m}$ was randomly perturbed with small increments or decrements and then the objective function was recomputed upon each perturbation. The role of the MML regularizer was intended to suppress any noise and "over-fitting" from the resulting estimates, but (as above had little effect). At the end of the annealing process, the estimated solution vector will be found. How well the estimate corresponds to the actual dipole component vector can then be observed.

4 Results

Simulation 1. A concurrent minimisation of the terms in (2) was carried out to find the best trade-off that leads to total minimisation rather than optimum results for each term in isolation. The ideal scenario would be if a certain vector $\hat{\mathbf{m}}_k$ results in the minimisation of both the first and second terms in (2) after the k-th iteration. It was observed that after 10 perturbations to the vector $\mathbf{m}$, the MML regularization term which we used did not deviate more than ± 0.3 nits. This occurred no matter how large the magnitude of perturbations introduced to the individual components in $\mathbf{m}$ was.

Simulation 2. The first term of (2) was minimized while the regularization term was kept as a constant to find out if there were any effects on the search. In Table 1, $\mathbf{m}$ denotes the true dipole component vector, $\hat{\mathbf{m}_0}$ is the best solution before optimisation and $\hat{\mathbf{m}}$ is the final estimate after convergence of the multi-dimensional search on the 12 variables x_1 to z_4. The estimates do not differ largely from the true values. However, this simulation was conducted without considering the effect of regularization. Results for this simulation are in Table 1. The last row of each column is the objective function of (2).

Simulation 3. In an attempt to optimise the search, the same vector (as in Simulation 2) was used but with different search parameters. The starting temperature, T, for the simulated annealing process was increased from 50 to 80

Table 1. Simulation 2 results from multi-dimensional search after perturbing variables

Component variables	Actual $\mathbf{m}$	Initial $\hat{\mathbf{m}}_0$	Estimate $\hat{\mathbf{m}}$
x1	1.0000	0.9010	0.9004
y1	0.0000	-0.0253	-0.0118
z1	0.0000	0.0253	0.0150
x2	1.0000	1.1196	1.0980
y2	0.0000	0.0653	0.1208
z2	1.0000	0.9426	0.9019
x3	0.0000	-0.0824	-0.0820
y3	0.0000	0.0761	0.0270
z3	0.0000	-0.1116	-0.1035
x4	0.0000	0.0090	0.0194
y4	1.0000	0.8600	0.8566
z4	1.0000	1.4020	1.3880
Expression 2	2.2361	2.3885	2.3501

and the number of perturbations to the vector per thermal equilibrium was also increased tenfold – from 10 to 100 perturbations. Despite the longer search duration, only mild improvements were seen. Resulting estimate vector = {0.9122 -0.0124 0.0151 1.0881 -0.0821 0.9091 -0.1820 0.2651 0.1031 -0.03645 0.95656 1.3547}. It was noted that increasing the number of perturbations did not really justify the increased computational time (3 seconds to complete the search in simulation 1 and 21 seconds to search in simulation 2).

5 Discussion and Further Work

Our preliminary results do not (yet) provide a clear indication whether the particular way we have used MML is effective as a regularization term in source localisation. Other factors that might contribute to these results include the use of an extremely simple 4-voxel head model. The fact that the problem was transformed from an underdetermined system to an overdetermined one might also be relevant. We recall the negligible effect that our initial MML regularization has had. This aspect could be further investigated. We could also investigate the effects of larger σ and more noise. A better head model that accomodates a greater number of voxels would assist. Previous works using MML on spatially correlated data [11,12] might help address whether the voxels might exhibit spatial correlation.

We thank Drs Timur Gureyev and Yakov Nesterets from CSIRO, Australia, for helpful discussions and for provision of the EEG lead field matrix used in our numeric experiments.

References

1. Baillet, S., Mosher, J.C., Leahy, R.M.: Electromagnetic Brain Mapping. IEEE Signal Processing Magazine 18(6), 14–30 (2001)
2. Boulton, D.M., Wallace, C.S.: An information measure for hierarchic classification. The Computer Journal 16(3), 254–261 (1973)
3. Dowe, D.L.: MML, hybrid Bayesian network graphical models, statistical consistency, invariance and uniqueness. In: Bandyopadhyay, P.S., Forster, M.R. (eds.) Handbook of the Philosophy of Science. Philosophy of Statistics, vol. 7, pp. 901–982. Elsevier (2011)
4. Edwards, R.T., Dowe, D.L.: Single factor analysis in MML mixture modelling. In: Wu, X., Kotagiri, R., Korb, K.B. (eds.) PAKDD 1998. LNCS, vol. 1394, pp. 96–109. Springer, Heidelberg (1998)
5. Grech, R., Cassar, T., Muscat, J., Camilleri, K., Fabri, S., Zervakis, M., Xanthopoulos, P., Sakkalis, V., Vanrumste, B.: Review on solving the Inverse problem in EEG source analysis. Journal of NeuroEngineering and Rehabilitation 5(1), 25 (2008)
6. Hallez, H., Vanrumste, B., Grech, R., Muscat, J., Clercq, W.D., Vergult, A., D'Asseler, Y., Camilleri, K.P., Fabri, S.G., Van Huffel, S.: Review on solving the forward problem in EEG source analysis. Journal of NeuroEngineering and Rehabilitation (1), 46 (2007)
7. Nunez, P.L., Srinivasan, R.: Electric Fields of the Brain: The neurophysics of EEG. Oxford Scholarship Online (May 2009)
8. Pascual-Marqui, R.D.: Standardized low resolution brain electromagnetic tomography (sLORETA): technical details. Technical report, The KEY Institute for Brain-Mind Research, University Hospital of Psychiatry Lenggstr, 31, CH-8029 Zurich, Switzerland (2002)
9. Phan, T.G., Gureyev, T.E., Nesterets, Y., Ma, H., Thyagarajan, D.: Novel application of EEG Source Localization in the assessment of the penumbra. Cerebrovascular Diseases 33(4), 405–407 (2012)
10. Ramirez, R.R.: Source Localization. Scholarpedia 3(11), 1733 (2008)
11. Visser, G., Dowe, D.L.: Minimum Message Length Clustering of Spatially-Correlated Data with Varying Inter-Class Penalties. In: Proc. 6th IEEE International Conf. on Computer and Information Science (ICIS) 2007, pp. 17–22 (July 2007)
12. Wallace, C.S.: Intrinsic Classification of Spatially Correlated Data. The Computer Journal 41(8), 602–611 (1998)
13. Wallace, C.S.: Statistical and Inductive Inference by Minimum Message Length. Springer (May 2005)
14. Wallace, C.S., Boulton, D.M.: An Information Measure for Classification. The Computer Journal 11(2), 185–194 (1968)
15. Wallace, C.S., Dowe, D.L.: Intrinsic classification by MML the snob program. In: Proc. Seventh Australian Joint Conf. Artificial Intelligence, pp. 37–44. World Scientific (1994)
16. Wallace, C.S., Dowe, D.L.: MML clustering of multi-state, Poisson, von Mises circular and Gaussian distributions. Statistics and Computing 10(1), 73–83 (2000)
17. Wu, S.C., Swindlehurst, A.L.: Matching Pursuit and Source Deflation for Sparse EEG/MEG Dipole Moment Estimation. IEEE Transactions on Biomedical Engineering PP(99), 1–1 (2013)

DEPTH: A Novel Algorithm for Feature Ranking with Application to Genome-Wide Association Studies

Enes Makalic, Daniel F. Schmidt, and John L. Hopper

The University of Melbourne
Centre for MEGA Epidemiology
Carlton VIC 3053, Australia
{emakalic,dschmidt,johnlh}@unimelb.edu.au

Abstract. Variable selection is a common problem in regression modelling with a myriad of applications. This paper proposes a new feature ranking algorithm (DEPTH) for variable selection in parametric regression based on permutation statistics and stability selection. DEPTH is: (i) applicable to any parametric regression task, (ii) designed to be run in a parallel environment, and (iii) adapts naturally to the correlation structure of the predictors. DEPTH was applied to a genome-wide association study of breast cancer and found evidence that there are variants in a pathway of candidate genes that are associated with a common subtype of breast cancer, a finding which would not have been discovered by conventional analyses.

1 Introduction

The problem of selecting predictor variables from a possibly large set of candidate variables occurs in many areas of science. An important recent example is the parametric regression model of genome-wide association studies (GWAS). GWA studies [1] measure thousands of genetic markers, typically single nucleotide polymorphisms (SNPs), for people affected by the disease of interest (cases) and people that are not affected by the disease (controls). The aim of a GWAS is to identify which SNPs, if any, are truly associated with risk of disease.

In the context of a GWAS, selecting potentially interesting variables is challenging due to: (i) the large number of SNPs measured, (ii) the correlation between SNPs, and (iii) the fact that the disease causing variants may not have been measured. The conventional strategy for finding disease associated SNPs is to test each SNP independently of all other measured SNPs using standard hypothesis testing methods. This approach yields a frequentist p-value for each measured SNP which is an indication of the strength of evidence for the association. The p-values are then adjusted for multiple testing [2] using, for example, the Bonferroni procedure and all SNPs whose p-values are less than some pre-specified threshold are in effect considered to be true associations; all the remaining SNPs are effectively discarded.

S. Cranefield and A. Nayak (Eds.): AI 2013, LNAI 8272, pp. 80–85, 2013.

This paper proposes a new algorithm (see Section 2) for discovering predictors using a regression model based on permutation statistics and stability selection. The basic idea is to rank each variable in terms of evidence for association and then measure the stability of the corresponding ranking by re-sampling the data. Intuitively, one expects the ranking of variables with little or no associations to be highly unstable under minor perturbations of the data. This is because their associations are essentially random and practically indistinguishable. In comparison, the ranking of stronger predictors should remain relatively stable under data permutation, even when there are groups of predictor variables that are highly correlated. The algorithm also adds random noise predictors and ranks these variables alongside the measured variables. Statistics computed for the noise variables correspond to an empirical null distribution and this is used to determine the relative importance of all variables as predictors.

In the context of a GWAS, the algorithm can be used for all SNPs in the genome or for any subset of SNPs. We have found that the algorithm shows good performance when compared to several established procedures using simulated data. When applied to a breast cancer GWAS data set, the proposed algorithm found evidence that there are variants in a pathway of candidate genes that are associated with a common subtype of breast cancer, a finding which would not have been discovered by conventional analyses (see Section 3).

2 Stability Selection Algorithm

Consider a data set $D = \{(\mathbf{x}_1, \mathbf{z}_1, y_1), (\mathbf{x}_2, \mathbf{z}_2, y_2), \ldots, (\mathbf{x}_n, \mathbf{z}_n, y_n)\}$, where $\mathbf{x}_i \in \mathbb{R}^p$, $\mathbf{z}_i \in \mathbb{R}^q$ and $y_i \in \mathbb{R}$ $(i = 1, 2, \ldots, n)$, assumed to be generated by the regression model

$$y_i = f_i(\mathbf{x}_i, \mathbf{z}_i; \boldsymbol{\theta}) + \varepsilon_i \tag{1}$$

parameterised by $\boldsymbol{\theta} \in \mathbb{R}^k$ with disturbances $\varepsilon_i \sim \pi(\cdot)$. This setup includes common linear as well as non-linear classification and regression models. The task is to rank the p regressor variables $\mathbf{x}$ in terms of the evidence for their strength of association with the target variable $\mathbf{y}$ and thus effectively select which of the p variables constitute signal and which variables are noise. Note that the q variables $\mathbf{z}$ are pre-selected and included in each candidate model.

This paper introduces a novel feature ranking algorithm for parametric regression called `DEPTH` (DEPendency of Association on the number of Top Hits) which is based on re-sampling techniques and stability selection. Briefly, the idea is to first rank all p variables based on their marginal contribution, adjusting for the fixed regressors $\mathbf{z}$, and then evaluate the stability of the corresponding ranking by re-sampling the data. The re-sampling is without replacement and repeated over many iterations. Statistics recorded during each iteration of sampling are then used to automatically select the predictors that are associated with the target. A detailed description of `DEPTH` is given in Algorithm 1.

Algorithm 1. A description of the DEPTH algorithm

Require: Data $D = \{(\mathbf{x}_1, \mathbf{z}_1, y_1), (\mathbf{x}_2, \mathbf{z}_2, y_2), \ldots, (\mathbf{x}_n, \mathbf{z}_n, y_n)\}$, $\mathbf{x}_i \in \mathbb{R}^p$, $\mathbf{z}_i \in \mathbb{R}^q$, $y_i \in \mathbb{R}$, number of iterations $T > 0$

1: $\mathbf{y}^p \leftarrow$ random permutation of target variable $\mathbf{y} \in \mathbb{R}^n$
2: $\mathbf{r}_0 \leftarrow$ initial ranking of variables using data D {see Algorithm 2}
3: $\mathbf{r}_0^p \leftarrow$ initial ranking of variables using D with $\mathbf{y}^p$ instead of $\mathbf{y}$ {see Algorithm 2}
4: $\mathbf{c} \leftarrow \mathbf{0}_p$, $\mathbf{c}^p \leftarrow \mathbf{0}_p$ {measure of variable significance}
5: **for** $t = 1$ to T **do**
6: $\quad D_* \leftarrow$ re-sample data D without replacement
7: $\quad D_*^p \leftarrow D_*$ where the target vector is randomly permuted
8: $\quad$ Append an extra p columns of noisy variables to data D_* and D_*^p
9: $\quad \mathbf{r}_t \leftarrow$ new ranking based on re-sampled data D_*
10: $\quad \mathbf{r}_t^p \leftarrow$ new ranking based on permuted data D_*^p
11: $\quad c_j \leftarrow c_j + 1$, $\forall \mathbf{x}$ variables in D_* ranked before the best ranked noise variable
12: $\quad c_j^p \leftarrow c_j^p + 1$, $\forall \mathbf{x}$ variables in D_*^p ranked before the best ranked noise variable
13: $\quad \mathbf{o}_t \leftarrow$ ranking overlap between $\mathbf{r}_0$ and $\mathbf{r}_t$ {see text}
14: $\quad \mathbf{o}_t^p \leftarrow$ ranking overlap between $\mathbf{r}_0^p$ and $\mathbf{r}_t^p$
15: **end for**
16: Ranking stability plot based on $\{\mathbf{o}_1, \ldots, \mathbf{o}_T\}$ and $\{\mathbf{o}_1^p, \ldots, \mathbf{o}_T^p\}$
17: $\mathbf{R} \leftarrow$ index of $\mathbf{x}$ variables obtained by sorting $\mathbf{c}$ in descending order
18: **return** final ranking of the p $\mathbf{x}$ variables $\mathbf{R}$

2.1 DEPTH Algorithm

DEPTH first creates a copy of the data set $\mathbf{D}$ where the target vector $\mathbf{y}$ is randomly permuted (Step 1). The new data set is denoted by D^p and corresponds to an empirical null distribution. Since the target vector in D^p is essentially random, all DEPTH statistics will be compared to the corresponding statistics obtained using this random data. DEPTH then calculates statistics from the data which are compared to the empirical null distribution to minimise false positive findings.

DEPTH ranks the p variables $\mathbf{x}$ in D and the $\mathbf{x}$ variables in D^p (Steps 2–3). DEPTH employs marginal variable ranking where each variable is examined independently of all other variables to be ranked. The ranking algorithm is described in Algorithm 2. For each of the p variables (Steps 3–7, Algorithm 2), the ranking function fits a regression model using one $\mathbf{x}$ variable at a time (Step 4, Algorithm 2), adjusting for all the $\mathbf{z}$ variables, and computes the corresponding log-likelihood (Step 5, Algorithm 2). The regression model is fitted using maximum likelihood estimation, though in principle another estimation technique can be used. Following maximum likelihood fitting, a ranking statistic is computed for each of the p $\mathbf{x}$ variables (Step 6, Algorithm 2). The ranking statistic is the difference between the log-likelihood of a model with one regressor x_j (and q regressors $\mathbf{z}$) and the log-likelihood of a model with only the q regressors $\mathbf{z}$. All p $\mathbf{x}$ variables are then ranked in ascending order of the ranking statistic (Step 8, Algorithm 2); the regressor x_j that results in the best improvement to the log-likelihood over the model with $\mathbf{z}$ regressors only is ranked first; the second

Algorithm 2. A description of the marginal feature ranking function

Require: Data $D_* = \{(\mathbf{x}_1, \mathbf{z}_1, y_1), (\mathbf{x}_2, \mathbf{z}_2, y_2), \ldots, (\mathbf{x}_{n_*}, \mathbf{z}_{n_*}, y_{n_*})\}$, $\mathbf{x}_i \in \mathbb{R}^{p_*}$, $\mathbf{z}_i \in \mathbb{R}^q$, $y_i \in \mathbb{R}$
Initialise score vector $\mathbf{s} = (s_1, s_2, \ldots, s_{p_*})' = \mathbf{0}_{p_*}$
$s_0 \leftarrow$ log-likelihood of model with regressors $\mathbf{z}_i$ $(i = 1, 2, \ldots, n_*)$
for $j = 1$ to p_* **do**
  Fit regression model using data $(x_{ij}, \mathbf{z}_i, y_i)$ $(i = 1, 2, \ldots, n_*)$
  $l_j \leftarrow$ log-likelihood of the fitted model
  $s_j \leftarrow l_j - s_0$ {difference in log-likelihood}
end for
$\mathbf{r} \leftarrow$ sort all variables in descending order of $\mathbf{s}$
return $\mathbf{r}$ {ranked list of all p_* variables}

best regressor $x_{j'}$ $(j \neq j')$ is the one that results in the second best improvement in log-likelihood, etc.

DEPTH performs $T > 0$ steps of data re-sampling and re-ranking in order to assess variable selection stability (Steps 5–15). DEPTH keeps two count vectors $\{\mathbf{c} \in \mathbb{R}^p, \mathbf{c}^p \in \mathbb{R}^p\}$ for each of the p variables in D and D^p which are used to measure variable importance. During each sampling iteration, DEPTH creates a new data sample D_* by sampling the original data without replacement (Step 6); the new data contains 66% of the original data points. In a similar fashion to Step 1, a copy of D_* with the target vector randomly permuted is stored in D_*^p (Step 7). Additional p columns of (noisy) data are then appended to D_* and D_*^p (Step 8). These extra columns will be used to determine the total number of significant $\mathbf{x}$ variables (a similar idea was mentioned in passing by Miller [3]). The data sets D_* and D_*^p therefore have $n_* = \lfloor 2n/3 \rfloor$ samples and $p_* = 2p$ variables that will be used for ranking. Algorithm 2 is used to compute a new ranking list r_t for the p_* variables in D_* (Step 9); the same procedure is also applied to the permuted data D_*^p resulting in $\mathbf{r}_t^p$ (Step 10). The count vector $\mathbf{c}$ is updated in Step 11 as follows: for each $\mathbf{x}$ variable j $(1 \leq j \leq p)$, if the variable ranks ahead of the best ranked noisy variable, add one to the count c_j, otherwise proceed to step 12. The counts $\mathbf{c}^p$ are updated in the same fashion using ranking list $\mathbf{r}_t^p$ instead of $\mathbf{r}_t$ (Step 13).

DEPTH uses the percentage of overlap between ranking vectors r_0 (based on data D) and r_t (based on re-sampled data D_*) as a metric of variable selection stability. Further, the percentage of overlap between $\mathbf{r}_0^p$ and $\mathbf{r}_t^p$ is used to estimate the empirical null distribution of overlap. The procedure to compute percentage overlap between two lists of ranks, say $\mathbf{r}_0$ and $\mathbf{r}_t$, is as follows (Steps 13–14). First, the p extra (noisy) variables that were added in Step 8 are removed. We then compute the intersection between the first j components of r_0 and r_t for all $(j = 1, 2, \ldots, p)$. The number of variables that remain in the intersection set for each j is stored in $\mathbf{o}_t$ and is a metric of ranking repeatability. As an example, consider two lists of rankings $\{3, 2, 5, 4, 1\}$ and $\{2,3,4,1,5\}$. Following the calculation of overlap percentage, the vector $\mathbf{o}_t$ for these two lists would be $\{0, 1, 0.67, 0.75, 1\}$. The first entry of $\mathbf{o}_t$ is 0 as the top ranked variable is different in the two lists. Similarly, the second entry of $\mathbf{o}_t$ is 1 since both lists rank variables $\{2, 3\}$ as most significant.

Following T re-sampling iterations, DEPTH produces: (i) an estimate of the number of significant predictors in the data γ ($0 \leq \gamma \leq p$), (ii) a plot of variable selection stability (Step 16), and (3) a ranking of all $\mathbf{x}$ variables in terms of their strength of association with the target $\mathbf{y}$. The total number of significant predictors is estimated as

$$\gamma = \frac{1}{T} \sum_{j=1}^{p} \max(c_j - c_j^p, 0), \quad (0 \leq \gamma \leq p) \tag{2}$$

where γ is the mean number of $\mathbf{x}$ variables ranked below the noise variables over T sampling iterations. Note that the number of significant variables in the permuted data D_p^*, which is an estimate of the empirical null distribution under the assumption that all variables are noise, is subtracted from the estimate obtained using D_*. This ensures that γ controls the type I error rate by recording the number of significant variables above what would be expected by chance.

The ranking stability plot is obtained from the overlap vectors $\{\mathbf{o}_1, \ldots, \mathbf{o}_T\}$ and $\{\mathbf{o}_1^p, \ldots, \mathbf{o}_T^p\}$. DEPTH overlays the median percentage overlap computed from data sets D_* and D_*^p in one figure. This gives the experimenter the ability to compare median replicability between the real data rankings and the random data rankings over T iterations of re-sampling. The area between the two median curves is a surrogate statistic for the amount of signal present in the data. If the area is small (that is, the curves are virtually overlapping), DEPTH may not able to distinguish between signal and noise variables. In contrast, a large area between two median curves may indicate presence of strong signal variables.

Finally, DEPTH produces a ranking of all $\mathbf{x}$ variables which is used in conjunction with the estimate of the number of significant predictors (2) for model selection. The DEPTH ranking is computed from the count vectors $\mathbf{c}$ and $\mathbf{c}^p$. The variable j with the largest count $(c_j - c_j^p)$ is ranked first; the variable j' with the second largest count $(c_{j'} - c_{j'}^p)$ is ranked second, etc. As the DEPTH ranking is an average over T re-sampling iterations, it is expected to be more stable than a single ranking.

3 Application to Breast Cancer GWAS Data

This section examines the application of DEPTH to a real GWAS of 204 women with breast cancer obtained from the Australian Breast Cancer Family Study [4] and 287 controls from the Australian Mammographic Density Twins and Sisters Study [5]. All women were genotyped using a Human610-Quad beadchip array resulting in over 600,000 SNPs per woman. Recommended GWAS data cleaning and quality control procedures (e.g., checks for SNP missingness, duplicate relatedness, population outliers [6]) were performed prior to analysis. We selected SNPs for DEPTH analysis from genes encoding a candidate susceptibility pathway. All SNPs were validated in the Caucasian population and were downloaded from the HapMap Consortium [7]. This particular pathway was chosen due to biological considerations and because previous GWAS research in the pathway has detected potentially interesting SNPs. The final data set consisted of 366 SNPs selected from a genomic region of approximately two million base pairs. The correlation structure was approximately block-diagonal where blocks of highly correlated SNPs are interspersed with regions of low correlation.

DEPTH ranking of the data was done using $T = 1,000$ re-sampling iterations and logistic regression for marginal variable ranking. The difference in area between two overlap curves showed the possible presence of multiple risk-associated SNPs. DEPTH selected five SNPs ($\gamma = 5.11$) as important. To examine whether there is any difference in SNP rankings across different types of breast cancer, we stratified the GWAS data into two groups. Breast cancer type was determined from the collected cancer pathology data and DEPTH ranking was then performed for all SNPs in the two subgroups. DEPTH showed that SNPs in the pathway are only associated with one common type of breast cancer and not the other. This is an important discovery which we are currently attempting to replicate using a large, independent breast cancer GWAS data set.

Due to time constraints, initial DEPTH tests have concentrated on subsets of the human genome, chosen by biological consideration. The DEPTH algorithm is now being implemented on the IBM BlueGene/Q supercomputer, a Victorian government initiative in partnership with the University of Melbourne and the IBM Research Collaboratory for Life Sciences. The BlueGene/Q comprises 4,096 compute nodes with 65,536 user processor cores in four racks. The authors have been granted a significant amount of compute time on the supercomputer and have been funded by the National Health and Medical Research Council to perform DEPTH analyses of GWAS data. A manuscript detailing DEPTH analyses of a large international breast cancer GWAS data set, obtained through the Breast Cancer Association Consortium, is currently in preparation.

References

1. Manolio, T.A.: Genomewide association studies and assessment of the risk of disease. The New England Journal of Medicine 363(2), 166–176 (2010)
2. Dudoit, S., Shaffer, J.P., Boldrick, J.C.: Multiple hypothesis testing in microarray experiments. Statistical Science 18(1), 71–103 (2003)
3. Miller, A.J.: Selection of subsets of regression variables. Journal of the Royal Statistical Society (Series A) 147(3), 389–425 (1984)
4. Dite, G., Jenkins, M., Southey, M., Hocking, J., Giles, G., McCredie, M., Venter, D., Hopper, J.: Familial risks, early-onset breast cancer, and BRCA1 and BRCA2 germline mutations. J. Natl. Cancer Inst. 95, 448–457 (2003)
5. Odefrey, F., Gurrin, L., Byrnes, G., Apicella, C., Dite, G.: Common genetic variants associated with breast cancer and mammographic density measures that predict disease. Cancer Research 70, 1449–1458 (2010)
6. Weale, M.: Quality control for genome-wide association studies. Methods Mol. Biol. 628, 341–372 (2010)
7. Consortium, I.H.: A second generation human haplotype map of over 3.1 million snps. Nature 449, 851–861 (2007)

Evidence for Response Consistency Supports Polychronous Neural Groups as an Underlying Mechanism for Representation and Memory

Mira Guise, Alistair Knott, and Lubica Benuskova

Dept of Computer Science, University of Otago, Dunedin, New Zealand
{mguise,alik,lubica}@cs.otago.ac.nz

Abstract. Izhikevich [6] has proposed that certain strongly connected groups of neurons known as polychronous neural groups (or PNGs) might provide the neural basis for representation and memory. Polychronous groups exist in large numbers within the connection graph of a spiking neural network, providing a large repertoire of structures that can potentially match an external stimulus [6,8]. In this paper we examine some of the requirements of a representational system and test the idea of PNGs as the underlying mechanism against one of these requirements, the requirement for consistency in the neural response to stimuli. The results provide preliminary evidence for consistency of PNG activation in response to known stimuli, although these results are limited by problems with the current methods for detecting PNG activation.

Keywords: spiking network, polychronous neural group, activation, representation, memory.

1 Introduction

It is widely assumed that synaptic plasticity provides the neural basis for long-term memory in the brain [1,2,9] although the precise nature of the underlying representation is still unclear [3]. Izhikevich [6] has proposed that certain strongly connected groups of neurons known as polychronous neural groups (or PNGs) might provide this representational mechanism. An understanding of this underlying mechanism is particularly relevant to the developing field of neuromorphic computing, but is also of interest to researchers in machine learning, or even information retrieval [5]. In this report we examine some of the requirements of a representational system and test the idea of PNGs as a mechanism of representation against one of these requirements, the requirement for consistency in the neural response to stimuli.

Polychronous groups arise from an interaction between the precise firing times of spatio-temporal input patterns and the variability of axonal transmission delays between neurons. Figure 1 shows a schematic example of such an interaction. The input stimulus is composed of a sequence of firing events, each representing the firing of a specific neuron at a precise point in time. The stimulus in this example forms an ascending spatio-temporal pattern as shown in Fig. 1 (unfilled circles).

S. Cranefield and A. Nayak (Eds.): AI 2013, LNAI 8272, pp. 86–97, 2013.

In this model network there exists a polychronous group whose intra-group axonal delays are congruent with the input stimulus (gray-filled circles). As shown in panels A, B and C, the spatio-temporal arrangement of three of the firing events that make up the stimulus (filled black circles in Fig. 1) interacts with the axonal delays, producing convergent input to group neurons. This firing event triplet acts as a trigger for PNG activation, producing a wave of neural firing that propagates throughout the polychronous group (only the first step is shown). Without this convergent input the neurons in the group would fail to reach the firing threshold and the input stimulus would not propagate.

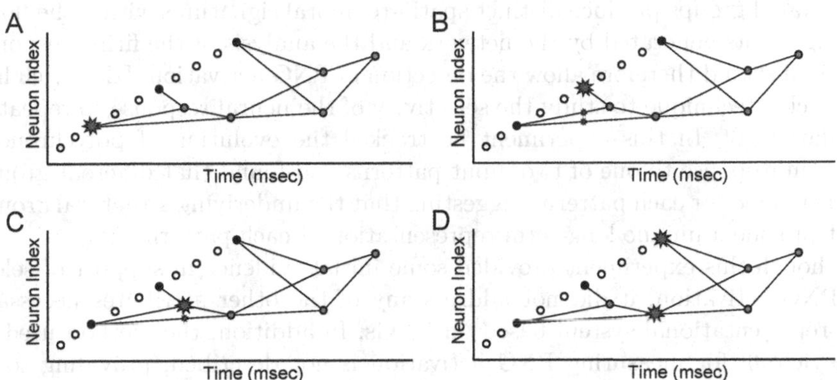

Fig. 1. Polychronous group activation following stimulation with an ascending firing pattern (unfilled circles). The precise timing of three of the firing events in the pattern (denoted by filled black circles) matches the axonal delays between group neurons, producing activation of the polychronous group. Other firing event combinations might produce additional group activations (not shown). Firing events resulting from PNG activation are shown with gray-filled circles. The first of the three firing events is fired in panel A, producing spikes that take time to propagate to PNG neurons. In panel B, the convergence of the propagating spike and the second of the triplet firing events is sufficient for a PNG neuron to reach the firing threshold (panel C). Further group firing events are supported by the axonal delays between group neurons (panel D).

This propagation of neural firing across the group is called group *activation*. When activated, the neurons in the polychronous group are said to *polychronize* in a causal chain of firing events that is both precisely timed and reproducible [6,8]. However, not all PNGs are capable of activation. *Structural* PNGs are defined purely topologically, as groups of neurons with connection latencies commensurate with a given input stimulus [10]. For polychronization to occur the synaptic connections converging on each group neuron must be sufficiently strong to allow the post-synaptic neuron to reach the firing threshold. Polychronous groups with compatible synaptic weights can activate when presented with a triggering stimulus at which point they are known as *activated* PNGs.

Izhikevich [6] observed that the number of structural PNGs in a network is typically many times larger than the number of neurons. Given this large repertoire of structural PNGs, how might we use it to build a representational system? Several attributes present themselves as necessary for a robust system and we will refer to these with the terms *selectivity*, *consistency*, *stability* and *capacity*. A *selective* system produces PNG activations in response to a stimulus that are sufficiently specific to allow the unique identification of the stimulus. A *consistent* system is able to dependably produce PNG activations on every presentation of the stimulus. A *stable* system is able to maintain long-term representations in the form of structural PNGs that are capable of activation, and a system with good *capacity* allows a biologically plausible number of these structural representations.

Activated groups produce distinct spatio-temporal signatures within the flood of firing events generated by the network and the analysis of the firing response to stimuli should therefore allow the detection of PNG activation. Izhikevich has used such a technique to study the selectivity of the neural response to repeated stimulation [6]. In this experiment he tracked the evolution of polychronous groups in response to one of two input patterns and found that different groups were activated for each pattern, suggesting that the underlying structural groups might provide a unique long-term representation of each pattern.

Although this experiment provided some initial evidence in support of selective PNG activation, it did not address any of the other attributes necessary for a representational system based on PNGs. In addition, the method used in Izhikevich [6] for measuring PNG activation is not described, providing some hurdles to the reproduction of these results. The experiments described below employ a template matching technique for detecting PNG activation (methods outlined below and in more detail in a separate technical report [4]).

In the remainder of this report we will focus on the requirement for a *consistent* representational system, using the pattern-specific activation of polychronous groups to measure the dependability of the neural response to known stimuli. Polychronous groups exist in a competitive medium in which the group affiliation of individual neurons is constantly fought over [8] and this dynamic environment therefore calls into question the reliability of a representational system based on PNG activation. Although PNG activation is often described as "stereotypical" and "reproducible", a specific PNG will not necessarily activate on every presentation of a triggering stimulus [6]. However, other PNG activations may result from the same stimulus and a stimulus-specific neural response consisting of some subset of the set of all stimulus-specific PNG activations may therefore occur with some consistency. In the following experiments we will assess the empirical probability of this stimulus-specific neural response given the presentation of a known stimulus. Does every stimulus presentation produce a relevant group activation, or only some presentations?

2 Methods

Twenty independent networks were created for these experiments, each composed of 1000 Izhikevich neurons (800 excitatory and 200 inhibitory) with parameters as described in [6]. The networks were matured for two hours by exposure to 1 Hz random input generated by a Poisson process. Following maturation, the networks were trained on one of two input patterns or were left untrained. The current experiments reproduce the few known details of the repeated stimulation experiment described in [6], namely a twenty minute training period, and the use of an *ascending* or *descending* input pattern as the stimulus (see Fig. 2).

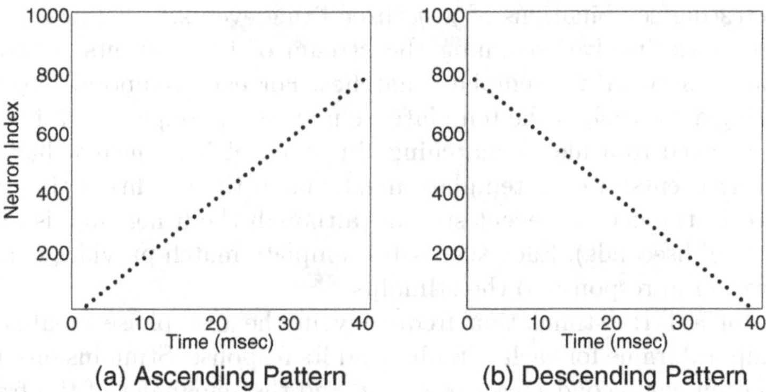

Fig. 2. The ascending and descending patterns: each spatio-temporal input pattern is composed of 40 firing events. Both patterns share the same neurons, differing only in the temporal order of their firing events.

The technique used by Izhikevich [6] for detecting PNG activations in the firing data was not described and therefore needed to be redeveloped for the current experiments. It was clear that this technique needed to discriminate pattern-specific PNG activations from unrelated PNG activations, and from other spiking events generated by the network. The original method was assumed to make use of the Izhikevich search algorithms [7] (see also [10]) to find structural PNGs in the network, suggesting the use of a template matching technique for the detection of PNG activation. The default behavior of these algorithms is to initiate a PNG search based on all combinations of three neurons in the network. However, we created a small modification that limits the search to triplet combinations that occur only in the training patterns. Any polychronous groups found by this modified algorithm are referred to as pattern-specific PNGs, as the activation of these groups is initiated by a firing event triplet that occurs in the input pattern.

The assumed template matching technique involves isolating PNGs from a trained network and using them as templates to probe for group activation. The technique is reproduced as follows: first, a network is trained with a specified input pattern and pattern-specific structural PNGs are isolated from the network at regular intervals; the isolated PNGs are then used as spatio-temporal templates to match the firing data. For convenience, the experiment is split into two phases: in an initial *training phase*, the network is repeatedly stimulated with the ascending or descending pattern at 5 or 25 Hz for twenty minutes; in the following *test phase* of the experiment, the network is stimulated with the same ascending or descending pattern at 1 Hz, and pattern-specific templates isolated during the training phase are used to probe for group activation.

At one minute intervals throughout the training phase a search is initiated for structural PNGs that can act as pattern-specific templates. The search involves testing all triplet combinations (i.e. combinations of three firing events) from the input pattern for their ability to discover structural PNGs in the network [4]. However, not all PNGs will be found as the algorithm is limited for performance reasons to testing combinations of just three firing events.

The test phase involves scanning the stream of firing events generated by the stimulated network for template matches. For each temporal offset in the network firing data, each of the templates is matched in sequence and successful matches are saved to a file. A matching threshold of 50% means that at least half the firing events in each template must match the the firing time and the neuron fired in the network event stream (although the firing time is allowed a jitter of $\pm$ 2 milliseconds). Each successful template match provides evidence of PNG activation in response to the stimulus.

The use of a 1 Hz stimulation frequency in the test phase creates a well-defined temporal frame for each stimulus and its response. Stimulus onset occurs at $t = 0$ in each one second *response frame*, and the remainder of the frame has sufficient temporal length to include all of the firing events in the resulting neural response. A 1 Hz random background pattern is also presented throughout each test period. For a more detailed description of the methods see the accompanying technical report [4].

3 Results

Together the training and testing phases of the experiment produce a large set of data that supports multiple analyses. Training phase data provides a view of the evolution of structural PNGs in response to the stimulus, while test phase data provides a snapshot of the process of PNG activation. Figure 3 uses a combination of both datasets to show a selection of three matching templates following low-intensity (1 Hz) test stimulation of a network. These matching templates are sampled from a larger pool of pattern-specific templates that match PNG activations triggered by some triplet combination from the ascending input pattern. The first few firing events in each of the templates in Fig. 3 are therefore upward-sloping, reflecting the isolation of the template from a network trained on the ascending pattern. Each group consists of multiple convergent connections that support the propagation of neural firing across the members of the group before terminating at an inhibitory neuron (gray-filled circles).

Temporal alignment of just these first few firing events for all matching templates (and with all other firing events removed) produces sloping firing patterns that can be seen in Fig. 4. Recall that the first few firing events in each template (the initial triplet) correspond to the stimulus trigger that leads to PNG activation. The gray-scale intensity in this figure encodes the number of times the corresponding firing event acted as a trigger for the initiation of PNG activation, where activation was measured by the number of matching templates

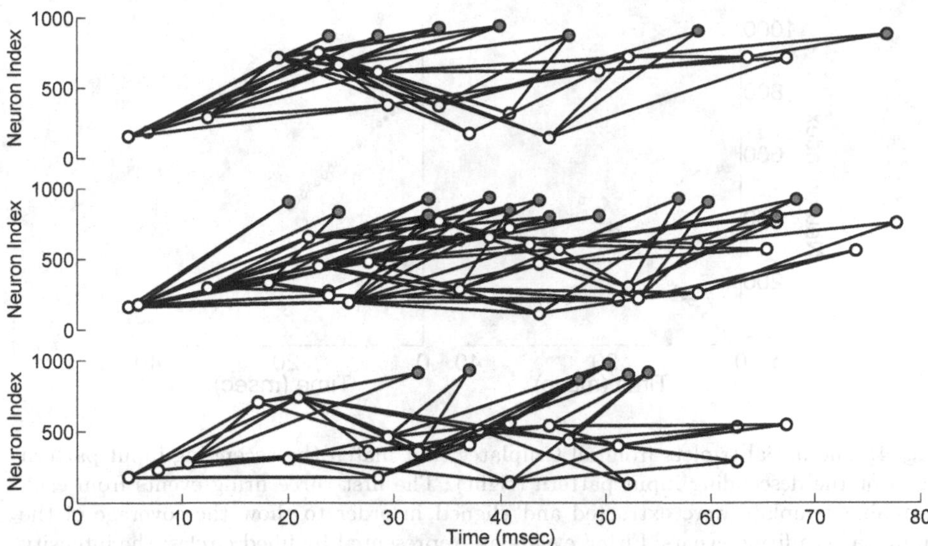

Fig. 3. A selection of three templates that match the firing data following stimulation with the ascending input pattern. The x- and y-axes for each template represent time in milliseconds and neuron index respectively (the y-axis is ordered so that inhibitory neurons are at the top of the graph). Nodes depict firing events generated by excitatory or inhibitory neurons and are drawn using either open circles (excitatory neurons) or gray-filled circles (inhibitory neurons). Lines between nodes represent causal connections between firing events.

accumulated across twenty independent networks. The figure therefore provides a picture of which of the input pattern firing events succeeded or failed at initiating PNG activation. Many of the forty firing events that make up each input pattern failed to initiate a responding group over the ten minutes (six hundred response frames) of the testing phase. Significantly, the majority of these failures are clustered in the later stages of the input pattern, suggesting that group response is concentrated on the early part of each stimulus presentation.

Nevertheless, the PNG activation response as a whole exhibits a high degree of consistency. Figure 5 shows the activation response of 40 networks (20 trained on the ascending pattern and twenty untrained networks) in the first 100 seconds of the 10 minute test run (only the first 100 of 600 response frames are shown in Fig. 5). The stimulus is presented at the start of each frame and any templates that match the firing events in the remainder of the frame are taken as evidence of PNG activation. Each row in Fig. 5 represents a single network and is divided into one hundred segments representing each of the one hundred response frames. The presence of a filled circle in each segment indicates the detection of a PNG activation response in the corresponding response frame. If there was no response, or the method was unable to detect the response, the segment is left empty.

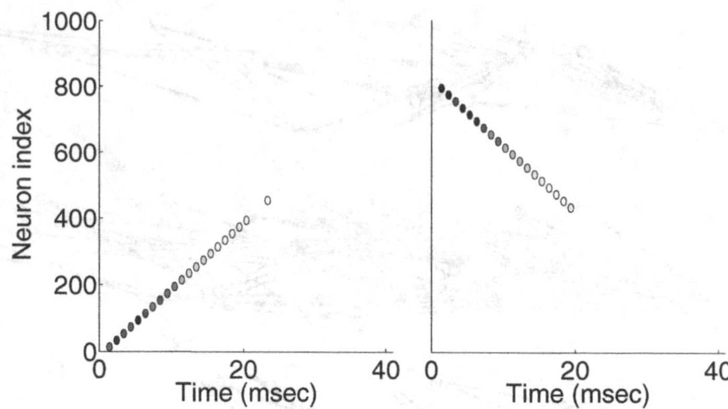

Fig. 4. The initial triplets from all templates that match the ascending input pattern (left) or the descending input pattern (right). The first three firing events from each matching template were extracted and aligned in order to show the coverage of the input pattern firing events. Firing events are represented by filled circles; the intensity of the fill color for each firing event represents the number of templates that matched PNG activations triggered by the event. This number, accumulated across twenty independent networks, is greatest in the early stages of each input pattern (darker fill color) and decreases in later stages of the input pattern (lighter fill color). The missing firing events in the later stages correspond to input pattern firing events that failed to initiate a group response during the test period.

The first 25 frames in this experiment used the ascending pattern, the next 25 used the descending pattern, the third group of 25 frames repeated the use of the ascending pattern, and in the final 25 frames no input pattern was provided (the null pattern). Using a combined pool of all templates to measure the PNG activation response, the twenty trained networks at the top of Fig. 5 show a consistent response to the ascending pattern but little or no response to the descending pattern or the null pattern. In contrast, the twenty untrained networks at the bottom of Fig. 5 show only sporadic activation and no apparent correlation with the type of input pattern. Comparing the activation response of the trained networks with the response of the untrained networks, we see a high degree of consistency in the response to the ascending pattern only where the network has been previously trained on the ascending pattern.

The PNG activation response to each stimulus presentation is assumed to occur in the early portion of each response frame, shortly after stimulus presentation at $t = 0$. Over this period, one of more PNG activations triggered by the stimulus have the opportunity to match the pool of PNG templates. Some insight into the temporal evolution of PNG activation is provided by computing the proportion of matches that occur at each temporal offset within the frame (the *template match ratio*) to produce an empirical measure of the likelihood of PNG activation at each offset. Firstly, each one second response frame is sliced into 1000 consecutive sub-frames and the number of template matches at each

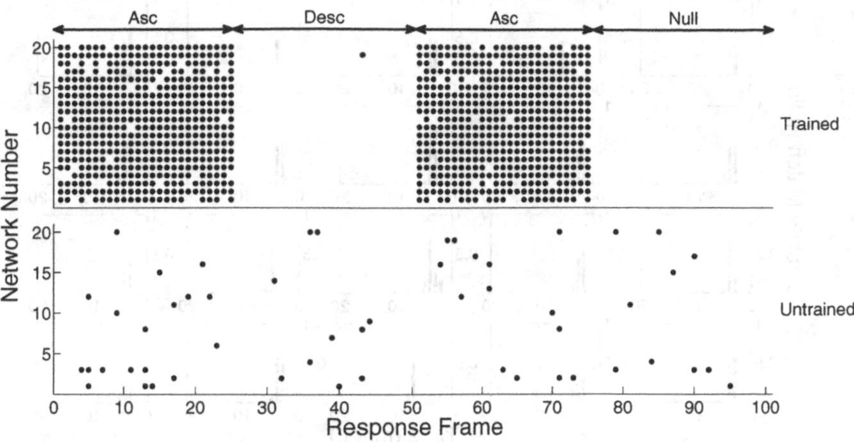

Fig. 5. The PNG activation response of twenty trained networks and twenty untrained networks over one hundred response frames. Trained networks were trained on the ascending pattern. A filled circle represents a positive response to the stimulus while an empty space denotes a lack of response. The stimulus for the first and third quarter of the one hundred frames was the ascending pattern and the stimulus for the second quarter was the descending pattern. No stimulus was provided in the fourth quarter (null pattern). The top figure shows the measured response for a network trained on the ascending pattern at 5 Hz and the bottom figure shows the result using an untrained network. The trained networks in the top figure were derived from the untrained networks in the corresponding row of the bottom figure.

one millisecond sub-frame is counted. The template match ratio for each offset is then computed by aggregating the number of matches for each offset across all response frames. Using this procedure we expect to see an isolated peak in the number of matches at a short delay following the stimulus at time $t = 0$, reflecting the transient activation of a responding PNG. However, due to limitations in the template matching method the delay can only be calculated to within half the length of each template (i.e. ± 15 milliseconds), depending on where on each template the match occurs.

Figure 6 shows the template match ratios for each network distributed over the first twenty sub-frames of each response frame. As predicted there is an isolated peak that consistently occurs in the first ten milliseconds following the stimulus. Within this small temporal window the likelihood of a template match typically reaches 50% or more, indicating that PNG activation is in full swing. As PNG activation comes to an end, the likelihood of a template match decreases to zero and remains at zero for the remainder of the response frame.

Although these positive results support the consistency of PNG activation, it is worth noting that the majority of templates are ineffective in matching the firing data. Here, we define an effective template as one that is able to match the firing data at least once during a ten minute period of stimulation with the corresponding input pattern. On average, just 32% of ascending templates and

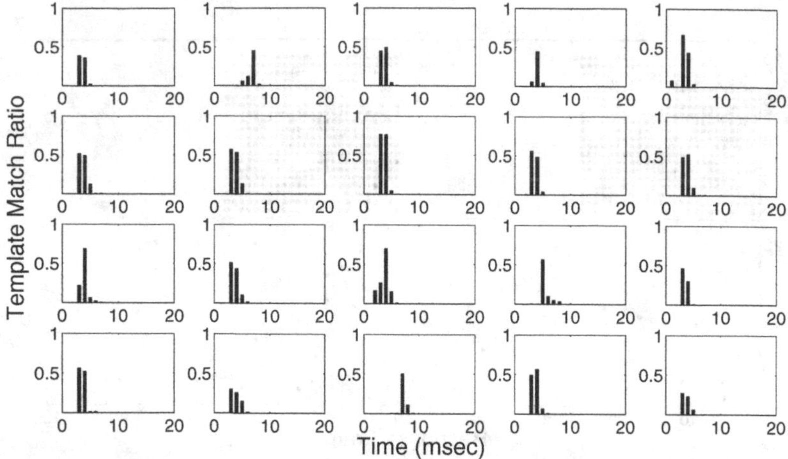

Fig. 6. Template match ratios distributed over each one second frame for each of twenty independent networks. The template match ratio was computed for each one millisecond slot in the response frame, accumulated over multiple frames. The response for each network is confined to the first ten milliseconds following the stimulus and therefore only the first twenty milliseconds of the frame are shown.

43% of descending templates were effective at finding a match (averaged across twenty independent networks). The template matching performance between networks is also very variable with some networks averaging as few as three matches in each response frame. In some frames, the evidence for PNG activation is based on a single template match suggesting that the template matching method is near to its sensitivity limit for some networks.

4 Discussion

The template matching method attempts to match spatio-temporal templates derived from the structural PNGs found in a trained network with the sequence of firing events that are produced when the network is stimulated with the same pattern. To ensure that template matches were pattern-specific, the selected templates were restricted to structural PNGs that were triggered by triplet combinations of the input pattern firing events. We can imagine that structural groups exist in the network that require larger, more complex, triggering patterns although it seems likely that the probability of finding groups with larger triggers decreases with the size of the triggering pattern. Templates that match the firing data such as those shown in Fig. 3 provide an impression of the corresponding PNG activations that occur in the milliseconds following each stimulus. However, looking at a selection of matching templates creates only a partial picture of the complex pattern of neural firing in response to spatio-temporal stimuli. Visualization of all of the PNG activations that are initiated by combinations of firing events from the input pattern produces a complex graph in which individual PNGs interact and merge (results not shown).

Izhikevich [8] has proposed that competitive interactions occur between the polychronous groups in the network, with neurons that are shared by multiple PNGs synchronising their firing times with different polychronising pathways at different times. However, cooperative interactions are also possible in which firing events generated by separate PNG activations together produce the required spatio-temporal initiators for additional PNG activations. The emerging picture is one in which the activation response to complex stimuli is a composition of individual PNG activations that interact and merge in a complex manner.

Interestingly, all of the templates that found a match in the neural response were initiated by firing event triplets from just the early portion of the input pattern. This effect was found across all networks and for both the ascending and descending input patterns. A possible explanation is that competition during PNG formation for use of shared neurons creates an interference effect between early PNG activations and those that come later, with the earlier activating groups forming first and therefore dominating the available neural resources.

This explanation has implications for the maximum number of simultaneous activations that a network of a given size is able to support, and might in turn impact the maximum number of representations that can simultaneously be "held in mind" in a representational system based on polychronous groups. However, note that this explanation does not contradict the extraordinary potential capacity of a PNG representational system [6] because any potential limitation in the number of *simultaneous* activations supported by a representational system does not necessarily affect the network capacity i.e. the total number of representations that can be stored within the network.

Despite any interference caused by interactions between simultaneous activations, the template matching method provides good support for the consistency of a PNG-based representational system. Using a combined pool of all templates, one or more template matches are detected in almost every response frame, suggesting a consistent PNG activation response following each stimulus presentation. The best single template for each network is also able to show quite a high degree of consistency, although most individual templates match only rarely.

Computing the template match ratio for each one millisecond time-slot in the response frame shows that all matches are confined to a narrow temporal window following each stimulus presentation (see Fig. 6). This strong interaction between the time of the stimulus and the time of template matching supports the view that template matching reflects the causal relationship between stimulus onset and subsequent PNG activation. The template match ratio can also be computed at frame level (i.e. the proportion of matching frames), producing a value that reflects the empirical likelihood of PNG activation given the stimulus. With the combined templates, this likelihood value approaches certainty for many networks, although there is considerable inter-network variation in performance.

Together these results indicate a high degree of consistency in the PNG activation response following a stimulus. However, despite this consistent response there are occasional response frames where no neural response is detected, despite the presence of a known stimulus. The lack of a detectable response does not mean that PNG activation did not occur and may instead be due to limitations in the template matching method. Examination of the precise timing of the firing events in consecutive response frames shows considerable jitter in the spike times of PNG neurons between frames (results not shown). Competition for neural resources between activating groups may increase this jitter to the point where the corresponding template fails to match.

The lack of tolerance of the template matching method to temporal jitter is just one of the flaws of this method for detecting PNG activations. Although this technique is able to respond selectively to substantially different stimuli (e.g. discriminating between the ascending and descending patterns, or the ascending and null patterns in Fig. 5), the low matching threshold used in these experiments potentially allows templates to match unrelated spatio-temporal patterns. The template matching method may therefore have difficulty in resolving stimuli that are too closely related.

Another problem with the template matching method is that it treats matching as a local process when it is likely to be a global one. The neural response to a complex stimulus is a unique *set of PNG activations*; it is therefore the set as a whole and not individual activations that provide a unique signature of the stimulus. Given a set-oriented view of the neural response, if a single template happens to match a single PNG activation, does this provide good evidence of the presence of the stimulus? For example, two stimuli with partial overlap in their spatio-temporal firing patterns could both match the same template and may therefore not be individually resolvable. In recognition of a set-oriented view of the neural response, the template matching method makes use of a pool of templates that are able to detect multiple PNG activations. However, this method does not take into account the number of unique matches in each response frame and is therefore unable to counter the problem of overlapping stimuli.

Each of the templates generated in the training phase contribute to the time it takes to scan the firing data in the testing phase. It is therefore a problem that the majority of templates are ineffective, with less than half of the templates ever able to generate a match. Although the single best template for each network matches the neural response very consistently, the majority of templates that match at all do so only rarely. In addition, the number of matches in each response frame is sometimes very low suggesting that this method is close to the threshold for maximum sensitivity for some networks.

It is likely that Izhikevich [6] used a similar technique to show selectivity in the activation response, despite the flaws of the template matching method. The issues with this method, while limiting the scope and accuracy of the current results, do not invalidate our overall finding. Here we provide preliminary evidence for the consistency of PNG activation in response to stimuli, suggesting that polychronous groups may be able to meet at least one of the necessary criteria for a representational system. The neural response to complex stimuli appears to involve multiple interacting PNG activations suggesting that an alternative method for measuring the neural response must treat any single PNG activation as only partial evidence in favor of a particular stimulus. Work is in progress on such an alternative technique that will address these limitations.

References

1. Abraham, W.C.: Metaplasticity: tuning synapses and networks for plasticity. Nature Reviews Neuroscience 9(5), 387–387 (2008)
2. Caporale, N., Dan, Y.: Spike timing-dependent plasticity: a hebbian learning rule. Annu. Rev. Neurosci. 31, 25–46 (2008)
3. Caroni, P., Donato, F., Muller, D.: Structural plasticity upon learning: regulation and functions. Nature Reviews Neuroscience 13(7), 478–490 (2012)
4. Guise, M., Knott, A., Benuskova, L.: Consistency of polychronous neural group activation supports a role as an underlying mechanism for representation and memory: detailed methods and results. Tech. rep., Dept of Computer Science, University of Otago, Dunedin (2013)
5. Hoffmann, H., Howard, M.D., Daily, M.J.: Fast pattern matching with time-delay neural networks. In: The 2011 International Joint Conference on Neural Networks (IJCNN), pp. 2424–2429. IEEE (2011)
6. Izhikevich, E.M.: Polychronization: computation with spikes. Neural Computation 18(2), 245–282 (2006)
7. Izhikevich, E.M.: Reference software implementation for the Izhikevich model: minimal spiking network that can polychronize (2006), `http://www.izhikevich.org/publications/spnet.htm`
8. Izhikevich, E.M., Gally, J.A., Edelman, G.M.: Spike-timing dynamics of neuronal groups. Cerebral Cortex 14(8), 933–944 (2004)
9. Martin, S., Grimwood, P., Morris, R.: Synaptic plasticity and memory: an evaluation of the hypothesis. Annual Review of Neuroscience 23(1), 649–711 (2000)
10. Martinez, R., Paugam-Moisy, H.: Algorithms for structural and dynamical polychronous groups detection. In: Alippi, C., Polycarpou, M., Panayiotou, C., Ellinas, G. (eds.) ICANN 2009, Part II. LNCS, vol. 5769, pp. 75–84. Springer, Heidelberg (2009)

A Neural Network Model of Visual Attention and Group Classification, and Its Performance in a Visual Search Task

Hayden Walles, Anthony Robins, and Alistair Knott

Dept of Computer Science, University of Otago, New Zealand

Abstract. Humans can attend to and categorise objects individually, but also as groups. We present a computational model of how visual attention is allocated to single objects and groups of objects, and how single objects and groups are classified. We illustrate the model with a novel account of the role of stimulus similarity in visual search tasks, as identified by Duncan and Humphreys [1].

1 Introduction

Humans can represent objects individually, but also collectively, as groups. We can attend to and categorise individual objects, but we can also attend to several objects as a group—and if the objects are all of the same type, we can classify them collectively as being of that type. In fact there is evidence that the visual object classification system is relatively insensitive to the number of items in a group. In monkeys, Nieder and Miller [2] showed that neurons in the inferotemporal (IT) cortex are sensitive to the type of objects in a group but relatively insensitive to their cardinality, while neurons in the intraparietal sulcus show the opposite pattern. A similar sensitivity to type but not number has been found in imaging studies of human IT, using a habituation paradigm where either the type of objects in a group or the size of the group was selectively changed (e.g. [3]). In previous work [4] we coined the term **cardinality blindness** to describe this phenomenon. We showed that a classifier called a **convolutional neural network** (CNN) shows cardinality blindness, and argued that this property also characterises the object classifier in the IT cortex of humans and other primates. Our classifier assigns the same class ('X') to a single visually presented X shape and to a homogeneous group of X shapes. However, when it is presented with a group of objects with different shapes (a heterogeneous group), it typically refuses to make a classification at all.

If the classifier in IT is cardinality blind, this may be expected to have consequences for the design of the attentional system that selects spatial regions to be classified [5,6]. For one thing, attention should be able to deliver *homogeneous groups* to the classifier as well as single objects, so that the objects in these groups can be classified in parallel. There should also be a system that acts in parallel with object classification, to compute the number information which is not provided by the classifier. In this paper we present a computational model of visual attention and object classification, in which the attentional system selects individuals and groups for the classifier. We also describe the performance of this model in a visual search task.

S. Cranefield and A. Nayak (Eds.): AI 2013, LNAI 8272, pp. 98–103, 2013.

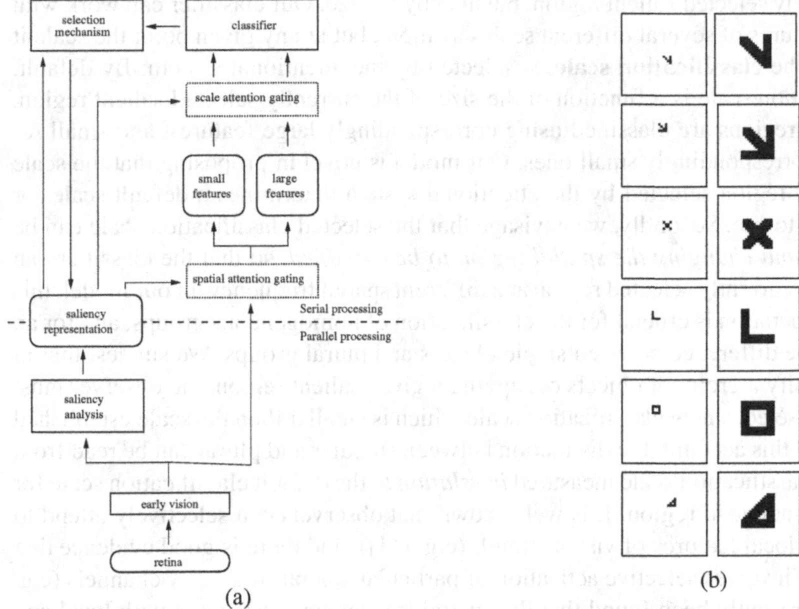

Fig. 1. (a) Our model of the visual attention system (left-hand pathway) and the object classification system (right-hand pathway). (b) The shapes used to train the model.

2 The Model of Visual Attention and Object Classification

The structure of our model of visual attention and classification is outlined in Figure 1a. The **attentional subsystem** (dorsal pathway) determines the salient regions on the retina, and activates these regions one at a time. The **classification subsystem** (ventral pathway) categorises the retinal stimulus in the currently activated region; its output changes as different regions are selected [7].

In the attentional subsystem in our model, the saliency of a region is determined by two factors: one is local contrast (how different it is from the surrounding region), the other is homogeneity (how similar its texture elements are). Salient regions can contain isolated visual features which contrast from their surroundings, but also regions containing repeated visual features. Computations of saliency are performed at multiple spatial frequencies, so salient regions containing isolated visual features can be of different sizes. Salient regions containing repeated visual features (i.e. homogeneous textures) can also be of different sizes.

There are several existing computational models of saliency that detect salient regions of different sizes (e.g. [8]), and numerous models of texture identification that detect regions containing repeated visual features (e.g. [9]). There are also many existing computational models of classification that allow objects of different sizes to be classified, by taking as input primitive visual features at a range of different scales (e.g. [10]). The main innovations in our model are in how the saliency mechanism interacts with the classifier. Firstly, in our system, classification is influenced not only by the location of the currently selected salient region, but also by its size. Our classifier can work with primitive features of several different scales as input, but at any given point the scale it uses, called the **classification scale**, is selected by the attentional system. By default, the classification scale is a function of the size of the currently selected salient region, so that large regions are classified using correspondingly large features, and small regions with correspondingly small ones. Our model is novel in proposing that the scale of the salient region selected by the attentional system determines a default scale for the classifier to use. Secondly, we envisage that the selected classification scale can be changed *without changing the spatial region to be classified*, so that the classifier can reanalyse the currently selected region at a different spatial frequency. In our model, this attentional operation is crucial for the classification of homogeneous groups, and for an account of the difference between single objects and plural groups. We suggest that in order to classify a group of objects occupying a given salient region, the observer must attentionally select a new classification scale which is smaller than the scale established by default. In this account, the distinction between singular and plural can be read from the current classification scale measured *in relation to* the default classification scale for the currently attended region. It is well known that observers can selectively attend to the global or local features of visual stimuli (e.g. [11]), and there is good evidence that this attention involves selective activation of particular spatial frequency channels (e.g. [12]). It has recently been found that the spatial frequencies associated with local and global features of an object are defined in relative not absolute terms ([13]). Our model makes use of this notion of relative classification scale to support an account of group classification and of the distinction between singular and plural in the visual system.

2.1 The Classification Subsystem

The visual classification subsystem is modelled by a convolutional neural network (CNN) previously described [4]. The classifier takes, as input, retinotopic maps of simple oriented visual features at two different spatial frequencies, or scales: one of these scales is selected by the attentional system. The classifier was trained with six shapes at each spatial frequency (see Figure 1b). The classifier has seven output units: six of these provide localist encodings of the six shape categories and the seventh encodes the verdict 'unknown category'. The units have activations ranging from zero to one. We define the classifier's decision to be the strongest output over 0.5. If no unit's activation exceeds 0.5 the classifier's decision is assumed to be 'unknown category'. In summary, the classifier provides two pieces of information: first, whether classification is possible and, if so, what that classification is.

The classifier exhibits two types of invariance which have been observed in IT [14] and are generally acknowledged to be crucial for a model of vision [10], namely location (or translation) invariance and scale invariance. Location invariance is a result of the architecture of the CNN, which intersperses feature combination layers with layers that abstract over space [4]. Scale invariance depends on the input having been prefiltered for the desired frequency: the small shapes must be classified with the high-frequency visual features, and the large ones with the low-frequency features. Importantly for the current paper, the classifier is also blind to the cardinality of homogeneous groups of small shapes: its accuracy varies from 95% for a single shape to 97% for a homogeneous group of five shapes. Interestingly, these results show a redundancy gain effect similar to that found in humans: the classifier's performance improves the more instances of a type it classifies.

2.2 The Attentional Subsystem

As shown in Figure 1a, the attentional subsystem can be divided into two interacting stages: a preattentive, or parallel, stage and an attentive, or serial, stage.

The preattentive stage includes an operation called saliency analysis. The job of saliency analysis is to analyse the local contrast and texture homogeneity of the input in parallel. These are used to implement the Gestalt grouping properties of proximity and similarity respectively. The result of this is a saliency representation, or saliency map [15]. This representation is the point of communication between saliency analysis and the selection mechanism. The selection mechanism uses the saliency representation to decide how best to deploy attention. Once processing of attended stimuli is complete the representation is updated and then used to redeploy attention.

The saliency representation is also used to gate the input to the classifier. Input is gated in two different ways. It is gated by location, which is a well-known idea [6,16]). And it is also independently gated by scale, which is a new idea in our model. The initial scale selected by the attentional system is the **default classification scale** for the selected region. In order to recognise a figure within a region, the primitive visual features which the classifier uses must be of an appropriate spatial scale—not too large and not too small (see Sowden and Schyns [17]). If they are too large, they cannot be combined to represent a complex shape within the region. And if they are too small, then their combinations are not guaranteed to represent the global form of the figure occupying the region. A novel idea in our model is that a selected region can first be classified at the default classification scale, and then subsequently at a finer classification scale. If the classifier returns a result in this second case, it is identifying the type of objects in a homogeneous *group* occupying the selected region. In the remainder of this section we will provide more details about the attentional subsystem and its interaction with the classifier. Full technical details are given in [18].

3 Performance of the System in a Visual Search Task

In this section we describe two experiments investigating the behaviour of our complete system in the domain of visual search There are well-known similarity and grouping effects in search, which our model may be able to explain.

In a visual search task, a subject searches for a target stimulus in a field of distractors. The search time is a function of the number of distractors, but also on the visual properties of the target and distractor stimuli. The earliest visual search experiments reported a discrete difference between 'parallel search', in which search time is independent of the number of distractors, and 'serial search', where search time is linearly proportional to the number of distractors (Treisman and Gelade [5]). In the original model explaining this finding, feature integration theory (FIT), parallel search is possible if there is a single 'visual feature' that the target possesses and the distractors do not, allowing it

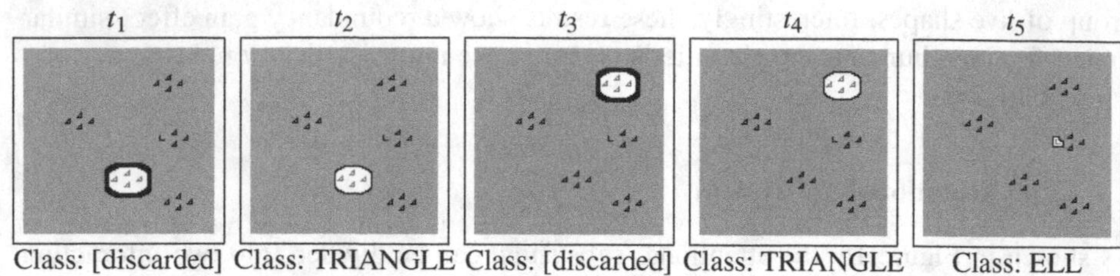

Fig. 2. An example sequence of operations during simple search. At t_1 the input is presented and at subsequent time steps attention is directed as shown until the target (ell) is found. Thick borders around a region indicate attention to the low spatial frequency, thin borders attention to the high spatial frequency.

to 'pop out' of the field of distractors; if the target is distinguished from the distractors by a specific conjunction of visual features, items in the visual field must be attended serially, to allow their features to be integrated.

Later experiments uncovered more complex patterns of visual search performance. Treisman [19] found that perceptual grouping affects search because subjects serially scan groups of items where possible, not just individual items. Treisman and Gormican's group scanning theory [20] drew this finding into the FIT model. In group scanning theory, when parallel search fails because the target cannot be discriminated from the distractors, attention is used to limit the spatial scope of the parallel search to a region where parallel search by feature discrimination *can* work. Parallel search then continues inside the attended area.

Duncan and Humphreys [1] presented results that challenged the basic assumption of a simple dichotomy between parallel and serial search. They gave subjects search tasks where the similarity between targets and distractors, and the similarity between distractors (i.e. the homogeneity of the set of distractors) were varied continuously. They found that increasing the degree of similarity between target and distractors progressively increases the slope of the search graph, and that increasing the similarity between distractors has the opposite effect. In Duncan and Humphreys' stimulus similarity theory (SST), pop-out and item-by-item serial search are opposite ends of a continuum of search processes, rather than discrete alternatives.

Our model of visual attention and classification is able to identify homogeneous groups and classify their elements in single operations; it therefore has some interest as a model of visual search. In this section we examine its performance on search tasks where target-distractor similarity and distractor-distractor similarity are varied, as in the experiment of Duncan and Humphreys.

To test the search performance of our model, we created four different search tasks, defined by varying two independent binary parameters based on those used by Duncan and Humphreys: target-distractor similarity (with values 't-d similar' and 'td-different') and distractor-distractor similarity (with values 'd-d similar' and 'd-d different'). Details of these tasks are given in [18]. We presented displays of each type to the model, and recorded how many serial attentional steps were taken for it to find the target. Figure 2 shows the steps taken by the system during a td-different/dd-similar search. We found that different search tasks have different slopes. Our simulation reproduces

Duncan and Humphreys' main experimental results: when targets are dissimilar to distractors but distractors are similar to one another the search slope is close to flat, and when targets are similar to distractors the slopes are highest. Details of these findings, and a comparison with other computational models of visual search, are given in [18].

References

1. Duncan, J., Humphreys, G.W.: Visual search and stimulus similarity. Psychological Review 91(3), 433–458 (1989)
2. Nieder, A., Miller, E.K.: A parieto-frontal network for visual numerical information in the monkey. Proceedings of the National Acadamey of Sciences 101(19), 7457–7462 (2004)
3. Izard, V., Dehaene-Lambertz, G., Dehaene, S.: Distinct cerebral pathways for object identity and number in human infants. PLoS Biology 6(2), 275–285 (2008)
4. Walles, H., Knott, A., Robins, A.: A model of cardinality blindness in inferotemporal cortex. Biological Cybernetics 98(5), 427–437 (2008)
5. Treisman, A.M., Gelade, G.: A feature-integration theory of attention. Cognitive Psychology 12, 97–136 (1980)
6. Moran, J., Desimone, R.: Selective attention gates visual processing in the extrastriate cortex. Science 229(4715), 782–784 (1985)
7. Zhang, Y., Meyers, E., Bichot, N., Serre, T., Poggio, T., Desimone, R.: Object decoding with attention in inferior temporal cortex. Proceedings of the National Academy of Sciences of the USA 108(21), 8850–8855 (2011)
8. Kadir, T., Brady, M.: Saliency, scale and image description. International Journal of Computer Vision 45(2), 83–105 (2001)
9. Kadir, T., Hobson, P., Brady, M.: From salient features to scene description. In: Workshop on Image Analysis for Multimedia Interactive Services (2005)
10. Riesenhuber, M., Poggio, T.: Hierarchical models of object recognition in cortex. Nature Neuroscience 2(11), 1019–1025 (1999)
11. Fink, G., Halligan, P., Marshall, J., Frith, C., Frackowiak, R., Dolan, R.: Where in the brain does visual attention select the forest and the trees. Nature 382, 626–628 (1996)
12. Flevaris, A., Bentin, S., Robertson, L.: Local or global? attentional selection of spatial frequencies binds shapes to hierarchical levels. Psychological Science 21(3), 424–431 (2010)
13. Flevaris, A., Bentin, S., Robertson, L.: Attention to hierarchical level influences attentional selection of spatial scale. Journal of Experimental Psychology: Human Perception and Performance 37(1), 12–22 (2011)
14. Logothetis, N.K., Sheinberg, D.L.: Visual object recognition. Annual Review of Neuroscience 19, 577–621 (1996)
15. Itti, L., Koch, C.: A saliency-based search mechanism for overt and covert shifts of visual attention. Vision Research 40, 1489–1506 (2000)
16. Moore, T., Armstrong, K.M.: Selective gating of visual signals by microstimulation of frontal cortex. Nature 421, 370–373 (2003)
17. Sowden, P., Schyns, P.: Channel surfing in the visual brain. Trends in Cognitive Sciences 10(12), 538–545 (2006)
18. Walles, H., Robins, A., Knott, A.: A neural network model of visual attention and object classification: technical details. Technical Report OUCS-2013-09, Dept of Computer Science, University of Otago (2013)
19. Treisman, A.: Perceptual grouping and attention in visual search for features and for objects. Journal of Experimental Psychology: HPP 8(2), 194–214 (1982)
20. Treisman, A., Gormican, S.: Feature analysis in early vision: Evidence from search asymmetries. Psychological Review 95(1), 15–48 (1988)

Affect Detection from Virtual Drama

Li Zhang[1], John Barnden[2], and Alamgir Hossain[1]

[1] Department of Computer Science and Digital Technologies,
Faculty of Engineering and Environment, Northumbria University, Newcastle, NE1 8ST, UK
[2] School of Computer Science, University of Birmingham, Birmingham, B15 2TT, UK
li.zhang@northumbria.ac.uk

Abstract. We have developed an intelligent agent to engage with users in virtual drama improvisation previously. The intelligent agent was able to perform sentence-level affect detection especially from user inputs with strong emotional indicators. However, we noticed that emotional expressions are diverse and many inputs with weak or no affect indicators also contain emotional indications but were regarded as neutral expressions by the previous processing. In this paper, we employ latent semantic analysis (LSA) to perform topic detection and intended audience identification for such inputs. Then we also discuss how affect is detected for such inputs without strong emotional linguistic features with the consideration of emotions expressed by the most intended audiences and interpersonal relationships between speakers and audiences. Moreover, uncertainty-based active learning is also employed in this research in order to deal with more open-ended and imbalanced affect detection tasks within or beyond the selected scenarios. Overall, this research enables the intelligent agent to derive the underlying semantic structures embedded in emotional expressions and deal with challenging issues in affect detection tasks.

Keywords: Affect detection, latent semantic analysis, and dialogue contexts.

1 Introduction

There has been significant progress for human computer interaction research to build human-like computer interfaces. This endeavour has given rise to agent-based user interfaces [1, 2]. Moreover, we believe it will make intelligent agents possess human-like behaviour and narrow the communicative gap between machines and human-beings if they are equipped to interpret human emotions during social interaction. Thus in this research[1], we focus on the production of intelligent agents with the abilities of interpreting dialogue contexts semantically to inform affect detection.

We previously developed an online multi-user role-play virtual drama framework, which allows school children to perform drama performance training. In this platform

[1] This work is supported by Northumbria Alumni Funding and TSB grant AK014L. The previous background research was supported by grant RES-328-25-0009 from the ESRC under the ESRC/EPSRC/DTI _PACCIT programme. It was also partially supported by EPSRC grant EP/C538943/1.

S. Cranefield and A. Nayak (Eds.): AI 2013, LNAI 8272, pp. 104–109, 2013.

young people could interact online in a 3D virtual drama stage with others. An intelligent agent is also involved in improvisation. It included an affect detection component, which detected affect from human characters' each individual turn-taking input. The agent also made responses based on the detected affect to stimulate the improvisation. This original affect detection was mainly built using pattern-matching rules that looked for simple grammatical patterns partially involving specific words. From the analysis of the transcripts, the previous affect detection without any contextual inference proved to be effective enough for those inputs containing strong emotional indictors such as 'haha', 'thanks' etc. There are also situations that users' inputs contain very weak affect signals, thus contextual inference is needed to further derive the affect conveyed in such inputs. This research especially deals with such challenges and discusses how LSA is used to perform affect detection from contexts. It will either detect affect using emotion contexts of the most intended audiences and relationships between characters when dealing with scenario related improvisation. Or it will detect affect using a min-margin based active learning when handling open-ended inputs beyond constraints of scenarios.

2 Related Work

Affect detection from texts has attracted great attention in recent years. ConceptNet [3] was developed as a toolkit to provide practical textual reasoning for affect sensing, text summarization and topic extraction. Ptaszynski et al. [4] employed context-sensitive affect detection with the integration of a web-mining technique to detect affect from users' input and verify the contextual appropriateness of the detected emotions. However, their system targeted interaction only between an AI agent and one human user in non-role-playing situations, which reduced the complexity of the modelling of the interaction context. Although Façade [5] included shallow natural language processing for characters' open-ended utterances, the detection of major emotions, rudeness and value judgments was not mentioned. Cavazza et al. [6] reported an AI agent embodied in a robot to provide suggestions for users on a healthy living life-style. Their system seemed not able to cope well with open-ended interactions based on the selected planning techniques. Endrass et al. [1] carried out study on the culture-related differences in the domain of small talk behaviour. Their agents were equipped to generate culture specific dialogues.

3 Affect Detection Based on Semantic Analysis of Contexts

In this section, we first of all discuss the development of latent semantic analysis based topic detection. Then we introduce how rule-based inference is used to interpret emotions from interaction contexts with the consideration of the detected topics and relationships between characters within the improvisation of the chosen scenarios. Finally, we discuss how a min-margin based active learning is developed to deal with imbalanced affect detection and affect classification beyond pre-defined scenarios.

In order to build a reliable and robust analyser of affect it is necessary to undertake several diverse forms of analysis and to enable these to work together to build stronger interpretations. Thus in this study, we integrate semantic interpretation of social contexts to inform affect analysis. In this section, we discuss our approaches of using latent semantic analysis [7] for terms and documents comparison to recover discussion themes and audiences for those inputs without strong affect indicators.

Latent semantic analysis generally identifies relationships between a set of documents and the terms they contain by producing a set of concepts related to the documents and terms. In order to compare the concepts behind the words, LSA maps both words and documents into a 'concept' space and performs comparison in this space. In detail, LSA assumed that there is some underlying semantic structure in the data which is partially obscured by the randomness of the word choice. This random choice of words also introduces noise into the word-concept relationship. LSA aims to find the smallest set of concepts that spans all the documents. It uses a statistical technique, called singular value decomposition, to estimate the hidden concept space and to remove the noise. This concept space associates syntactically different but semantically similar terms and documents. We use these transformed terms and documents in the concept space for retrieval rather than the original ones.

In our work, we employ the semantic vectors package [8] to perform LSA and calculate similarities between documents. This package provides APIs for concept space creation. It applies concept mapping algorithms to term-document matrices using Apache Lucene, a powerful search engine library [8]. We integrate this package with our AI agent's affect detection component to calculate semantic similarities between user inputs and training documents with clear themes. In this paper, we employ transcripts of the Crohn's disease[2] scenario for context-based affect analysis.

In order to compare the improvisational inputs with documents from different topic categories, we have to collect some sample training documents with strong themes. Personal articles from the Experience project (www.experienceproject.com) are borrowed to construct training documents. Since we intend to perform topic detection for the transcripts of the Crohn's disease scenario, we extracted sample articles close enough to this scenario including articles of Crohn's disease (5 articles), school bullying (5), family care for children (5), food choice (3), school life including school uniform (10) and school lunch (10). Phrase and sentence level expressions implying 'disagreement' and 'suggestion' are also gathered from several other articles published on this website. Thus we have training documents with eight discussion themes including 'Crohn's disease', 'bullying', 'family care', 'food choice', 'school lunch', 'school uniform', 'suggestions' and 'disagreement'. In order to detect some metaphorical phenomena, we include five types of metaphors published on online resources in our training corpus. These include mental (ideas), cooking, family, weather, and farm metaphors. All the sample files of the above 13 categories (including five types of metaphors) are regarded as training files. We use the following example interaction produced by subjects to demonstrate how to detect topics for those inputs with weak affect indicators.

[2] Peter has Crohn's disease and has the option to undergo a life-changing but dangerous surgery. He needs to discuss the pros and cons with friends and family. Janet (Mum) wants Peter to have the operation. Matthew (younger brother) is against it. Arnold (Dad) is not able to face the situation. Dave (the best friend) mediates the discussion.

1. Arnold: lets order some food. Peter *drop the subject*! [disapproval]
2. Dave: Arnold, Peter needs ur support and u can't just ignore it. [*played by the AI agent*]
3. Peter: how would u all feel if u were in my situation? [*Topic themes: family care, disease and disagreement; Target audience: Arnold; Rhetorical question: angry*]
4. Janet: *nobody cares* peter. [angry]
5. Arnold: we know peter now *stop talking* about it. [disapproval]
6. Janet: well you have got mine, peter. [*Topic themes: family care and disease; Target audience: Peter; Emotion: caring*]
7. Peter: *help me* daddy. [sad]
8. Arnold: *not now* son. Wife, ur blood pressure will get too high. [disapproval]
9. Peter: *im confused and hurt*. [sad]
10. Janet: *nobody cares* wat u think dad. [angry]
11. Dave: Arnold, y u don't want 2 b involved in? Peter is ur son. [*played by the AI agent*]
12. Arnold: dave I said bring me da menus! [*Topic themes: 'food' and 'bullying'; Target audience: Dave; Emotion: angry*]
13. Janet: *whos Arnold*. [neutral; *Small talk*]
14. Peter: Natalie. [neutral; *Small talk*]
15. Arnold: I just *don't want* to talk about it. I do care about peter. [disapproval]

First of all, the previous affect detection provides affect annotation for those inputs with strong emotion signals in the above example. The emotion indicators are also illustrated in italics in the above example. The inputs without an affect label followed straightaway are those with weak affect indicators (3rd, 6th, and 12th inputs). Therefore further processing is needed to recover their discussion themes and identify their most likely audiences in order to identify implied emotions more accurately. The general idea for the topic detection is to use LSA to calculate similarities between each test input and all the training files with clear themes. Semantic similarities between the test input and the 13 topic terms (such as 'disease') are also calculated. The final detected topics are derived from the integration of the similarity outputs from the above two channels. We start with the 3rd input to demonstrate the topic detection.

In order to produce a concept space, the corresponding semantic vector APIs are used to create a Lucene index for all the training samples and the test file as the first step. This generated index is also used to create term and document vectors, i.e. the concept space. We first provide rankings for all the training files and the test input based on their semantic similarities to a topic term. We achieve this by searching for document vectors closest to the vector of a specific term (e.g. 'disease'). The 3rd input thus obtains the highest semantic similarity to the topic, 'family care', among the 13 topics with a similarity score of 0.797. Then we also find semantic similarities between training and test documents by using the CompareTerms semantic vector API. It shows there are three training files (crohn3.txt (0.788), family_care3.txt (0.783) and disagree1.txt (0.779)) semantically most similar to the test file. These training files respectively recommend the following three themes: 'disease', 'family care' and 'disagreement'. With the integration of the results obtained from both of the above two semantic-based processings, the system concludes that the 3rd input from Peter relates most closely to the topics of 'family care', 'disease' and 'disagreement'.

In order to identify the 3rd input's audiences, we conduct topic theme detection for the previous two inputs and retrieve step-by-step until we find the input sharing at least one topic with the current input. The above topic detection identifies the 1st input from Arnold shares one theme (i.e. disapproval) with the 3rd input. Thus Arnold is regarded as the intended audience of the 3rd input.

In our application domain, one character's manifestations of emotion can thus influence others. Thus interpersonal relationships (such as friendly or hostile) are also employed to advise the affect detection in social contexts. In this example, since Peter (the speaker) and Arnold (the audience) have a negative relationship and Arnold showed 'disapproval' in the most related context with an identified 'bullying' intention, Peter is most likely to indicate a resentful 'angry' emotion in the 3rd input. Appraisal rules reflecting the above reasoning are generated to derive affect from contexts. The rules accept the most recent emotions expressed by the target audiences identified by the topic detection and relationships between the audiences and the speaker for affect interpretation. Four transcripts from employed scenarios are used for the rule generation. The topic processing also identifies the 6th input from Janet conveys a 'caring' emotion from mum, Janet, towards Peter and the 12th input from Arnold is more likely to express 'anger' towards the AI agent, Dave. It also shows great potential in distinguishing between scenario driven and small talk dialogues (e.g. 13th and 14th inputs) to effectively guide the responding regime of the AI agent.

In order to improve the robustness of our system and deal with imbalanced and open-ended affect classifications beyond scenarios, we have also used a min-margin based active learning method for affect detection. Supervised learning algorithms usually require a large number of labelled training data and most of them suffered from the lack of sufficient training data or heavily imbalanced classification samples. Thus, active learning becomes a promising method to solve such bottleneck labelling problems [9, 10]. The active learning algorithm we employ is able to effectively inform the classifier with the best queries. Providing the true labels for those examples with the most ambiguous affect annotations has improved the system's performance greatly for each learning process. Initial experiments also indicated that the algorithm showed impressive performances (reaching >90% accuracy after several learning iterations) when tested with 180 heavily imbalanced and open-ended inputs.

4 Evaluation and Conclusion

The overall system is able to perform affect detection in real-time as the development of each improvisation. The rule-based reasoning inferences emotions from contexts with the consideration of emotions expressed by the audiences and the relationships between speakers and the audiences. The detected emotions and topic themes also enable the AI agent to make appropriate responses during improvisation. We select 265 inputs of the Crohn's disease scenario with agreed topic annotations provided by two human judges to evaluate the efficiency of the context-based affect and topic detection. A keyword pattern matching baseline system for topic detection is used to compare the performance with that of the LSA. The LSA based analysis achieves the precision and recall scores respectively 0.736 and 0.733 for the detection of 13 topics while the baseline system with precision and recall results respectively 0.603 and

0.583. The two judges also annotated the 265 examples with 15 emotions including 'neutral', 'approval', 'disapproval', 'angry', 'grateful', 'regretful', 'happy', 'sad', 'worried', 'stressful', 'sympathetic', 'embarrassed', 'praising', 'threatening' and 'caring'. The inter-agreement between human judge A/B for affect annotation is 0.63. While the previous version of affect detection achieves 0.46, the new version achieves inter-agreements scores respectively 0.56 and 0.58. The inter-agreements achieved by the updated system become very close to the agreement level between the two judges.

Then 120 example inputs from another scenario, school bullying and 26 articles from the Experience website are also used to further prove the robustness of topic and affect detection. Comparing with the annotations provided by one human judge, the LSA-based topic detection achieves an 83% accuracy rate for the annotation of the bullying inputs and 86% for the annotation of the online articles beyond scenarios.

In this research, we make initial explorations on developing a semantic-based approach and active learning in order to provide solutions for challenging issues such as affect detection from inputs without strong linguistic affective indicators, open-ended and imbalanced affect categorizations. The proposed system achieves impressive performances. In future work, we also intend to extend the emotion modeling with the consideration of personality and culture. We also aim to equip the AI agent with culturally related small talk behavior in order to ease the interaction and further contribute to natural human agent interaction.

References

1. Endrass, B., Rehm, M., André, E.: Planning Small Talk Behavior with Cultural Influences for Multiagent Systems. Computer Speech and Language 25(2), 158–174 (2011)
2. Zhang, L., Jiang, M., Farid, D., Hossain, A.M.: Intelligent Facial Emotion Recognition and Semantic-based Topic Detection for a Humanoid Robot. Expert Systems with Applications 40(13), 5160–5168 (2013)
3. Liu, H., Singh, P.: ConceptNet: A practical commonsense reasoning toolkit. BT Technology Journal 22 (2004)
4. Ptaszynski, M., Dybala, P., Shi, W., Rzepka, R., Araki, K.: Towards Context Aware Emotional Intelligence in Machines: Computing Contextual Appropriateness of Affective States. In: Proceeding of IJCAI (2009)
5. Mateas, M.: Ph.D. Thesis. Interactive Drama, Art and Artificial Intelligence. School of Computer Science, Carnegie Mellon University (2002)
6. Cavazza, M., Smith, C., Charlton, D., Zhang, L., Turunen, M., Hakulinen, J.A.: 'Companion' ECA with Planning and Activity Modelling. In: Proc. of 7th Int. Conf. on Autonomous Agents and Multiagent Systems (AAMAS), Estoril, Portugal, pp. 1281–1284 (2008)
7. Landauer, T.K., Dumais, S.: Latent semantic analysis. Scholarpedia 3(11), 4356 (2008)
8. Widdows, D., Cohen, T.: The Semantic Vectors Package: New Algorithms and Public Tools for Distributional Semantics. In: IEEE Int. Conference on Semantic Computing (2010)
9. Olsson, F.: Bootstrapping Named Entity Recognition by Means of Active Machine Learning. PhD thesis, University of Gothenburg (2008)
10. Zhang, L.: Contextual and Active Learning-based Affect-sensing from Virtual Drama Improvisation. ACM Transactions on Speech and Language Processing (TSLP) 9(4), Article No. 8 (2013)

A One-Shot Learning Approach to Image Classification Using Genetic Programming

Harith Al-Sahaf, Mengjie Zhang, and Mark Johnston

Evolutionary Computation Research Group,
Victoria University of Wellington, P.O. Box 600, Wellington, New Zealand
{harith.al-sahaf,mengjie.zhang}@ecs.vuw.ac.nz,
{mark.johnston}@msor.vuw.ac.nz

Abstract. In machine learning, it is common to require a large number of instances to train a model for classification. In many cases, it is hard or expensive to acquire a large number of instances. In this paper, we propose a novel genetic programming (GP) based method to the problem of automatic image classification via adopting a *one-shot learning* approach. The proposed method relies on the combination of GP and *Local Binary Patterns* (LBP) techniques to detect a predefined number of informative regions that aim at maximising the *between-class* scatter and minimising the *within-class* scatter. Moreover, the proposed method uses only two instances of each class to evolve a classifier. To test the effectiveness of the proposed method, four different texture data sets are used and the performance is compared against two other GP-based methods namely *Conventional GP* and *Two-tier GP*. The experiments revealed that the proposed method outperforms these two methods on all the data sets. Moreover, a better performance has been achieved by Naïve Bayes, Support Vector Machine, and Decision Trees (J48) methods when extracted features by the proposed method have been used compared to the use of domain-specific and Two-tier GP extracted features.

Keywords: Genetic Programming, Local Binary Patterns, Image Classification, One-shot Learning.

1 Introduction

The ability of recognising objects surrounding us represents one of the supreme tasks of human brains, specifically the visual system. Different parts of our bodies (i.e. eyes, hands, tongue, and brain) cooperate with each other in order to learn new objects. Humans are heavily relying on the visual system to capture the variety of object characteristics such as colour, shape, size, and distance. One study [3] shows that the brain of a six year child can recognise objects from more than 10^4 categories, and the learning process continues throughout life. Furthermore, the human brain has the ability to organise learnt objects into different informative groups.

Image classification is mainly concerned with the task of grouping images based on the similarity of its contents, which represents an important task in a

S. Cranefield and A. Nayak (Eds.): AI 2013, LNAI 8272, pp. 110–122, 2013.

variety of fields such as automatic face recognition, disease detection, and machine vision. The importance of this operation has attracted many researchers over the last three decades; and a rich set of different methods have been proposed to the problems of object classification, detection, and recognition. However, performing image classification by machines remains difficult and not as easy as it is by humans.

Genetic Programming (GP) is an evolutionary computation method based on Darwinian principles of natural selection [10]. The promising results achieved using GP techniques to solve a variety of problems in different domains represent a major reason that motivated researchers to investigate those techniques even more over decades. However, the high computational cost of such techniques represents its major drawback.

Local Binary Patterns (LBP) aims at extracting image descriptors based on the relation between each pixel value in an image and its 8 (3×3 window) neighbouring pixels [16]. Since 1995, a number of LBP variants have been introduced and investigated in the literature. In Section 2 of this paper we will provide more details about this operator.

Generally, the task of learning or evolving models requires tuning a large number of parameters in order to capture features covering a diversity of different objects. It has been observed that a large number of training instances are required to adjust or estimate the models' parameters values [7], [22], [23], [24]. In many cases, the task of acquiring a large number of instances can be difficult, expensive or infeasible (e.g. ID-card identification and e-passport). Jain and Chandrasekaran [9] discussed the problem of the training set size in general. Raudys and Jain [20] investigated the effect of using a smaller training set on statistical pattern recognition and gave guidelines and recommendations for practitioners. Moreover, Duin [5] showed that one possible way to reduce the number of used training instances is via reducing the searching space size (i.e. number of features). The main difficulty of this approach is that it has to be handled by a domain expert with good background knowledge about the problem nature, which is in many cases hard and expensive.

Motivated by humans remarkable ability of learning relatively new objects using one or few images, researchers have tried to replicate this functionality in machines and termed it as *one-shot learning* [6]. This problem has been broadly researched and numerous methods are proposed in the field of, but not limited to, machine vision. To stimulate the ability of humans to rapidly learn numerous types of regularities and generalisations, Yip and Sussman [26] proposed a novel method towards fast learning in the domain of morphology. The method exploits the characteristics of sparse representations and forced constraints by a plausible hardware mechanism. A Bayesian-based method is proposed by Fei-Fei and Fergus [6] investigating the problem of object categories using the one-shot learning approach. The aim of their study was to use only one or very few of images to learn much information about a category. Their results show the system can effectively use information gathered during the learning phase to discriminate between unseen instances. Lake et al. [12] proposed a generative model motivated by the concepts of one-shot learning. The method shows how obtaining knowledge from previously learnt characters can be relied on to infer the use of strokes for different characters composition. A hierarchical Bayesian model has been developed in [21] that uses a single training example to learn informative information about a complete category.

In this paper, we propose a hybrid GP based method that adopts the one-shot learning approach to the problem of image classification. The proposed method relies on the combination of GP and a LBP operator to handle the task of automatic binary classification in images using raw pixel values. We are interested in investigating the following objectives:

- To develop a program structure that has the ability to capture informative information of the two classes;
- To find a suitable design of a fitness function that minimises the *within-class* scatter and maximises the *between-class* scatter;
- To test the performance of the system against other GP based methods for automatic image classification (i.e. Two-tier GP [2]), and conventional GP using hand-crafted domain-specific features; and
- To investigate the ability of the proposed method for feature extraction by comparing the features extracted by the proposed method with domain-specific features and those extracted by Two-tier GP on three commonly used methods: Naïve Bayes, Support Vector Machines and Decision Trees (J48).

The rest of this paper is structured as follows. Section 2 briefly explains the Local Binary Patterns (LBP) operator and LBP histogram (LBPH). A detailed description of the proposed method is given in Section 3. Experimental design, data sets, baseline methods, and parameter settings are described in Section 4. Results are shown in Section 5. Section 6 concludes the paper.

2 Local Binary Patterns

The Local Binary Patterns (LBP) operator was originally proposed by Ojala et al. [16] in which the authors aimed at calculating each pixel value of an image based on the values of its neighbouring pixels. The basic LBP operator works in a 3×3 window, which consists of three steps: (1) Assign the value of a neighbouring pixel to 0 if it is less than the centre value of the window and 1 otherwise; (2) the values of the resulted matrix are then multiplied by the power of two in a clockwise direction; and (3) the value of the centre pixel is then replaced by the summation of the resulted values as shown in Figure 1. Formally, the LBP operator can be defined as

$$LBP_{n,r}(x_c, y_c) = \sum_{j=0}^{n-1} 2^j s(V_j - V_c) \tag{1}$$

where r is the radius and n is the number of neighbouring pixels. The values of x_c and y_c represent the coordinate of the centre pixel of the current window. The jth pixel value is denoted as V_j whilst the value of the centre pixel is V_c. The

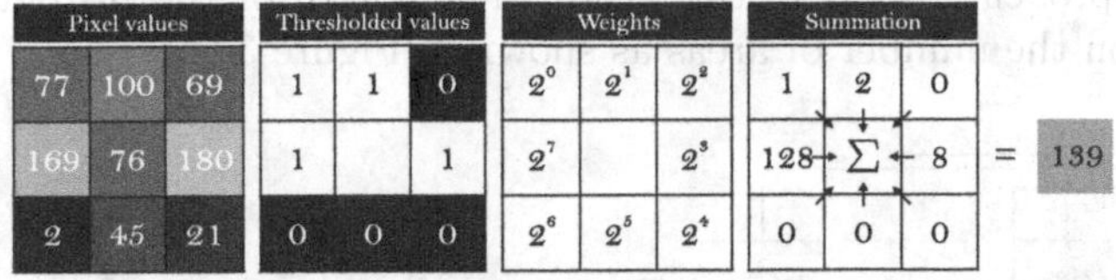

Fig. 1. An example of the required steps to extract an LBP code

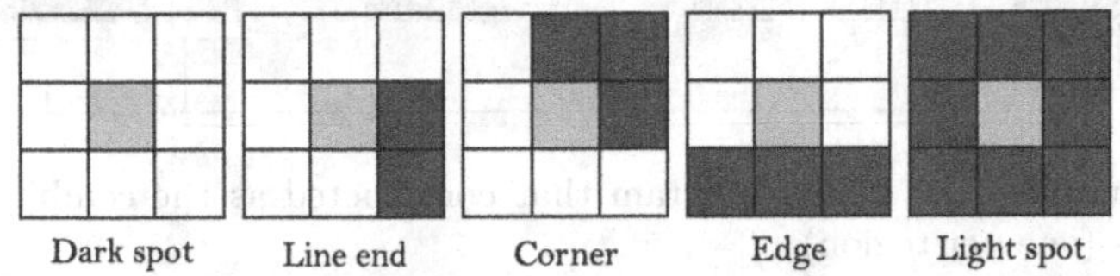

Fig. 2. Some examples of texture primitives of different uniform LBP codes

function $s(x)$ returns 1 if $x \geq 0$ and zero otherwise. A variety of LBP operators have been proposed in the literature that differ in the way of thresholding the neighbouring pixel values, the size of the window and radius, and calculating the final value (more details can be found in [18]). An important extension of basic LBP is known as *uniform patterns* [17] that is denoted as $LBP^{u2}_{n,r}$. A circularly traversed pattern is considered to be *uniform* if bits equal 1 are consequent. For example, if we have the code 00000000, then 00011000, 00110000, and 00111000 are all uniform codes. Uniform codes are important for two reasons: (1) the frequency of uniform codes is higher than non-uniform ones [1]; hence, omitting non-uniform patterns reduces the number of possible LBP codes significantly; and (2) uniform codes can be used to detect different texture primitives as shown in Figure 2.

In [13], a grey-scale invariant intensity-based descriptor is proposed named *NI-LBP*. The main difference between basic LBP and NI-LBP is that the latter uses the mean value of all pixels in a window as the threshold instead of only the value of the centre pixel. In this study, we use the same descriptor (NI-LBP) to extract the pattern of each pixel.

Traditionally, the frequencies of LBP codes appearance are used to form Local Binary Pattern Histogram (LBPH) [8]. For example, if we have an 8-bit codes then we can label 256 different labels starting from 0 (00000000) up to 255 (11111111). Each label of the LBPH is considered as a bin that accumulates the number of occurrences of a specific value or label. By omitting non-uniform codes, there will be only 59 bins (58 uniform codes plus one bin for all non-uniform ones). Let $LBP_{n,r}(i,j)$ identify the calculated LBP code of the $pixel(i,j)$ where $0 \leq i < N$ and $0 \leq j < M$ of an $N \times M$ image; then the histogram h of length L of the entire image can be formally defined as

$$h(l) = \sum_{i=0}^{N-1} \sum_{j=0}^{M-1} (LBP_{n,r}(x_i, y_j) = l) \qquad l = 0, 1, ..., L-1 \tag{2}$$

LBP histograms can either be calculated over the entire image or combine multiple histograms that are obtained from different areas (mostly non-overlapping).

The latter approach is used in this study. The length of the histogram vector varies based on the number of areas as shown in Figure 3.

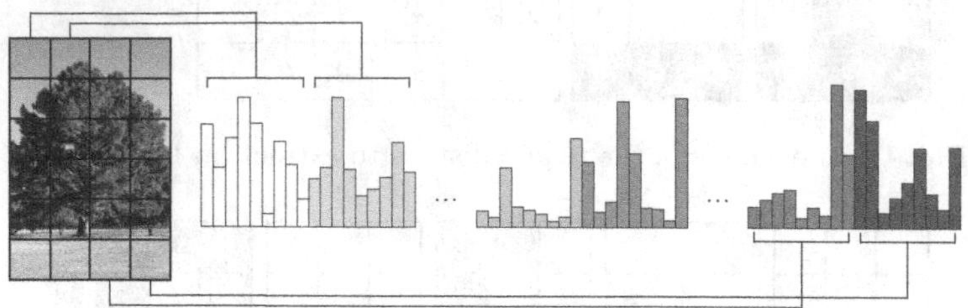

Fig. 3. An example of an LBP histogram that constructed as the combination of 24 sub-histograms (one per region)

3 The New Method

The representation of a new *one-shot* GP-based method is described in this section. The design of the GP individual that extracts LBP histograms from a raw image pixel value and calculates the similarity of two histograms occupies the first part of this section. The rest of the section gives the terminal and function sets, and the fitness function.

3.1 One-Shot GP and Program Structure

An individual is made up of three types of non-terminal nodes: (1) controller node; (2) histogram node; and (3) area node. Figure 4 shows a general structure of an individual for binary classification. Each individual has a set of *controlling instances*, one for each class notated as $controller^X$ where X represents the class label of that instance. The training process starts by extracting the histogram of each controller instance depending on the detected areas by the current individual. Hence, in our case of binary classification we have $histogram^A$ and $histogram^B$ where A and B are the class labels of the two classes. The controller histogram will be compared with the histogram of each and every instance in the training and test sets. The system then iterates over all instances of the training set and for each instance there will be H histograms (equal to the maximum number of classes) one from each branch of the individual tree. The distance between each of the corresponding histograms (controller and instance resulted from the same branch) is then calculated and passed over to the *controller* node (more details of this step in the fitness function subsection). To meet the condition of *one-shot learning* (only one or a few number of instances), the training set is made up of two randomly selected instances of each class. One is used as a controller and the other is used to reflect the goodness of the evolved individual on unseen data (validation set).

The size of the evolved individual under this design is fixed in terms of depth and width of the tree. The number of branches (children of the controller node) increases only if more classes are added which will increase the size of the tree horizontally. The number of detected areas has the same impact and will result in a wider tree. Hence, the positions and sizes of the detected areas represent the only dynamic part of the evolved individual.

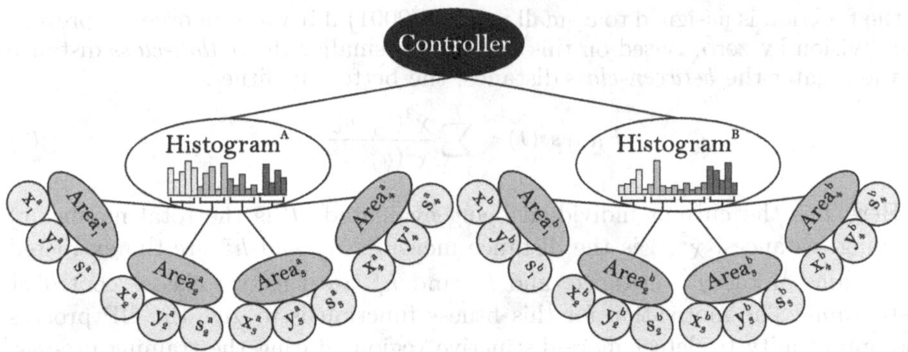

Fig. 4. The program structure of a *one-shot GP*, which consists of two *histogram* nodes that each has four *area* nodes

3.2 Function Set

The function set in this study is *strongly-typed* [15], and is composed of three different types of nodes in which each type has its own functionality. The *controller* node represents the first type, which restricted to be the root of the individual tree. Hence, each individual has one and only one node of this type. The output of this node is a vector of double-precision values for each instance. The length of the resulted vector is fixed and equivalent to the number of classes. The type of input values of this node is the extracted LBP histograms from its children. The second type is the *histogram* node that represents the root node of each branch of the controller node children. *Histogram* nodes are responsible for extracting the LBP histograms by combining the received sub-histograms from each of its children. Each histogram node has a predefined number of *area* nodes that represent the third type of the function set nodes. Each *area* node detects an area of the image, calculates the LBP histogram, and passes it over to its parent (histogram node).

3.3 Terminal Set

The terminal set is also made of three nodes: 1) X-Coordinate; 2) Y-Coordinate; and 3) Window size. Each of these nodes is of type *integer* value and has its own restrictions. For example, *x-cord* and *y-cord* nodes take values in the range $[1, N-2]$ and $[1, M-2]$ respectively, where N and M represent the width and height of an image respectively. The *size* node is restricted to be in the range between 3 and 15.

3.4 Fitness Function

The fitness function used here aims at minimising the *within-class* scatter and maximising the *between-class* ones as shown in Equation (3). The fitness value is proportional to the distance of the same class instances and inversely proportional to distance between instances of different class. However, the denominator of the function is assigned to a small value (0.0001) if it was 0 in order to prevent the division by zero. Based on this design, the smaller the *within-class* distance or the greater the *between-class* distance, the better the fitness.

$$Fitness(I) = \sum_{j=1}^{T} \frac{\bar{\chi}^2(h_j^x, h_c^x)}{\bar{\chi}^2(\bar{h}_j^x, h_c^{\bar{x}})} \tag{3}$$

Here I is the current individual being evaluated, T is the total number of training instances, $\bar{\chi}^2(.)$ is the distance measure, h_j^x and $\bar{h}_j^x$ are the extracted histograms of the jth instance, and h_c^x and $h_c^{\bar{x}}$ are the two classes controller histograms. The motivation for this fitness function is to give the GP process the opportunity to detect more distinctive regions during the training process. Note that the use of *accuracy* makes the system suffer from the problem of over-fitting[1].

In [25], the χ^2 distance measurement shown in Equation (4) is used to measure the similarity between two histograms (bin-by-bin approach). To give the system more flexibility, a shape based version of this formula is used to make it shifting invariant (unbinned approach) via using the *mean* and *standard deviation* as shown in Equation (5) [19].

$$\chi^2(h^a, h^b) = \frac{1}{2} \sum_{i=0}^{\bar{B}-1} \frac{(h_i^a - h_i^b)^2}{h_i^a + h_i^b} \tag{4}$$

$$\bar{\chi}^2(h^a, h^b) = \frac{(\mu^a - \mu^b)^2}{\sigma^a + \sigma^b} \tag{5}$$

where h^a and h^b are the two histograms, $\bar{B}$ is the total number of bins, μ^a and μ^b are the mean of histogram h^a and h^b respectively, and σ^a and σ^b are the standard deviation of histogram h^a and h^b respectively.

4 Experimental Design

The main focus of this section is to highlight the parameter settings and data sets that were used to evaluate the proposed method.

4.1 Data Sets

In order to test the effectiveness of the proposed method, four different texture image-based data sets are used. Six different classes of the *Kylberg Texture*

[1] The GP process can easily detects areas that can be relied on to discriminate between a small number of instances. However, exposing such a model to a large set of unseen instances can result in a very poor performance.

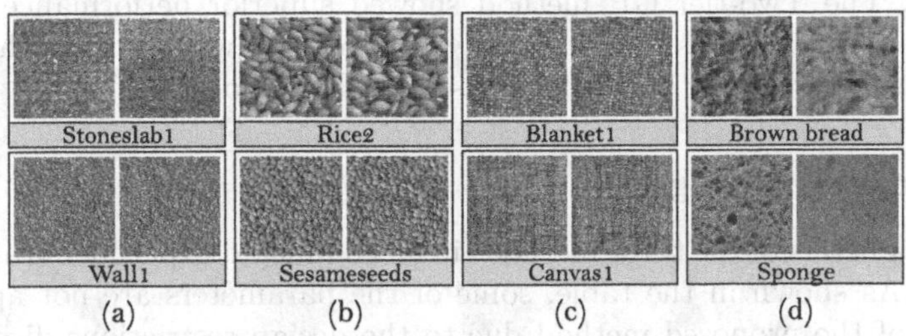

Fig. 5. Samples of the four texture data sets

Dataset [11] are selected to form three data sets in this paper. Originally, this data set is composed of 28 different classes that each consists of 160 instances. Each instance of this data set is a grey-scale image of size 576 × 576 pixels that was resampled to 57 × 57 pixel in our experiments. Furthermore, we equally divided the total number of instances of each class between the training and test sets. Hence, each set is made of 160 instances in total (2 classes × 80 instances).

The *stoneslab1* and *wall1* classes are picked to represents the first data set in this study and we named it as *Textures-1*. Figure 5(a) shows a sample of these two classes. *Textures-2* represents the second set that is made of the *rice2* and *sesameseeds* classes as shown in Figure 5(b). The training and test sets of the third set are made of the *blanket1* and *canvas1* class instances which we refer to as *Textures-3* in this paper and some of its instances are shown in Figure 5(c).

The instances of the fourth data set were taken from the *Columbia-Utrecht Reflectance and Texture* (CUReT) data set [4]. The CUReT data set is made of 61 classes in total that only *brown bread* and *sponge* classes are selected to form the fourth data set that we call it *Textures-4* in this study as shown in Figure 5(d). The size of each instance is a 200 × 200 pixel and there are 81 instances in each class. Hence, the training and test sets consist of 40 and 41 instances of each class respectively.

4.2 Baseline GP Methods

The performance of the proposed system has been compared against two other GP-based methods. The details of those two methods are discussed below.

Conventional GP. The conventional GP method is applied on a set of hand-crafted pre-extracted features. The *mean* and *standard deviation* of the entire image, the four quadrants, and the centre part of each instance have been calculated. The extracted 12 values are then stored in a text file which is later fed to a GP package to evolve a model. Then the evolved model is tested on the test set that was created in a similar way to the training set.

Two-tier GP. Al-Sahaf et al. [2] have developed a GP-based method for automatic feature extraction and selection, and image classification named it as *Two-tier GP*. The system automatically detects areas of different shapes and sizes, and uses different functions to extract the features from pixel values of those areas. The Two-tier GP method showed superior performance over all other competitive methods in that study [2]. Hence, we will compare the performance of the *one-shot GP* with this Two-tier GP method.

4.3 Parameter Settings

The GP parameters of the three methods in all conducted experiments are shown in Table 1. As shown in the table, some of the parameters are not applicable in the case of the proposed method due to the design restrictions discussed in Section 3. The rates of crossover, mutation and reproduction are 0.80, 0.19, and 0.01 respectively. The *tournament* selection method of size 7 is used to maintain population diversity.

4.4 Evaluation

Two different set of experiments are conducted that each aims at testing a different objective. In the first set, the focus was toward investigating the performance of the proposed method against the two baseline (GP) methods. Hence, each of the three methods (two baseline methods and the proposed one) has been evaluated on the four data sets described at the beginning of this section. Only four instances (two of each class which represents the smallest number based on the current design) are randomly selected to play the role of training set. The best evolved individual of each run is then tested against the unseen instances (test set) and the accuracy is reported. In the case of both baseline methods, the value of 0 is used to divide the results space such that all negative values and 0 are considered representing one class while the other class is considered having positive values. However, the instance is evaluated as belonging to the branch having a smaller distance of the evolved program in the case of the proposed method.

This process is repeated for 30 independent runs using the same training and test sets. The average performance of the best evolved programs of the 30 runs on the test set is then recorded. The use of different instances to evolve the model has large impact on the performance of the resulted program. Hence, the entire procedure is repeated 10 times ($10 \times 30 = 300$ runs) using different instances in the training set each time while test set kept the same. At the end of all 10 repetitions the highest and lowest average performances are reported, and the mean and standard deviation statistics are calculated as shown in Table 2. The same procedure is used to evaluate all three methods using exactly the same training and test instances each time.

In the case of the second experiment, the best evolved program at the end of each of the 30 runs of the first experiment is used to extract features of the detected areas in the case of the proposed and Two-tier GP methods. Hence, there will be 10 different individuals as the first experiment is repeated 10 times. In the case of Two-tier GP method, the calculated values of the *aggregation* nodes are used to represents the extracted features similar to the work in [2]. In the case of one-shot GP method, the calculated differences between the two

Table 1. The GP Parameters of all experiments

Parameter	Value	Conve-GP	Two-tier GP	One-shot GP
Generations	20	✓	✓	✓
Population Size	1000	✓	✓	✓
Tree min-depth	2	✓	✓	✗
Tree max-depth	10	✓	✓	✗
Crossover min-depth	2	✓	✓	✗
Crossover max-depth	10	✓	✓	✗
Mutation min-depth	2	✓	✓	✗
Mutation max-depth	10	✓	✓	✗
Initial Population	Ramped half-and-half	✓	✓	✗

controller histograms and resulted histograms of each instance are considered to be the extracted features. This process is also repeated 10 times for each individual due to having 10 different training sets (as mentioned in above). It is important to notice that, extracted features by any two individuals of the test set are different due to different detected areas in terms of position, size, and/or number. In addition to feature sets extracted by those two methods, domain-specific features are also used in this experiment (as stated earlier). The extracted features by each of the three methods are then fed to three different classification methods, i.e., Naïve Bayes (NB), Support Vector Machines (SVM) and Decision Trees (J48). The goal of this set of experiments is to investigate the capability of the new method for automatic feature extraction.

The proposed method, as well as the two other GP-based methods, were all implemented using the platform provided by *Evolutionary Computing Java-based* (ECJ) package [14]. This is mainly because implementing strongly-typed GP in this package is relatively easy.

5 Results and Discussions

This section highlights the results of the experiments. Here, t-tests have been used to compare the difference between the performances of the proposed method and each of the two baseline methods. The "†" and "§" signs in the tables appear if the performance of the proposed method is significantly different than conventional GP and Two-tier GP methods respectively. The bolded numbers in the tables represent the highest mean value amongst the three methods.

Table 2 shows the average performance of the 10 repetitions (each with 30 independent runs) gained from the first experiment on the four data sets. The proposed method has significantly outperformed both conventional GP and Two-tier GP methods on all of the data sets. Moreover, the proposed method, in its worst case, shows better performance than the best pérformance of the Two-tier GP method on all experimented data sets. However, in the case of conventional GP this property does not hold on the Textures-3 and Textures-4 data sets, but the best performance of the proposed one-shot GP is 16% and 30% higher than the conventional GP and Two-tier GP respectively.

The performance statistics of the NB, SVM and DT (J48) on the four data sets using three different sets of features are shown in Table 3. The three classifiers show significantly better performance on all data sets using the features

Table 2. Comparison between conventional GP, Two-tier GP, and One-shot GP

	Conventional GP (%)			Two-tier GP (%)			One-shot GP (%)		
	Min	Max	$\bar{x} \pm s$	Min	Max	$\bar{x} \pm s$	Min	Max	$\bar{x} \pm s$
Texture-1	52.73	66.37	57.10 ± 4.43	49.29	53.56	51.52 ± 1.23	82.37	92.23	**87.76** ± 3.65 †§
Texture-2	47.92	63.98	55.50 ± 4.76	50.29	53.33	51.30 ± 1.05	98.88	99.81	**99.38** ± 0.34 †§
Texture-3	50.77	66.48	57.75 ± 4.85	48.29	54.06	51.62 ± 1.69	64.92	82.50	**76.95** ± 6.09 †§
Texture-4	44.72	63.21	56.35 ± 5.94	50.41	55.08	53.05 ± 1.49	54.11	93.25	**81.12** ± 14.64 †§

Table 3. The performance of Naïve Bayes, Support Vector Machine, and Decision Trees (J48) classification methods on the four texture data sets using domain-specific features, and features extracted by each of the Two-tier GP and One-shot GP methods

		Domain-specific (%)			Two-tier GP (%)			One-shot GP (%)		
		Min	Max	$\bar{x} \pm s$	Min	Max	$\bar{x} \pm s$	Min	Max	$\bar{x} \pm s$
Tex-1	NB	53.75	92.50	70.00 ± 12.18	47.50	66.25	55.56 ± 5.65	83.75	96.25	**91.07** ± 4.29 †§
	SVM	47.50	94.38	74.64 ± 13.67	50.00	65.63	55.25 ± 5.44	88.13	98.75	**92.69** ± 3.72 †§
	J48	66.25	97.50	**88.13** ± 10.87	46.88	64.38	53.82 ± 6.55	51.88	90.00	79.00 ± 12.64 §
Tex-2	NB	96.25	100.0	98.50 ± 1.36	46.88	69.38	55.13 ± 6.35	93.75	100.0	**98.69** ± 2.07 §
	SVM	83.75	100.0	96.75 ± 4.98	48.75	58.13	52.50 ± 3.85	93.13	100.0	**99.13** ± 2.15 §
	J48	43.13	92.50	**71.50** ± 15.78	50.00	58.75	53.63 ± 3.59	53.75	96.88	69.63 ± 16.16 §
Tex-3	NB	61.25	92.50	**79.00** ± 12.39	36.25	63.75	53.57 ± 8.06	60.00	85.00	76.82 ± 8.56 §
	SVM	55.00	96.25	**85.75** ± 12.47	55.00	65.63	55.44 ± 6.05	78.75	86.25	82.94 ± 2.65 §
	J48	56.88	93.75	**76.25** ± 12.12	39.38	67.50	53.50 ± 8.17	52.50	86.25	69.56 ± 12.25 §
Tex-4	NB	42.68	82.93	61.22 ± 11.82	46.34	71.95	64.27 ± 8.81	36.59	98.78	**82.44** ± 19.22 †§
	SVM	42.68	51.71	61.59 ± 16.61	50.00	74.39	60.00 ± 7.53	67.07	100.0	**88.17** ± 10.94 †§
	J48	14.63	86.59	60.37 ± 19.89	29.27	76.83	55.98 ± 14.67	50.00	95.12	**75.00** ± 16.90 §

extracted by the proposed method over using those extracted by the Two-tier GP. However, this property holds for NB and SVM in Textures-1 and Textures-4 data sets in the case of comparing the use of domain-specific features and the features extracted by the proposed method. In Textures-3 data set, the three classification methods have achieved better results using domain-specific features over features extracted by the other two methods, which is opposite to the case of Textures-4. We can observe that the features extracted by the new method have positive influence on the performances of both NB and SVM (scored first in three out of four data sets). This is also true in the case of DT (J48) in the Textures-4 data set. In all other cases, these methods have the second rank using the set of extracted features by the proposed method.

6 Conclusions

In this paper, a one-shot learning approach has been adopted to the problem of automatic image classification. The proposed method uses the combination of GP and LBP techniques to evolve a classifier. Moreover, the fitness function has been designed to maximises the distance of *between-class* and minimises the *within-class* distance. We used four texture data sets to evaluate the performance of the proposed method. The conventional GP and Two-tier GP methods have been used as competitive methods. Two experiments have been conducted that aim at investigating different objectives. The performance of the proposed method is investigated in the first experiment and compared to each of the two baseline methods. The results of this experiment show superior performance of the new method over the other two competitive methods. The second experiment aims at investigating the effectiveness of the extracted features by the proposed method on the performance of Naïve Bayes, Support Vector Machines, and Decision Trees (J48) classification methods. The resulted performances are also compared against the use of both domain-specific features and features extracted by the Two-tier GP method. The results of this experiment show that significantly better or at least comparable performance can be achieved when features extracted by the proposed method are used.

References

1. Ahonen, T., Hadid, A., Pietikainen, M.: Face description with local binary patterns: Application to face recognition. IEEE Transactions on Pattern Analysis and Machine Intelligence 28(12), 2037–2041 (2006)
2. Al-Sahaf, H., Song, A., Neshatian, K., Zhang, M.: Extracting image features for classification by two-tier genetic programming. In: IEEE Congress on Evolutionary Computation, pp. 1–8. IEEE (2012)
3. Biederman, I.: Recognition-by-components: A theory of human image understanding. Psychological Review 94, 115–147 (1987)
4. Dana, K.J., van Ginneken, B., Nayar, S.K., Koenderink, J.J.: Reflectance and texture of real-world surfaces. ACM Transactions on Graphics 18(1), 1–34 (1999)
5. Duin, R.P.: Small sample size generalization. In: Proceedings of the Ninth Scandinavian Conference on Image Analysis, Uppsala, Sweden, vol. 2, pp. 957–964 (1995)
6. Fei-Fei, L., Fergus, R., Perona, P.: One-shot learning of object categories. IEEE Transactions on Pattern Analysis and Machine Intelligence 28(4), 594–611 (2006)
7. Fergus, R., Perona, P., Zisserman, A.: Object class recognition by unsupervised scale-invariant learning. In: Proceedings of the IEEE Conference on Computer Vision and Pattern Recognition, vol. 2, pp. 264–271 (June 2003)
8. Hegenbart, S., Maimone, S., Uhl, A., Vécsei, A., Wimmer, G.: Customised frequency pre-filtering in a local binary pattern-based classification of gastrointestinal images. In: Greenspan, H., Müller, H., Syeda-Mahmood, T. (eds.) MCBR-CDS 2012. LNCS, vol. 7723, pp. 99–109. Springer, Heidelberg (2013)
9. Jain, A.K., Chandrasekaran, B.: Dimensionality and sample size considerations in pattern recognition practice. In: Classification Pattern Recognition and Reduction of Dimensionality, vol. 2, pp. 835–855. Elsevier (1982)
10. Koza, J.R.: Genetic Programming: On the Programming of Computers by Means of Natural Selection. MIT Press, Cambridge (1992)
11. Kylberg, G.: The Kylberg texture dataset v. 1.0. External report (Blue series) 35, Centre for Image Analysis, Swedish University of Agricultural Sciences and Uppsala University, Uppsala, Sweden (2011)
12. Lake, B.M., Salakhutdinov, R., Gross, J., Tenenbaum, J.B.: One shot learning of simple visual concepts. In: Proceedings of the 33rd Annual Conference of the Cognitive Science Society, Austin, TX, pp. 2568–2573 (2011)
13. Liu, L., Zhao, L., Long, Y., Kuang, G., Fieguth, P.: Extended local binary patterns for texture classification. Image and Vision Computing 30(2), 86–99 (2012)

14. Luke, S.: Essentials of Metaheuristics, 2nd edn. Lulu (2013), `http://cs.gmu.edu/~sean/book/metaheuristics/`
15. Montana, D.J.: Strongly typed genetic programming. Evolutionary Computation 3(2), 199–230 (1995)
16. Ojala, T., Pietikäinen, M., Harwood, D.: A comparative study of texture measures with classification based on feature distributions. Pattern Recognition 29(1), 51–59 (1996)
17. Ojala, T., Pietikäinen, M., Mäenpää, T.: Multiresolution gray-scale and rotation invariant texture classification with local binary patterns. IEEE Transactions on Pattern Analysis and Machine Intelligence 24(7), 971–987 (2002)
18. Pietikäinen, M., Hadid, A., Zhao, G., Ahonen, T.: Local binary patterns for still images. In: Computer Vision Using Local Binary Patterns. Computational Imaging and Vision, vol. 40, pp. 13–47. Springer London (2011)
19. Porter, F.C.: Testing Consistency of Two Histograms. ArXiv e-prints, pp. 1–35 (2008)
20. Raudys, S.J., Jain, A.K.: Small sample size effects in statistical pattern recognition: Recommendations for practitioners. IEEE Transactions on Pattern Analysis and Machine Intelligence 13(3), 252–264 (1991)
21. Salakhutdinov, R., Tenenbaum, J.B., Torralba, A.: One-shot learning with a hierarchical nonparametric bayesian model. Journal of Machine Learning Research - Proceedings Track 27, 195–206 (2012)
22. Schneiderman, H., Kanade, T.: A statistical method for 3d object detection applied to faces and cars. In: Proceedings of Computer Vision and Pattern Recognition, pp. 1746–1759. IEEE Computer Society (2000)
23. Viola, P.A., Jones, M.J.: Rapid object detection using a boosted cascade of simple features. In: Proceeding of Computer Vision and Pattern Recognition, pp. 511–518. IEEE Computer Society (2001)
24. Weber, M., Welling, M., Perona, P.: Unsupervised learning of models for recognition. In: Vernon, D. (ed.) ECCV 2000. LNCS, vol. 1842, pp. 18–32. Springer, Heidelberg (2000)
25. Xie, J., Zhang, D., You, J., Zhang, D.: Texture classification via patch-based sparse texton learning. In: IEEE International Conference on Image Processing (ICIP), pp. 2737–2740 (2010)
26. Yip, K., Sussman, G.J.: Sparse representations for fast, one-shot learning. In: Proceedings of the Fourteenth National Conference on Artificial Intelligence, pp. 521–527. AAAI Press / The MIT Press (1997)

Event Detection Using Quantized Binary Code and Spatial-Temporal Locality Preserving Projections

Hanhe Lin, Jeremiah D. Deng, and Brendon J. Woodford

Department of Information Science, University of Otago
P.O. Box 56, Dunedin 9054, New Zealand
{hanhe.lin,jeremiah.deng,brendon.woodford}@otago.ac.nz

Abstract. We propose a new video manifold learning method for event recognition and anomaly detection in crowd scenes. A novel feature descriptor is proposed to encode regional optical flow features of video frames, where quantization and binarization of the feature code are employed to improve the differentiation of crowd motion patterns. Based on the new feature code, we introduce a new linear dimensionality reduction algorithm called "Spatial-Temporal Locality Preserving Projections" (STLPP). The generated low-dimensional video manifolds preserve both intrinsic spatial and temporal properties. Extensive experiments have been carried out on two benchmark datasets and our results compare favourably with the state of the art.

Keywords: manifold learning, event recognition, anomaly detection.

1 Introduction

Recent advances in imaging, multimedia compression and storage technologies have led to the rapid expansion of the use of crowd surveillance systems. The unprecedented availability of large amounts of video data demand new technologies and tools for efficient and automatic surveillance video content analysis. Indeed, crowd scene analysis and classification has been catching growing attention in computer vision research.

Among others, the high density of objects in a video scene remains a challenge to crowd scenes analysis in surveillance videos. A conventional approach would normally aim at tracking individual objects (e.g., [1]), but this cannot cope with frequently occurring conditions such as severe occlusions, small object sizes, and strong similarity among the objects. To overcome these difficulties, various approaches based on features that characterize crowd motions have been proposed for crowd event recognition and abnormal crowd event detection in surveillance videos. These features used for crowd motion modeling include optical flow [2,3], spatial-temporal gradient [4], and volumetric shape matching [5].

On the other hand, a social force model was employed in [6] to estimate the interaction force between pedestrians. Video frames as normal and abnormal are

S. Cranefield and A. Nayak (Eds.): AI 2013, LNAI 8272, pp. 123–134, 2013.

classified by using a bag-of-words approach. Ref. [7] used the largest Lyapunov exponents and correlation dimensions to examine the chaotic dynamics of particle trajectories for anomaly detection in crowd scenes. In [8], testing samples are classified as abnormal or normal based on their sparse reconstruction cost.

Recently, learning low-dimension video manifolds using dimension reduction algorithms has become a popular approach [9,10]. In [9], unusual events were detected by combining Laplacian Eigenmaps (LE) [11] with temporal information. Focusing on the analysis of crowd scenes, a framework was proposed for event detection using optical flow and spatio-temporal Laplacian Eigenmap [10,12]. However, LE is a nonlinear dimension reduction algorithm that works in batch mode; it is unclear how to embed new testing frames, hence it is unsuitable for real-time applications.

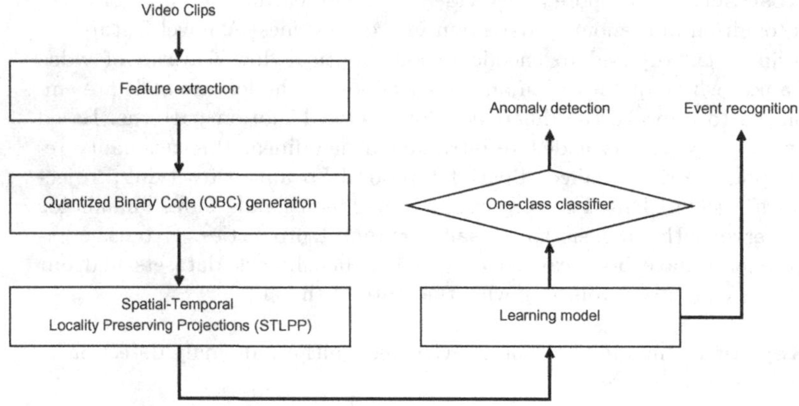

Fig. 1. The computational framework for crowded event recognition and anomaly detection

In this paper, we propose a framework for crowd event recognition and abnormal crowd event detection in surveillance videos based on learning video manifolds. A new feature code design by applying adaptive quantization and binarization is adopted so as to increase the dissimilarity between motion patterns. We propose a linear dimensionality reduction algorithm that considers both spatial and temporal similarities between frames when generating the video manifold. Experimental results demonstrate that our new method outperforms the state-of-the-art methods.

2 The Computational Framework

Our computational framework is shown in Figure 1. Motion features based on optical flows are first extracted between two successive frames. Using these features, we generate the Quantized Binary Code (QBC) for every frame. Then the QBCs are embedded into a low-dimensional manifold using a new algorithm called

"Spatial-Temporal Locality Preserving Projections" (STLPP). Event recognition and anomaly detection are eventually conducted on the low-dimensional manifolds. Details of these algorithmic steps are explained as follows.

2.1 Feature Extraction

We adopt a matrix of size $W \times H \times L$ for a video clip, where W and H are the width and height of the video frame respectively, and L is the total number of frames of the clip. Optical flow between successive frames is estimated according to [13]. The optical flow vector of the frame at time t is denoted as: $\{(f_{x,t}, f_{y,t})\}$. Then, insignificant values due to camera motion or noise are removed based on a predefined threshold.

The orientation θ of a nonzero optical flow vector is determined as:

$$\theta = \begin{cases} \tan^{-1}(\frac{f_y}{f_x}) & f_x > 0 \\ \tan^{-1}(\frac{f_y}{f_x}) + \pi & f_y \geq 0, f_x < 0 \\ \tan^{-1}(\frac{f_y}{f_x}) - \pi & f_y < 0, f_x < 0 \\ +\frac{\pi}{2} & f_y > 0, f_x = 0 \\ -\frac{\pi}{2} & f_y < 0, f_x = 0 \end{cases} \tag{1}$$

with $-\pi < \theta < \pi$. We then construct an 8-bin histogram for θ.

We compute the average magnitude of optical flows in each bin, and a histogram stacking average flows within $m \times n$ regions is formed. Figure 2 shows an example of the feature extraction process. We partition a frame into 48 (i.e., 6×8) regions (shown on the left), and every region has 8 bins. This results in a 2-D histogram (shown on the right). The corresponding direction with high magnitudes in the histogram indicates the main directions of crowd motions, and the region index indicates the position of the crowd in the frame.

2.2 Quantized Binary Code

Based on the 2-D optical flow histogram, we next propose a novel feature scheme called *quantized binary code* (QBC). Rather than using the optical flow histogram vectors directly for further computational procedures, we transform the histogram code through quantization and binarization. Our intention is to make the transformed feature code more discriminative for different motion types.

For each region we have a 8-dim flow vector f after feature extraction. This is expanded into a QBC vector as a 2-tuple $\mathbf{c} = (\mathbf{cl}, \mathbf{cr})$, where $\mathbf{cl}$ and $\mathbf{cr}$ are both of length 8. The binary element values in the tuple are assigned with the help of a binarization threshold T_b:

$$\begin{aligned} \mathrm{cl}_i &= 1, \text{ IF } 0 < f_i < T_b; \\ \mathrm{cr}_i &= 1, \text{ IF } f_i \geq T_b; \end{aligned} \tag{2}$$

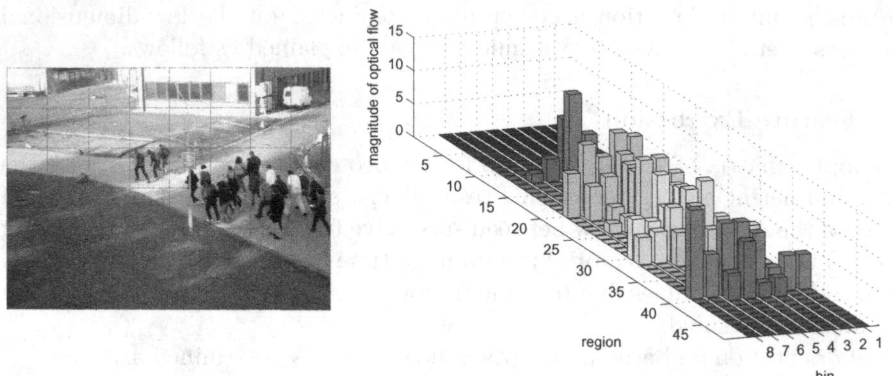

Fig. 2. The splitting of frames into regions and the generation of optical flow histogram

where $i = 1, 2, \cdots, 8$. Otherwise, these elements will remain 0 as initialized.

The generation of QBC is demonstrated in Figure 3. The QBC has 16 binary elements, the first 8 of them corresponding to **cl** and the next 8 elements corresponding to **cr**. Threshold T_b is indicated by the red dotted line in the diagram. Despite the quantization and binarization operations, the QBC feature still preserves both the directional and magnitude information of local motions.

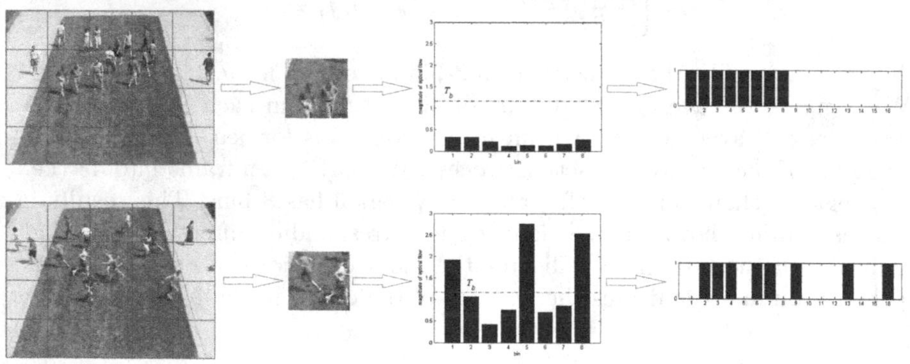

Fig. 3. Generation of the quantized binary code from optical flow histograms

The threshold T_b can be adaptively determined from a k-means clustering process ($k = 2$) on the optical flow vectors. Identical activities in different regions usually do not generate the same scale of flow values due to camera perspectives. Hence we obtain a local T_b threshold for each region separately.

2.3 Spatial-Temproal Locality Preserving Projections

The proposed Spatial-Temporal Locality Preserving Projections (STLPP) is based on Locality Preserving Projections (LPP) [14]. Unlike the original LPP algorithm, however, STLPP utilizes both spatial and temporal information for manifold learning. The procedure works as follows.

Constructing the Weight Matrix. Let W be a symmetric $m \times m$ (m is the total number of frames) matrix, with the weight between Frame i and Frame j given as:

$$w_{ij} = SS_{ij} \times TS_{ij}. \tag{3}$$

Here SS_{ij} denotes the spatial similarity between $\mathbf{c}_i$ and $\mathbf{c}_j$, i.e., the QBC vectors of Frame i and Frame j respectively:

$$SS_{ij} = \frac{\mathbf{c}_i \cdot \mathbf{c}_j}{\| \mathbf{c}_i \| \| \mathbf{c}_j \|}, \tag{4}$$

and TS_{ij} denotes the temporal similarity between the two frames:

$$TS_{ij} = e^{-\frac{(i-j)^2}{\sigma^2}}, \tag{5}$$

where σ is a parameter controlling the effective scope for temporal similarity.

Generating Eigenmaps. Having had the similarity matrix W, we deal with the generalized eigenanalysis problem [15]:

$$XLX^T \boldsymbol{v} = \lambda XDX^T \boldsymbol{v}, \tag{6}$$

where D is a diagonal matrix whose entries are column (or row) sums of W, i.e., $d_{ii} = \sum_j w_{ij}$; $L = D - W$ is the Laplacian matrix; and X denotes the data matrix whose i-th column x_i corresponds to the QBC vector of Frame i.

Let column vectors $\boldsymbol{v}_0, \cdots, \boldsymbol{v}_{l-1}$ be the solutions of Eq. (6), with the corresponding eigenvalues in ascending order: $\lambda_0 < \cdots < \lambda_{l-1}$. Then, the l-dimensional embedding vector y_i corresponding to x_i is estimated by:

$$y_i = V^T x_i, V = (\boldsymbol{v}_0, \boldsymbol{v}_1, \cdots, \boldsymbol{v}_{l-1}) \tag{7}$$

2.4 Event Recognition and Anomaly Detection

Now the high-dimensional video frames are embedded into a low-dimensional manifold. A frame in a video clip is represented as a data point in the embedded space. A trajectory $\mathbf{S}_i$ is constructed as $\mathbf{S}_i = \{y_t^i | t = 1, 2, \cdots, T\}$, where T denotes the temporal window size, and y_t is the l-dimensional embedding vector obtained in Eq. (7). Machine learning algorithms can then be employed on the embedded manifolds for event classification and anomaly detection.

Event Recognition. To recognize events in crowd scenes, first we need to measure the similarity between the reference trajectory and the test trajectory in the low-dimensional embedding space. Specifically, given two trajectories $\mathbf{S}_1$ and $\mathbf{S}_2$ we use the Hausdorff Distance to compute their similarity:

$$H(\mathbf{S}_1, \mathbf{S}_2) = \max(h(\mathbf{S}_1, \mathbf{S}_2), h(\mathbf{S}_2, \mathbf{S}_1)), \tag{8}$$

where

$$h(\mathbf{S}_1, \mathbf{S}_2) = \max_{y_i^1 \in \mathbf{S}_1} \min_{y_j^2 \in \mathbf{S}_2} \|y_i^1 - y_j^2\|, \tag{9}$$

and $\| \cdot \|$ is the Euclidean norm.

Crowd events are classified by the 1-Nearest Neighbour (1-NN) algorithm. Denote the training set of trajectories as TR. For a test trajectory $\mathbf{S}_{te}$, it is classified as Class c if it finds the nearest match in TR with a class label c:

$$c = \arg\min_c H(\mathbf{S}_{te}, \mathbf{S}^c), \ \forall \ \mathbf{S}^c \in \text{TR}, \text{class}(\mathbf{S}^c) == c. \tag{10}$$

Note that a more sophisticated classifier could be employed, but here we concentrate on evaluating the 1-NN discriminative ability of QBC and STLPP.

Anomaly Detection. We adopt the One-Class Support Vector Machine (OC-SVM) [16] for anomaly detection. Given a sample of normal trajectories $\{\mathbf{S}_i\}_{i=1}^n$, OC-SVM maps these trajectories into an inner product space and finds the optimal boundary that encloses them. A new trajectory that falls within the boundary is labeled as "normal", otherwise as "abnormal". More specifically, let Φ be a feature map $\mathbf{S}_i \rightarrow F$, OC-SVM solves the following quadratic program:

$$\min_{w \in F, \boldsymbol{\xi} \in R^l, \rho \in R} \frac{1}{2} \parallel w \parallel + \frac{1}{vn} \sum_{i=1}^{n} \xi_i - \rho \tag{11}$$

subject to $(w \cdot \Phi(\mathbf{S}_i)) \geqslant \rho - \xi_i$, $\xi_i \geqslant 0$. Here w, ξ, and ρ are the solution of this problem, v is a specified prior, and the function $f(x) = \text{sgn}((w \cdot \Phi(\mathbf{S}_i)) - \rho)$ determines whether a new trajectory is "in" (normal) or "out" (abnormal).

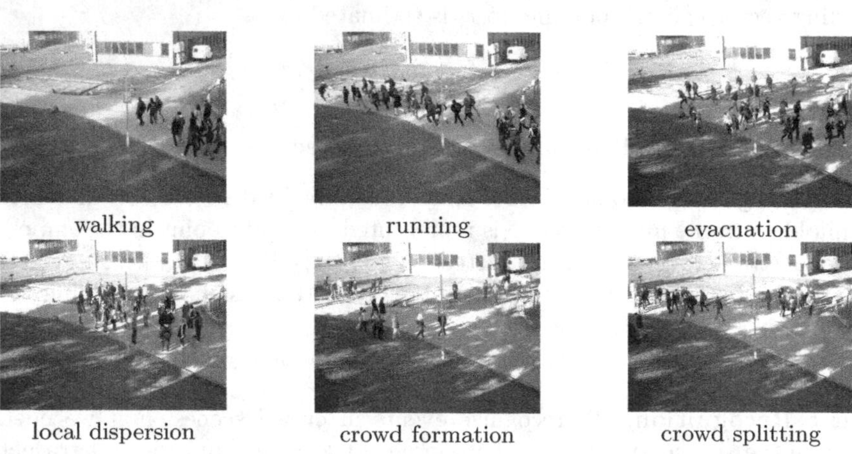

Fig. 4. Sample frames in the PETS dataset, each representing a crowd event

3 Experiments

3.1 Crowd Event Recognition

The PETS 2013 dataset [1] is used for crowd event recognition. It contains four video sequences (1059 frames in total) with timestamps 14-16, 14-27, 14-31 and 14-33. As shown in Figure 4, crowd events vary over time as follows: walking, running, evacuation (rapid dispersion), local dispersion, crowd formation and splitting. We use the same ground truth as in [12], shown in Table 1.

Evaluation of QBC. Binarization thresholds are derived individually for each region, based on which the QBC for each frame is obtained.

To acquire the optimal σ value, we first set the embedding dimension l and the temporal window size T both to 3, obtaining trajectories of 9 dimensions; then a 4-fold cross-validation is employed on these trajectories to compute recognition accuracy with σ increasing exponentially ($\sigma = 2^{-5}, 2^{-4}, \cdots, 2^{9}$), and the scale with the best cross-validation accuracy is chosen. Meanwhile, the QBC and plain optical flow histogram are tested respectively to verify the effectiveness of the QBC. The outcome is illustrated in Figure 5. It shows that, over a large range of σ values (from $\frac{1}{4}$ to 64), the performance of the QBC is stable and better.

Table 1. Ground truth for crowd event recognition in the PETS dataset

Classes	**Timestamp** [frames]
Walking	14-16 [0-36, 108-161] 14-31[0-50]
Running	14-16 [37-107, 162-223]
Local dispersion	14-16 [0-184, 280-333]
Local movement	14-33 [197-339] 14-27 [185-279]
Crowd splitting	14-31 [51-130]
Crowd formation	14-33 [0-196]
Evacuation	14-33 [340-377]

Evaluation of STLPP. Figure 6 shows the 3-D manifolds of four dimension reduction algorithms: LPP [14], principal component analysis (PCA), ST-LE [10], and STLPP. In STLPP, σ is set as $16 (i.e., 2^4)$. It can be seen that STLPP and ST-LE show better visual clustering effects. However, the effect of temporal information in ST-LE is too strong, shaping the generated video manifold almost into a time series. In our STLPP, similar motion pattern still cluster together despite integrating temporal similarity.

To evaluate the classification performance, a training/testing split ratio is set to 1/2 for the dataset (the same as in [12]). The average results are reported from 10 randomized runs. The confusion matrix corresponding to the highest recognition accuracy, with the embedding dimension and temporal window both set to 5, is shown in Table 2. Clearly most crowd events have very good recognition

[1] Available from URL `http://pets2013.net/`

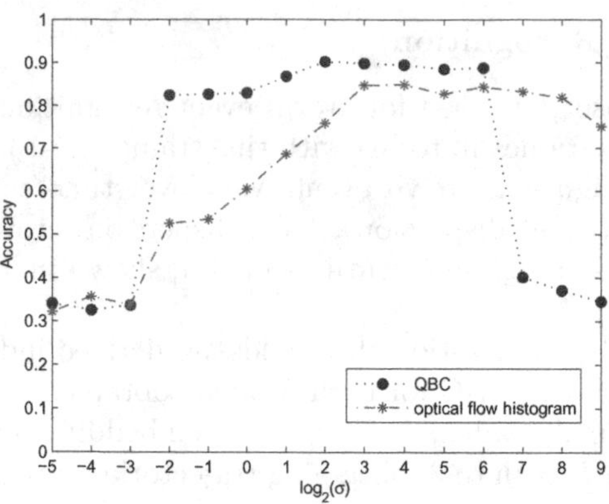

Fig. 5. QBC vs optical flow histogram

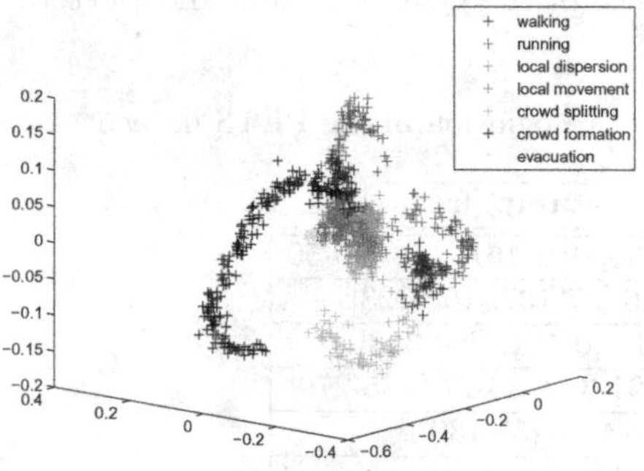

(a) STLPP

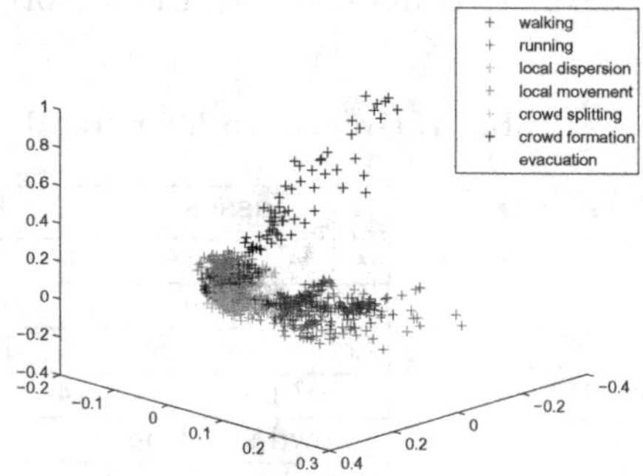

((b) LPP

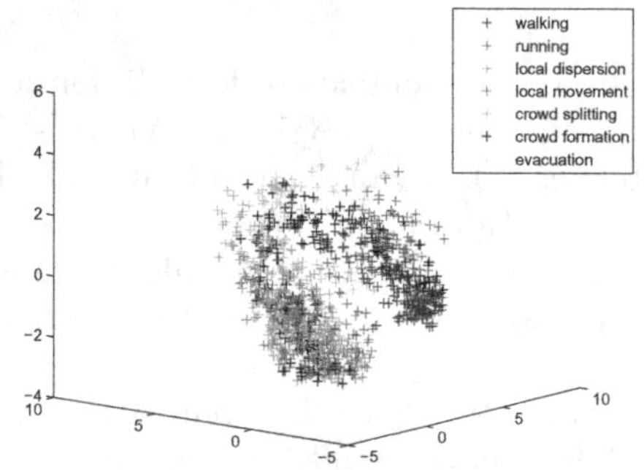

(c) PCA

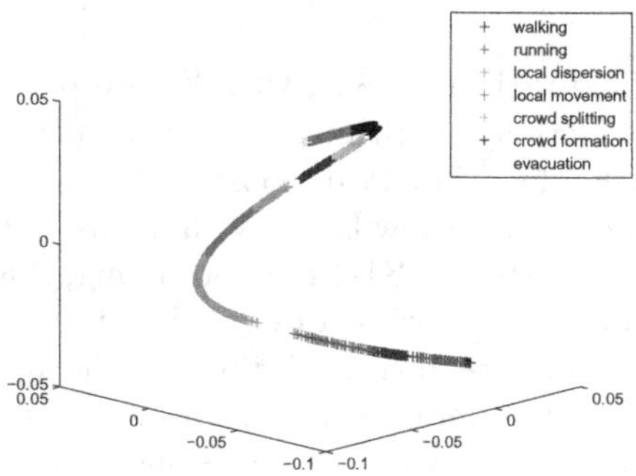

(d) ST-LE

Fig. 6. 3-D manifolds of crowd events generated by different algorithms

Table 2. Confusion matrix for event recognition in the PETS dataset ($l = 5, T = 5$)

Walking (W)	1.00	0	0	0	0	0	0
Running (R)	0.03	0.97	0	0	0	0	0
Local dispersion (LD)	0	0	0.95	0.05	0	0	0
Local movement (LM)	0	0	0.07	0.92	0	0.01	0
Splitting (S)	0.03	0	0	0	0.97	0	0
Formation (F)	0	0	0	0.02	0	0.98	0
Evacuation (E)	0	0	0	0.01	0	0	0.99
Classified as →	W	R	LD	LM	S	F	E

Table 3. Comparison of event recognition accuracy on the PETS dataset

Method	Ref.[17]	Ref.[18]	Ref.[12]	Ours
Accuracy	0.83	0.84	0.90	**0.967**

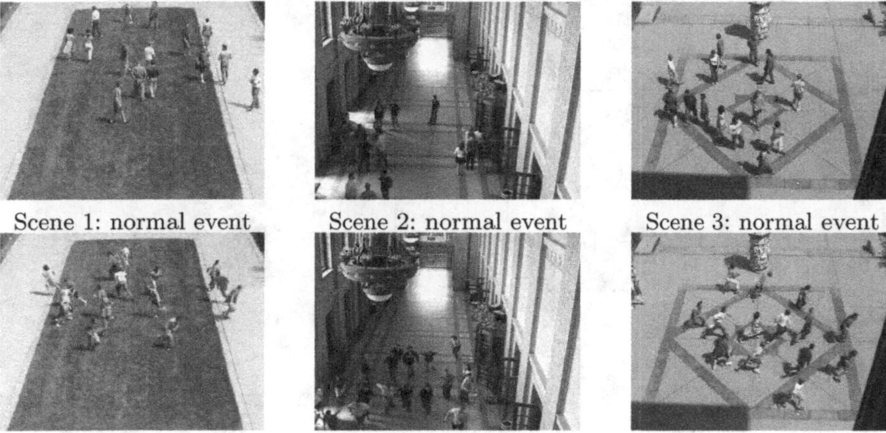

Scene 1: normal event Scene 2: normal event Scene 3: normal event

Scene 1: abnormal event Scece 2: abnormal event Scene 3: abnormal event

Fig. 7. Sample scene frames in the UMN dataset

performance. The high similarity between local movement (LM) and local dispersion (LD) contributes to a significant confusion, which is however consistent with the manifold shown in Figure 6(a). Recognition accuracy of our proposed method is compared with the state of the art in Table 3, which indicates that we have increased the recognition accuracy by around 7%.

3.2 Anomaly Detection

We use the University of Minnesota dataset [2] for crowd anomaly detection. It contains eleven video clips of three different scenes: two clips of Scene 1 (outdoor), six clips of Scene 2 (indoor), and three clips of Scene 3 (outdoor).

[2] Available from URL `http://mha.cs.umn.edu`

Table 4. Comparison of average AUC on the UMN dataset

Methods	Ref.[7]	Ref.[6]	Ref.[8]	Ref.[12]	Ref.[19]	Ours
AUC	0.99	0.96	0.98	0.97	0.89	**0.99**

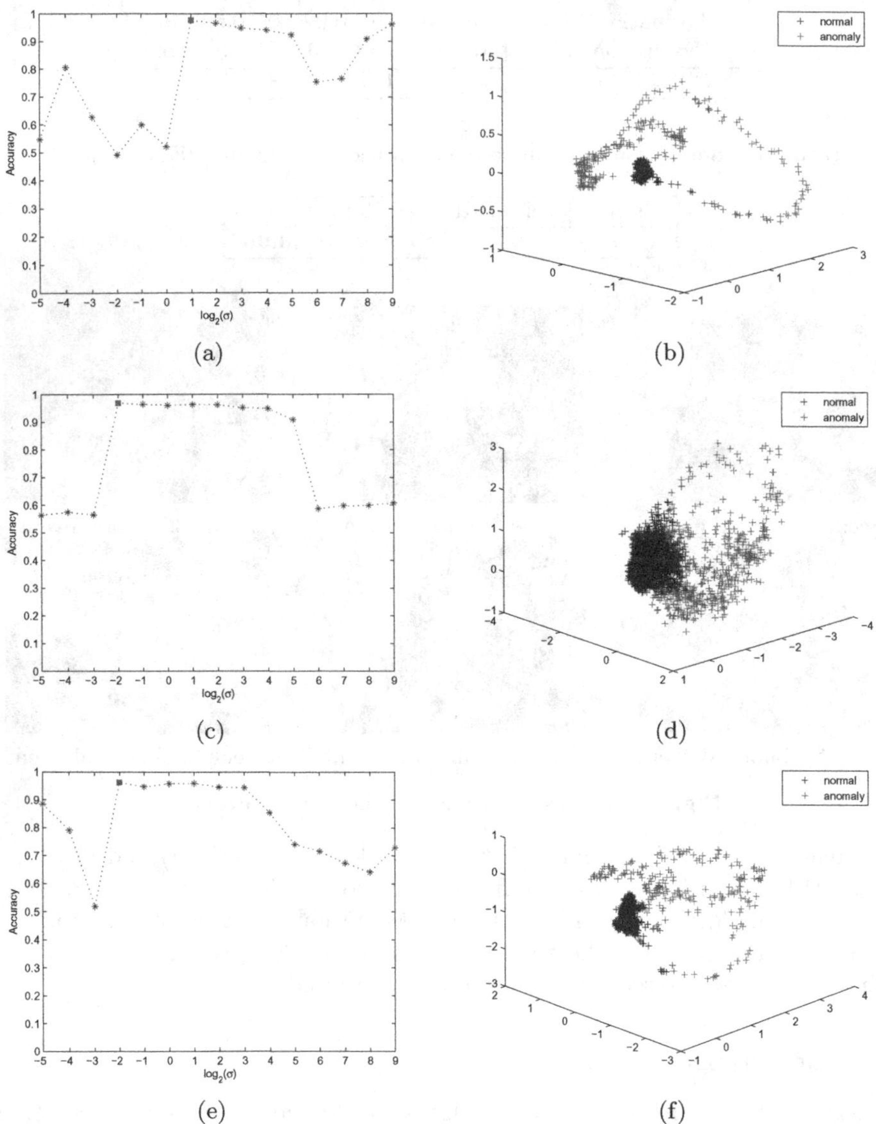

Fig. 8. STLPP manifolds for the UMN scenes (shown on the right) generated from optimal σ values (indicated by red square markers shown on the left).(a)(b) Scene 1; (c)(d) Scene 2; (e)(f) Scene 3.

Each clip starts with normal activities and ends with some anomalies – see Figure 7. Motion patterns in different clips vary. For comparison purposes, we follow the same usage as in other works: 3/4 normal trajectories are selected randomly for training and the rest trajectories (normal and abnormal) are used for testing. The results reported are averaged from 10 randomized runs. In this experiment, we split each frame (240×320) into 20 (4×4) regions.

Similar to experiments in Section 3.1, we first fix the embedding dimension and temporal window size (both to 3) and compute the best detection accuracy while increasing σ exponentially. The corresponding results are shown in Figure 8. The σ with the best detection accuracy in three scenes are 2, 2^{-2} and 2^{-2} respectively. Their corresponding 3-dimension manifolds demonstrate that our proposed method separate normal and anomaly events clearly in the low-dimensional embedding space.

Eventually, we compare our method using QBC and OC-SVM with six other state-of-the-art methods in Table 4, where the area under the ROC (AUC) values are reported. The performance of our method is competitive, being the same as [7], and outperforming others [6,8,12,19].

4 Conclusion and Future Work

In this paper we have proposed a new method for event recognition and anomaly detection in crowd scenes based on video manifold learning. Optical flow features are first encoded through adaptive quantization and binarization. Codes are embedded in a low-dimensional space using a manifold projection algorithm which is improved by integrating both spatial and temporal similarities among frames. The proposed approach is able to generate manifolds with well-shaped motion pattern clusters. Experimental results have verified that our proposed method enhances the discriminative ability of optical flow features, and is effective for event recognition and anomaly detection in crowd scenes.

Despite the competitive results, STLPP works in batch-mode, hence cannot satisfy the demand of applications where data are received incrementally from online video streams. To address this limitation, we intend to explore an incremental learning model in the future.

References

1. Nguyen, H.T., Ji, Q., Smeulders, A.W.: Spatio-temporal context for robust multi-target tracking. IEEE TPAMI 29(1), 52–64 (2007)
2. Andrade, E.L., Blunsden, S., Fisher, R.B.: Modelling crowd scenes for event detection. In: Proc. ICPR 2006, vol. 1, pp. 175–178 (2006)
3. Adam, A., Rivlin, E., Shimshoni, I., Reinitz, D.: Robust real-time unusual event detection using multiple fixed-location monitors. IEEE TPAMI 30(3), 555–560 (2008)
4. Kratz, L., Nishino, K.: Anomaly detection in extremely crowded scenes using spatio-temporal motion pattern models. In: Proc. CVPR 2009, pp. 1446–1453 (2009)

5. Ke, Y., Sukthankar, R., Hebert, M.: Event detection in crowded videos. In: Proc. ICCV 2007, pp. 1–8 (2007)
6. Mehran, R., Oyama, A., Shah, M.: Abnormal crowd behavior detection using social force model. In: Proc. CVPR 2009, pp. 935–942 (2009)
7. Wu, S., Moore, B.E., Shah, M.: Chaotic invariants of lagrangian particle trajectories for anomaly detection in crowded scenes. In: Proc. CVPR 2010, pp. 2054–2060 (2010)
8. Cong, Y., Yuan, J., Liu, J.: Sparse reconstruction cost for abnormal event detection. In: Proc. CVPR 2011, pp. 3449–3456 (2011)
9. Tziakos, I., Cavallaro, A., Xu, L.Q.: Event monitoring via local motion abnormality detection in non-linear subspace. Neurocomputing 73(10), 1881–1891 (2010)
10. Thida, M., Eng, H.-L., Dorothy, M., Remagnino, P.: Learning video manifold for segmenting crowd events and abnormality detection. In: Kimmel, R., Klette, R., Sugimoto, A. (eds.) ACCV 2010, Part I. LNCS, vol. 6492, pp. 439–449. Springer, Heidelberg (2011)
11. Belkin, M., Niyogi, P.: Laplacian eigenmaps for dimensionality reduction and data representation. Neural Computation 15(6), 1373–1396 (2003)
12. Thida, M., Eng, H.L., Monekosso, D.N., Remagnino, P.: Learning video manifolds for content analysis of crowded scenes. IPSJ Transactions on Computer Vision and Applications 4, 71–77 (2012)
13. Liu, C., Freeman, W.T., Adelson, E.H., Weiss, Y.: Human-assisted motion annotation. In: Proc. CVPR 2008, pp. 1–8 (2008)
14. Niyogi, X.: Locality preserving projections. Neural Information Processing Systems 16, 153 (2004)
15. Golub, G.H., van Loan, C.F.: Matrix computations (1996)
16. Schölkopf, B., Platt, J.C., Shawe-Taylor, J., Smola, A.J., Williamson, R.C.: Estimating the support of a high-dimensional distribution. Neural Computation 13(7), 1443–1471 (2001)
17. Garate, C., Bilinsky, P., Bremond, F.: Crowd event recognition using hog tracker. In: Proc. PETS-Winter 2009, pp. 1–6 (2009)
18. Chan, A.B., Morrow, M., Vasconcelos, N.: Analysis of crowded scenes using holistic properties. In: Proc. PETS-Winter 2009, pp. 101–108 (2009)
19. Shi, Y., Gao, Y., Wang, R.: Real-time abnormal event detection in complicated scenes. In: Proc. ICPR 2010, pp. 3653–3656 (2010)

Growing Neural Gas Video Background Model (GNG-BM)

Munir Shah, Jeremiah D. Deng, and Brendon J. Woodford

Department of Information Science, University of Otago, Dunedin, New Zealand
{munir.shah,jeremiah.deng,brendon.woodford}otago.ac.nz

Abstract. This paper presents a novel growing neural gas based background model (GNG-BM) for foreground detection in videos. We proposed a pixel-level background model, where the GNG algorithm is modified for clustering the input pixel data and a new algorithm for initial training is introduced. Also, a new method is introduced for foreground-background classification and online model update. The proposed model is rigorously validated and compared with previous models.

Keywords: Background subtraction, online learning, growing neural gas, video processing and Gaussian mixture model.

1 Introduction

Precise localization of foreground objects is one of the most important building blocks of computer vision applications such as smart video surveillance, automatic sports video analysis and interactive gaming [1]. However, accurate foreground detection for complex visual scenes in real time is a difficult task due to the intrinsic complexities of real-world scenarios. Some of the key challenges are: dynamic background, shadows, sudden illumination changes and foreground aperture [1, 2].

Recently, significant research efforts have been made in developing methods for detecting foreground objects in complex video scenes [1, 3–5]. The mixture of Gaussians (MoG) is one of the most popular background models, due to its ability in handling multi-model backgrounds, and robustness to gradual illumination changes [1, 3]. Thus, it is widely adopted as a basic framework in many subsequent models. However, MoG based models have some common limitations. Firstly, these models assume that the video data follow a Gaussian or normal distribution, which is not always true [6]. Secondly, during the maintenance process the distribution tails are not updated, i.e., the distribution is only updated when a new pixel value is within its variance range. Consequently, the standard deviation can be underestimated and values in the tail may be misclassified [7]. This motivated us to explore a growing neural gas based approach for video background modeling.

The main contributions of this study are: firstly, a pixel-level background model, where the GNG algorithm is modified for clustering the input pixel data.

S. Cranefield and A. Nayak (Eds.): AI 2013, LNAI 8272, pp. 135–147, 2013.

Secondly, a new algorithm for initial training, which helps the model to become more stabilized and better approximate the distribution of the training set. Thirdly, a new criteria to differentiate between foreground and background nodes. Finally, an online method to update the background model.

2 Growing Neural Gas (GNG)

The concept of GNG was introduced by [8] as an extension of Neural Gas, which is an unsupervised incremental clustering algorithm [9]. The GNG incrementally constructs a graph of nodes for a given n-dimensional input data distribution in a R^n space. This algorithm starts with two nodes and constructs a graph, where neighboring nodes are connected by an edge. The nodes in the network compete for determining the node with the highest similarity measured in Euclidean distance to the input datum using competitive Hebbian learning (CHL) [10]. The CHL algorithm directs the local adaptation of nodes and insertion of new nodes, where local error is maintained during the learning process to determine the location of new nodes. A new node is inserted between the two nodes having the highest and the second highest accumulated error.

One advantage of GNG is its ability of adapting itself to a slowly changing input distribution, i.e., it move the nodes to cover new distributions. Furthermore, GNG add nodes incrementally during execution, therefore there is no need to define the number of nodes *a priori*. The insertion of nodes stops if a maximum network size is reached or some user defined criteria is met.

The detail procedure of GNG is given in Algorithm 1. The GNG contains a set of nodes and edges connecting them in a topological structure. Each node k consists of a reference vector $\bar{w}_k$ in R^n, a local accumulated error variable error_k and a set of edges defining the neighbors of node k. The $\bar{w}_k$ is the position of a node k in the input space. The local accumulated error is a statistical measure used to determine insertion points for the new nodes. Furthermore, an age variable is maintained for each edge, which is used to remove obsolete edges in ordered to keep the topology updated.

Figure 1 pictorially illustrates the working of GNG algorithm, where GNG is initialized with two random nodes. The accumulated error, node movements, edges and topology update, and nodes insertion are the important parts of the GNG algorithm. Each of these components are discussed as follows.

2.1 The Local Accumulated Error

GNG computes a local accumulated error for each node for each input signal. The local error is a statistical measure reflecting the portion of the input distribution covered by that node. The large coverage means larger updates of the local error values because inputs at greater distances are mapped to the node. Since the GNG tries to minimize the errors, knowing where the error is large is helpful to find the location for a new node at its insertion time. The local error for each node s is updated as in Eq. (1).

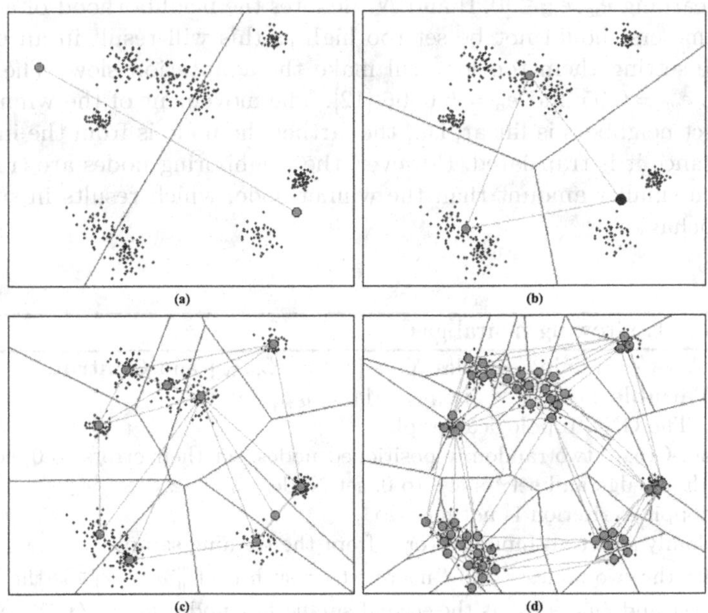

Fig. 1. An illustration of the GNG algorithm. (a) The initialization of the GNG algorithm by creating two nodes at random location; (b) The state of the GNG algorithm after 1000 iterations, a third node is inserted as depicted in blue color; (c) After 9000 iterations, GNG has 10 nodes covering data space; (d) After 50000 iterations, 50 nodes are spread out over the data space.

$$\text{error}_s \leftarrow \text{error}_s + \left\|\overline{w}_s - x\right\|^2, \tag{1}$$

where the error is updated with the squared distance to the input. This provides a way to detect all nodes that cover a larger proportion of the input distribution. Finally, errors for all nodes are decreased which keeps the local errors from growing out of proportion and at the same time, it gives more importance to the recent errors.

$$\forall_{j \in S} \ \text{error}_j \leftarrow \text{error}_j - \beta \times \text{error}_j, \tag{2}$$

where β is a learning rate used to reduce over all error. In the default settings $\beta = 0.0005$ [11].

2.2 Node Movements

When a new input x is presented, GNG updates the winning node s and its direct topological neighbors:

$$\overline{w}_s \leftarrow \overline{w}_s + e_w(x - \overline{w}_s), \tag{3}$$

$$\overline{w}_n \leftarrow \overline{w}_n + e_n(x - \overline{w}_n), \ \forall n \in N_s, \tag{4}$$

where the learning $e_w, e_n \in [0, 1]$ and N_s denotes the neighborhood of a node s. These parameters should not be set too high as this will result in an unstable graph while setting them too low will make the adaptation slow. The default settings are $e_w = 0.05$ and $e_n = 0.0006$ [12]. The movement of the winner node and its direct neighbors is linear, i.e., the farther the node is from the input the greater distance it is translated. However, the neighboring nodes are translated with a much smaller amount than the winner-node, which results in smoother clustering behavior.

Algorithm 1. Growing neural gas

Input: A set of training samples $X \leftarrow \{x_1, x_2, ..., x_N\}$ and **Control Variables:** a_{max}, λ, Max_{iter}, Max_{nodes}.

Output: The GNG topological graph

Initialize: Create two randomly positioned nodes, set their errors to 0, connect them with an edge and set its age to 0, set $\text{Node}_{count} \leftarrow 2$.

while Stopping criterion is not met **do**

- Randomly select an input vector x from the training samples
- Locate the two nodes s and t nearest to x such that $\|\overline{w}_s - x\|^2$ is the smallest and $\|\overline{w}_t - x\|^2$ is the second smallest, $\forall$ nodes k. /* $\overline{w}_s$ and $\overline{w}_t$ are reference vectors for s and t. */
- Update the local error for the winner node s using Eq. (1).
- Move s and its direct topological neighbors towards x by using Eq. (3) and (4) and increment the age of all edges from node s to its neighbors.
- **if** s and t are connected by an edge **then**
  - Set the age of that edge to 0.
- **end**
- **else**
  - Create an edge between them.
- **end**
- Remove edges having age greater than a_{max}, remove nodes with no edges and decrement Node_{count}.
- **if** Current iteration is an integer multiple of λ and $\text{Node}_{count} < \text{Max}_{nodes}$ **then**
  - Determined the node u with largest accumulated error and among its neighbors, determine the node v with the largest error.
  - Insert the new node r between u and v as $\overline{w}_r \leftarrow \frac{\overline{w}_u + \overline{w}_v}{2}$
  - Create edges between u and r, and v and r. and remove edge between u and v.
  - Update errors for u, v using Eq. (5) and (6) and set error for r using Eq. (7).
  - $\text{Node}_{count} \leftarrow \text{Node}_{count} + 1$.
- **end**
- Decrease the errors for all nodes j by a factor β using Eq. (2).

end

2.3 Update Edges and Topology

The movement of the nodes in the direction of input may cause the current construction of Delaunay triangulation to become invalid. Therefore, GNG uses a local aging process to remove old edges that should not be a part of the Delaunay triangulation. More specifically, the edges for which CHL has not detected any activity in a_{max} steps are considered as invalid edges and thus removed from the graph. In its default setting $a_{max} = 100$ [12]. This removal of edges may result in a node with no edges. This type of nodes are called dead-nodes and thus removed from the graph. Furthermore, the GNG uses CHL to create and update the edges. The age variable for the recently accessed edges is reset to 0, which prevents them from being removed immediately.

2.4 Node Insertion

In the GNG algorithm, a new node r is inserted after a predefined number of training epochs. Since GNG tries to minimize the error, r is placed in the middle position of the node u with the largest accumulated error and the nodes v among the neighbors of u with the largest accumulated error. This decreases the coverage area (Voronoi regions) of both u and v, which leads to the minimization of their future errors. Thus the errors for u and v are decreased as in Eq. (5) and (6).

$$\text{error}_u \leftarrow \alpha \times \text{error}_u, \tag{5}$$

$$\text{error}_v \leftarrow \alpha \times \text{error}_v, \tag{6}$$

where the value of α determines the amount of reduction in the error. Usually it is set to some high value ($\alpha = 0.5$). However, how much it should be decreased depends on the particular application.

The main motivation for reducing errors for u and v comes from the argument that a new node r inserted in the middle of them covers some of the input distribution from both u and v. Therefore, it makes sense to reduce their error variable because the error in essence represents the coverage of the input distribution. Furthermore, decreasing the errors for u and v prevents the insertion of the next node in the same region.

Also, the error for the new node r is set as a mean of errors of u and v as:

$$\text{error}_r \leftarrow \frac{\text{error}_u + \text{error}_v}{2}. \tag{7}$$

Initializing the error in this way gives the approximation of the error for a new node.

Furthermore, a fixed insertion rate scheme (λ) might not always be desirable because it may lead to unnecessary or untimely insertions. Normally, $\lambda = 600$ and setting λ too low will result in poor initial distribution of nodes because the local errors will be approximated badly. Also, it may lead to the risk of nodes becoming inactive because they are not close enough to the inputs, thus not adapted. On the other hand, setting the λ parameter too high will result in slow growth, which means a large number of iterations are required. However, nodes will be well distributed. The desirable insertion scheme can be based on the global mean error, i.e., a threshold constant could be used on the mean squared error to trigger node insertion.

Furthermore, due to the incremental nature of the GNG, it is not required to specify the number of nodes *a priori.* This is a very important characteristic of GNG compared with other clustering algorithms such as k-means, which makes it an attractive choice for problems where the input distribution is unknown. Thus the GNG algorithm is suitable for online or real-time clustering algorithms.

We explore the GNG for modeling video backgrounds. Since GNG gives better clustering capabilities and has less sensitive parameters [11], it can be a good choice for online background modeling. The proposed GNG background model is explained in the next section.

3 Growing Neural Gas Background Model (GNG-BM)

In the propose model, values of a pixel in a video are modeled as a GNG graph. Suppose we have a set of training samples $X = \{x_1, x_2, ..., x_N\}$ for a pixel consisting of N YCbCr vectors (3-dimensional). The proposed model maintains the GNG topological graph for the training samples. In this algorithm, each node is represented by a reference vector in the R^3 space. It can be seen as a topological graph in a three dimensional plane, where each dimension corresponds to individual color channel in YCbCr color space. Here the maximum value for each dimension, i.e., color channel, is 255, therefore the size of the plane is $255 \times 255 \times 255$. Also, we maintain the winning frequency of each node, which will be used for differentiating between background and foreground nodes. The complete algorithm for GNG training is listed in Algorithm 2.

3.1 Initial Training

The proposed algorithm starts with two randomly created nodes. In each training epoch, a sample is randomly selected from the training set and the winner and runner-up nodes are determined based on their Euclidean distance from the input vector x. Some of the initial video frames are used to generate training set for each pixel in the frame. Empirically, we found that the 100 to 300 frames would be sufficient for initial training. Particularly, when input vector is selected randomly from the training set for a specified number of iterations, some training samples will be repeated several times, which gives sufficient time for the GNG algorithm to adapt itself to the data distribution. In the proposed model, the number of iterations are set to 10 times the number of samples N in the training set. In this way, the samples will be presented or selected a sufficiently large number of times, which helps the GNG algorithm to achieve more stabilized topological approximation of the data distribution.

In this algorithm, for each selected vector the winner node and runner-up node is determined based on their Euclidean distance from the input vector x. After

Algorithm 2. Growing neural gas background model (Training)

Input: A set of training samples $X \leftarrow \{x_1, x_2, ..., x_N\}$ for a pixel, where N is the number of samples and **Control Variables:** a_{max}, λ, Max_{iter}, Max_{nodes}, $\text{win}_{counter}$ counts the number of wins for a node.

Output: Background model for a pixel

1 **Initialize:** Create two randomly positioned nodes, set their errors to 0, $\text{win}_{counter}$ to 1, connect them with an edge and set its age to 0, and set $\text{Node}_{count} \leftarrow 2$.
2 **while** iter < Max_{iter} **do**
3   Randomly select input $x \leftarrow \{Y, Cb, Cr\}$ from the training samples.
4   Update the network topology as in Step 4 to 13 of Algorithm 1 and increment the $\text{win}_{counter}$ for the winning node s.
5   **if** s does not satisfies the criteria in Eq. (8) OR iter is an integer multiple of λ **then**
6   Insert new node r as in Step 15 to 19 of Algorithm 1 and set its $\text{win}_{counter}$ to 1.
7   **end**
8   Decrease errors for all nodes j by a factor β using Eq. (2).
9   iter $\leftarrow$ iter + 1
10 **end**

that the winner node, its topological neighbors and topology of the graph are updated, as in Step 4 of Algorithm 2. The winner counter $win_{counter}$ that counts the number of wins by a node is updated each time node wins. The $win_{counter}$ reflects the portion of data covered by a particular node.

In the proposed model, the pixel value $x \leftarrow \{Y, Cb, Cr\}$ is a 3-D vector, where Y represents the intensity information, Cb and Cr represent color information. In the realistic scenarios, Y channel (intensity) experience larger noise or variations than color channels, especially in case of shadows and highlights. Therefore, matching criteria used to decide whether the insertion of a new node is required or not is slightly modified in the proposed model.

$$\text{MATCH} = \begin{cases} \textbf{true}, & \text{if } \|\overline{w}_s(Y) - x(Y)\| < T_i \text{ AND } \|\overline{w}_s(Cb) - x(Cb)\| < T_c \\ & \qquad \text{AND } \|\overline{w}_s(Cr) - x(Cr)\| < T_c, \\ \textbf{false}, & \text{otherwise,} \end{cases} \tag{8}$$

where T_i and T_c are thresholds for the intensity and the color channels respectively. In the proposed model, T_i is set higher than T_c, so that it can cover higher variation in the intensity channel than the color channels. Empirically, we found that $T_i = 10$ and $T_c = 5$ are good values for these thresholds.

If the winner node s does not satisfy the criteria given in Eq. (8), or the current iteration is an integer multiple of λ, a new node is inserted in the graph as in Step 5 of Algorithm 2. It should be noted that the new node is inserted near the winner node instead of near the node with the largest error as in the original GNG model. This makes sense because a new data point is closer to the winner node than any other node, thus the node created for this data point should be topological neighbor of the winner node. Although the original GNG algorithm may also adapt this node to move it closer to the winning nodes over the time, but it takes time.

Further, if an edge is older than a_{max} (maximum age of the edge), i.e., it is not updated for a long time, that edge is considered as an obsolete edge and therefore can be safely removed from the graph. After that, a node with no edges should also be removed. This is necessary, especially when input data is erroneous or data distribution is changing over the time. In case of erroneous input data, the GNG creates nodes and edges for it, but these inputs do not occur frequently. Thus, the corresponding nodes unnecessarily occupy the resources. On the other hand, in case of changing data distribution in an online environment, previous nodes become invalid. Thus old edges and nodes are removed from the network, which keeps the GNG graph compact.

Moreover, for background modeling, the foreground objects can be considered as noise and thus nodes created for foreground objects should be removed from the graph. No foreground object stays at the same location for more than 300 frames with the exception of stationary foreground objects. Therefore, in the proposed model $a_{max} = 300$ is used for all of our experiments. This means that an edge which is not accessed for more than 300 frames is removed from the graph.

Algorithm 3. Foreground detection and online model update

Input: Pixel value $x \leftarrow \{Y, Cb, Cr\}$ at time tt and a background model M.
Output: Detection result and updated background model

Update the network topology as in Step 4 to 13 of Algorithm 1 and increment the win$_{counter}$ for the winning node s.
if s satisfies the condition as in Eq. (8) and $win_{counter} > T_{win}$ **then**
    Classify the pixel as **background**
end
else
    Classify the pixel as **foreground**
    if s does not satisfy the condition as in Eq. (8) **then**
        Insert new node r as in Step 15 to 19 of Algorithm 1 and set its win$_{counter}$ to 1.
    end
end
Decrease errors for all nodes j by a factor β using Eq. (2).

3.2 Foreground Detection and Online Update

After the initial training phase, a new incoming pixel value x is compared against the background model M, and the winner node s and runner-up node t is determined. If the winner nodes s satisfied the condition as in Eq. (8) and its $win_{counter} > T_{win}$, this pixel is classified as background, otherwise it is classified as foreground. The $win_{counter}$ of a node reflects the portion of the data modeled by that node. As the background surfaces appear more frequently than the foreground surfaces, the nodes corresponding to the background will have higher $win_{counter}$ compared with the nodes corresponding to the foreground. In the proposed model, the threshold value of $T_{win} = 300$ is empirically found a reasonable setting for a variety of scenes.

Furthermore, in real-world scenarios the geometry of the scene may change over a period of time. For instance, a new chair is brought into the room. Also, lighting conditions may also change, e.g, clouds appearing in front of the sun change the brightness condition of the outdoor scene, To handle such situations, the model is adapted online. The online adaptive algorithm is listed in Algorithm 3. If a pixel is classified as foreground and the winning node s does not satisfy the condition in Eq. (8), that means no node is existing in the model that corresponds to the current pixel. Thus, a new node r is inserted in the network using the same procedure as explained in Algorithm 2.

4 Experimental Results

In this section, the proposed model GNG-BM (growing neural gas based background model) is validated and compared with the MoG [3], KMoG [4], SMoG [13], KDE [14] and PBAS [15] background models. The parameters values for the proposed model are set as mentioned in the original papers respectively. These models are compared both qualitatively and quantitatively on the CDnet dataset [16]. This dataset contains 31 indoor and outdoor video sequences, which covers various challenging real life sceneries. Importantly, for precise validation, this dataset provides ground truth at pixel resolution for each frame in a video. This makes detailed quantitative analysis and comparison possible. Due to the space limitation we will present example results on "DynamicBackground" and "shadow" categories, however similar trend is experienced for other categories.

Fig. 2. Comparative qualitative results on the "fall" video sequence in the "DynamicBackground" category

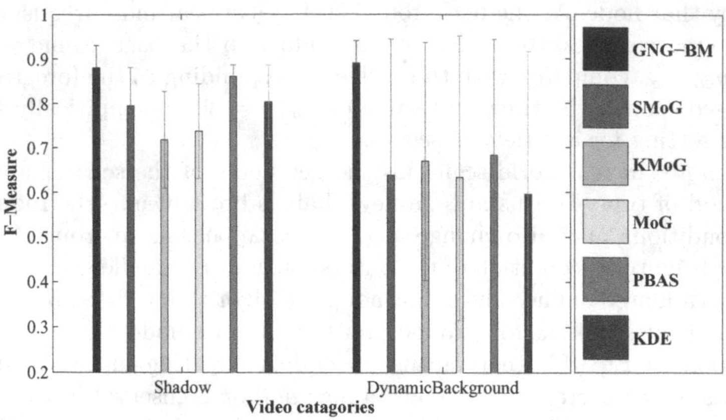

Fig. 3. Comparative quantitative results

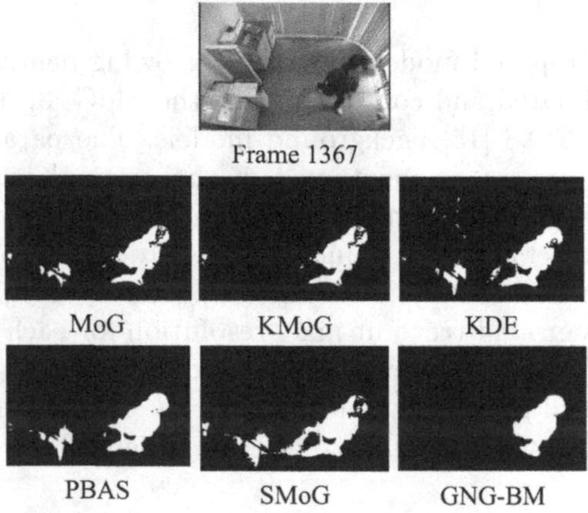

Fig. 4. Comparative qualitative results on the "copyMachine" video sequence in the "shadow" category

Figure 2 presents an example result for the "fall" video sequence in the CDnet dataset. This video is one of the most complex videos for dynamic background modeling. Particularly, the tree in the middle sways irregularly due to the strong wind, which generates multiple background surfaces in pixel values. Also, this tree sways in a large area. As one can see in Figure 2, GNG-BM gives better results than other proposed models. These results are confirmed by the quantitative results presented in Figure 3. This figure presents an average F-measure results for six videos in the "DynamicBackground" category. The F-measure takes into consideration both recall and precision, therefore it is a better choice to present results concisely. As shown in the figure, GNG-BM obtained a better F-measure than other methods.

Furthermore, Figure 4 presents the visual results for the "copyMachine" video sequence in the CDnet dataset. This is a very challenging video sequence for local illumination changes handling techniques. It can be seen in the figure that the GNG-BM gives almost ideal results, whereas MoG gives some false positive results. The same trend is confirmed from the quantitative results presented in Figure 3. This figure plots the average F-measure obtained by the background modeling algorithms for six videos in the "Shadow" category in the CDnet dataset. The GNG-BM achieved a better overall F-measure than other models.

To measure the processing speed, we implemented all the algorithms in C++ using the OpenCV library and ran them on an Intel machine with 2.6 GHz CPU, 3 GB main memory and the Microsoft Windows XP operating System. We measure the processing speed averaged over 10 video sequences of resolution 320×240 (Width $\times$ Height). The proposed model gives processing speed equal to 16 frames per second.

4.1 Discussion

The proposed model is validated and compared with the previous models. The GNG-BM gives significantly better results than other MoG based models both for dynamic backgrounds and illumination changes. This improvement is due to the better clustering abilities of the GNG algorithm and new the training procedure. Especially, the random selection of data samples for training over the specified number of iterations stabilized the model and thus helps in achieving better segmentation. Also, the GNG based background model is good for dynamic backgrounds because it updates the age for the recently or frequently accessed nodes. In this way, even if the background surfaces occurs less frequently, they will have their edges updated regularly and thus they will remain in the model, which enables the proposed model to handle less regular background surfaces. Also, it gives a better solution for the illumination changes, especially for shadows because it creates more nodes near the wining node, which increases the chances of having active nodes corresponding to background surface shifted due to the illumination changes.

Furthermore, it provides a number of advantages compared to traditional methods. Firstly, the GNG based algorithm has the ability to grow and shrink itself according to the data distribution. Therefore, there is no need to specify a number of nodes in advance. Secondly, the inherently adaptive mechanism can model changing data distributions. Thirdly, it has less sensitive parameters [11]. Finally, it has better online clustering capabilities than its competitors. These are the desirable characteristics for online data modeling algorithms. Therefore, it has shown great potential for robust background modeling in complex environments.

However, it requires some data samples for initial training and one the pass learning algorithm is not sufficient to achieve good approximation of the data distribution. Therefore, the initial learning procedure has to be repeated for a number of iterations, which takes time. The GNG-BM gives a processing speed lower than what is required for real-time processing, but it gives better segmentation accuracy than other methods.

5 Conclusion

We proposed a pixel-level background model, where the GNG algorithm is modified for clustering the input pixel data. An algorithm is introduced for initial training, which helps the model to become more stabilized and better approximate the distribution of the training set. Also, a new criteria is introduced to differentiate between foreground and background nodes. Finally, the proposed online method to update the background model helps the model to cover changing distribution of the data. Overall, the proposed model gives more than 20% and 10% improvements for the "dynamicBackground" and the "shadow" category respectively. Although the proposed model achieved significantly better segmentation accuracy than the previous model, it gives a processing speed lower than what is required for real-time processing. However, the structure of the proposed model is inherently parallelizable. We are intending to propose a GPU based model to get better processing speed.

References

1. Bouwmans, T., Baf, F.E., Vachon, B.: Statistical Background Modeling for Foreground Detection: A Survey, pp. 181–199. World Scientific Publishing (2010)
2. Toyama, K., Krumm, J., Brumitt, B., Meyers, B.: Wallflower: Principles and Practice of Background Maintenance. In: Proceedings of the IEEE International Conference on Computer Vision, ICCV 1999, vol. 1, p. 255. IEEE Computer Society (1999)
3. Stauffer, C., Grimson, W.: Adaptive Background Mixture Models for Real-Time Tracking. In: Proceedings of the IEEE Computer Society Conference on Computer Vision and Pattern Recognition, CVPR 1999, vol. 2, pp. 246–252. IEEE Computer Society (1999)
4. KaewTraKulPong, P., Rowden, R.: An improved adaptive background mixture model for real-time tracking with shadow detection. In: Proceedings of the Second European Workshop on Advanced Video Based Surveillance Systems, pp. 149–158 (2001)
5. Shah, M., Deng, J., Woodford., B.: Enhancing the Mixture of Gaussians background model with local matching and local adaptive learning. In: Proceedings of the 27th Conference on Image and Vision Computing New Zealand (IVCNZ 2012), pp. 103–108. ACM (2012)
6. Heikkila, M., Pietikainen, M.: A Texture-Based Method for Modeling the Background and Detecting Moving Objects. IEEE Transaction on Pattern Analysis and Machine Intelligence 28(4), 657–662 (2006)
7. Bouwmans, T., Baf, F.E., Vachon, B.: Background Modeling using Mixture of Gaussians for Foreground Detection - A Survey. Recent Patents on Computer Science 1(3), 219–237 (2008)

8. Fritzke, B.: A Growing Neural Gas Network Learns Topologies. In: Proceedings of the Advances in Neural Information Processing Systems, pp. 625–632. MIT Press (1995)
9. Martinetz, T., Berkovich, S., Schulten, K.: 'Neural-gas' network for vector quantization and its application to time-series prediction. IEEE Transactions on Neural Networks 4(4), 558–569 (1993)
10. Martinetz, T.: Competitive hebbian learning rule forms perfectly topology preserving maps. In: Proceedings of the International Conference on Artificial Neural Networks (ICANN 1993), pp. 427–434. Springer London (1993)
11. Qin, A.K., Suganthan, P.N.: Robust growing neural gas algorithm with application in cluster analysis. Neural Networks 17(8-9), 1135–1148 (2004)
12. Fritzke, B.: A self-organizing network that can follow non-stationary distributions. In: Gerstner, W., Hasler, M., Germond, A., Nicoud, J.-D. (eds.) ICANN 1997. LNCS, vol. 1327, pp. 613–618. Springer, Heidelberg (1997)
13. Evangelio, R.H., Ptzold, M., Sikora, T.: Splitting Gaussians in Mixture Models. In: Proceedings of the 9th IEEE International Conference on Advanced Video and Signal-Based Surveillance, AVSS 2012, pp. 300–305. IEEE Computer Society (2012)
14. Elgammal, A., Duraiswami, R., Harwood, D., Davis, L.: Background and foreground modeling using nonparametric kernel density estimation for visual surveillance. Proceedings of the IEEE 90(7), 1151–1163 (2002)
15. Hofmann, M., Tiefenbacher, P., Rigoll, G.: Background Segmentation with Feedback: The Pixel-Based Adaptive Segmenter. In: Proceedings of the IEEE Conference on Computer Vision and Pattern Recognition -Change Detection Workshop, pp. 38–43. IEEE (2012)
16. Goyette, N., Jodoin, P., Porikli, F., Konrad, J., Ishwar, P.: Changedetection.net: A new change detection benchmark dataset. In: Proceedings of the IEEE Computer Society Conference on Computer Vision and Pattern Recognition (CVPR) -Workshops, pp. 1–8. IEEE Computer Society (2012)

Sparse Principal Component Analysis via Joint $L_{2,1}$-Norm Penalty

Shi Xiaoshuang, Lai Zhihui, Guo Zhenhua*, Wan Minghua, Zhao Cairong, and Kong Heng

Shenzhen Key Laboratory of Broadband Network & Multimedia
Graduate School at Shenzhen, Tsinghua University
Shenzhen, 518055, China
cszguo@gmail.com

Abstract. Sparse principal component analysis (SPCA) is a popular method to get the sparse loadings of principal component analysis(PCA), it represents PCA as a regression model by using lasso constraint, but the selected features of SPCA are independent and generally different with each principal component (PC). Therefore, we modify the regression model by replacing the elastic net with $L_{2,1}$-norm, which encourages row-sparsity that can get rid of the same features in different PCs, and utilize this new "self-contained" regression model to present a new framework for graph embedding methods, which can get sparse loadings via $L_{2,1}$-norm. Experiment on Pitprop data illustrates the row-sparsity of this modified regression model for PCA and experiment on YaleB face database demonstrates the effectiveness of this model for PCA in graph embedding.

Keywords: SPCA, $L_{2,1}$-norm, row-sparsity.

1 Introduction

High dimensionality of data may bring a big issue to science and engineering applications, therefore, many methods have been proposed by scientists to reduce data dimensions [1]. PCA [2] is one popular technique for dimension reduction. It has been used successfully because it can find a range of linearly independent basis by minimizing the loss of information. However, PCA also has obvious drawback such as each of linearly independent basis is constructed by nonzero elements, which make it hard to explain what variables play a major role in the data. In some high-dimensional data analysis, people always want to know what variables play a major role, which variables do not work, and the unimportant variables can be ignored to reduce the workload. In addition, people wish to use the least components to explain data. Yet, PCA cannot solve these problems.

In order to compensate the shortcoming, many methods have been proposed. Rotation techniques are commonly used to help practitioner to interpret principal components [3]; simple principal components which restrict the loadings to

* Corresponding author.

S. Cranefield and A. Nayak (Eds.): AI 2013, LNAI 8272, pp. 148–159, 2013.

take values from a set of candidate integers such as 0, 1 and -1 [4]; SCoTLASS was proposed to get modified principle components with possible zero loadings [5]; sparse principal component analysis (SPCA) [6], one of famous algorithms was also been proposed. In SPCA, PCA is decomposed by the singular value decomposition (SVD) [7] and its objective function is expressed as a "self-contained" regression-type optimization problem, and the lasso [8], [18] with L_1-norm is integrated into the regression criterion, which can lead to the sparse loadings [9]. However, the selected features by SPCA are independent and generally different with each PC.

Recently, $L_{2,1}$-norm has been proposed, which can encourage the loadings row-sparsity [10], [16], each row of projection matrix corresponds to a feature. In addition, researchers have successfully applied $L_{2,1}$-norm in Group Lasso [12], multi-task feature learning [13], joint covariate selection and joint subspace learning [14]. Moreover, $L_{2,1}$-norm also was used in graph embedding [15], which is a famous framework including PCA for reducing the dimension of data [11]. However, the motivations between graph embedding and SPCA are different and their optimization models are also different, thus, the optimization process of SPCA is different from graph embedding.

Based on the above motivation, in this paper, we employ $L_{2,1}$-norm to modify the regression model of SPCA firstly, in which $L_{2,1}$-norm is used to minimize the loss function and regularization instead of using the L_2-norm and L_1-norm. Then, we unify the optimization problem of graph embedding into the new "self-contained" regression model, which can get orthogonal solution of graph embedding, and use this modified model to get the sparse loadings of PCA in graph embedding.

The rest of this paper is organized as following: Section 2 briefly reviews SPCA and graph embedding. In Section 3, we modify the model of SPCA via $L_{2,1}$-norm and unify the optimization model of graph embedding into the new model. Experiments on different datasets are showed in Section 4. Finally, a conclusion is drawn and future work is pointed out in Section 5.

2 A Brief Review of SPCA and Graph Embedding

2.1 Notations and Definitions

For matrix $M = (m_{ij}) \in R^{p \times d}$, m_i and m^j respectively present its i-th row and j-th column. The L_p-norm of vector $v \in R^n$ is defined as $\|v\|_p = (\sum_{i=1}^{n} |v_i|^p)^{\frac{1}{p}}$. The L_1-norm of the matrix is defined as $\|M\|_1 = \sum_{i=1}^{n} \|m_i\|_1$, the L_2-norm of M is defined as $\|M\|_2^2 = \sum_{i=1}^{p} \|m_i\|_2^2$ and $L_{2,1}$-norm of Matrix M is defined as $\|M\|_{2,1} = \sum_{i=1}^{p} \|m_i\|_2$.

2.2 SPCA

Given training data set $X \in R^{n \times p}$, which has n observations with p predictors, the mean of $X_i \in X$ is zero. $Y \in R^{n \times k}$ is the data after dimension reduction,

$\beta \in R^{p\times k}$ is a projection matrix, which can be obtained by the following least squares regression [17]

$$\min_{\beta} \sum_{i=1}^{n} \|X_i\beta - Y_i\|_2^2 \quad (1)$$

It is also viewed as a loss function. By adding a penalty term, it can be written as following [19]

$$\min_{\beta} \sum_{i=1}^{n} \|X_i\beta - Y_i\|_2^2 + \gamma\Phi(\beta) \quad (2)$$

If the penalty term $\Phi(\beta)$ equals $\frac{1}{\gamma}(\lambda_1 \|\beta\|_2^2 + \sum_{j=1}^{k} \lambda_{2,j} \|\beta(:,j)\|_1)$, then above optimization problem is transformed into the following elastic net problem [20]

$$\min_{\beta} \sum_{i=1}^{n} \|X_i\beta - Y_i\|_2^2 + \lambda_1 \|\beta\|_2^2 + \sum_{j=1}^{k} \lambda_{2,j} \|\beta(:,j)\|_1 \quad (3)$$

PCA can be done via singular value decomposition (SVD) of X, $X = UDV^T$, $Y_i = U_iD_i$ is the i-th principal component. Therefore, in SPCA, above regression model is expressed as the following optimization problem [6]

$$(\alpha^*, \beta^*) = arg \min_{\alpha,\beta} \sum_{i=1}^{n} \left\|X_i\beta\alpha^T - X_i\right\|_2^2 + \lambda_1 \|\beta\|_2^2 + \sum_{j=1}^{k} \lambda_{2,j} \|\beta(:,j)\|_1$$
$$s.t.\ \alpha^T\alpha = I \quad (4)$$

Then, $\beta^* = \frac{\beta(:,j)}{|\beta(:,j)|}$, for $j = 1, 2, \cdots, k$. $\alpha \in R^{p\times k}$.
This formula successfully integrates PCA into a regression type problem. By solving this model, the sparse projection matrix β can be obtained.

2.3 Graph Embedding

Graph embedding can be seen as a general framework, which can interpret many dimensionality reduction methods [15]. In graph embedding, a data graph G can be constructed, whose vertices correspond to $X = \{x_1, x_2, \cdots, x_n\} \in R^{p\times n}$. $W \in R^{n\times n}$ is a symmetric matrix to measure the similarity of any two different vertices. The purpose of graph embedding is to find low-dimensional vector $Y = \{y_1, y_2, \cdots, y_n\}^T \in R^{n\times k}$ representing vertices, for the linear dimensional reduction, that is, $X^TA = Y$, $A \in R^{p\times k}$, the optimal A is given by the following optimization problem

$$\min_{A} Tr\, A^TXLX^TA$$
$$s.t.\ A^TXDX^TA = I \quad (5)$$

where $D_{ii} = \sum_{i=1}^{n} W_{ij}$ is a diagonal matrix, $L = D - W$ is Laplacian matrix.

Different linear dimensional reduction methods have different W in graph embedding framework. For PCA, supposing the intrinsic graph connecting all the vertices with equal weights and constrained by scale normalization on the projector vector, W is defined as

$$W_{ij} = \begin{cases} \frac{1}{n} & i \neq j \\ 0 & otherwise \end{cases} \tag{6}$$

The first optimization model (5) with $L_{2,1}$-norm has been solved by transforming it into two following optimization problems [11]:

$$\begin{aligned} &\min_{Y} Tr\, Y^T LY \\ &s.t.\, Y^T DY = I \end{aligned} \tag{7}$$

$$\begin{aligned} &\min_{A} \|A\|_{2,1} \\ &s.t.\; \left\|X^T A - Y\right\|_2^2 \leq \delta^2 \end{aligned} \tag{8}$$

By using (6) to get W and obtaining L based on $L = D - W$, then substituting L into (7) and (8), the sparse loadings of PCA can be got. In this paper, this method using graph embedding for sparse PCA via joint $L_{2,1}$-norm is called $L_{2,1}$GSPCA.

3 Sparse Principal Component Analysis via Joint $L_{2,1}$-Norm

3.1 Sparse PCA via Joint $L_{2,1}$-Norm

In SPCA, the number of zero loadings of projection matrix is controlled by L_1-norm, but the zero loadings in each PC are different, which means the selected features in each PC are independent. Meanwhile, since each row of projection matrix corresponds to a feature in data matrix and $L_{2,1}$-norm can make some rows of projection matrix zeros. In addition, the coefficient λ_1 of L_2-norm in (4) does not influence the correlation between β^* and V in (4) after normalization of β^* [6]. These motivates us modify the regression model of SPCA. By using $\frac{\lambda}{\gamma}\|\beta\|_{2,1}$ to replace $\Phi(\beta)$ of (2), then the model becomes as following:

$$\min_{\beta} \sum_{i=1}^{n} \|X_i\beta - Y_i\|_2^2 + \lambda \|\beta\|_{2,1} \tag{9}$$

Therefore, formula (4) can be translated into the following optimization problem

$$\begin{aligned} (\alpha^*, \beta^*) = arg \min_{\alpha,\beta} \textstyle\sum_{i=1}^{n} \left\|X_i\beta\alpha^T - X_i\right\|_2^2 + \lambda \|\beta\|_{2,1} \\ s.t.\; \alpha^T\alpha = I \end{aligned} \tag{10}$$

Then $\beta^* = \frac{\beta(:,j)}{|\beta(:,j)|}$, for $j = 1, 2, \cdots, k$. $\alpha \in R^{p\times k}$.
From (10), it can get

$$\begin{aligned} &\textstyle\sum_{i=1}^{n} \left\|X_i\beta\alpha^T - X_i\right\|_2^2 + \lambda \|\beta\|_{2,1} \\ &= TrX^TX + Tr\beta^T(X^TX + \lambda G)\beta - 2Tr\alpha^T X^T X\beta \end{aligned} \tag{11}$$

where G is a diagonal matrix and the i-th diagonal element is

$$g_{ii} = \begin{cases} 0 & if\ \beta_i = 0 \\ \frac{1}{2\|\beta_i\|_2^{\frac{1}{2}}} & otherwise \end{cases} \tag{12}$$

For fixed α, the formula (11) is minimized at

$$\beta = (X^T X + \lambda G)^{-1} X^T X \alpha \tag{13}$$

Substituting (13) into (11), we get

$$\begin{aligned} \max_{\alpha} Tr\, \alpha^T X^T X (X^T X + \lambda G)^{-1} X^T X \alpha \\ s.t.\ \alpha^T \alpha = I \end{aligned} \tag{14}$$

From (13) and (14), we can see that if $\lambda = 0$, $\alpha = \beta = V$.
Based on the Theorem 3 and 4 in SPCA [6], the solution of α can be obtained by SVD of $X^T X \beta$, $X^T X \beta = U D V^T$, the solution of α is

$$\alpha = U V^T. \tag{15}$$

In summary, we present the algorithm for optimizing (10) in Algorithm 1.

Algorithm 1. Sparse principal components analysis via joint $L_{2,1}$-norm ($L_{2,1}$SPCA)

Initialize: $G_0 = I$, $t = 0$;
α_0 start at $V(:, 1 : k)$, the loadings of first k principal components.
Repeat
  Given fixed α_t, compute β_{t+1} by $(X^T X + \lambda G_t)^{-1} X^T X \alpha_t$;
  For each fixed β_{t+1} , do the SVD of $X^T X \beta_{t+1} = U D V^T$, then $\alpha_{t+1} = U V^T$;
  Compute G_{t+1} based on β_{t+1};
  $t = t + 1$;
until β converge
Normalization: $\beta^*_{:,j} = \frac{\beta_{:,j}}{|\beta_{:,j}|}$, $j = 1, 2, \cdots, k$.

3.2 Time Complexity

In Algorithm 1, the first step is to compute the matrix $\sum = X^T X$ requires np^2 operations. Then the computational cost of G is $O(pk)$, and the inverse of the matrix $\hat{\sum} = \sum + \lambda G$ is of order $O(p^3)$. In addition, computing $X^T X \alpha$ and $X^T X \beta$ need $p^2 k$ and pJk operations respectively, J is the number of nonzero loadings. The SVD of $X^T X \beta$ is of order $O(pJk)$. Generally, k and J is smaller than n and p, therefore, if $p > n$, the total computational cost is $mO(p^3)$, otherwise, it is $mO(np^2)$, where m is the number of iterations before convergence.

3.3 Joint $L_{2,1}$-Norm Orthogonal Regression in Graph Embedding

Except the objection function (5), the objection function to get the optimal A in graph embedding also can be written as following:

$$\begin{aligned} &\min_{A} Tr\, A^T XLX^T A \\ &s.t.\ A^T A = I \end{aligned} \tag{16}$$

The optimization model (16) is different from (5) and we solve it by "self-contained" regression type in this section.

Theorem 1: Denote $\forall$ matrix X, X_i is its i-th row vector. If $X^T X = S$, for any λ, let

$$\begin{aligned} (\alpha^*, \beta^*) = arg \min_{\alpha,\beta} \textstyle\sum_{i=1}^{n} \left\| X_i \beta \alpha^T - X_i \right\|_2^2 + \lambda \left\| \beta \right\|_2^1 \\ s.t.\ \alpha^T \alpha = I \end{aligned}$$

Then, for any X with product S, the PCs are invariable.

Proof: Based on (13) and $X^T X = S$. Then, the solution of β can be written as

$$\beta = (S + \lambda G)^{-1} S\alpha \tag{17}$$

In addition, based on $X^T X = S$, (14) can be expressed as following

$$\begin{aligned} &\max_{\alpha} Tr\, \alpha^T S (S + \lambda G)^{-1} S\alpha \\ &s.t.\ \alpha^T \alpha = I \end{aligned} \tag{18}$$

Therefore, the solution of α can be obtained by SVD of $S\beta$, $S\beta{=}UDV^T$, then $\alpha{=}UV^T$.

Theorem 2: For any matrix X and L, if L is a real symmetric positive definite matrix, let

$$H = X^T LX$$

Then, $\exists$ matrix $\Phi(X)$, such that $\Phi(X^T)\Phi(X) = H$.

Proof: Due to L is a real symmetric positive definite matrix, thus, based on the Cholesky Decomposition [21]:
$L = RR^T$.
Therefore, $XLX^T{=}XRR^T X^T$.
Let $\Phi(X) = R^T X^T$, then
$XLX^T{=}\Phi(X)^T \Phi(X) = H$.

Based on Theorem 1 and Theorem 2, by using the model of $L_{2,1}$SPCA, we write (16) as the following "self-contained" regression model

$$\begin{aligned} (\alpha^*, \beta^*) = arg \min_{\alpha,\beta} \textstyle\sum_{i=1}^{n} \left\| \Phi(X_i) \beta \alpha^T - \Phi(X_i) \right\|_2^2 + \lambda \left\| \beta \right\|_2^1 \\ s.t.\ \alpha^T \alpha = I \end{aligned} \tag{19}$$

Then $\beta^* = \frac{\beta(:,j)}{|\beta(:,j)|}$, for $j = 1, 2, \cdots, k$.

Solving (19), the solution of β is similar with (17) and can be written as $\beta = (H + \lambda G)^{-1} H\alpha$.

The solution of β can be obtained by SVD of $H\beta$. H$\beta = UDV^T$, the solution of α is $\alpha = UV^T$.

In summary, this algorithm to solve the optimization problem of (16) can be presented in Algorithm 2.

Algorithm 2. Joint $L_{2,1}$-norm orthogonal regression in graph embedding ($L_{2,1}$ORGE)

Initialize: $XLX^T = H$, $G_0 = I$, $t = 0$;
α_0 start at $V(:, 1 : k)$, the loadings of first k principal components.
Repeat
  Given fixed α_t, compute β_{t+1} by $(H + \lambda G_t)^{-1} H\alpha_t$;
  For each fixed β_{t+1} , do the SVD of $H\beta_{t+1}$=UDV^T, then $\alpha_{t+1} = UV^T$;
  Compute G_{t+1} based on β_{t+1};
  t=t+1;
until β converge
Normalization: $\beta^*_{:,j} = \frac{\beta_{:,j}}{|\beta_{:,j}|}$, $j = 1, 2, \cdots, k$.

For obtaining the sparse loadings of PCA, use (6) to get W and obtain L by $L = D - W$, then substitute L into Algorithm 2. As a side note, (16) is a criterion for minimum, but SCPA is opposite, it should be noted for β^* in the selection in Algorithm 2.

4 Experiments

4.1 Pitprop Data

Pitprop data is a classic data set for PCA analysis. It was first used by ScoTLASS [5] and also used in SPCA. In order to compare with SPCA easily, we also use this data set and try to interpret the first 6 PCs to evaluate the performance of our algorithm.

In the model of $L_{2,1}$SPCA, the penalty coefficient λ can influence the performance including the variance and the number of zero loadings, which can be showed in Fig. 1. Fig. 1(a) displays the variance changes of 6 PCs with different λ and Fig. 1(b) shows the total number of zero loadings changes of 6 PCs (the total number of zero loadings is the sum of the number of each PC's zero loadings) with different λ. All the variance of 6 PCs are not changing until λ near 3.4 in Fig. 1(a), correspondingly, the total number of zero loadings is increasing in Fig. 1(b). When λ is from 3.4 to 5.3, all the variance of 6 PCs are nearly still, but the total number of zero loadings is still slightly increasing till largest. As λ is larger than 5.3, all variance of 6 PCs increase and reach stable quickly, but the total number of zero loading decreases till nearly zero. Therefore, the suitable value of λ is between 3.4 and 5.3.

Table 1 displays the sparse effect of $L_{2,1}$SPCA, in which the absolute value of loadings less than 0.01 is set to 0. To better evaluate the sparse effect by

$L_{2,1}$SPCA, the sparse result of two famous sparse algorithms ScoTLASS and SPCA are showed in Table 2, in which all of them used the same low limitation as Table 1. From Table 2, it can be seen that SPCA has the strong performance on sparse loadings but weak on row-sparsity, on the contrary, $L_{2,1}$SPCA has strong performance on row-sparsity. In detail, it can be seen that 6 features do not work at all and three features, which include testsg, bowmax and bowdist, work little in 6 PCs in Table 1. It shows the strong performance of $L_{2,1}$SPCA on explaining the features which play an important role in total variance.

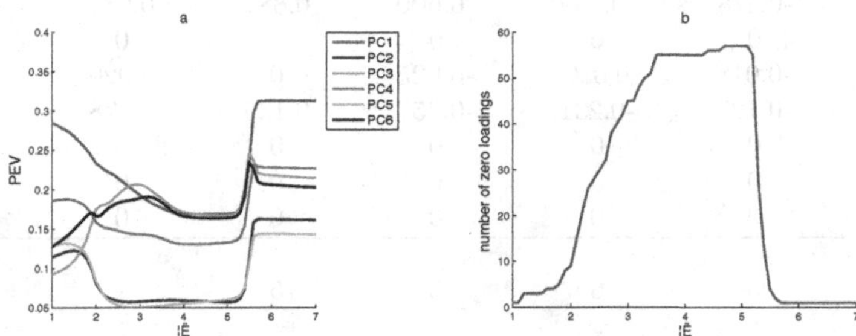

Fig. 1. Pitprops data: the sequence of sparse approximations to the first 6 principal components. (a) The percentage of explained variance (PEV) by different λ (b)The total number of zero loadings by different λ.

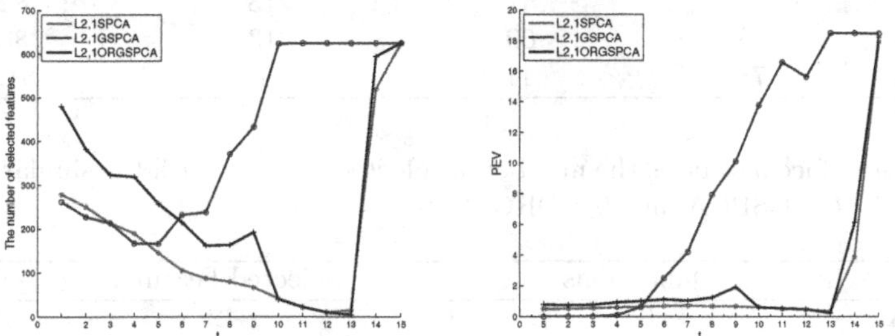

Fig. 2. YaleB face database: the correlation among the number of selected features, PEV and λ. (a) The number of selected features by different λ (b)The percentage of explained variance (PEV) by different λ, $\lambda = 0.1 * 2^t$.

4.2 YaleB Face Database

Besides using $L_{2,1}$SPCA to obtain the row-sparsity of loadings, there are still two situations including (5) and (16) in graph embedding via $L_{2,1}$-norm, we name the model (5) via $L_{2,1}$-norm for sparse PCA as $L_{2,1}$GSPCA and the model (16) via

Table 1. Pitprops data: loadings of the first 6 modified PCs by $L_{2,1}$SPCA

$\lambda = 4$						
Variable	PC1	PC2	PC3	PC4	PC5	PC6
topdiam	-0.618	0.124	-0.472	0.289	-0.064	0.118
length	-0.611	0.222	-0.537	0.332	0.029	0.162
moist	0	0	0	0	0	0
testsg	0	0	0	0	-0.019	-0.014
ovensg	0	0	0	0	0	0
ringtop	0	-0.019	0	0.022	0.013	-0.025
ringbut	-0.478	-0.939	-0.600	0.887	0.981	-0.969
bowmax	0	0	0	0	0	0
bowdist	-0.018	0.030	-0.025	0	0.020	0
whorls	-0.127	-0.231	-0.357	0.135	0.178	-0.140
clear	0	0	0	0	0	0
knots	0	0	0	0	0	0
diaknot	0	0	0	0	0	0
Number of nonzeros loadings	5	6	5	5	7	6
Variance	20.00	7.23	21.83	15.62	9.48	7.27

Table 2. Pitprops data: performance of the first 6 modified PCs by ScoTLASS, SPCA and $L_{2,1}$SPCA

Algorithm	Total loadings	Zero loadings	Selected Features	Total Variance
ScoTLASS	78	37	13	82.8
SPCA	78	60	13	80.5
$L_{2,1}$**SPCA**	78	44	7	81.43

Table 3. YaleB face database: the number of selected features with the similar variance by $L_{2,1}$SPCA, $L_{2,1}$GSPCA and $L_{2,1}$ORGSPCA

Algorithm	Dimensions	t	Selected Features	Variance
$L_{2,1}$**SPCA**	64	10	42	0.6051
L21GSPCA	64	5	167	0.6265
$L_{2,1}$**ORGSPCA**	64	13	40	0.5952

$L_{2,1}$-norm for sparse PCA as $L_{2,1}$GSVDSPCA in this paper. To compare their sparse performance, we use the face images of YaleB database, there are including 38 human subjects with 65 face images of each person and the resolution of each image is 25*25, which means each face image can be represented by a 625-dimensional vector. The correlation among the selected features, PEV and the parameter λ is showed in Fig. 2, in which the 625-dimensional vector was reduced to 64.

In Fig. 2, λ is growing exponentially, thus, we use t to represent it, $\lambda = 0.1 * 2^t$. Based on Fig. 2, it shows the different λ can lead to huge difference between the results, which means the suitable value of λ is very important. The smallest number of selected features is 10, 167 and 5 in $L_{2,1}$SPCA, $L_{2,1}$GSPCA and $L_{2,1}$ORGSPCA when t is 12, 5 and 13 respectively, which suggests that the $L_{2,1}$ORGSPCA can obtain the smallest number of selected features among three methods. However, we also can see that the number of selected features of $L_{2,1}$SPCA is smaller than $L_{2,1}$ORGSPCA before t reaching 10 in Fig. 2(a). Moreover, the PEV of $L_{2,1}$SPCA is more stable than $L_{2,1}$ORGSPCA in Fig. 2(b) before t growing to 10. Therefore, the overall performance of $L_{2,1}$SPCA is better than $L_{2,1}$ORGSPCA.

As the number of selected features is the smallest in $L_{2,1}$GSPCA, its variance is 0.6265. To better compare these three algorithms' performance, we choose the similar variance for $L_{2,1}$SPCA and $L_{2,1}$ORGSPCA, and the comparison result is showed in Table 3, which infers that $L_{2,1}$SPCA and $L_{2,1}$ORGSPCA have smaller number of selected features than $L_{2,1}$GSPCA as their variance are similar. To be clearer, the selected features represented by white pixel in face image are showed in Fig. 3. Similar phenomena can be observed when the 625-dimensional vector is reduced to other dimensions which are smaller than 64. For the space limit, we do not show them.

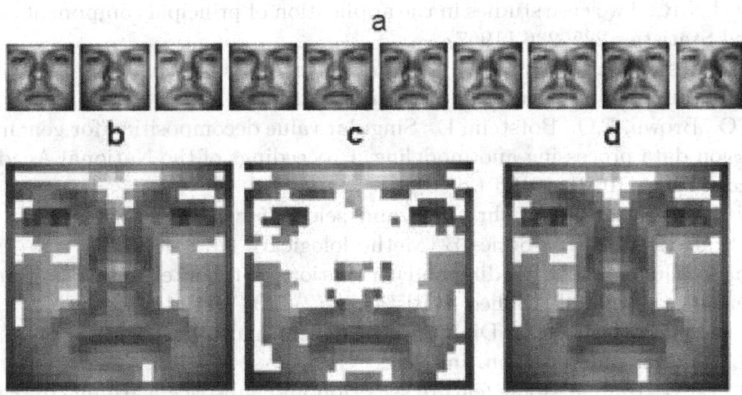

Fig. 3. YaleB face database: the selected features of different sparse algorithms. (a) Original images (b) $L_{2,1}$SPCA (c) $L_{2,1}$GSPCA (d) $L_{2,1}$ORGSPCA.

5 Conclusion

In this paper, we propose two methods $L_{2,1}$SPCA and $L_{2,1}$ORGE. $L_{2,1}$SCPA is obtained by modifying the regression model of SPCA through replacing the penalty term L_2-norm and L_1-norm with $L_{2,1}$-norm, and $L_{2,1}$ORGE is got via this modified model to unify the optimization problem of graph embedding for obtaining the sparse loadings. Experiment on Pitprop data illustrates $L_{2,1}$SPCA can get row-sparse loadings of PCA and experiment on Yale database demonstrates the effectiveness of $L_{2,1}$ORGSPCA, which is one application of $L_{2,1}$ORGE, for obtaining row-sparse loadings of PCA. However, there are still some problems unsolved, for instance, how to choose a stable λ to get the sparest solution. In our future work, we should continue to study these problems and apply $L_{2,1}$ORGE on other methods of graph embedding.

Acknowledgments. The work was supported by Nature Science Foundation of China (NSFC), No. 61101150, 61203376 and 61375012, and Shenzhen research fund: JCYJ20120831165730901.

References

1. Yang, J., Zhang, D., Frangi, A.F., Yang, J.Y.: Two-dimensional, P.C.A.: a new approach to appearance-based face representation and recognition. IEEE Transactions on Pattern Analysis and Machine Intelligence 26(1), 131–137 (2004)
2. Jolliffe, I.: Principal component analysis. John Wiley & Sons Ltd. (2005)
3. Richman, M.B.: Rotation of principal components. Journal of Climatology 6(3), 293–335 (1986)
4. Vines, S.K.: Simple principal components. Journal of the Royal Statistical Society: Series C (Applied Statistics) 49(4), 441–451 (2000)
5. Jeffers, J.N.R.: Two case studies in the application of principal component analysis. Applied Statistics 225–236 (1967)
6. Zou, H., Hastie, T., Tibshirani, R.: Sparse principal component analysis. Journal of Computational and Graphical Statistics 15(2), 265–286 (2006)
7. Alter, O., Brown, P.O., Botstein, D.: Singular value decomposition for genome-wide expression data processing and modeling. Proceedings of the National Academy of Sciences 97(18), 10101–10106 (2000)
8. Tibshirani, R.: Regression shrinkage and selection via the lasso. Journal of the Royal Statistical Society. Series B (Methodological), 267–288 (1996)
9. Cadima, J., Jolliffe, I.T.: Loading and correlations in the interpretation of principle compenents. Journal of Applied Statistics 22(2), 203–214 (1995)
10. Nie, F., Huang, H., Cai, X., Ding, C.H.: Efficient and robust feature selection via joint $L_{2,1}$-norms minimization. In: Proc. NIPS, pp. 1813–1821 (2010)
11. Gu, Q., Li, Z., Han, J.: Joint feature selection and subspace learning. In: Proceedings of the Twenty-Second International Joint Conference on Artificial Intelligence, vol. 2. AAAI Press (2011)
12. Yuan, M., Lin, Y.: Model selection and estimation in regression with grouped variables. Journal of the Royal Statistical Society: Series B (Statistical Methodology) 68(1), 49–67 (2006)
13. Argyriou, A., Evgeniou, T., Pontil, M.: Convex multi-task feature learning. Machine Learning 73(3), 243–272 (2008)
14. Obozinski, G., Taskar, B., Jordan, M.I.: Joint covariate selection and joint subspace selection for multiple classification problems. Statistics and Computing 20(2), 231–252 (2010)
15. Yan, S., Xu, D., Zhang, B., Zhang, H.J., Yang, Q., Lin, S.: Graph embedding and extensions: a general framework for dimensionality reduction. IEEE Transactions on Pattern Analysis and Machine Intelligence 29(1), 40–51 (2007)

16. Hou, C., Nie, F., Yi, D., Wu, Y.: Feature selection via joint embedding learning and sparse regression. In: Proceedings of the Twenty-Second International Joint Conference on Artificial Intelligence, vol. 2. AAAI Press (2011)
17. Efron, B., Hastie, T., Johnstone, I., Tibshirani, R.: Least angle regression. The Annals of Statistics 32(2), 407–499 (2004)
18. Journe, M., Nesterov, Y., Richtrik, P., Sepulchre, R.: Generalized power method for sparse principal component analysis. The Journal of Machine Learning Research 11, 517–553 (2010)
19. Jenatton, R., Audibert, J.Y., Bach, F.: Structured variable selection with sparsity-inducing norms. The Journal of Machine Learning Research 12, 2777–2824 (2011)
20. Zou, H., Hastie, T.: Regression shrinkage and selection via the elastic net, with applications to microarrays. Journal of the Royal Statistical Society: Series B 67, 301–320 (2003)
21. Trefethen, L.N., Bau III, D.: Numerical linear algebra. SIAM (1997)

Image Segmentation with Adaptive Sparse Grids

Benjamin Peherstorfer[1], Julius Adorf[1],
Dirk Pflüger[2], and Hans-Joachim Bungartz[1]

[1] Department of Informatics, Technische Universität München, Germany
[2] Institute for Parallel and Distributed Systems, University of Stuttgart, Germany

Abstract. We present a novel adaptive sparse grid method for unsupervised image segmentation. The method is based on spectral clustering. The use of adaptive sparse grids achieves that the dimensions of the involved eigensystem do not depend on the number of pixels. In contrast to classical spectral clustering, our sparse-grid variant is therefore able to segment larger images. We evaluate the method on real-world images from the Berkeley Segmentation Dataset. The results indicate that images with 150,000 pixels can be segmented by solving an eigenvalue system of dimensions 500×500 instead of $150,000 \times 150,000$.

Keywords: sparse grids, image segmentation, out-of-sample extension.

1 Introduction

In unsupervised image segmentation, the objective is to partition images into disjunct regions. In this paper, we focus on segmenting natural images from the Berkeley Segmentation Dataset [1]. Our key idea is to combine the strengths of both spectral clustering and adaptive sparse grids. This is accomplished by adapting a recent clustering method [2] to the image segmentation problem.

Spectral clustering methods often outperform classical methods such as k-means [3], but they are computationally expensive. For an image with M pixels, the computational complexity for solving the involved generalized eigenproblem is in $O(M^3)$. To make things worse, the number of pixels M grows quadratically with the image resolution.

In our method, we address this problem by employing a so-called *out-of-sample extension* based on sparse grids [2]. We learn sparse grid functions by solving an eigensystem whose dimensions depend on the number of sparse grid points rather than the number of image pixels. The sparse grid functions are evaluated for all image pixels to produce the input for a final k-means clustering step.

The proposed method is related to the two out-of-sample extensions presented in [4,5]. Fowlkes et al. [4] use the Nyström method, where eigenfunctions are learned directly on a subsample of the image pixels. As an alternative, Alzate and Suykens [5] formulate spectral clustering in terms of weighted kernel PCA. The principal components are computed for a subsample. The remaining out-of-sample pixels can be projected onto these principal components. Both methods [4,5] reduce computational costs in comparison to classical spectral

S. Cranefield and A. Nayak (Eds.): AI 2013, LNAI 8272, pp. 160–165, 2013.

clustering but have the disadvantage that the dimensions of the involved eigensystems still depend on the number of sample pixels. This motivates our choice to use adaptive sparse grids.

2 Spectral Clustering

Let $S = \{x_1, \ldots, x_M\} \subset \mathbb{R}^d$ be the data set that we want to cluster with spectral clustering. We define a weighted graph $G = (S, E)$, where the weight W_{ij} of the edge connecting the data point x_i with x_j is a measure for the similarity between the two data points. Spectral clustering minimizes the so-called normalized cut between the different partitions. For that we solve the generalized eigenproblem $Ly = \lambda Dy$ with where $L = D - W$ is the so-called graph Laplacian with the diagonal matrix $D_{ii} = \sum_j W_{ij}$, see [6]. The components $\{y_1, \ldots, y_M\}$ of the solution vector y are a low-dimensional embedding of the data points in S. The advantage of this low-dimensional embedding is that the cluster structure becomes more evident. That is why the components $\{y_1, \ldots, y_M\}$ are clustered with k-means to determine the cluster assignment of the data points. We refer to [3,6] for more details.

3 Sparse-Grid-Based Out-of-Sample Extension

In this section, we describe how to formulate the classical spectral clustering problem with sparse grid functions [2]. The sparse-grid-based out-of-sample extension provides an explicit clustering model because once the sparse grid function is learned, it can be evaluated on the whole dataset. Thus, it allows us to solve the spectral clustering eigenproblem for a couple of training points only and to interpolate cheaply the cluster assignment for all other points.

We represent a function $f(x) = \sum_{i=1}^{N} \alpha_i \phi_i(x)$ as a linear combination of basis functions $\phi_1, \ldots, \phi_N$ stemming from a grid with N grid points. The costs for evaluating such a function are independent from the number of data points M and depend linearly on the number of grid points N. Unfortunately, a straightforward discretization with 2^l grid points in each direction suffers from the curse of dimensionality – the number of grid points would increase exponentially with the number of dimensions d. That is why we employ so-called sparse grids that consist of $\mathcal{O}(2^l d^{l-1})$ grid points instead of $\mathcal{O}(2^{ld})$ grid points as with ordinary discretizations, see Fig. 1. Under certain smoothness assumptions, the sparse grid discretizations leads to similar accuracies as the classical discretization but with orders of magnitude fewer grid points [7]. To further reduce the number of sparse grid points, we can employ adaptivity and tailor the grid structure to the current problem at hand [8,7]. To achieve that we require an error indicator or adaptivity criterion. In Sect. 4 we will introduce an error indicator that is well-suited for image segmentation.

For our sparse-grid-based out-of-sample extension, we want to construct a sparse grid function f that approximates the eigenfunctions corresponding to the eigenproblem $Ly = \lambda Dy$ of the previous section. This is achieved by solving

Fig. 1. On the left a full grid, in the middle a regular sparse grid, and on the right an adaptive sparse grid [8]

$B^T L B\alpha = \lambda B^T D B\alpha$ where α is the vector containing the hierarchical coefficients of f and the matrix $B \in \mathbb{R}^{M \times N}$ has the entries $B_{ij} = \phi_j(x_i)$, see [2] for details. Note that we can evaluate the sparse grid function at the data points in $S = \{x_1, \ldots, x_M\}$ with the matrix-vector product $B \cdot \alpha$. It is important that the dimensions of the eigenproblem are $N \times N$ where N is the number of sparse grid points. Thus, in contrast to other out-of-sample extensions [4,5,9], the dimensions are completely independent from the number of data points M.

4 Image Segmentation

The proposed sparse-grid-based image segmentation algorithm consists of extracting features, setting up and solving the eigensystem, optional adaptive refinement, and a final clustering step with k-means (see Fig. 2).

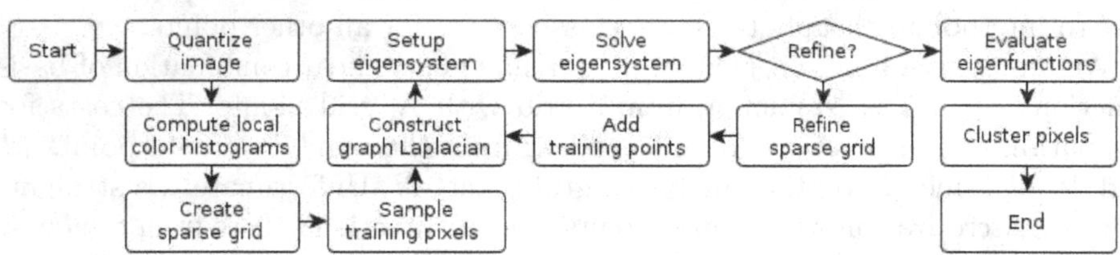

Fig. 2. Flowchart of the image segmentation algorithm with sparse grids. This chart provides a rough overview on how an image is preprocessed, how spectral decomposition on a sparse grid is performed in one or more iterations, and how the final clustering is obtained.

Constructing the graph Laplacian for a subsample of M training pixels is the first step. As in [4,5], we use local color histograms as pixel features. These are computed within a 5×5 neighborhood on an image reduced to C colors with minimum-variance quantization [10]. The value h_{ic} denotes the c-th bin value for the i-th pixel. We define the distance t_{ij} between the i-th and the j-th pixel as

$$t_{ij} = \frac{1}{2} \sum_{c=1}^{C} \frac{(h_{ic} - h_{jc})^2}{h_{ic} + h_{jc}}$$

A radial basis function provides the weights $w_{ij} = \exp\left(-t_{ij}/\sigma^2\right)$ for the graph Laplacian. Then the spectral clustering eigensystem is solved. In order to achieve this, we use a combination of two freely available software libraries: the `SG++` [8] toolbox for sparse grids, and `GSL`[1] for solving the eigensystem. Only the first $k-1$ eigenvectors $\{\alpha^{(r)}\}_{r=1}^{k-1}$ are taken into account, where k is the number of desired clusters. Each eigenvector contains the *hierarchical coefficients* of one sparse grid function. Evaluating the $k-1$ sparse grid functions on all image pixels yields the input for the final k-means clustering.

With the optional adaptive refinement, the choice of sparse grid and the sampling strategy can be adapted to the image. We use the hierarchical coefficients as an error indicator [7]. For each of the $k-1$ sparse grid functions defined by the hierarchical coefficients $\alpha^{(r)}$, we refine a fixed percentage of grid points with the largest hierarchical coefficient. Refining a grid point here means that four grid points are added within its support. Adding more grid points increases the degrees of freedom in the sparse grid. Therefore, we also extend the set of training pixels by sampling an additional fixed number of pixels uniformly from the whole image.

Fig. 3. Five images (top row) from the Berkeley Segmentation Dataset 300. The corresponding segmentations (bottom row) have been obtained on regular sparse grids with levels 7, 8, 7, 7, and 7 (from left to right), and 8000 training pixels each.

5 Results

We show qualitative results on five images of the Berkeley Segmentation Dataset 300 [1], which has also been used for evaluation in [4,5]. The five images are shown in the top row of Fig. 3, and are coined *elephant*, *bird*, *pyramids*, *surfer*, and *parade* for easier reference. The segmentation results obtained with our method are given in the bottom row of Fig. 3. We discuss results on regular sparse grids first and evaluate adaptive sparse grids next. Note that throughout the evaluation, we use linear boundaries for the sparse grids. A more detailed treatment of the choice of parameters can be found in [11].

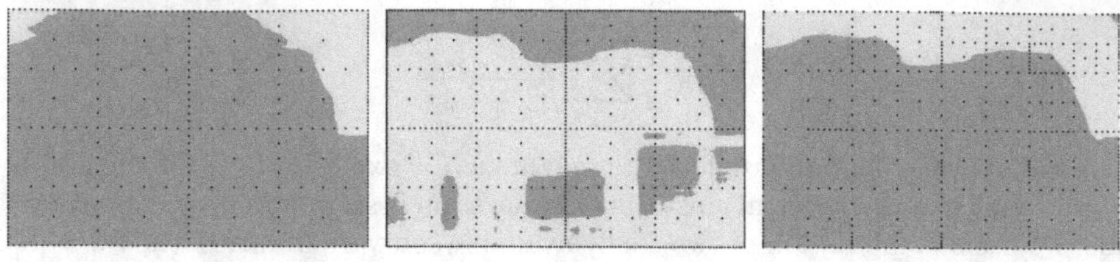

Fig. 4. Three segmentations of image *elephant*, obtained with regular sparse grids of level 6 (left) and level 7 (middle), and with an adaptive sparse grid (right)

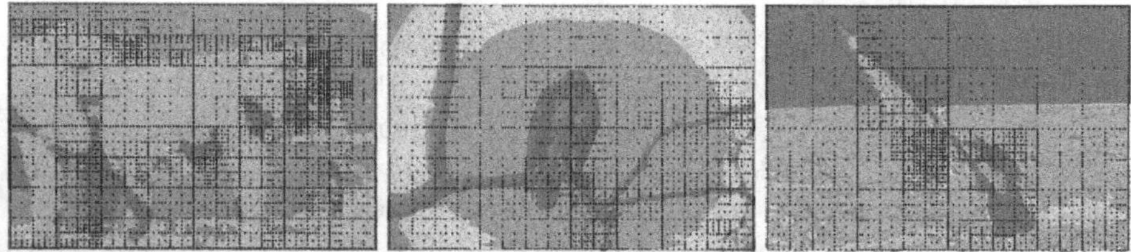

Fig. 5. Three segmentations obtained on adaptive sparse grids. They illustrate that the grids are adaptively refined mostly in regions containing natural boundaries.

The choice of the sparse grid, together with the similarity measure, decide about the amount of details in an image that can be resolved by the segmentation method. Figure 4 shows the segmentations of *elephant* for regular sparse grids (see Fig. 1) of level 6 and 7. The regular sparse grid at level 6 is not sufficient to obtain good segmentation boundaries. However, we obtain satisfactory results already with the regular sparse grid of level 7 (833 grid points).

Figure 5 shows segmentations on adaptive sparse grids for *elephant*, *bird*, and *surfer*. Only few grid points fall into relatively homogeneous regions, but many grid points are spent near natural contours in the images. This is desirable because sparse grid points are only placed in image regions that require higher resolution.

Adaptive sparse grids can achieve similar segmentations as regular sparse grids with considerably fewer grid points. Figure 4 (right) shows the adaptive segmentation of *elephant* with 499 grid points, achieving a similar result as with a regular sparse grid of level 7 (833 grid points). At the same time, the adaptive sparse grid clearly improves upon the regular sparse grid of level 6 (385 grid points).

6 Conclusion

We employed a sparse-grid-based out-of-sample extension to segment images of the Berkeley Segmentation Dataset with spectral clustering. The advantage of

[1] `http://www.gnu.org/software/gsl`

our approach is that the dimensions of the corresponding eigensystem do not dependent on the number of data points.

The results indicate that even in the absence of a sophisticated similarity measure, we are able to segment real-world images. We showed how the segmentation is affected by the choice of grid level and concluded that by using adaptive sparse grids, as few as 500 grid points can be sufficient. This confirms that the adaptivity criterion indeed refines only near cluster boundaries where many grid points are required to achieve a sufficient approximation and that only few grid points are spent in homogeneous regions of the images. Overall, for the presented examples, we can reduce the dimension of the eigensystem involved in the spectral clustering method from about 150,000 to 500 only.

References

1. Martin, D., Fowlkes, C., Tal, D., Malik, J.: A Database of Human Segmented Natural Images and its Application to Evaluating Segmentation Algorithms and Measuring Ecological Statistics. In: Proceedings of the 2001 IEEE International Conference on Computer Vision, vol. 2, pp. 416–423 (2001)
2. Peherstorfer, B., Pflüger, D., Bungartz, H.-J.: A Sparse-Grid-Based Out-of-Sample Extension for Dimensionality Reduction and Clustering with Laplacian Eigenmaps. In: Wang, D., Reynolds, M. (eds.) AI 2011. LNCS, vol. 7106, pp. 112–121. Springer, Heidelberg (2011)
3. von Luxburg, U.: A Tutorial on Spectral Clustering. Statistics and Computing 17(4), 395–416 (2007)
4. Fowlkes, C., Belongie, S., Chung, F., Malik, J.: Spectral Grouping Using the Nystrom Method. IEEE Transactions on Pattern Analysis and Machine Intelligence 26(2), 214–225 (2004)
5. Alzate, C., Suykens, J.A.K.: Multiway Spectral Clustering with Out-of-Sample Extensions through Weighted Kernel PCA. IEEE Transactions on Pattern Analysis and Machine Intelligence 32(2), 335–347 (2010)
6. Belkin, M., Niyogi, P.: Laplacian eigenmaps for dimensionality reduction and data representation. Neural Computation 15(6), 1373–1396 (2003)
7. Bungartz, H.J., Griebel, M.: Sparse grids. Acta Numerica 13, 147–269 (2004)
8. Pflüger, D.: Spatially Adaptive Sparse Grids for High-Dimensional Problems. Verlag Dr. Hut, München (August 2010)
9. Bengio, Y., Paiement, J.F., Vincent, P.: Out-of-sample extensions for LLE, isomap, MDS, eigenmaps, and spectral clustering. In: Advances in Neural Information Processing Systems, pp. 177–184. MIT Press (2003)
10. Wu, X.: Efficient Statistical Computations for Optimal Color Quantization. In: Arvo, J. (ed.) Graphics Gems II, pp. 126–133. Academic Press (1991)
11. Adorf, J.: Nonlinear Clustering on Sparse Grids. Studienarbeit/SEP/IDP, Institut für Informatik, Technische Universität München (August 2012)

Diversify Intensification Phases in Local Search for SAT with a New Probability Distribution

Thach-Thao Duong, Duc-Nghia Pham, and Abdul Sattar

Queensland Research Laboratory, NICTA and
Institute for Integrated and Intelligent Systems, Griffith University,
QLD, Australia
{t.duong,d.pham,a.sattar}@griffith.edu.au

Abstract. A key challenge in developing efficient local search solvers is to intelligently balance diversification and intensification. This study proposes a heuristic that integrates a new dynamic scoring function and two different diversification criteria: variable weights and stagnation weights. Our new dynamic scoring function is formulated to enhance the diversification capability in intensification phases using a user-defined diversification parameter. The formulation of the new scoring function is based on a probability distribution to adjust the selecting priorities of the selection between greediness on scores and diversification on variable properties. The probability distribution of variables on greediness is constructed to guarantee the synchronization between the probability distribution functions and score values. Additionally, the new dynamic scoring function is integrated with the two diversification criteria. The experiments show that the new heuristic is efficient on verification benchmark, crafted and random instances.

1 Introduction

Stochastic Local Search (SLS) is a competitive and an efficient approach to find the optimal solution or the approximately optimal solution for very large and complex combinatorial problems. Some examples of practical combinatorial problem instances that have been solved efficiently by SLS under the Satisfiability (SAT) framework are hardware verification and planning. Despite this significant progress, SLS solvers still have limitations compared with systematic solvers in practical and structured SAT problems as evident through the series of SAT competitions. Because structured and practical SAT problems have tighter constraints than randomized SAT problems, SLS algorithms are easily trapped in local minima and have difficulty to escape from stagnation. This problem does not exist in systematic search algorithms because of the nature of complete searching strategies.

Since the introduction of the GSAT algorithm [15], there have been huge improvements in developing efficient SLS algorithms for SAT. These improvements need to properly regulate diversification and intensification in local search. There are some common techniques to boost diversification such as random walk [10]

S. Cranefield and A. Nayak (Eds.): AI 2013, LNAI 8272, pp. 166–177, 2013.

and Novelty$^+$ [8]. In addition, some diversification boosting methods includes variable weighting [13] or clause weighting [16,12,6]. As reported in [14], the clause weighting scheme is able to escape the local minima by focusing on satisfying long-time unsatisfied clauses. The Hybrid [18] and TNM [17] algorithms exploit variable weight distribution to regulate the two noise heuristics but do not directly exploit variable weights to select variables. Recently, stagnation weights were introduced as a new diversification criterion to avoid local minima in gNovelty$^+$PCL [11]. Stagnation weights can be considered as an extension to variable weights because they record frequencies of variables involved in stagnation paths [5]. Sparrow2011 [1], the winner of the SAT competition 2011, is based on gNovelty$^+$ [12] framework but instead of using the Novelty$^+$ strategy at stagnation phases, it employs its own dynamic scoring function to escape from local minima. Recently, CCASat [3], the winner of SAT Challenge 2012, heuristically switches between two greedy modes and one diversification mode, and uses configuration checking to prevent the blind unreasonable greedy search.

In terms of intensification enhancement, the majority of local search solvers greedily explore the search space. More specifically, a local search will choose the most decreasing variable (i.e. a variable that leads to the most decrease in the number of unsatisfied clauses if being flipped). If there is more than one best decreasing variable, the algorithm prefers the variable with better diversification (i.c the least recently flipped or the lowest variable weight). The diversification mode is invoked when the search cannot greedily explore the search space. During the intensification mode, the scoring function (i.e. objective function) is very important. The drawback of most scoring function and variable selection methods is a lack of compromise between scores and tiebreak criteria. Mostly the tiebreak criteria are variable properties such as variable ages in most SLS solvers and variable weights in VW2 or stagnation weights in gNovelty$^+$PCL. The tiebreak criteria are considered as diversification boosting properties. Despite the fact that the current gradient-based scoring function (i.e. score in G2WSAT) in greedy phases works efficiently with current SLS solvers, a more advanced scoring function is needed to balance the score and the diversification tiebreaks. It motivates us to develop a single scoring function that combines greedy scores and diversification criteria.

In this work, we present a new SLS solver which uses an integration of a new probability-based dynamic scoring function as the objective function and two diversification criteria. The proposed dynamic function is controlled by a diversification noise α and is designed as a combination of clause-weighting score function and diversification criteria. The remainder of this paper is structured as follows. Section 2 summaries the background of SLS in developing objective functions. The motivation and construction of the probability-based dynamic formula are presented in section 3. Section 4 describes our algorithm, named PCF. The experiments on verification, crafted and random instances of SAT competitions are reported in section 5. This section also discusses the coverage of optimal diversification parameters. Section 6 concludes the paper and outlines the future work.

2 Preliminaries

Most modern SLS solvers operate in two modes : greedy (or intensification) mode and random (or diversification) mode. Starting from a randomized candidate solution, the solver computes the objective function for each variable. The function reflects the improvements in regards to the decrease in the number of the number of unsatisfied clauses when a variable is flipped. The most basic objective function (or scoring function) is computed as the decrease in the number of unsatisfied clauses. While this traditional score was used by TNM, sattime2011, the most up-to-date dynamic scoring function is the additive clause-weighting score used by many SLS solvers (e.g. gNovelty$^+$, Sparrow, EagleUP, CCASat).

In the greedy mode, if there exist variables with positive scores, the solver will select the variable with the highest score, breaking ties by the least recently flipped variables. Otherwise, the local search resides in a local minimum in which there is no possible greedy move. In such cases, it selects variables according to the random mode. There are numerous heuristics for the variable selection at the random mode. Most of them randomly pick an unsatisfied clause and select variables within that clause. Random walk is the simplest way of selecting variables randomly from the selected clauses. Novelty$^+$ is the most common and efficient scheme integrated in adaptG2WSAT, gNovelty$^+$.

2.1 Basic Scoring Function

The most basic scoring function for SAT is proposed by GSAT. It defines the number of unsatisfied clauses. Afterwards, the score is computed in an alternative objective function of the decrease in the number of unsatisfied clauses in G2WSAT. Eq. 1 expresses the score computed in G2WSAT.

$$score(v) = \sum\nolimits_{c} (Cls'(c, \alpha, v) - Cls(c, \alpha)) \tag{1}$$

where $score(v)$ is the decrease in the number of unsatisfied clauses if variable v is flipped. $Cls(c, \alpha)$ is the value of clause c under the candidate solution assignment α. If clause c is satisfied, $Cls(c, \alpha) = 1$, else $Cls(c, \alpha) = 0$. Given an circumstance of assignment α, $Cls'(c, \alpha, v)$ is the value of clause c after variable v is flipped.

2.2 Dynamic Scoring Function

To prevent the search from getting trapped in local minima, other dynamic penalties of clauses are integrated into the objective function of the search. The SLS using dynamic scoring functions is named the dynamic local search. This method is based on modifying the scoring function at each search step to re-evaluate the objective function in conjunction with changed circumstances. The purpose of the dynamic scoring function is to adjust the circumstance of local minima from the static method of computing an objective function. Because the dynamic scoring function can adjust the objective function, it assists the local search to dynamically avoid failing into previous stagnation.

Clause-Weighting. The state-of-the-art dynamic scoring function is based on the clause weighting scheme. This scheme typically associates weights with clauses. At each step, clause weights are adjusted according to truth value of corresponding clauses. Then instead of minimising the number of false clauses, the algorithms minimise the sum of clause weights. Eq. 2 expresses the clause-weighting score.

$$score_w(v) = \sum_c Wgh(c) \times (Cls'(c, \alpha, v) - Cls(c, \alpha)) \quad (2)$$

where $Wgh(c)$ is the clause weight of clause c.

VW2. Another way of computing dynamic scoring function based on variable weights is firstly proposed in VW1 and VW2 [13]. The dynamic scoring function of VW2 uses variable weights as diversification properties involved in the score. The scoring function of VW2 is computed as follows:

$$score_{VW2}(v) = \Big(\sum_c break(c, \alpha, v)\Big) + \Big(b \times (vw(v) - \overline{vw})\Big) \quad (3)$$

where $break(c, \alpha, v)$ is set to one if clause c becomes unsatisfied when variable v is flipped in the candidate solution α; otherwise its value is set to zero. $vw(v)$ is the weight of variable v. $\overline{vw}$ denotes the average of variable weights across all variables. b is a pre-defined parameter. The update and continuous smoothing procedure on variable weights vw is described in [13].

Sparrow. introduced a dynamic scoring function to overcome local minima. Its scoring function is modeled under a probability distribution and computed based on $score_w$ and variable ages. The variable selection is still based on the $score_w$ in the greedy mode. However, this dynamic scoring function is employed at stagnation phases only. The solver selects randomly one of the yet unsatisfied clauses at random. The selected clause is notated as $u_i = (x_{i_1} \vee .. \vee x_{i_k})$. The probability distribution to select variables in clause u_i is computed as the Eq. 4.

$$p(x_{i_j}) = \frac{p_s(x_{i_j}) \times p_a(x_{i_j})}{\sum_{l=1}^{k} p_s(x_{i_l}) \times p_a(x_{i_l})} \quad (4)$$

with $p_s(x_{i_j}) = a_1^{score_w(x_{i_l})}$, and $p_a(x_{i_j}) = (\frac{age(x_{i_l})}{a_3})^{a_2}$
The constant a_1, a_2, a_3 are experimentally determined and reported in [1].

3 Probability-Based Dynamic Scoring Function

3.1 Motivation

As mentioned in the previous sessions, few solvers have addressed the issue of combining greediness and diversification criteria into a single function. One drawback of variable selections of most SLS solvers is the fact that algorithms greedily select the most promising variable in terms of scores as the first priority. In case two variables have the same score, the algorithms will consider about tiebreak criteria (e.g. variable age). The scores and variable ages are considered separately in variable selection. Moreover, scores have greater priority than variable ages.

In order to construct a single scoring function as a trade-off between scores and tiebreaks, we decided to use probability knowledge to formulate a new dynamic scoring function. The idea of using a probability-based dynamic scoring function was firstly introduced in Sparrow [1]. Sparrow2011 and EagleUP [7] won the first and third places respectively in the SAT 2011 competition in Random track. These two solvers are efficient on random instances, which was attributed to their probability-based scoring function. However, their probability-based scoring function is restricted to stagnation phases whereas the conventional clause-weighting score is still used during intensification phases. We decided to approach the problem differently from the Sparrow formula by using additive formulation opposed to multiplicative formulation (Eq. 4). One reason for creating additive formulation is to modify the function gradually instead of adjusting rapidly as the multiplicative formulation. Additionally, we preferred to employ fewer parameters for users to regulate the scoring function more easily.

3.2 Defining a Probability Distribution

The proposed dynamic scoring function is formulated as Eq. 5. The scoring function is a summary of greediness and diversification probability distribution. The probability of diversification contribution is regulated by a user-defined parameter α. Thus, greediness probability contribution is regulated by $(1 - \alpha)$.

$$P(v_i) = (1 - \alpha) \times P_g(v_i) + \alpha \times P_d(v_i) \tag{5}$$

where v_i is the i-th variable and $P(v_i)$ is the probability of selecting variable v_i. The higher the value of $P(v_i)$, the more likelihood v_i is selected. $P_g(v_i)$ is the probability of greediness and $P_d(v_i)$ is the probability of diversification for variable v_i. In accordance to the fact that $P(v_i)$, $P_g(v_i)$, $P_d(v_i)$ are probability distribution functions (pdf), their values are scaled in the range of [0,1].

Probability Distribution on Greediness. In this work, we chose $score_w$ in Eq. 2 to compute $P_g(v_i)$. $score_w(v_i)$ is firstly scaled into the range [0,max_{score_w} - min_{score_w}] to satisfy the condition that the probability distribution function is non-negative. The adjusted scoring function for v_i is calculated by Eq. 6.

$$score'_w(v_i) = score_w(v_i) - min_{score_w} \tag{6}$$

where min_{score_w} and max_{score_w} is the minimum and maximum $score_w$ across all variables. Afterwards, the scoring function is normalized to satisfy the condition of a probability distribution function (i.e. probability distribution functions are in the range [0,1] and summarized to one) as follows:

$$P_g(v_i) = \mathcal{N}(score'_w(v_i)) = \frac{score'_w(v_i)}{\sum_j score'_w(v_j)} \tag{7}$$

Probability Distribution on Diversification. In this work, we chose two diversification properties separately to compute the probability distribution of diversification $P_d(v_i)$ in order to investigate the effect of different diversification properties. These properties are variable weights and stagnation weights.

Variable weights are first used in the work VW1 and VW2. To improve the diversification capacity, the variables with low flipped frequencies (i.e. low variable weights) are preferred to be selected. It was reported that variable weights improved the diversification capacity of SLS solvers [13].

Stagnation weights are presented in the work [12] as diversification criteria with the purpose of preventing local minima. The stagnation weight of a variable is computed as the frequency of its occurrences in stagnation paths. A stagnation path within a given tenure k is defined as a list of k consecutively flipped variables leading to a local minimum [4].

In order to compute $P_{d_v}(v_i)$, $vw(v_i)$ and $sw(v_i)$ are scaled into $[0, max_{vw} - min_{vw}]$, $[0, max_{sw} - min_{sw}]$ and transferred to $vw'(v_i)$, $sw'(v_i)$ respectively as follows:

$$vw'(v_i) = max_{vw} - vw(v_i) \tag{8}$$

$$sw'(v_i) = max_{sw} - sw(v_i) \tag{9}$$

where $vw(v_i)$ and $sw(v_i)$ are variable weights and stagnation weights respectively. min_{vw} and max_{vw} are the minimum and maximum variable weights. min_{sw} and max_{sw} are the minimum and maximum stagnation weights.

The functions $P_{d_{vw}}(v_i)$ and $P_{d_{sw}}(v_i)$ are the diversification probability distributions of variable weights and stagnation weights respectively. The formulas to compute the scaled diversification criteria $P_{d_{vw}}(v_i)$ and $P_{d_{sw}}(v_i)$ are presented as follows:

$$P_{d_{vw}}(v_i) = \mathcal{N}(vw(v_i)) = \frac{vw'(v_i)}{\sum_j vw'(v_j)} \tag{10}$$

$$P_{d_{sw}}(v_i) = \mathcal{N}(sw(v_i)) = \frac{sw'(v_i)}{\sum_j sw'(v_j)} \tag{11}$$

The probability distributions $P_{d_{vw}}(v_i)$ and $P_{d_{sw}}(v_i)$ grant bigger values to variables with higher scores and low variable weights and low stagnation weights respectively (e.g. preferring the least frequent flipped variables and least stagnated variables).

3.3 Diversification Parameter α

According to the probabilistic scoring function (Eq. 5), α is a pre-defined diversification parameter. It specifies the contribution of distribution probabilities in the scoring function in Eq. 5, whereas $(1 - \alpha)$ takes charge of the degree of intensification. In conventional clause-weighting, the objective function $score_w$ is maximized in order to greedily exploit the current searching position. In this case, the degree of intensification is 100% and the diversification parameter α is zero.

4 The PCF Algorithm

This section describes our proposed SLS solver, named PCF. It is based on gNovelty$^+$ and employs the new probability-based scoring function. gNovelty$^+$ is the state-of-the-art framework for SLS solvers and some currently superior SLS algorithms for SAT (e.g. Sparrow, CCASat) are tightly correlated with gNovelty$^+$ framework. The adjustment of PCF with the gNovelty$^+$ are listed below:

- Scoring function:
  - Instead of using the $score_w$, PCF applies the new scoring function P described in section 3.2 for both greedy and diversification phases.
  - Re-usage of the additive weighting-scheme for the greediness distribution function P_g
- Diversification criteria:
  - Applying variable weights and stagnation weights as tiebreaks of the scoring function.
  - Variable weights and stagnation weights are contributed in computing the scoring function.

The algorithm PCF is presented in Algorithm 1. PCF utilities a probabilistic objective function to determine search directions. The new heuristics PCF has one extra parameter, the diversification probability α. We use the clause-weighting scoring function in Eq. 2 as the greediness function. According to the diversification criteria of variable weights and stagnation weights respectively, we named the two variants of PCF as $PCF_v$and PCF_s.

At the initialization stage, clause weights are initiated to one; variable weights and stagnation weights are set at zero (line 2). If promising variables exist, the promising variable with the maximum probabilistic distribution value P is selected to be flipped, breaking ties by diversification criteria (line 9). Promising variables are defined as the variables whose $score_w$ are positive. If there is no promising variable, the Novelty strategy is invoked to escape from local minima. The procedure of updating stagnation weights is performed according to the original work gNovelty$^+$PCL [12,5] (line 11). Afterwards, weights of unsatisfied clauses are increased by one according to the additive weighting-scheme in gNovelty$^+$. More specifically, with probability sp, weights of weighted clauses are decreased by one. Weighted clauses are declared as clauses whose weights are larger than one [12]. The variable weight of the selected variable var is increased by one (line 17).

Algorithm 1. PCF(Θ, sp)

```
Input  : A formula Θ, wp = 0.01, diversification parameter α, smooth probability sp
Output: Solution σ (if found) or TIMEOUT
randomly generate a candidate solution σ;
initiate all clause weights to 1, stagnation weights and variable weights to 0;
while not timetout do
    if σ satisfied the formula Θ then  return σ ;
    if within the random walk probability wp then
        var = Random Walk in an unsatisfied clause;
    else
        if there exists promising variable then
            var = variable maximized P function, breaking ties by diversification
            criteria;
        else
            Update stagnation weights;
            var = Novelty Escape with P function, breaking ties by diversification
            criteria;
            Increase the weights of unsatisfied clauses by 1;
            if within smooth probability sp then
                Decrease clause weights of weighted clauses;
    Update candidate solution σ with the selected variable var;
    Increase the variable weight of var and adapt Novelty noise;
return TIMEOUT;
```

5 Experiments

The experiments were conducted on cbmc[1](a set of software verification problems), crafted instances of the SAT 2012 competition and medium-sized random instances of the SAT 2011 competition[2]. In our experiments, the time limit was set at 600 seconds. The number of runs per solver are 50 times for cbmc and 10 times for SAT Competition instances. The two PCF variants in the experiments are PCF_v and PCF_s, which employ variable weights and stagnation weights as the diversification criteria. The experiments were conducted on Griffith University Gowonda HPC Cluster Intel(R) Xeon(R) CPU X5650 2.67GHz. Our proposed algorithms PCF_v and PCF_s were compared with seven common SLS solvers:

- VW2, gNovelty$^+$PCL: originally uses variable weights and stagnation weights.
- gNovelty$^+$, sattime2011, EagleUP, Sparrow2011 [2], CCASat: are the top-3 best solvers of SAT competitions.

Table 1 and Table 2 present the results on structured instances (cbmc and crafted 2012) and medium-sized random instances of SAT 2011. Performance of a solver for a specific dataset are reported in three rows. The first and second rows indicate the success rate and the average of median CPU times. The third row specifies the number of flips in thousands. The number of flips of VW2, Sparrow2011 and CCASat are not reported because of the over-flown counted number of flipped in the VW2 , Sparrow2011; and CCASat did not provide that information in the output.

[1] `http://people.cs.ubc.ca/davet/papers/sat10-dave-instances.zip`
[2] `http://www.satcompetition.org`

Table 1. Results on the cbmc, Crafted 2012

Instances	VW2	gNovelty$^+$	sattime2011	EagleUP	Sparrow2011	gNovelty$^+$PCL	CCASat	PCF$_v$	PCF$_s$
cbmc	31%	85%	54%	0%	51%	100%	54%	100%	100%
(39)	439.128	247.997	322.528	600.000	384.359	1.453	276.196	**0.634**	1.013
		230,030	354,874	544,647	-	1,287	-	**523**	782
Crafted	0%	88%	84%	23%	70%	88%	82%	93%	**95%**
(74)	600.000	91.896	107.680	464.004	230.463	128.590	115.210	73.552	**70.377**
	-	22,476	22,784	142,440	-	38,067	-	13,725	**13,234**

As presented in Table 1, gNovelty$^+$PCL , PCF$_v$ and PCF$_s$ are the three solvers gaining a success rate of 100%. Among them, two variants of PCF performed better than gNovelty$^+$PCL. In regards to crafted instances, the two PCF variants gained better results than other solvers in terms of success rate, CPU time and flips.

Table 2. Results on SAT2011 Medium size

Instances	VW2	gNovelty$^+$	sattime2011	EagleUP	Sparrow2011	gNovelty$^+$PCL	CCASat	PCF$_v$	PCF$_s$
3-SAT	1%	50%	100%	100%	100%	93%	98%	100%	100%
(100)	595.033	336.962	**0.880**	6.471	20.498	52.893	9.691	1.378	1.966
	-	445,280	**1,504**	9,631	-	98,343	-	1,863	2,688
5-SAT	36%	98%	100%	100%	100%	98%	100%	100%	100%
(50)	410.750	48.136	3.815	30.260	55.846	27.300	7.199	**2.889**	3.277
	-	20,391	2,275	14,925	-	15,134	-	**1,280**	1,473
7-SAT	37%	100%	100%	100%	98%	100%	94%	100%	100%
(51)	400.219	20.026	9.454	23.917	65.350	22.152	51.236	7.944	**7.439**
	-	3,360	2,255	5,301	-	4,874	-	1,250	**1,203**
Random	19%	75%	100%	100%	100%	96%	98%	100%	100%
Medium	499.761	184.698	3.785	16.815	40.671	38.726	19.613	**3.420**	3.681
(201)	-	227,457	1,887	9,849	-	53,928	-	**1,562**	2,009

The medium-sized random instances in Table 2 are categorized into three subsets: 3-SAT, 5-SAT and 7-SAT [3]. On random 3-SAT instances, sattime2011, Sparrow2011, PCF$_v$ and PCF$_s$ are the four solvers having a success rate of 100%. Among the four solvers, sattime2011 is the best solver in terms of the average time of 0.88 seconds and the number of flips. Although PCF is not the best solver on 3-SAT instances, the two PCF variants performed well compared with other solvers on 5-SAT and 7-SAT. More specifically, in random 5-SAT, PCF$_v$ is the best solver with an average time of 2.899 seconds. On the other hand, PCF$_s$ performed better than PCF$_v$ with an average time of 7.439 seconds. For the whole dataset, PCF$_v$ is considered better than PCF$_s$ in terms of average CPU time of 3.42 and 3.681 respectively. It is clear from the Table 2, the two PCF variants generally performed well in medium-sized random instances compared with other solvers in the experiments.

[3] Random k-SAT instances consistently have k variables in every clause.

Figure 1 illustrates the comparison of solvers for cbmc and SAT 2011 Random Medium and SAT 2012 Crafted instances. The comparison is plotted in the log-log scale cactus presenting the distribution of the number of solved runs when the time limit increases. A run of an instance with a solver is defined as solved if it produces a solution within the given time limit. The x-axis corresponds to the time limit in seconds and the y-axis presents the number of solved runs within the corresponding time limit. The data points in these figures are plotted in every 50 seconds.

According to Figure 1(a) on the cbmc dataset, PCF_v, PCF_s and gNovelty$^+$PCL are outperformed other solvers. In the SAT 2011 Random Medium dataset, PCF_s, PCF_v and sattime2011 were consistently better than other solvers as displayed in Figure 1(b). The plot displayed in Figure 1(c) indicates that PCF_s and PCF_v steadily improved upon other solvers in crafted instances of the SAT 2012 competition. More specifically, PCF_s was not as good as PCF_v until 150 seconds but surpasses PCF_v thereafter.

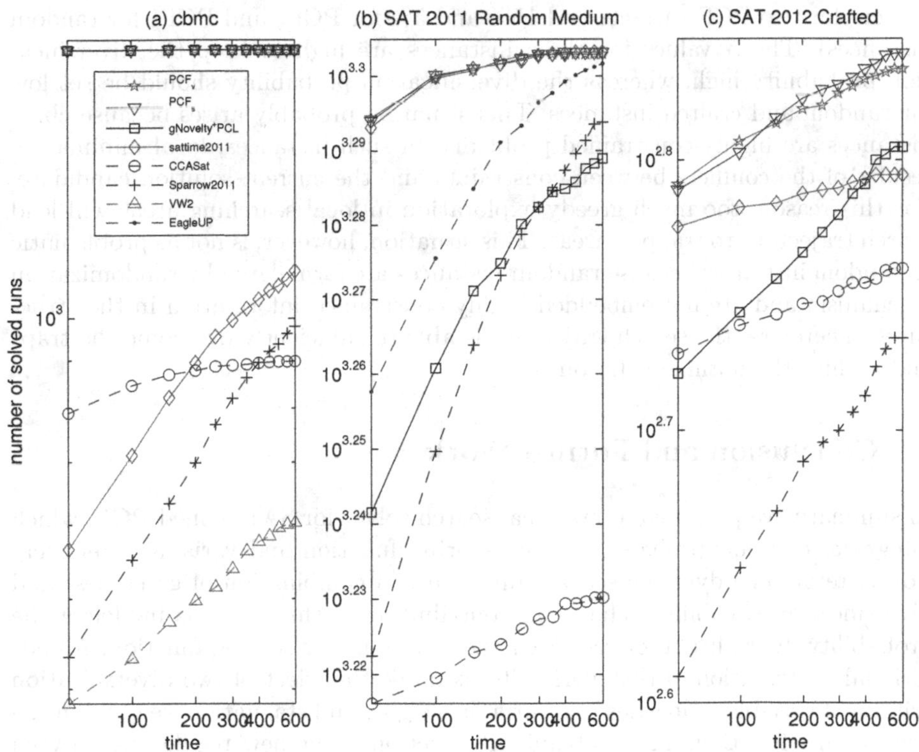

Fig. 1. Log-log scale plot of distribution graph of solved runs over time of cbmc, SAT 2011 medium size random and SAT 2012 Crafted

5.1 Analysis of Parameter Configurations

The parameter configurations for PCF variants were optimized separately on each instance set within {0.5;1;1.5;2;2.5;3} days by ParamILS with its default setting. ParamILS is a local search optimization tool for parameterized algorithms [9]. The best parameter settings of PCF variants are reported in Table 3. The sets for training and testing were divided half and separately from the original sets.

Table 3. PCF parameters trained by ParamILS

Solvers	cbmc			Crafted			Random		
	tenure	α	sp	tenure	α	sp	tenure	α	sp
PCF_v	n/a	0.5	0.1	n/a	0.05	0.30	n/a	0.15	0.30
PCF_s	30	0.5	0.1	30	0.05	0.3	15	0.05	0.35

As can be seen in Table 3 for cbmc, the α values are high. In contrast, for crafted and random instances, α converges to low values (i.e. 5% for crafted instances for two PCF variants and 15% and 5% for PCF_v and PCF_s for random instances). The α values for cbmc instances are high to keep the diversification probability high, whereas the diversification probability should be set low for random and crafted instances. This situation probably arises because cbmc instances are highly-constrained problems. In such instances, local minima are results of the conflicts between constraints and the current solution candidate. For this reason, too much greedy exploration in local searching areas will lead search trajectory to trapped areas. This situation, however, is not as problematic as random instances because random instances are formulated by randomization mechanism and are not embedded highly-constrained information in the structures. Therefore, the search trajectory is able to sufficiently overcome the traps and archive the optimal solution.

6 Conclusion and Future Work

In summary, we proposed a new local search solver for SAT named PCF, which integrates our new probability-based scoring function and variable diversification criteria. The dynamic scoring function is a combination of greediness and diversification capability. The main contribution of this study is modeling the probability distribution for each variable as a dynamic scoring function. An additional contribution of this work is to examine the effect of two diversification criteria of variable properties (e.g variable weights and stagnation weights) in the new scoring function. The probabilistic function is the new regulation between intensification in terms of the clause-weighting score and diversification in terms of variable properties.

The experiments showed that the new solver PCF significantly improved on the performance of other solvers. Comparative experiments demonstrated that the proposed approach outperformed original solvers (e.g. gNovelty$^+$, VW2 and gNovelty$^+$PCL). Furthermore, PCF outperformed other contemporary solvers on structure problems and medium-sized random instances. Our observation from optimization parameters suggests that the diversification parameter α for the verification benchmark and structure instances should be assigned high. In contrast, for random and crafted instances, diversification probability should be defined low.

In future, we plan to apply the probability-based scoring function to other solvers and self-tune the parameters α. Additionally, the intuitive research on the different performance of PCF_v and PCF_s should be investigated thoroughly on different benchmarks.

References

1. Balint, A., Fröhlich, A.: Improving stochastic local search for SAT with a new probability distribution. In: Strichman, O., Szeider, S. (eds.) SAT 2010. LNCS, vol. 6175, pp. 10–15. Springer, Heidelberg (2010)
2. Balint, A., Fröhlich, A., Tompkins, D.A., Hoos, H.H.: Sparrow 2011. In: Booklet of SAT 2011 Competition (2011)
3. Cai, S., Su, K.: Configuration checking with aspiration in local search for sat. In: AAAI (2012)
4. Duong, T.-T., Pham, D.N., Sattar, A.: A method to avoid duplicative flipping in local search for SAT. In: Thielscher, M., Zhang, D. (eds.) AI 2012. LNCS, vol. 7691, pp. 218–229. Springer, Heidelberg (2012)
5. Duong, T.T., Pham, D.N., Sattar, A.: A study of local minimum avoidance heuristics for sat. In: ECAI, pp. 300–305 (2012)
6. Duong, T.T.N., Pham, D.N., Sattar, A., Newton, M.A.H.: Weight-enhanced diversification in stochastic local search for satisfiability. In: IJCAI, pp. 524–530 (2013)
7. Gableske, O., Heule, M.J.H.: EagleUP: Solving random 3-SAT using SLS with unit propagation. In: Sakallah, K.A., Simon, L. (eds.) SAT 2011. LNCS, vol. 6695, pp. 367–368. Springer, Heidelberg (2011)
8. Hoos, H.H.: An adaptive noise mechanism for WalkSAT. In: Proceedings of AAAI 2002, pp. 635–660 (2002)
9. Hutter, F., Hoos, H.H., Leyton-Brown, K., Stützle, T.: Paramils: An automatic algorithm configuration framework. J. Artif. Intell. Res (JAIR) 36, 267–306 (2009)
10. McAllester, D.A., Selman, B., Kautz, H.A.: Evidence for invariants in local search. In: AAAI/IAAI, pp. 321–326 (1997)
11. Pham, D.N., Duong, T.T., Sattar, A.: Trap avoidance in local search using pseudo-conflict learning. In: AAAI, pp. 542–548 (2012)
12. Pham, D.N., Thornton, J., Gretton, C., Sattar, A.: Combining adaptive and dynamic local search for satisfiability. JSAT 4(2-4), 149–172 (2008)
13. Prestwich, S.D.: Random walk with continuously smoothed variable weights. In: Bacchus, F., Walsh, T. (eds.) SAT 2005. LNCS, vol. 3569, pp. 203–215. Springer, Heidelberg (2005)
14. Selman, B., Kautz, H.A.: Domain-independent extensions to gsat: Solving large structured satisfiability problems. In: IJCAI, pp. 290–295 (1993)
15. Selman, B., Levesque, H.J., Mitchell, D.G.: A new method for solving hard satisfiability problems. In: AAAI, pp. 440–446 (1992)
16. Thornton, J.R., Pham, D.N., Bain, S., Ferreira Jr., V.: Additive versus multiplicative clause weighting for SAT. In: Proceedings of AAAI 2004, pp. 191–196 (2004)
17. Wei, W., Li, C.M.: Switching between two adaptive noise mechanisms in localsearch. In: Booklet of the 2009 SAT Competition (2009)
18. Wei, W., Li, C.M., Zhang, H.: A switching criterion for intensification and diversification in local search for SAT. JSAT 4(2-4), 219–237 (2008)

A New Efficient In Situ Sampling Model for Heuristic Selection in Optimal Search

Santiago Franco, Michael W. Barley, and Patricia J. Riddle

Department of Computer Science, University of Auckland
santiago.franco@gmail.com, {barley,pat}@cs.auckland.ac.nz

Abstract. Techniques exist that enable problem-solvers to automatically generate an almost unlimited number of heuristics for any given problem. Since they are generated for a specific problem, the cost of selecting a heuristic must be included in the cost of solving the problem. This involves a tradeoff between the cost of selecting the heuristic and the benefits of using that specific heuristic over using a default heuristic. The question we investigate in this paper is how many heuristics can we handle when selecting from a large number of heuristics and still have the benefits outweigh the costs. The techniques we present in this paper allow our system to handle several million candidate heuristics.

1 Introduction

Techniques exist that enable problem-solvers to automatically generate an almost unlimited number of heuristics for any given problem. These heuristics are tailored to a given problem, therefore the cost of determining which heuristic to use must be included in the cost of solving the problem. The system must weigh the costs of selecting a heuristic against the savings resulting from using that better heuristic instead of a default heuristic. As runtime is an important criteria for evaluating problem-solvers, costs and savings should be expressed in terms of the impact on the system's total runtime.

Most of these techniques create new heuristics by combining heuristics. There are three standard ways of combining heuristics: maxing over them, randomly picking one of them, and summing their values together. Maxing is currently the most common way of combining heuristics. In this paper, we call the heuristics that are being combined *base* heuristics and the combined heuristic is a *heuristic combination*. Our system, RIDA* (Reconfigurable IDA*), is given a set of heuristics from which it finds a "good" combination to max over.

A parametric model for heuristic search is at the heart of the system's reasoning about runtimes of heuristic combinations. The model has variables that specify the problem solver's per node runtime costs as well as variables specifying the number of nodes in the search tree. Some of these variables (e.g., average branching factor (ABF)) cannot be determined *a priori*. Our system estimates these values by sampling the behavior of the different heuristics while expanding early portions of the problem's search tree. The most expensive calculation is the

S. Cranefield and A. Nayak (Eds.): AI 2013, LNAI 8272, pp. 178–189, 2013.

ABF of the heuristic combinations. This paper presents techniques for reducing the runtime costs for determining the ABF.

These techniques allow RIDA* to handle several million heuristic combinations and still show a reduction in total runtime over simply using the default heuristic, maxing over the set of base heuristics. We have tested RIDA* using two sliding tile puzzles (the 15-puzzle and 24-puzzle) with IDA* and the Towers of Hanoi (ToH) with A*. The base heuristics used were PDB-based heuristics.

2 Parametric Model

$$Time(comb, iter) = Nodes(comb, iter) * t(comb) \tag{1}$$

$$Nodes(comb, iter) = SampledNodes(comb) * ABF(comb)^D \tag{2}$$

$$ABF(comb) = \frac{Nodes(comb, LastSampledIteration)}{Nodes(comb, PenultimateSampledIteration)} \tag{3}$$

$$D = F(CurrentIteration) - F(LastSampledIteration) \tag{4}$$

Equation 1 is used by RIDA* to estimate the time performance for every heuristic combination it considers. *t(comb)* is the sum of the average cost of generating a node plus the average cost of evaluating the node for each of the base heuristics in *comb*. The evaluation of the base heuristics is the average per node evaluation time of a base heuristic. This is determined *a priori*, i.e., before solving any problems, for the domain. Equation 1 needs to know the number of nodes generated for the previous iteration. Equation 2 estimates the number of nodes generated for the current iteration based on two parameters: the number of nodes generated for the combination on the previous iteration (SampledNodes) and the Average Branching Factor(ABF). D stands for the difference between the previous iteration's f-limit and this iteration's f-limit, IDA*'s f-limits need not go up by one for successive iterations. For example, the f-limits for the 8-puzzle using Manhattan Distance heuristic, usually goes up by 2 for each iteration. RIDA* assumes that if a sufficiently large part of the search space has been sampled, then ABF behaves asymptotically for the remaining iterations until a solution is found. F is the f-limit for an iteration.

The proposed parametric model requires RIDA* to sample enough IDA* iterations to be able to make good predictions for each combination, assuming the ABF behaves asymptotically. Doing this naively, i.e. sampling each combination by growing a separate search tree, is too costly. The next section describes how we reduce the sampling costs so that we can efficiently sample each combination.

3 Techniques for Reducing Sampling Costs

RIDA*'s parametric model requires the number of nodes generated in the last sampled iteration and an average branching factor. There are two possible approaches to estimate these parameters, one is to calculate it as the ratio of generated nodes between the last two sampled iterations. The other is to use models which are based on the probability distribution of heuristic values. The latter is too memory intensive, see Section 4.

To calculate the average branching factor we need to count how many nodes a heuristic combination generated on the current iteration compared to how many it generated on the previous iteration. The naive way of doing this is, for each heuristic combination, grow a search tree deep enough to get a useful estimate of that combination's average branching factor. However, there is a lot of redundancy in growing the search trees. Many of the nodes are going to be common to many of the search trees.

To minimize the costs of sampling multiple heuristic combinations we created a data structure called the Heuristic Union Search Tree(*HUST*). A regular heuristic search tree, which we call a *maxTree*, assigns to each node the maximum heuristic value. The HUST can be seen as the union of all the maxTrees that would have resulted if we had grown a separate search tree for each combination.

Nodes in the HUST are pruned only if all heuristics prune them. By contrast, maxTrees will prune a node if any of the available heuristics prune it. For each iteration there is a maxTree for each heuristic combination. There is only one HUST for each iteration and each of the maxTrees is a subtree of the HUST.

The HUST is the union of all possible maxTrees. It is more efficient than doing all possible maxTrees because nodes common to two or more heuristic combinations are represented as one node in the HUST. Hence the HUST of an iteration is a lossless compression of all the maxTrees of that iteration.

HUSTs are substantially larger than any single maxTree for the same iteration. In order to achieve our objective of keeping the sampling costs low we need to stop growing the HUST at least one iteration before the solution for the current problem instance. If the problem instance is solved while we are generating the HUST, the resulting time performance will be as bad, if not worse, than the worst available heuristic combination.

Note that in order to generate the HUST we do not need to evaluate all heuristics in the set for each node. Once a heuristic prunes a node, the heuristic is suspended until IDA* backtracks past this node. There is a small associated bookkeeping cost, but its negligible compared to the resulting savings.

3.1 The Credit Assignment Problem

The HUST reduces the sampling costs via eliminating the generation of redundant nodes. Any node which is reachable by any heuristic combination is present once and only once in the HUST. Hence, creating a HUST reduces node redundancy from up to 2^H to 1 where H is the number of base heuristics in the set.

This merging of search trees removes some of the redundant work in growing multiple maxTrees, however, we can go further. Even if nodes are being generated only once for multiple heuristic combinations, we need to account for which combinations were responsible for generating each of the HUST's nodes. We call this the "credit assignment problem". In a naive implementation of the HUST, we would still treat each heuristic combination separately. We would evaluate each node with every combination of heuristics and would count how many nodes would be expanded by each combination. This means that every node is evaluated by every heuristic combination, and then added to the corresponding combination counters. This is an extremely high overhead per node.

We need to reduce these per node costs. Ideally we would like to reduce the evaluation costs from using every heuristic combination to only using every base heuristic We can certainly do that by using the HUST's expansion rule, i.e. expand any node if any base heuristic would expand it. This way we know that some combination would expand that node and therefore we want to expand that node in the HUST. However, this makes the proper accounting of how many nodes each heuristic combination would expand more complicated. How do we keep the accounting straight?

Ideally, we would like to reduce the number of additions per node to one, but is that possible? While it is not strictly possible we have come close to that ideal. We have split the counting into two phases. One phase is while growing the HUST. The second phase is after growing the HUST when we compute the generated nodes and branching factors for the different heuristic combinations.

The first phase calculates which base heuristics are "guilty" of expanding a node. We call this heuristic a "culprit". For each expanded node, we associate a "*culprit id*" (*CI*) which identifies all the heuristics which would expand the node. Examples of CIs are shown in Figure 1, e.g., the root's CI is 111, which means that all three heuristics expanded the root node. We have a counter, called a "*culprit counter*" (*CC*), for each CI. When we expand a node, we update its CC (the one associated with its CI). An example of a CC is in Figure 1 where the CC for CI 100 is shown as six. Therefore we only do 1 counter update per expanded node, while growing the tree.

In the second phase we compute how many nodes a particular heuristic combination would have generated (storing this number in *Heuristic Combination Counters* (HCC)), if we had grown its search tree (maximizing all heuristics in the combination). For example, in Figure 1 the HCC for CI 100 is nine. We do this using the CCs we updated while growing the HUST. We need to clarify the relation between what nodes were expanded by a given CI in the HUST and what nodes a given combination of heuristics would have expanded in its own search tree. Happily, this relation is straightforward. Given a heuristic combination, it only expands a node if all of its heuristics would expand that node. So given a CI, we would add its associated CC to the HCC only if the "guilty" heuristics in the CI are a superset of the heuristics in the combination, we call this a *ContributingCC*. This means we can compute the nodes that would have been generated by a given heuristic combination by summing together all of the CCs for CIs which represent supersets of the heuristics in that combination. Note that we only keep track of non-zero culprit counters. Also note, that depending upon the number of base heuristics, there could have been far fewer nodes expanded in the HUST than the number of possible heuristic combinations. Therefore if we have a smart way of selecting interesting combinations, we can calculate those combinations without having to explicitly deal with all possible combinations.

In the worst case, in the second phase we have a separate culprit counter for each node in the HUST, but this is unlikely.

3.2 Culprit Counters Example

We will be referring to an example search tree (Figure 1) to explain RIDA*'s HUST and CCs. The HUST's objective is to cheaply measure the number of generated nodes, for any heuristic combination, up to the current IDA*'s f-limit[1]. RIDA* uses the estimated generated nodes for each combination, together with the sampled average branching factor, to predict the number of generated nodes and ultimately search time for the next f-limit. Figure 1 is a HUST tree. Each node is identified by its path, e.g. root's left child is called "L" and L's right child is called "LR". Note that the node also contains which heuristics expanded it (1) and which prune it (0), e.g. 100 means that only h_1 expands this node.

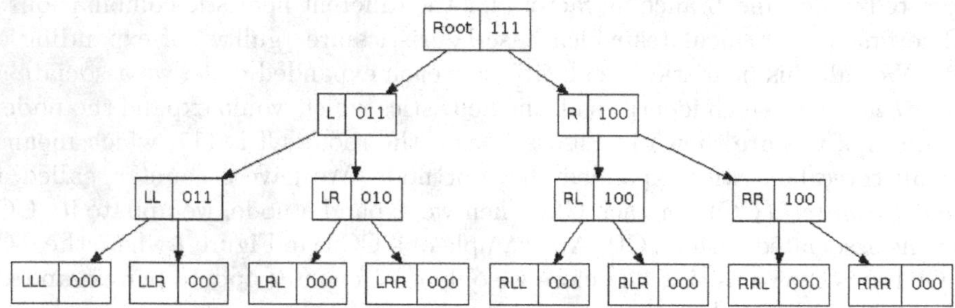

Fig. 1. Example HUST with 3 heuristics (h1,h2,h3)

Table 1. CCs and Heuristic Combination Counters(HCC) for Example Fig. 1

Populated Culprit Counters	
CI	*Associated Children Nodes*
111	$CC(111)=2$
011	$CC(011)=4$
100	$CC(100)=6$
010	$CC(010)=2$
Generated Nodes When Using Maximization	
Heuristic Combinations	*Generated Nodes*
$h1\&h2\&h3$	$HCC(111) = 1 + \sum CC(111) = 3$
$h1\&h2$	$HCC(110) = 1 + \sum CC(11X) = 3$
$h1$	$HCC(100) = 1 + \sum CC(1XX) = 9$
...	...

[1] IDA* iteratively expands its search tree down to a given f-value, called its f-limit.

RIDA* solves the credit assignment problem by using the Culprit Counters(CC). For instance, given a set of three heuristics, $CC(100) = 6$ means that only six nodes in the HUST were generated by parent nodes on which the first heuristic was below the f-limit but pruned by the other two (their f-value was higher than the current f-limit). Table 1 shows how RIDA* uses CCs to calculate the number of nodes generated by any heuristic combination by summing the corresponding CCs, e.g. to calculate how many nodes were generated by maximizing heuristics h1 and h2, sum all CCs whose first and second bit indexes are one plus the root node. For example, HCC(100) = 1 (for the root node) + CC(111) + CC(100) = 9.

The CC solution has four main strengths. Firstly, it reduces the per node costs from exponential to linear as a function of the number of base heuristics. Secondly, the on-line bookkeeping costs are reduced further by only updating the corresponding CC when a node is expanded. Pruned nodes have no bookkeeping costs. Thirdly, CCs are memory efficient, we only generate a CC if the specific CI generated a node in the HUST. For example, in Figure 1 there are no nodes that were only expanded by heuristic *h3*, therefore there are no culprit counters for the culprit id 001. Note a large collection of HUST nodes can be represented by a single CC. Lastly, it enables RIDA* to only solve the credit assignment problem once per iteration instead of once per generated node. Without a solution like CCs, sampling a large number of heuristic combinations would be impractical due to large bookkeeping costs.

3.3 Managing Large Powersets

Unfortunately, for the post-sampling phase, the number of heuristic combinations grows exponentially as a function of the number of base heuristics. For example, for a set of 100 heuristics there are $2^{100} = 1.27 * 10^{30}$ heuristic combinations. In this paper the biggest set is even larger (120 heuristics for ToH). The time it takes to solve the credit assignment problem should be small enough to keep some of the HUST's overall savings. We now describe two techniques used to reduce the cost of the post-sampling phase.

Limiting the Combination Degree. The first technique is to limit the Maximum Combination Degree (MCD). Taking advantage of the diminishing returns property, we focus the sampling on the lower combination degrees by ignoring all heuristic combinations above a predetermined degree. This reduces the number of heuristic combinations which need to be computed from 2^H to $\sum_{i=1}^{i=MCD} \binom{H}{i}$, H being the number of base heuristics. The maximum practical MCD depends on the number of base heuristics. In our experiments we show that it works quite well for up to 120 heuristics. Since the total number of heuristic combinations is the number of base heuristics choose MCD, this means that the larger the set of base heuristics, the lower the MCD must be, otherwise there would be too many counters to fit into memory. However, RIDA*'s objective is not to select the best possible combination, but to balance the *in situ* prediction costs against their potential savings.

Sparse Representation of the Culprit Counter Lattice. The search trees we are creating in this paper's experiments can be several hundred million nodes. But, as previously said, the number of possible heuristic combinations can be many orders of magnitude larger. Evidence from our experiments with the 100 heuristic set in the 24-puzzle domain shows that for the HUST tree, which has a few million nodes, there are less than a million CCs. Given 100 base heuristics, there could theoretically be up to 10^30 CCs which is approximately 24 orders of magnitude greater than the number of CCs that were actually created. The more base heuristics the sparser the CC lattice is. The intuitive reason is that as only one CC, at most, can be populated for each expanded HUST node, and the number of nodes generated is orders of magnitude smaller than the amount of possible heuristic combinations, the CC lattice must be very sparsely populated for large sets of base heuristics.

Equation 5 calculates the number of additions, per populated CC, to calculate the number of generated nodes for all heuristic combinations up to the maximum combination degree(MCD). Each CC contributes to those combinations where its set of generating heuristics is a superset of the combination's heuristics. Hence, the maximum combination degree each CC contributes to ActvDegree is the smallest of either the MCD or the number of contributing heuristics (called in the equation the NumContributingCC). For instance, if the culprit counter has two expanding heuristics then it only contributes to heuristic combinations of degree one or two, any degree combinations of three or higher would not have generated the CC's associated nodes because at least one base heuristic would prune them.

$$TotalCCsAdditions \leq \sum_{\forall\ CC} \sum_{i=1}^{ActvDegree} \binom{H}{i} \tag{5}$$

$$ActvDegree = min(MCD, NumContributingCC)$$

3.4 Capping Limit

The purpose of the HUST in RIDA* is to sample enough nodes for each heuristic combination so that we can make good predictions of their branching factors, but not so many nodes that the sampling cost is too expensive. The worst heuristic combinations can create maxTrees whose size is orders of magnitude larger than their more informed counterparts for the same f-limit. Hence we are potentially investing much more sampling effort on the worst heuristics while making lower quality predictions, due to their low sampling frequency, on the better heuristic combinations. We solve this by capping the HUST.

Capping the HUST means that once any of the base heuristics expands more nodes than a predetermined number, we call the *capping limit*, we will stop expanding any heuristic combinations using this heuristic for future HUST iterations. Hence sampling effort is more evenly split over all heuristic combinations. In Section 5 we show that by using the right capping limits we can achieve a good prediction quality while keeping sampling costs reasonable. Once all the base heuristics are capped, RIDA* makes its selection and solves the rest of the problem using maximization.

3.5 Summary

Using the HUST and CCs reduces several sources of redundancy from repeatedly regrowing parts of the search tree, repeated base heuristic evaluations of a node (base heuristics are repeated within different combinations), and aggregating counts by CIs so that we have many fewer additions to perform overall.

4 Related Research

Parametric models In the last decade there has been significant research into improving heuristic performance prediction. This paper introduces a new parametric model to predict the performance of heuristics. Lopez and Junghanns[1] introduced a parametric model of perimeter search which enabled his system to predict the optimal perimeter depth. Their work is the only other work that actually uses a runtime formula to modify their problem solving behavior.

Selective Maxing Using Bayesian Classifiers. The only other system which makes efficient *in situ* heuristic selections in the context of optimal search, given a set of heuristics, is [5]. It is based on using Bayesian classifiers to select heuristics on a state-by-state basis. The first important difference between [5] and RIDA* is that their system has on-line meta-reasoning costs for every generated node. RIDA* instead uses the early IDA* iterations of a problem to gather data up to a sampling cap. Once RIDA* has made a selection for the current problem, there are no more meta-reasoning costs. This also means that RIDA* cannot adapt its choice if its initial sampling is not representative of the whole problem.

Secondly, their experiments used heuristics which have large on-line overheads. Note that their meta-reasoning increases the average overhead for all generated nodes. It would be harder for such a system to be as efficient when using pre-calculated heuristics. RIDA*'s sampling mechanism is designed to be efficient even when using heuristics with a small on-line overhead.

Finally, their heuristic set contains only two heuristics. Their paper states that the number of classifiers it needs to update on-line, and presumably the overhead costs, grow quadratically as a function of the number of base heuristics. This means that for the larger heuristic sets used in our experiments their on-line meta-reasoning costs would increase by 4 orders of magnitude. That would make their listed 2% average meta-reasoning costs per node too expensive. RIDA*'s on-line meta-reasoning costs grow linearly instead of quadratically.

Choosing Heuristics Based on the Number of Generated Nodes. There are several papers on how to generate the best abstraction-based heuristics, e.g. iPDB [7], Merge-and-Shrink [8], etc. There are surprisingly few examples attempting to efficiently and automatically select from different types of admissible heuristics *in situ* in the context of optimal search.

All these heuristic generation procedures do a restricted on-line heuristic selection, to determine which abstractions will reduce the search space the most. The main difference with our approach is that RIDA* is using a parametric model to estimate which heuristic combination will solve the problem faster. Generating the lowest amount of nodes does not guarantee the fastest solving time, otherwise maximizing all available heuristics would always be the fastest option. Even though it is not done in this paper, RIDA* handles heuristics with very different evaluation times.

5 Experiments

For the 15-puzzle we used a manually selected set of 5 PDB-based heuristics for a suite of 1,000 random problems. The heuristics were manually generated to complement each other.

For the 24-puzzle we decided to automatically generate sets of heuristics. Each pattern was selected randomly, albeit with the condition that the patterns must be "neighours" in the goal description, and also no pattern is repeated. The main difference with the manual selection of PDBs for the 15-puzzle is that we did not take into account how well each generated heuristic will complement the other heuristics in the set. This makes the heuristic set more amenable to RIDA*, i.e. RIDA* finds the heuristic subset which it estimates to be most complementary while balancing this with the overhead costs to select good heuristic subsets. We created a set of 100 heuristics. We only included 9 random problems for the 24-puzzle set. The problems take about a month to solve when evaluating performance against maximizing over the whole set.

The ToH problem is a classic search problem, consisting of 3 pegs and a set of disks. We used 4 pegs because there is no known deterministic algorithm which guarantees an optimal solution for the 4 peg problem[9]. The heuristic set was created using the same pattern database as in [9]. They calculated the largest pattern which can fit in memory, then use the domain symmetries to generate a large set of a hundred and twenty heuristics [9].

5.1 Sampling Cost Savings

Table 2 shows how increasing the number of combinations generally increases redundancy compression and time compression. The number of populated Culprit Counters is also a factor on the HUST's savings ratio.

In general, the more heuristic combinations in the set, the bigger the redundancy reduction (second column). Note that the number of combinations RIDA*

Table 2. Sampling Savings w/wo Credit Assignment(CA)

15-puzzle			
Heuristic Set	Redundancy[a]	Time Compress[b]	TC after CA[c]
5 heuristics, MCD=5, 32 combos	4.04	3.85	3.85
24-puzzle			
Heuristic Set	Redundancy	Time Compress	TC after CA
8 heurs, MCD=8, 256 combos	3.01	2.12	2.12
25 heurs, MCD=4, 1.53E+004 combos	47.41	22.37	21.7
25 heurs, MCD=5, 6.84E+004 combos	121.43	65.98	58.79
100 heurs, MCD=3, 1.67E+005 combos	367.26	52.50	32.67
50 heurs, MCD=4, 2.51E+005 combos	463.58	143.70	68.91
50 heurs, MCD=5, 2.37E+006 combos	2326.40	910.89	85.43
100 heurs, MCD=4, 4.09E+006 combos	2496.51	703.10	47.59
Towers of Hanoi			
Heuristic Set	Redundancy	Time Compress	TC after CA
120, MCD=3, 2.88E+005 combos	10,173	1,696.51	768.35
120, MCD=4, 8.50+006 combos	204,589	34,900	709.25

[a] $Redundancy\ Reduction = \frac{NodesGenerated\ by\ AllPossibleSubsets}{Nodes\ generated\ in\ HUST}$
[b] $Time\ Compression = \frac{Time\ to\ Generate\ All\ Possible\ Subsets}{Time\ to\ Generate\ HUST}$
[c] HUST's Time Compression after solving credit assignment problem.

considers is a function of the number of heuristics in the set and the maximum combination degree being considered. Solving the credit assignment can significantly reduce the overall time savings.

5.2 Overall Runtime Results

The first part of Table 3 shows that RIDA* was only slightly faster than simply maximizing the heuristic set on the 15-puzzle. The reason for the poor performance improvement is that the heuristic set is made of a small number of base heuristics which were generated manually to complement each other. RIDA* could not find significant savings by selecting a subset of these heuristics. For the 24-puzzle, Table 3 shows the overall speed up achieved when using RIDA* compared to simply maximizing over 25, 50 and 100 heuristics sets respectively. The overall times for each set is calculated by summing over all 9 problems. The best results are achieved for the 100 heuristic set, where RIDA* is 3.91 times faster than using the default combination. Note that the overall times being compared include all RIDA* sampling and prediction costs. It is interesting to note that for the two biggest heuristic sets (50 and 100 heuristics) the better Maximum Combination Degree(MCD) is getting smaller. The reason for the poorer performance of larger MCDs is the high costs associated with solving the credit assignment problem for large heuristic sets. However, note that even when RIDA* is dealing with 4 million heuristic combinations (100 base heuristics and MCD = 4), its speed-up ratio is 3.54.

Table 3. RIDA*'s Speed-up ratio vs Maximization $Speed\_up\ Ratio(SR) = \frac{Maximize\ Whole\ Set\ Runtime}{RIDA*\ Overall\ Runtime}$

<table>
<tr><td colspan="6">Fifteen Puzzle
5 Heuristic Set,MCD 5,Capping Limit=5,000 nodes.
SR=1.17</td></tr>
<tr><td colspan="6">Twenty-Four Puzzle
Capping Limit = 5 Million Nodes</td></tr>
<tr><td colspan="2">25 Heuristic Set</td><td colspan="2">50 Heuristic Set</td><td colspan="2">100 Heuristic Set</td></tr>
<tr><td>MCD 4</td><td>MCD 5</td><td>MCD 4</td><td>MCD 5</td><td>MCD 3</td><td>MCD 4</td></tr>
<tr><td>SR=1.83</td><td>SR=2.03</td><td>SR=3.10</td><td>SR=3.04</td><td>SR=3.9</td><td>SR=3.54</td></tr>
<tr><td colspan="6">Towers of Hanoi (16 Disks, 4 Pegs)
120 Heuristic Set, Capping Limit=100,000 nodes</td></tr>
<tr><td colspan="3">MCD 3
SR=4.12</td><td colspan="3">MCD 4
SR=1.36</td></tr>
</table>

The third part of Table 3 lists the speed-up ratio when using RIDA* with an MCD of 3 in the Towers of Hanoi (ToH) domain as 4.12 times faster but with an MCD of 4 RIDA* is only 1.36 times faster. The best results in our experiments were obtained in the ToH domain with a MCD of 3. The same reasoning as before applies for using this low MCD. Namely the high cost of solving the credit assignment problem for higher MCDs causes the speed-up ratio to drop significantly when RIDA* uses an MCD of 4. An MCD of 4 with 120 base heuristics created 8.5 million heuristic combinations. The cost of just computing the combination counters from the culprit counters for an MCD of 4 was twice the total cost of solving the problem when the MCD was 3. This illustrates that our current techniques do have limits on how many heuristic combinations they can handle.

6 Conclusions

The goal of this paper was to investigate how many heuristic combinations RIDA* can handle and still solve the problem faster than simply maxing over the set of base heuristics. In our experiments we showed that on the 24-puzzle even when handling 4 million heuristic combinations, RIDA* was significantly faster (speed-up ratio of 3.54) than the default heuristic. However, our experiments showed that on ToH 8.5 million heuristic combinations was not much faster (speed-up ratio of 1.36) than the default.

RIDA* uses 4 different techniques to reduce the cost of computing the quality of the heuristic combinations. Firstly, the HUST reduces the number of duplicated nodes produced during sampling. Secondly, the culprit counters reduce the number of heuristic evaluations of a state from the number of heuristic combinations to the number of base heuristics and reduce the number of counter updates per expanded node from the number of heuristic combinations to just 1. Thirdly, only non-zero culprit counters are stored which reduces the number of culprit counters needed from the number of heuristic combinations to a worst case of the number of internal nodes in the HUST. Lastly, the law of diminishing returns allows us to limit the number of base heuristics in any single combination, which reduces the number of combinations considered from 2^H to a few million.

This is the only study where so many heuristic combinations were evaluated on-line. Our conclusion is that using techniques, like the ones described in this paper, makes it possible to reduce the cost of evaluating combinations to the point where a system can evaluate relatively large numbers of heuristic combinations and still have the savings outweigh the cost.

Acknowledgements. This material is based on research sponsored by the Air Force Research Laboratory, under agreement number FA2386-12-1-4018. The U.S. Government is authorized to reproduce and distribute reprints for Governmental purposes notwithstanding any copyright notation thereon. The views and conclusions contained herein are those of the authors and should not be interpreted as necessarily representing the official policies or endorsements, either expressed or implied, of the Air Force Research Laboratory or the U.S. Government.

References

1. López, C.L., Junghanns, A.: Perimeter Search Performance. In: Schaeffer, J., Müller, M., Björnsson, Y. (eds.) CG 2002. LNCS, vol. 2883, pp. 345–359. Springer, Heidelberg (2003)
2. Korf, R.E., Reid, M., Edelkamp, S.: Time complexity of iterative-deepening-A*. Artificial Intelligence 129(1-2), 199–218 (2001)
3. Zahavi, U., Felner, A., Burch, N., Holte, R.C.: Predicting the performance of IDA* with conditional distributions. In: Fox, D., Gomes, C.P. (eds.) AAAI Conference on Artificial Intelligence (AAAI 2008), pp. 381–386. AAAI Press (2008)
4. Zahavi, U., Felner, A., Schaeffer, J., Sturtevant, N.R.: Inconsistent heuristics. In: AAAI Conference on Artificial Intelligence (AAAI 2007), pp. 1211–1216 (2007)
5. Domshlak, C., Karpas, E., Markovitch, S.: To max or not to max: Online learning for speeding up optimal planning. In: AAAI Conference on Artificial Intelligence (AAAI 2010), pp. 1701–1706 (2010)
6. Tolpin, D., Beja, T., Shimony, S.E., Felner, A., Karpas, E.: Towards rational deployment of multiple heuristics in a*. CoRR abs/1305.5030 (2013)
7. Haslum, P., Botea, A., Helmert, M., Bonet, B., Koenig, S.: Domain-independent construction of pattern database heuristics for cost-optimal planning. In: AAAI Conference on Artificial Intelligence (AAAI-2007), vol. 22(2), p. 1007. AAAI Press, MIT Press, Menlo Park, Cambridge (2007)
8. Helmert, M., Haslum, P., Hoffmann, J.: Flexible abstraction heuristics for optimal sequential planning. In: Proceedings ICAPS 2007, pp. 176–183 (2007)
9. Felner, A., Korf, R.E., Hanan, S.: Additive pattern database heuristics. Journal of Artificial Intelligence Research (JAIR) 22, 279–318 (2004)

A Framework for the Evaluation of Methods for Road Traffic Assignment

Syed Galib and Irene Moser

Swinburne University of Technology, Melbourne Victoria, Australia
{sgalib,imoser}@swin.edu.au

Abstract. Research in traffic assignment relies largely on experimentation. The outcomes of experiments using different traffic assignment methods on different road network scenarios may vary greatly. It is difficult for the reader to assess the quality of the results without prior knowledge about the difficulty of the road network. In the current work, we propose an approximate metric to characterise the difficulty of network scenarios which takes travellers' origins and destinations as well as link capacities into account rather than relying on the size of the network. As this metric considers number of overlapping routes between the origins and destinations of the travelles, a higher number in the metric would indicate a higher possibility of congestion in the road network scenario.

Keywords: Traffic Assignment, Evolutionary Optimisation, Network Complexity, Experimental Evaluation.

1 Introduction

The road network is used by travellers who make trips from origins to destinations. Given these origin-destination (OD) pairs, the travellers usually have a set of route options to choose from. The selection of a route for an OD pair is known as traffic assignment (TA). TA is the cause of the distribution of traffic on the road network. Researchers have used a variety of approaches to approximate traffic distributions. Among others, multiagent systems [1,2], evolutionary games and evolutionary algorithms such as genetic algorithms [3] and ant colony optimisation [5] have been applied to the problem. To test their approaches, many authors have performed experiments either on hypothetical road networks or road networks modelled on real geographical areas. However, the results of different TA approaches over different road networks cannot be compared easily, as there is no a priori indication of the results to be expected.

The success of road traffic optimisation is mostly measured in terms of the cost incurred by the traveller. Often, this cost is expressed as travel time. To assess the success of a TA approach, it would be helpful to have an indication how 'difficult' the optimised network scenario is. Larger road networks would appear more difficult for road traffic optimisation. However, in road traffic scenarios, the difficulty in achieving an optimal distribution of travellers arises not from the number of links available between an origin and a destination, but rather the number of travellers competing for road capacity. The deciding factor is where the travellers come from and where they wish to go, which forms the OD pair of a traveller. A network may be very simple for one set of OD pairs

S. Cranefield and A. Nayak (Eds.): AI 2013, LNAI 8272, pp. 190–195, 2013.

which have no or few overlapping routes, whereas the same network may become very complex for another set of OD pairs which have many overlapping routes. We refer to networks in combination with OD pairs as traffic scenarios.

The difficulty measure for road network experimentation proposed here is based on how many times each link is included in the alternative routes between each OD pair. Thus, practitioners can have an expectation of congestion in a road network scenario by the value of this metric.

2 Traffic Assignment Experimentation

In transport modelling, the most commonly applied state-of-the-art technique is the so-called four-step model: trip generation, trip distribution, mode choice and traffic assignment [6,7,8]. The last step, traffic assignment (TA) has been receiving much attention from a number of researchers internationally.

Chen and Ben-Akiva [9] attempted to achieve optimal traffic flows by applying game-theoretic formulations. The authors evaluated their methods on a simple hypothetical network with seven nodes, seven links and 3 OD pairs. Bazzan and Klügl [1] investigated the behaviour of agents under the effect of real-time information and thus studied how the agents change their route mid-way. Their experimental scenario was a 6x6 grid network with 36 nodes and 60 one-way links. The scenario had three main origins and one main destination as well as several other origins and destinations with some probabilities. Zhu et al. [2] proposed an agent-based route choice (ARC) model where nodes, links and travellers were modelled as agents. The agents achieve an equilibrium by exchanging experience of previous route choices. The authors reported the results in terms of link flows from experiments on the Sioux-Falls network, which has 24 nodes and 76 links. They also demonstrated their approach on the Chicago Sketch Network with 933 nodes and 2950 links.

Kitamura and Nakayama [10] investigated the effect of distributing information about travel time to the travellers. The experiments were performed on a network with only three nodes and three links with two OD pairs. The investigation was based on the Minority Game (MG) model introduced by Challet and Zhang [11] which shows emerging cooperation among the agents by means of self-organisation without precise information. Galib and Moser [12] proposed using modified MG for TA. The approach was demonstrated on a network with 10 nodes, 24 links and 9 OD pairs. The network scenario was chosen to represent a 'difficult' situation with many overlapping routes.

In an attempt to assess the degree of challenge posed by the network, Bazzan and Klügl [1] discussed the complexity of the 6x6 grid they used for experimentation. Their explanation of the network complexity considers the distribution of OD combinations and defines the capacity of the roads as non-homogeneous. The researchers except Bazzan and Klügl [1], mentioned in this section, did not attempt to measure or assess the difficulty of their road network scenarios.

Defining the difficulty of a network scenario or measuring network complexity has been of interest in various disciplines [13,14]. In graph theory, the complexity of a graph can be measured by the number of spanning trees the graph contains [15] or by the minimum number of union and intersection operations required to obtain the whole set of its edges starting from the simplest graph (a star graph) [16]. Kim and Wilhelm [17] and Da Costa, Rodrigues and Traviseo [18] have surveyed some approaches which measure the complexity of graphs. Dehmer et al. [13] described how such complexity measures can be applied to chemical structures. Strogatz [14] explored several complications different types of networks may have and investigated the complexity of networks in different scientific fields. However, methods based on graph theory do not lend themselves for road networks because of the presence of OD pairs. Therefore, we attempt to quantify the difficulty of road network scenarios using link cross-inclusion between the OD pairs.

3 Quantifying Network Difficulty

3.1 Cross-Inclusion Factor

Road networks can differ in the number of nodes/intersections, number of links/roads, and the average degree of the network i.e. the average number of roads connected to an intersection, capacity and free flow travel time. From the point of view of TA, a crucial factor is how quickly one can expect to travel from one node in the network to another. From this point of view, one of the main contributors to network difficulty is the distribution of the origins and destinations within the network. If the possible paths between one OD pair are entirely separate from another, the difficulty of route choice depends entirely on the driver having current information about the route alternatives and their usages and capacities.

However, when links have to be shared between routes, the allocation task becomes more difficult. Assuming, realistically, that most drivers with experience of travelling from an origin to a destination are likely to allocate themselves to a set of the fastest routes they are aware of, we are suggesting a difficulty measure which is based on common link inclusions in the path options between OD pairs. To calculate link cross-inclusion, we first apply a breadth-first graph traversal technique which determines the shortest path between each intermediate node to the destination. To compile the set of intermediate nodes between each OD pairs, we first traverse the road network as a graph from the origin to the destination and then from the destination to the origin, collecting the traversed nodes into two sets. The intersection of the sets results in a set of links that comprise feasible routes between the origins and destinations.

Let $G(V, E)$ be a road network where V is the set of nodes (intersections) and E is the set of links (roads). The set of OD pairs is denoted by M, its cardinality $|M|$. The links belonging to the shortest paths are comprised in the sets $S_1 - S_{|M|}$. We denote as η the total number of links included in all shortest paths between the OD pairs in M. Here, ω_ℓ and $\tau_{\ell,0}$ are the capacity and free flow travel time of the link ℓ, respectively. Considering x_i as the proportion of travellers assigned to OD-pair i, we can calculate a measure of difficulty C_G for a network scenario as shown in Eq. 1.

$$C_G = \frac{\sum_{i=1}^{|M|} x_i \sum_{\ell \in S_i} \left(1 - \frac{\omega_\ell}{\sum_{\ell \in S_i} \omega_\ell}\right) * \frac{\tau_{\ell,0}}{\sum_{\ell \in S_{1-|M|}} \tau_{\ell,0}}}{\eta} \quad (1)$$

The cross-inclusion factor C_G considers the cross-use of links in shortest paths between origins and destinations, relative to all links used in shortest paths η. The link capacity ω_ℓ is normalised and inverted, as an increased capacity is expected to lead to a shorter travel time. Eq. 1 also considers the free flow travel times of the links τ_ℓ, which has the opposite effect on the outcome, as larger free flow travel times are mostly - but not exclusively - indications of the lengths of the links which increase the experienced travel times. It can be argued that the inclusion of all path options between each node to the destination is not necessary. However, when travellers choose a route from a node to the destination, their decisions often change on the fly and second and third shortest routes are included ad hoc. For this reason, we calculate the link cross-inclusion within the shortest paths from each intermediate node to the destination.

The cross-inclusion factor x represents the number of drivers allocated to a particular OD pair as a proportion of all travellers. In traffic experimentation, the number of travellers is often distributed uniformly among the OD pairs. In such a situation, the factor x_i in eq. 1 can be omitted. In realistic traffic scenarios, different percentages of travellers will be travelling between the OD-pairs. Traffic authorities can be consulted for the actual proportions in this case. In simulations, researchers may wish to vary the proportions of travellers between the OD pairs.

3.2 Discussion

It is clear that the metric proposed is not without its limitations; a network scenario may have several near-optimal routes from an origin to a destination, making it easy to find alternative routes with low impact on the travel times in a case where some of the routes are used by another OD pair.

Also, the heavy cross-inclusion of a few common links scores similarly on the cross-usage metric as the sharing of a multitude of links between few OD pairs. The latter situation is arguably less likely to have a severe impact on travel times.

A high cross-inclusion factor can also indicate that one or several links are shared between OD pairs without other options. In such a case, the optimisation problem is not difficult, as there are no options to optimise, but the resulting travel times are guaranteed to by high. Therefore, cross-inclusion sometimes identifies unavoidable delays rather than optimisation challenges. Even so, cross-usage can help adjust our expectation of travel times resulting from TA experimentation.

4 Result

Results of TA experiments can be shown in terms of travellers' travel times. One of the most common alternatives for calculating travel times for each link in transportation research is the BPR (Bureau of Public Roads) Volume Delay Function (VDF) [7], shown in Eq. 2.

$$\tau_\ell = \tau_{\ell,0} + [1 + \alpha(\frac{\rho_\ell}{C_\ell})^\beta] \quad (2) \qquad \tau_n = \sum_{\ell \in \Re_n} \tau_\ell \quad (3)$$

where τ_ℓ is the travel time on link ℓ, $\tau_{\ell,0}$ is the free flow link travel time for link ℓ, C_ℓ is the capacity of link ℓ, ρ_ℓ is the traffic volume on link ℓ and α and β are parameters. The time to travel from an origin to a destination for a traveller can be calculated by Eq. 3.

We simulate 1260 travellers in a road network of 1000 nodes and 2459 links. The scenario has 9 OD pairs with 886 overlapped links between the OD pairs resulting in a score of 2.74 on the difficulty measure. We subsequently removed the most included link at a time and recalculated C_G, repeating this process 9 times to obtain the scenarios with the new cross-inclusion factors shown in the legend of Fig. 1.

Higher cross-inclusion generally predicts longer travel times. Removing the link which is included in the most routes should, intuitively, lower the cross-inclusion factor because the heaviest contributor has been removed. In practice, the cross-inclusion factor can rise or fall depending on the alternatives included in the routes instead. The magnitude of the link cross-inclusion factor depends not only on the number of inclusions but also on the relative free flow travel time and the relative capacity of the link removed and the links taking its place.

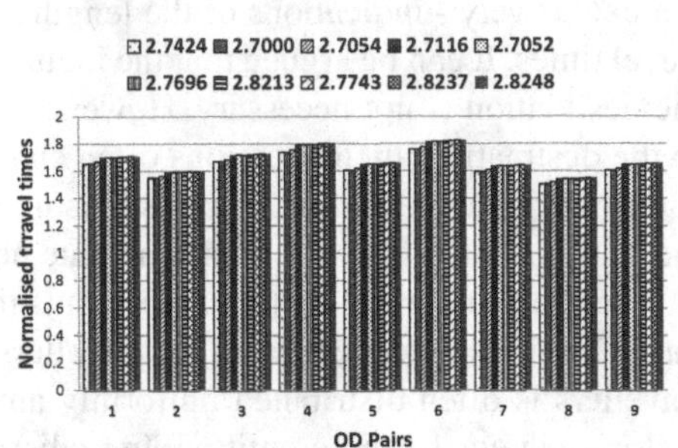

Fig. 1. Average travel times for OD pairs in a network where the most heavily included link is repeatedly removed

Fig. 1 illustrates that initially, links can be removed with the effect of lowering the link cross-inclusion factor, when there are still links which can be included instead of the removed link. However, as links are removed repeatedly, the cross-inclusion naturally has an increasing trend, as fewer links are available and the cross-use becomes more intense. Given the load is not adjusted, it is clear that the travel times also tend to rise.

5 Conclusion

Road traffic scenarios are largely dependent on the physical structure of the road network - the network topology, the distribution of the OD pairs and their geographical/physical locations. Traffic Assignment methods proposed by researchers mainly attempt to optimise the traffic distribution in the road networks. There has been a necessity to define the level of difficulty of a road network scenario. Therefore, in this paper, we have introduced a metric to quantify the difficulty level of a road network scenario. As the topology of the network and graph theoretic methods to measure network complexity do not sufficiently account for the difficulty of a road network scenario due to the presence of the OD pairs, we proposed that the number of overlapping routes between the OD pairs can be a good indication to determine the degree of difficulty of a road network scenario. A high score on the metric suggests that the network scenario is difficult for the travellers to avoid congestion than a scenario with a low score on the link-cross inclusion measure.

To evaluate a TA method, travellers using the method have to be simulated in realistic road network scenarios. The scenario can first be measured using the link cross-inclusion metric. This provides an idea about the difficulty level of the scenario that helps assess whether the TA method can easily avoid congestions. If there are many overlapping links, the scenario would have a high score indicating a higher possibility of congestion on some links even though an intelligent TA method is applied. Thus, the measure can help one reasonably expect the difficulty of optimising the traffic distribution in a road network scenario. Then the method can be simulated with a 'reasonable' or 'optimisable' traffic load in the road network scenario and finally the results of the method in terms of travel times can be compared with those of a benchmark method which has also been simulated with the same traffic load in the same scenario. Therefore, the proposed metric can act as a part of the framework to evaluate TA experimentations where proposing a benchmark TA approach with a traffic load categorisation technique is our future work.

References

1. Bazzan, A.L., Klügl, F.: Re-routing Agents in an Abstract Traffic Scenario (2008)
2. Zhu, S., Levinson, D., Zhang, L.: An agent-based route choice model. Working Papers 000089, University of Minnesota: Nexus Research Group (2007)
3. Sadek, A.W., Smith, B.L., Demetsky, M.J.: Dynamic traffic assignment: Genetic algorithms approach. Transportation Research Record: Journal of the Transportation Research Board 1588, 95–103 (1997)
4. Cruz, F., van Woensel, T., Smith, J.M., Lieckens, K.: On the system optimum of traffic assignment in state-dependent queueing networks. European Journal of Operational Research 201(1), 183–193 (2010)
5. D'Acierno, L., Montella, B., De Lucia, F.: A stochastic traffic assignment algorithm based on ant colony optimisation. In: Dorigo, M., Gambardella, L.M., Birattari, M., Martinoli, A., Poli, R., Stützle, T. (eds.) ANTS 2006. LNCS, vol. 4150, pp. 25–36. Springer, Heidelberg (2006)
6. de Dios Ortuzar, J., Willumsen, L.G.: Modelling Transport, 2nd edn. John Willey & Sons (1994)
7. Fricker, J.D., Whitford, R.K.: Fundamentals of transportation engineering: a multimodal approach. Pearson Prentice Hall, Upper Saddle River (2004)
8. Bazzan, A.L.: Traffic as a Complex System: Four Challenges for Computer Science and Engineering (2007)
9. Chen, O., Ben-Akiva, M.: Game-Theoretic Formulations of Interaction Between Dynamic Traffic Control and Dynamic Traffic Assignment. Transportation Research Record: Journal of the Transportation Research Board 1617(-1), 179–188 (1998)
10. Kitamura, R., Nakayama, S.: Can travel time information influence network flow? - Implications of the minority game. Transportation Research Record, 14–20 (2007)
11. Challet, D., Zhang, Y.C.: Emergence of cooperation and organization in an evolutionary game. Physica A 246(3-4), 12 (1997)
12. Galib, S.M., Moser, I.: Road traffic optimisation using an evolutionary game (2011)
13. Dehmer, M., Barbarini, N., Varmuza, K., Graber, A.: A Large Scale Analysis of Information-Theoretic Network Complexity Measures Using Chemical Structures. PLoS ONE 4(12), e8057 (2009)
14. Strogatz, S.H.: Exploring complex networks. Nature 410(6825), 268–276 (2001)
15. Constantine, G.: Graph complexity and the laplacian matrix in blocked experiments. Linear and Multilinear Algebra 28, 8 (1990)
16. Jukna, S.: On graph complexity. Combinatorics, Probability and Computing 15, 22 (2006)
17. Kim, J., Wilhelm, T.: What is a complex graph? Physica A: Statistical Mechanics and its Applications 387(11), 2637–2652 (2008)
18. da F. Costa, L., Rodrigues, F., Travieso, G.: Characterization of complex networks: A survey of measurements. Advances in Physics 56(1), 76 (2007)

Constraint Optimization for Timetabling Problems Using a Constraint Driven Solution Model

Anurag Sharma and Dharmendra Sharma

Information Technology and Engineering
University of Canberra, Australia
{Anurag.Sharma,Dharmendra.Sharma}@canberra.edu.au

Abstract. Many science and engineering applications require finding solutions to planning and optimization problems by satisfying a set of constraints. These constraint problems (CPs) are typically NP-complete and can be formalized as constraint satisfaction problems (CSPs) or constraint optimization problems (COPs). Evolutionary algorithms (EAs) are good solvers for optimization problems ubiquitous in various problem domains. A variation of EA - Intelligent constraint handling evolutionary algorithm (ICHEA) has been demonstrated to be a versatile constraints-guided EA for all forms of continuous constrained problems in our earlier works. In this paper we investigate an incremental approach through ICHEA in solving benchmark exam timetabling problems which is a classic discrete COP and compare its performance with other well-known EAs. Incremental and exploratory search in constraint solving has shown improvement in the quality of solutions.

Keywords: constraint satisfaction problems, constraint optimization problems, evolutionary algorithms, exam timetabling problems.

1 Introduction

Many engineering problems ranging from resource allocation and scheduling to fault diagnosis and design involve constraint satisfaction as an essential component that require finding solutions to satisfy a set of constraints over real numbers or discrete representation of constraints [12, 13, 19]. There are many classical algorithms that solve CSPs like branch and bound, backtrack algorithm, iterative forward search algorithm, local search but heuristic methods such as evolutionary algorithms (EAs) have mixed success and for many difficult problems these are the only available choice [2, 13, 17]. EAs however suffer from some of its inherent problems to solve CSPs as it does not make use of knowledge from constraints and blindly search in the vast solution space using its heuristic search mechanism. Constraints can reduce the search space and direct the evolutionary search towards feasible regions. Additionally *large* static CPs can be solved like a dynamic CP formulation where a subset of constraints is added incrementally. The incremental approach shows more effective results in terms of evaluation parameters of success rate (SR) and efficiency. SR is the rate of successful trials for each problem i.e. $SR = successful\ trials/\ total\ trials.$

S. Cranefield and A. Nayak (Eds.): AI 2013, LNAI 8272, pp. 196–201, 2013.

The main contribution of this paper is to show that incremental approach in solving constraints leads to better quality solutions. Incremental approach also helps in getting feasible solutions without any need to have a separate bespoke algorithm. We have enhanced the existing ICHEA [20, 21] to solve discrete COPs. Now ICHEA can be used to solve any form of a CP. The paper is organized as follows: Section 2 describes the cost function of the benchmark timetabling problems and Section 3 describes how the algorithm of ICHEA incrementally solves them. Section 4 demonstrates experiments on benchmark exam timetabling problems. Section 5 discusses the experimental results and Section 6 concludes the paper by summarizing the results confirming the claim against the established hypothesis and proposing some further possible extensions to the research.

2 Solving Exam Timetabling Problems

There are two types of constraints: if constraints are required to be satisfied under any circumstances to have an acceptable solution are known as *hard constraints*. Another type of constraints are called *soft constraints* that are considered to be desirable but not essential [3]. Solutions, which satisfy all the hard constraints, are often called *feasible* solutions. Soft constraints can have some degree of satisfaction or order of preferences for a particular problem. Soft constraints can be represented by penalty functions for COPs where higher weights demonstrate lower preferences and vice versa for higher preferences. We used University of Toronto benchmark exam timetabling problems (version I) given in [18, 23] where the given weights based on the spread of exams for each student is:

$$W_d = S_d 2^{4-d} \tag{1}$$

where d is the distance between two timeslots in the range [0 4], S_d is total corresponding students and W_d is the total corresponding weight. The cost function is the average weight corresponds to each student given as:

$$f = \frac{1}{S} \sum_{d=0}^{4} S_p 2^{4-d} \tag{2}$$

We mainly used our modified *Kempe* chain [8, 14] techniques for mutating timetables in ICHEA that also follows a reversible hill climbing technique that works like a backtracking algorithm. The details will be provided in the extended journal paper.

3 ICHEA Algorithm for Exam Timetabling Problems

Some *large* CPs like exam timetabling problems can be divided into several components (subsets of constraints) then each component can be solved incrementally. This divide and conquer approach solves a CP by taking a component to get feasible solutions before taking next component. In the literature, exam timetabling problems sort the constraints according to the largest degree (LD), saturation degree (SD), largest weighted degree (LWD), largest penalty (LP) or random Order (RO) [4, 7, 9]. LD and

SD are commonly used sorting order. In LD exams are ordered decreasingly according to the number of conflicts each exam has with others, and in SD the exams are ordered increasingly according to the number of remaining timeslots available to assign them without causing conflicts. The definition of other sorting orders can be found in [7]. ICHEA uses LD to sort all the exams based on clashes with other exams. It takes only 5% of the sorted exams in every increment and once a feasible solution is obtained the optimization operators are applied for G generations before taking next increment of exams. The value of G is 500 in our experiments.

Solving CPs incrementally has many advantages. Incrementality in solving CPs also comes handy when a new constraint is added or an existing constraint is changed. A by-product of incrementality in search is a set of generated partial solutions for each increment that can be stored separately and later reused, where a new constraint can be added or an existing constraint can be changed without making too much distortion to the current solution. More importantly incremental ICHEA does not have to define any problem specific algorithm to get feasible solutions as many other approaches like [1, 6, 14] use bespoke algorithms or SD graph-coloring heuristics to get the feasible solutions.

4 Experiments for Discrete COPs

Hyper-heuristics have been frequently used to solve benchmark exam timetabling problems which show promising results [8, 14]. ICHEA is a meta-heuristic algorithm that uses multiple mutation strategies to optimize a CP as described in Section 3. All the benchmark problems have been experimented on a Windows 7 machine with Pentium (R) i5 CPU 2.52 GHz and 3.24 GB RAM except the problem Pur93 which was run on a server machine (Intel Xeon CPU 2.90GHz and 128 GB RAM) because of its size and memory requirements. We ran all the problems overnight because of their size and complexity. Additionally, real world timetabling problem does not required to be solved within minutes or hours [1, 14]. Even though smaller sized problems like Hec92 and Sta83 can be solved within an hour or two; however problem Pur93 had to be run for almost 24 hours because of its huge size.

Table 1. Statistical summary of results from IICHEA and ICHEA

Instance	Best	Median	Worst	SD
Car91	4.91 (5.1)	5.04 (5.3)	5.16 (5.46)	0.01 (0.15)
Car92	4.08 (4.3)	4.1 (4.45)	4.2 (4.54)	0.05 (0.10)
Ear83	33.24 (33.6)	34.02 (34.69)	34.7 (37.39)	0.57 (1.24)
Hec92	10.13 (10.17)	10.33 (10.45)	10.61 (11.15)	0.15 (0.37)
Kfu93	13.58 (13.8)	13.8 (14.1)	14.21 (15.09)	0.20 (0.37)
Lse91	10.37 (10.95)	10.51 (11.34)	10.67 (11.8)	0.11 (0.27)
Pur93	4.67 (5.2)	4.78 (5.43)	4.99 (5.81)	0.12 (0.21)
Rye92	8.63 (9.07)	8.76 (9.4)	8.85 (9.7)	0.08 (0.19)
Sta83	157.03 (157.03)	157.03 (157.03)	157.03 (157.03)	0.0 (0.0)
Tre92	8.33 (8.8)	8.5 (9.3)	8.8 (9.6)	0.16 (0.28)
Uta92	3.28 (3.48)	3.41 (3.60)	3.57 (3.64)	0.07 (0.07)
Ute92	24.85(24.9)	24.9 (25.7)	25.1 (27.0)	0.10 (0.87)

All the experimental results have been verified through the standard evaluator program available in the dedicated website for research on benchmark exam timetabling problems [23]. We executed each problem for 10 trials to get SRs and establish statistical evaluation in Table 1.

Many times incremental approach to solve a complex static CP gives better results than solving entire constraints altogether. We used both approaches in the experiments to demonstrate supremacy of one approach over another. We observed that this incremental approach also helps in quickly providing feasible partial solutions and eventually feasible solutions at the SR of 100% for all the benchmark problems; whereas SRs of non-incremental ICHEA are very low for bigger problems like Car91 and Uta92 have only 0%-10% of SR, and 30%-70% for other problems of medium size. Non-incremental ICHEA also takes much longer duration to get the first feasible solution. The unpromising outcome from non-incremental ICHEA has led us to do the experiments with incremental ICHEA only. We first sort the constraints (exams clashes) according to LD then remove first 5% of the total exams as input for each increment in ICHEA. *Intermarriage* crossover constructs new partial feasible solutions which are then optimized using mutation strategies for feasible partial solutions. We used two instances of ICHEA for the experiments to demonstrate the validation of incrementality. The first and second instances of ICHEA optimize the partial solutions for 0 and 500 generations respectively. The only difference between these two instances is the first one does not apply optimization strategies to partial solutions while the other optimizes the partial solution for 500 generations. However, both instances get the feasible solutions incrementally. To distinguish the two instances the first one is called ICHEA and second one is called incremental ICHEA (IICHEA) as it fully exploits the notion of incrementality.

The statistical results of IICHEA and ICHEA on all the problems from University of Toronto benchmark exam timetabling problems (version I) from [18, 23] are shown in Table 1. We only used version I because it has been mostly reported in the literature. ICHEA results are in the brackets. We also compared our best solutions with other published results from [1, 5, 8, 10, 11, 14–16, 22] cited frequently in the literature in Table 2.

Table 2. Best results from the literature compared with IICHEA

Algorithms	Car91	Car92	Ear83	Hec92	Kfu93	Lse91	Pu93	Rye92	Sta83	Tre92	Uta92	Ute92	Yor83
IICHEA	4.9	4.1	33.2	10.1	13.6	10.4	4.7	8.6	157.0	8.3	3.3	24.8	36.2
[15]	7.1	6.2	36.4	10.8	14.0	10.5	3.9	7.3	161.5	9.6	3.5	25.8	41.7
[31]	5.1	4.3	35.1	10.6	13.5	10.5	-	8.4	157.3	8.4	3.5	25.1	37.4
[16]	5.4	4.4	34.8	10.8	14.1	14.7	-	-	*134.9*	8.7	-	25.4	37.5
[52]	4.5	3.9	33.7	10.8	13.8	10.4	-	8.5	158.4	7.9	3.1	25.4	36.4
[1]	5.2	4.4	34.9	10.3	13.5	10.2	-	8.7	159.2	8.4	3.6	26.0	36.2
[26]	5.2	4.3	36.8	11.1	14.5	11.3	-	9.8	157.3	8.6	3.5	26.4	39.4
[9]	4.6	3.8	32.7	10.1	12.8	9.9	4.3	7.9	157.0	7.7	3.2	27.8	34.8
[13]	4.9	4.1	33.2	10.3	13.2	10.4	-	-	156.9	8.3	3.3	24.9	36.3
[23]	4.5	3.8	32.5	10.0	12.9	10.0	5.7	8.1	157.0	7.7	3.1	24.8	34.6

5 Discussion

To solve a CSP ICHEA works with allele coupling only. So only the definition of constraints and the rules for coupling of two constraints need to be provided for *intermarriage* crossover. Experimental results for benchmark exam timetabling problems for COPs are very promising. Results for problems *Ear83*, *Hec92*, *Sta83*, *Tre92*, *Ute92* are in top three and other results are also in the upper half of the best results. It is noted that IICHEA has been giving consistent results for all the problems. It is noted that exam timetabling problems show good results with hyper-heuristics. Using the incrementality technique of IICHEA on these hyper-heuristics can produce even better results as shown in the comparative results between ICHEA and IICHEA. Incrementality in ICHEA produces better results than without incrementality. Consequently, IICHEA can also be used for real time discrete COPs. IICHEA also does not require having a separate problem specific algorithm to get feasible solutions as a preprocessor for constraint optimization. It has found feasible solutions for all the problems at the SR of 100%.

6 Conclusion

This paper focuses on incorporating ICHEA for solving discrete COPs. ICHEA has been designed as a generic framework for evolutionary search that extracts and exploits information from constraints. ICHEA has shown promising results experimented on benchmark exam timetabling problems. We proposed another version of *intermarriage* crossover operator for discrete CSPs to get the feasible solutions. Constraint optimization requires additional optimization techniques that are not all generic in its current form. ICHEA uses many problem specific mutation strategies to optimize exam timetabling problems. A major experimental observation was realizing the efficacy of incrementality in evolutionary search. Incrementality helps in getting feasible solutions with SR of 100% that also produces solutions of better quality. Incremental ICHEA can also be used for real time dynamic COPs in discrete domain. The competitive results from ICHEA shows its potential in making a generic evolutionary computational model that discovers information from constraints.

References

1. Abdullah, S., et al.: Investigating Ahuja–Orlin's large neighbourhood search approach for examination timetabling. Spectr. 29(2), 351–372 (2006)
2. Brailsford, S.: Constraint satisfaction problems: Algorithms and applications. Eur. J. Oper. Res. 119(3), 557–581 (1999)
3. Burke, E., et al.: A Time-Predefined Local Search Approach to Exam Timetabling Problems, vol. 1153, pp. 76–90 (2003)
4. Burke, E., et al.: Hybrid Graph Heuristics within a Hyper-Heuristic Approach to Exam Timetabling Problems (2005)

5. Burke, E., Bykov, Y.: A Late Acceptance Strategy in Hill-Climbing for Exam Timetabling Problems. Presented at the PATAT 2008 Proceedings of the 7th International Conference on the Practice and Theory of Automated Timetabling (2008)
6. Burke, E.K., Newall, J.P., Weare, R.F.: A memetic algorithm for university exam timetabling. In: Burke, E.K., Ross, P. (eds.) PATAT 1995. LNCS, vol. 1153, pp. 241–250. Springer, Heidelberg (1996)
7. Burke, E.K., et al.: Adaptive selection of heuristics for improving exam timetables. Ann. Oper. Res., 1–17 (2012)
8. Burke, E.K., et al.: Hybrid variable neighbourhood approaches to university exam timetabling. Eur. J. Oper. Res. 206(1), 46–53 (2010)
9. Caramia, M., Dell'Olmo, P., Italiano, G.F.: New Algorithms for Examination Timetabling. In: Näher, S., Wagner, D. (eds.) WAE 2000. LNCS, vol. 1982, pp. 230–241. Springer, Heidelberg (2001)
10. Carter, M.W., et al.: Examination Timetabling: Algorithmic Strategies and Applications. J. Oper. Res. Soc. 47(3), 373 (1996)
11. Casey, S., Thompson, J.: GRASPing the Examination Scheduling Problem. In: Burke, E.K., De Causmaecker, P. (eds.) PATAT 2002. LNCS, vol. 2740, pp. 232–244. Springer, Heidelberg (2003)
12. Craenen, B.G.W., et al.: Comparing evolutionary algorithms on binary constraint satisfaction problems. IEEE Trans. Evol. Comput. 7(5), 424–444 (2003)
13. Craenen, B.G.W.: Solving constraint satisfaction problems with evolutionary algorithms. Phd Dissertation, Vrije Universiteit (2005)
14. Demeester, P., et al.: A hyperheuristic approach to examination timetabling problems: benchmarks and a new problem from practice. J. Sched. 15(1), 83–103 (2012)
15. Eley, M.: Ant algorithms for the exam timetabling problem. In: Burke, E.K., Rudová, H. (eds.) PATAT 2007. LNCS, vol. 3867, pp. 364–382. Springer, Heidelberg (2007)
16. Merlot, L.T.G., Boland, N., Hughes, B.D., Stuckey, P.J.: A Hybrid Algorithm for the Examination Timetabling Problem. In: Burke, E.K., De Causmaecker, P. (eds.) PATAT 2002. LNCS, vol. 2740, pp. 207–231. Springer, Heidelberg (2003)
17. Müller, T.: Constraint-based Timetabling. PhD Dissertation, Charles University (2005)
18. Qu, R., et al.: A survey of search methodologies and automated system development for examination timetabling. J. Sched. 12(1), 55–89 (2008)
19. Shang, Y., Fromherz, M.P.J.: Experimental complexity analysis of continuous constraint satisfaction problems. Inf. Sci. 153, 1–36 (2003)
20. Sharma, A., Sharma, D.: ICHEA for Discrete Constraint Satisfaction Problems. In: Thielscher, M., Zhang, D. (eds.) AI 2012. LNCS, vol. 7691, pp. 242–253. Springer, Heidelberg (2012)
21. Sharma, A., Sharma, D.: Solving Dynamic Constraint Optimization Problems Using ICHEA. In: Huang, T., Zeng, Z., Li, C., Leung, C.S., et al. (eds.) ICONIP 2012, Part III. LNCS, vol. 7665, pp. 434–444. Springer, Heidelberg (2012)
22. Yang, Y., Petrovic, S.: A Novel Similarity Measure for Heuristic Selection in Examination Timetabling. In: Burke, E.K., Trick, M.A. (eds.) PATAT 2004. LNCS, vol. 3616, pp. 247–269. Springer, Heidelberg (2005)
23. Benchmark Exam Timetabling Datasets, `http://www.cs.nott.ac.uk/~rxq/data.htm`

Learning Risky Driver Behaviours from Multi-Channel Data Streams Using Genetic Programming

Feng Xie, Andy Song, Flora Salim, Athman Bouguettaya, Timos Sellis, and Doug Bradbrook

RMIT University, Mornington Peninsula Shire Council
{feng.xie,andy.song,flora.salim,athman.bouguettaya, timos.sellis}@rmit.edu.au, doug.bradbrook@mornpen.vic.gov.au

Abstract. Risky driver behaviours such as sudden braking, swerving, and excessive acceleration are a major risk to road safety. In this study, we present a learning method to recognize such behaviours from smartphone sensor input which can be considered as a type of multi-channel time series. Unlike other learning methods, this Genetic Programming (GP) based method does not require pre-processing and manually designed features. Hence domain knowledge and manual coding can be significantly reduced by this approach. This method can achieve accurate real-time recognition of risky driver behaviours on raw input and can outperform classic learning methods operating on features. In addition this GP-based method is general and suitable for detecting multiple types of driver behaviours.

1 Introduction

Road safety is a significant issue in modern society. Road fatalities and injuries cost the Australian economy 27 billion dollars a year [1]. More than 15,000 people have been killed in road accidents during 2002 to 2011 in Australia [2]. These tragic figures could be dramatically reduced if improper driving was instantaneously notified to the drivers and recorded for the stakeholders, as risky driving behaviours can cause road traumas. One cost-effective approach is performing driver behaviour recognition on smart phones which all have built-in accelerometer, gyroscope and magnetometer for sensing movements and directions.

Continuous digital signals produced by each sensor can be viewed as a time series. So the readings from all sensors can be considered as multi-channel time series. Therefore the task of risky driver behaviour recognition can be treated as a type of time series classification [3]. One difficulty of using this approach is to find a suitable set of features which can be used to differentiate risky behaviours from normal behaviours. Usually a feature set that is good for one problem is not suitable for a slightly different task. Furthermore, the duration or time-span of a target event should be defined beforehand although that might vary and even be unknown in real world scenarios. Even experienced road safety experts might not

S. Cranefield and A. Nayak (Eds.): AI 2013, LNAI 8272, pp. 202–213, 2013.

be able to provide a good set of features for various scenarios. The other approach which does not require features is to build a mathematical model for normal behaviours [4]. Any statistically significant change in the model would indicate a risky behaviour. However these modelling methods often are applicable on one variable, not suitable for multi-channel stream data. In addition, significant variations may exist even in the same type of driver behaviours. Beside these issues another important consideration is real-time performance.

In this study we aim to establish a Genetic Programming (GP) based learning method which can avoid the problem of finding suitable features. In particular the main goals are:

- How can a suitable method be established to learn patterns of risky driver behaviours from multi-channel smartphone data without using manually constructed features?
- How is this GP-based learning method compared to conventional learning methods on this time series classification problem, especially when multiple types of detection are required?
- How would the learnt detectors perform on real-time road tests?

To address the above questions three typical types of risky driver behaviours are studied: excessive acceleration, sudden braking and swerving (either left or right). Several classic algorithms were used for comparison including Decision Tree, Naïve Bayes, IB1, SVM and a boosting classifier. For these methods two sets of manually constructed features were used while our GP-based method directly operates on raw multi-channel stream data.

2 Related Work

Improving road safety by analyzing driver behaviours is an active research area. Wahlstrom et al. [5] used video camera to analyze the pupils of the driver to determine whether the driver is distracted or not. Oliver and Pentland used Coupled Hidden Markov Models (CHMMs) to predict driver behaviours which were captured by in-vehicle sensors such as video camera, face and gaze movement trackers, and the car internal state (speed, acceleration, steering wheel angle, gear, and brake) [6]. Their data was originated from a driving simulator rather than the real-world scenarios. In [7], methods to identify drunk driving behaviours in real-time were proposed. Two stages of ubiquitous data mining were applied, which is clustering and classification of driver behaviours. The major challenge was in linking the results of clustering models with the existing expert knowledge in the road safety field. Sensor data for the evaluation was generated using simulation based on an expert study, which categorised drunk driving behaviour into sober, borderline, drunk, and very drunk. The classification rules consisted of the following variables: number of correct responses, number of collisions, time over speed limit, reaction time, speed deviation, and lane deviation [7]. Due to the computational capability, smart phones have been widely used for behaviour or activity recognition [8]. Dai et al. proposed a method to detect drunk driving based on smart phone accelerometer [9]. In all above researches, domain knowledge is required so the solution can be problem-dependent. Our method can address this problem.

Another area of related work is on Genetic Program which has been demonstrated as a powerful problem solving mechanism especially for some complex tasks such as scheduling, structural design, object recognition and classification [10]. GP has been successfully applied in time series data analysis, including prediction [11,12] and pattern discovery [13]. GP has shown success in time series classification as well. Song and Pinto used GP on motion detection [14]. Xie et. al used GP for differentiating various events from time series data [15]. In the last two studies, GP were directly operating on raw input without any feature extraction process. These applications show the potential of GP being used in our study which is a complex time series classification task.

3 Data Collection

In this study sensor data was collected on an iPhone 5 when the subject is driving. The phone has three types of built-in sensors: accelerometer, gyroscope and magnetometer, all providing tri-axial measurements. The accelerometer read user acceleration as well as gravitational acceleration. The gyroscope measures the rotation of the device in rotation rates and the angles of the rotation in three dimensions as "roll", "pitch" and "yaw". The magnetometer reports the magnetic fields around the device. The readings will be from the earth magnetic field if there is no other detectable field. In total, these sensors provide 21 channels of input which are all used in this study (see Table 1). Without domain knowledge from a human expert, it is difficult to determine which channels are relevant or more important for detecting different types of driver behaviors.

We sample the 21-channel time series data at a frequency of $10Hz$. Figure 1(a) shows 500 time-intervals of original readings. It contains a swerving to the left and then to the right, and some gentle turning behaviors. Clearly the patterns of swerving are not obvious from the graph. After close examination of the data, we removed the readings of Yaw, Pitch, Roll (No. 10-12) and Magnetic Heading (No. 19-21) as shown in Figure 1(b). Then some patterns can be observed. Our GP-based learning method is expected to automatically capture these patterns while ignoring irrelevant data input.

Table 1. Sensor Channels available on iPhone

No.	Channel
1-3	Raw Acceleration X, Y, Z (raw accelerometer reading)
4-6	Gravity X, Y, Z
7-9	User Acceleration X, Y, Z
10 - 12	Yaw, Pitch, Roll
13 - 15	Raw Rotation Rate X, Y, Z (raw gyro reading)
16 - 18	Unbiased Rotation Rate X, Y, Z
19 - 21	Magnetic Heading X, Y, Z

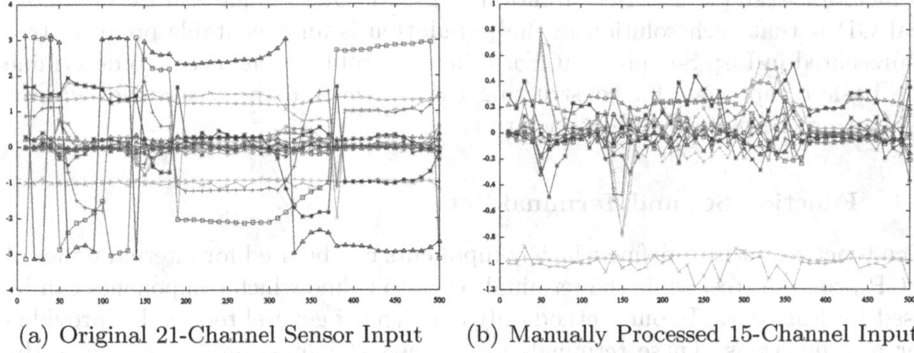

(a) Original 21-Channel Sensor Input (b) Manually Processed 15-Channel Input

Fig. 1. Examples of the Multi-Channel Sensor Input

During the data collection process, the phone was attached to the windshield in front of the driver. Because our approach is a type of supervised learning, all the data recorded are labelled for training and test purposes. Data labelling is done through voice command. When a person in the car feels uncomfortable with the impact of swerving, excessive acceleration or deceleration, the person will speak out and the data recorded at that time will be labelled as "*positive*". Otherwise the data will be labelled as "*negative*". One positive label indicates the occurrence of an event which is one of the three types of risky driver behaviors. The duration of an event is not recorded.

Table 2. Training and Test Data Set for Three Types of Driving Behaviours

	Training			Test		
	Total Instances	Positives	Negatives	Total	Positives	Negatives
1. Harsh Acceleration	1182	12	1160	690	9	681
2. Sudden Brake	1688	9	1679	947	6	941
3. Swerving	1206	12	1194	828	6	822

Table 2 presents the total number of instances, the number of positive instances and negative instances in the training and test datasets for detecting the three type of driver behaviors. As shown in the table, the data is highly unbalanced, much more negatives than positives. A suitable learning method should be able to handle such imbalance.

4 Methodology

As one of the main evolutionary computing techniques, the basic principle of Genetic Programming is also "survival of the fittest". A GP evolution process firstly generates a population of initial solutions for a particular problem, then reproduces the next generation of solutions by recombining or mutating the good solutions, and repeats this reproduction process until a certain stopping criteria is met, for example a perfect solution is found. One unique feature of canonical GP is that each solution in the population is an executable program tree represented in Lisp S-expression. For different problems one needs to determine the basic components for constructing these trees and the fitness measure for evaluating the performance of each tree.

4.1 Function Set and Terminal Set

The function set is to define which components can be used for internal nodes of a GP program tree, while the terminal set is to define which components can be used for leaf nodes. In our method only one type of general terminal is provided for any functions. These terminals are named "Channel_[m]" where m is the index of the channel. In addition some function nodes have their own specific terminals.

Table 3. Function Set

Function	Parameter		
	Parameter No	Type	Value
+	1	Double	[DOUBLE_MIN, DOUBLE_MAX]
	2	Double	[DOUBLE_MIN, DOUBLE_MAX]
−	1	Double	[DOUBLE_MIN, DOUBLE_MAX]
	2	Double	[DOUBLE_MIN, DOUBLE_MAX]
*	1	Double	[DOUBLE_MIN, DOUBLE_MAX]
	2	Double	[DOUBLE_MIN, DOUBLE_MAX]
/	1	Double	[DOUBLE_MIN, DOUBLE_MAX]
	2	Double	[DOUBLE_MIN, DOUBLE_MAX]
$Window$	1	Double	[DOUBLE_MIN,DOUBLE_MAX]
	2	Temporal-Index	$[1, 2^{window-size} - 1]$
	3	Temporal-Operation	AVG,STD,DIF,SKEWNESS
$Temporal\_Diff$	1	Double	[DOUBLE_MIN,DOUBLE_MAX]
$Multi\_Channel$	1	Double	[DOUBLE_MIN,DOUBLE_MAX]
	2	Channel-Index	$[1, 2^{num-of-channels} - 1]$
	3	Channel-Operation	AVG,STD,MED,RANGE

Table 3 shows all the function nodes designed for this task, including number of parameters, data type of each parameter and the range of parameter values for each function. All these functions return double value as output. Two functions are specifically designed for processing time series data, $Window$ and $Temporal\_Diff$. The last function $Multi\_Channel$ is designed to capture inter-variable dependency in multi-channel data. The details are explained below. The 2nd and 3rd parameters of $Window$ and $Multi\_Channel$ are special terminals.

Function *Window*. This function samples data points from a time series for analysis. It has three parameters: "Input", "Temporal Index" and "Temporal Operation". The first parameter records the current reading from the time series. The second parameter "Temporal Index" is for data point selection. This $Window$ function stores a sequence of data points of length specified in *window-size* (which is 12 in this study). Not every points should participate in the subsequent analysis. For example, if an event is short, then extra data points will be irrelevant and may even bias the results. In addition, using fewer points means less computational cost. "Temporal Index" is designed for this purpose. For example, if this number is set as 15 by the GP process, then the binary equivalent is 000000001111. So the last four data points will be selected for analysis as the last four digits are 1 while others are 0. Effectively it defines a sub-window size of 4 within the sampled data sequence. Furthermore, non-consecutive points can be selected in this way. For example an index value of 101010000000 (decimal 2688) would select the first, the third and the fifth data points for analysis.

The third parameter of *Window* is "Temporal Operation" which selects one of four operations: `AVG`, `STD`, `DIF` and `SKEWNESS`. The operation type is determined during GP evolution. They are terminals which will return the average value, the standard deviation, the sum of absolute differences and skewness of the points selected by parameter "Temporal Index". Presumably a poor selection of operation will result in poor accuracy, hence will not survive the evolutionary process.

Function *Temporal_Diff*. Function "Temporal_Diff" is similar to the standard temporal difference function. It returns the the deviation of two adjacent points on a time series. This function can be attached to other functions such as "Window" function. It can be nested so higher order of deviations can be extracted. This function only operates on two consecutive points therefore behaves similar to a "Window" function of size 2 with a DIF operator, despite the DIF operation returns absolute values.

Function *Multi_Channel*. Analyzing driver behaviours involves multiple channels. For example the pattern of swerving would appear in both accelerometer and gyroscope readings. Therefore capturing the dependences between channels would be helpful for detecting risky driver behaviours. So function "Multi_Channel" is introduced. It is similar to Function "Window", except it operates on data channels rather than data points inside of a window. The two parameters "Channel Index" and "Multi_Channel Operation" work the same way as the second and the third parameter of "Window" function. "Channel Index" has a value range between 1 and $2^M - 1$, where M is the number of channels. Moreover. The operations here are `AVG`, `STD`, `MED` and `RANGE` which return the average, the standard derivation, the median and the distance between the maximum and minimum input of a channel. It should be noted that this function works on the most recent data point. It does not record historical values.

4.2 Fitness Function

As illustrated in Section 3, it is natural that the behavioural data is dominated by negative instances. Hence accuracy is not a reliable measure of performance in this case. Instead we use AUC as the fitness measure.

AUC (Area under the ROC (Receiver Operating Characteristics) Curve) measures how far the two different classes can be separated. It is considered as better alternative to overcome the bias caused by data imbalance [16]. While AUC takes all possible thresholds into account to calculate the fitness value. In practice we expect the selected threshold can maximise True Positive Rate and minimise False Positive Rate at the same time. In this study we choose the threshold corresponding to the point on ROC curve with the shortest distance to the left top corner (the point representing 100% of True Positive and 0% of False Positive) [17].

5 Experiments

The parameters of our experiments on the GP method are listed in Table 4, which have not been tuned deliberately. Population is the number of solutions in each generation. A large population of 1000 is used to increase the chance of finding a good solution. The maximum and minimum depth of our GP tree are 8 and 2 respectively. Crossover rate is the probability of a solution being generated by swapping tree branches between two selected parents. Mutation rate is the probability of a solution being generated by mutating one parent. Elitism rate determines the proportion of good solutions which can be copied from the previous generation to the current generation. Each evolutionary process is repeated 10 times for each task. The best evolved programs are selected for testing. The stopping criteria is either the process has found an program which has 0% of recognition error, or the maximum generation 50 is reached.[1]

Table 4. GP Runtime Parameters

Population	1000
Generation	50
Maximum Depth	8
Minimum Depth	2
Mutation Rate	5%
Crossover Rate	85%
Elitism Rate	10%
Number of Runs	10

5.1 Methods for Comparison

Five classic classifiers were chosen for comparison purpose. They are *Decision Tree* in particular *J48* [2] [18], *Naïve Bayes (NB)* [19], *IB1* [20], *Support Vector Machine (SVM)* [21] and Adaboost [22]. The last one is a boosting classifier which takes the best of the other four classifiers as the base learner. For all these methods, two sets of features are manually created. There features are commonly used in time series classification such as Activity Recognition, which is very similar to our task which is to detect patterns from sensory data.

The first set of features are the temporal difference between two consecutive data points of each channel. The second set contains the average and the standard deviation of data points along the time series. A sliding window of size 12

[1] The conventional cross-validation is not particularly suitable for this streaming data, hence was not used in the experiments.

[2] The Java implementation of C4.5 in Weka.

is used to go through the time series and to return the two values for each time interval. The sliding window has the same size as the window size in "Window" function.

5.2 Results

The best programs generated from the GP evolutionary process are evaluated on test data. Table 5 shows the test results on raw sensor input data, including the accuracy, true positive rate (TP) and true negative rate (TN). All traditional classifiers performed very poorly on the three recognition tasks. Their high accuracies are due to the significantly unbalanced data. Their True Positive rate are quite low. The performance of classical classifiers are very poor in recognizing these three risky driver behaviours. Only J48 and Naïve Bayes obtained reasonable results on detecting swerving. In comparison our GP method identified all risky driving behaviours with minor false positives on all three tasks.

Table 5. Results on Raw Stream Data(%)

Tasks	J48	Naïve Bayes	IB1	SVM	AdaBoost	GP
1. Sudden Acceleration	95.73 TP : 33.3 TN : 96.6	77.76 TP : 66.7 TN : 77.9	98.23 TP : 33.3 TN : 99.1	98.53 TP : 0 TN : 99.9	98.23 TP : 33.3 TN : 99.1	**97.79** **TP : 100** **TN : 97.76**
2. Sudden Braking	99.36 TP : 50 TN : 99.7	99.36 TP : 0 TN : 100	99.36 TP : 0 TN : 100	99.36 TP : 0 TN : 100	99.36 TP : 50 TN : 99.7	**99.68** **TP : 100** **TN : 99.68**
3. Swerving	99.51 TP : 83.3 TN : 99.6	94.86 TP : 83.3 TN : 94.9	99.39 TP : 66.7 TN : 99.6	97.8 TP : 0 TN : 98.5	99.39 TP : 33.3 TN : 99.9	**99.02** **TP : 100** **TN : 99.01**

Table 6. Comparing with Conventional Methods on Pre-defined Feature Set 1

Tasks	J48	Naïve Bayes	IB1	SVM	AdaBoost	GP
1. Sudden Acceleration	98.09 TP : 33.3 TN : 99	47.28 TP : 100 TN : 46.6	96.9 TP : 0 TN : 98.2	98.67 TP : 0 TN : 100	98.67 TP : 0 TN : 100	**97.79** **TP : 100** **TN : 97.76**
2. Sudden Braking	69.8 TP : 67.3 TN : 71.4	39.96 TP : 50 TN : 39.9	99.36 TP : 0 TN : 100	99.36 TP : 0 TN : 100	99.47 TP : 16.7 TN : 100	**99.68** **TP : 100** **TN : 99.68**
3. Swerving	98.9 TP : 16.7 TN : 99.5	71.85 TP : 83.3 TN : 71.8	99.39 TP : 16.7 TN :100	99.27 TP : 0 TN : 100	99.27 TP : 0 TN : 100	**99.02** **TP : 100** **TN : 99.01**

To further compare the performance, we applied these traditional methods on extracted features which are manually designed. Table 6 and Table 7 show the results along with the GP results on raw data.

Feature set 1 did not bring much benefits, while feature set 2 did contribute towards much higher accuracies. J48 achieved the best performance over all the others on task 1 and task 3. Unfortunately, it did not perform well on detecting sudden braking. In comparison the performance of our GP method is consistent, and not statistically worse than J48 on Task 1 and 3. One may argue that with better features or combination of features, these traditional methods can achieve

Table 7. Comparing with Conventional Methods on Pre-defined Feature Set 2

Tasks	J48	Naïve Bayes	IB1	SVM	AdaBoost	GP
1. Sudden Acceleration	**99.26** **TP : 100** **TN : 99.3**	65.68 TP : 66.7 TN : 65.7	96.2 TP : 33.3 TN : 97	98.67 TP : 0 TN : 100	99.12 TP : 44.4 TN : 99.9	97.79 TP : 100 TN : 97.76
2. Sudden Braking	98.61 TP : 16.7 TN : 99.1	99.36 TP : 0 TN : 100	99.36 TP : 0 TN : 100	99.36 TP : 0 TN : 100	98.61 TP : 16.7 TN : 99.1	**99.68** **TP : 100** **TN : 99.68**
3. Swerving	**99.76** **TP : 100** **TN : 99.8**	94.86 TP : 100 TN : 94.8	99.39 TP : 100 TN : 99.4	99.27 TP : 0 TN : 100	99.51 TP : 100 TN : 99.5	99.02 TP : 100 TN : 99.01

good results. However that is exactly the point of this study. We would like to avoid this feature extraction step as a good feature set is often dependent on the problem and sensitive to the learning method.

6 Discussion and Analysis

To further test the detection performance, we selected the programs which obtained the highest test accuracy for each task and embedded them into an iphone app. Figure 2 shows the app in action. The left photo shows the phone detecting a sudden braking event and right photo shows the detection of a swerving event.

(a) Sudden Braking Detected

(b) Swerving Detected

Fig. 2. Evolved Detection Programs Running During Road Test

Our road test was performed on a slightly different condition compared to the original data collection exercises. The total driving time was over 20 minutes. The result from the road test is shown in Table 8. For three types of behaviors, there was no false negatives, meaning these trained GP programs did not miss any case of positives. For swerving, there was no false positives. There were some false

Table 8. Road Test

	True Positives	False Positives	False Negatives
1. Sudden Acceleration	3	2	0
2. Sudden Braking	2	3	0
3. Swerving	3	0	0

positives for detecting excessive acceleration and deceleration. These detection were actually caused by bumpy road surface. The person in the car could feel the movement at that time. This type of movement might generate a similar pattern as that of sudden acceleration and deceleration, hence triggered the detection program to report a positive. Inside the iphone app shown in Figure 2 there are

```
[1] Sudden Acceleration
Multi_Channel(AVG, 1779415) + Channe_14 /Window(STD, 751, Channe_12)
 - Window(AVG, 3847, Channel_10) - Window(STD, 580, Channel_18)

[2] Sudden Braking
Channel_8 + Channel_7 - Window(DIF, 1391, Channel_13 + Channe_13)

[3] Swerving
 - Window(STD, 1920, Channel_6) * (Temporal_Diff(Channel_4) +
 Channel_13) * (Channel_4 + Window(DIF, 1918, Channel_12)) +
 Channel_0 * Channel_13
```

Fig. 3. Examples of Evolved Programs for the Three Tasks

three evolved programs embedded. Each program is responsible for one type of driver behaviour. The sensor input streamed from 21 channels is processed by all of them. If one program reports positive, then the app shows and records the corresponding driver behaviour. Even with three programs running inside, the app can still achieve real-time performance because these GP-evolved programs are quite small. Examples are shown in Figure 3. They were for detecting sudden braking, sudden acceleration and swerving respectively. As we can see there are no more than 20 nodes on each program and there is no loop in any of them.

The exact algorithms evolved in these three programs are difficult to explain (which is an common issue in GP paradigm). However we can obtain some insights from these programs. The detecting program for sudden acceleration is a little complex. However it did not all 21 channel but 14 of them. Its Multiple Channel function takes input 1779415 of which the binary is 11011001001101101 0111. Most of the z-axis channels were not used.

The program for detecting sudden braking is simpler and only uses four channels: x-axis gravity, y-axis user acceleration, z-axis user acceleration and y-axis raw rotation rate. Its choice of channels seems understandable as three channels are related to acceleration and sudden braking will surely generate significant readings on acceleration. Moreover, this program did not use any magnetic sensors. For swerving, five channels are selected by the evolved program including x-axis raw acceleration, y-axis gravity, x-axis user acceleration, x-axis raw rotation rate, y-axis raw rotation rate. Intuitively gyro should play a more important role here.

7 Conclusions and Future Work

This study presents a GP-based methodology for learning the recognition of risky driver behaviours through smartphone sensor input. The advantage of this approach can be shown from the results in comparison with other classic methods. This approach can be directly applied on raw multi-channel sensor data without any pre-processing and manually designed features. This characteristic is very desirable in situations where domain knowledge is not clear or not available. By this approach, there is little need for road safety experts to manually determine what kind of accelerometer reading and other sensor readings should be considered as unsafe. Furthermore the evolved detection programs are small in size and low in complexity. They can achieve real-time performance on road tests. GP can provide a feasible solution for recognizing risky behaviours and help improve road safety.

This study can lead to a range of future work. We will continue collaborating with road safety experts and authorities to test our methods on more road conditions. Furthermore we will extend the recognition to more types of driver behaviours. Another future investigation is how to eliminate false positives caused by bumpy rides which may produce patterns similar to those of genuine risk driver behaviours. We also intent to further study the behaviours of these evolved GP programs to reveal the reasons behind their success. In addition, Hidden Markov approaches will be included in the comparison.

References

1. National road safety strategy 2011-2020. Australian Transport Council (May 2011)
2. A.D. of Infrastructure and Transport, "Road deaths australia 2011 statistical summary" (May 2012), `http://www.bitre.gov.au/publications/2012/files/RDA_Summary_2011.pdf`
3. Shahabi, C., Yan, D.: Real-time pattern isolation and recognition over immersive sensor data streams. In: Proceedings of the 9th International Conference on Multi-Media Modeling, pp. 93–113 (2003)
4. Guralnik, V., Srivastava, J.: Event detection from time series data. In: Proceedings of the fifth ACM SIGKDD International Conference on Knowledge Discovery and Data mining, KDD 1999, pp. 33–42. ACM, New York (1999)
5. Wahlstrom, E., Masoud, O., Papanikolopoulos, N.: Vision-based methods for driver monitoring. In: 2003 IEEE Proceedingsof Intelligent Transportation Systems., vol. 2, pp. 903–908. IEEE (2003)
6. Oliver, N., Pentland, A.P.: Graphical models for driver behavior recognition in a smartcar. In: Proceedings of the IEEE Intelligent Vehicles Symposium, IV 2000, pp. 7–12. IEEE (2000)

7. Horovitz, O., Krishnaswamy, S., Gaber, M.M.: A fuzzy approach for interpretation of ubiquitous data stream clustering and its application in road safety. Intelligent Data Analysis 11(1), 89–108 (2007)
8. Lu, H., Yang, J., Liu, Z., Lane, N., Choudhury, T., Campbell, A.: The jigsaw continuous sensing engine for mobile phone applications. In: Proceedings of the 8th ACM Conference on Embedded Networked Sensor Systems, pp. 71–84. ACM (2010)
9. Dai, J., Teng, J., Bai, X., Shen, Z., Xuan, D.: Mobile phone based drunk driving detection. In: 2010 4th International Conference on-NO PERMISSIONS Pervasive Computing Technologies for Healthcare (PervasiveHealth), pp. 1–8. IEEE (2010)
10. Poli, R., Langdon, W.B., McPhee, N.F.: A Field Guide to Genetic Programming. Lulu Enterprises, UK Ltd. (2008)
11. Wagner, N., Michalewicz, Z.: An analysis of adaptive windowing for time series forecasting in dynamic environments: further tests of the dyfor gp model. In: Proceedings of the 10th Annual Conference on Genetic and Evolutionary Computation, GECCO 2008, pp. 1657–1664. ACM, New York (2008)
12. Kaboudan, M.: Spatiotemporal forecasting of housing prices by use of genetic programming. In: The 16th Annual Meeting of the Association of Global Business (2004)
13. Hetland, M.L., Sætrom, P.: Temporal rule discovery using genetic programming and specialized hardware. Applications and Science in Soft Computing 24, 87 (2004)
14. Song, A., Pinto, B.: Study of gp representations for motion detection with unstable background. In: 2010 IEEE Congress on Evolutionary Computation (CEC), pp. 1–8. IEEE (2010)
15. Xie, F., Song, A., Ciesielski, V.: Event detection in time series by genetic programming. In: 2012 IEEE Congress on Evolutionary Computation (CEC), pp. 1–8 (June 2012)
16. Ling, C., Huang, J., Zhang, H.: Auc: a better measure than accuracy in comparing learning algorithms. In: Advances in Artificial Intelligence, pp. 991–991 (2003)
17. Liu, C., Berry, P.M., Dawson, T.P., Pearson, R.G.: Selecting thresholds of occurrence in the prediction of species distributions. Ecography 28(3), 385–393 (2005)
18. Quinlan, J.R.: C4. 5: programs for machine learning, vol. 1. Morgan Kaufmann (1993)
19. John, G.H., Langley, P.: Estimating continuous distributions in bayesian classifiers. In: Proceedings of the Eleventh Conference on Uncertainty in Artificial Intelligence, pp. 338–345. Morgan Kaufmann Publishers Inc. (1995)
20. Aha, D., Kibler, D., Albert, M.: Instance-based learning algorithms. Machine Learning 6(1), 37–66 (1991)
21. Hastie, T., Tibshirani, R.: Classification by pairwise coupling. In: Proceedings of the 1997 Conference on Advances in Neural Information Processing Systems 10, NIPS 1997, pp. 507–513. MIT Press, Cambridge (1998)
22. Freund, Y., Schapire, R.E., et al.: Experiments with a new boosting algorithm. In: ICML, vol. 96, pp. 148–156 (1996)

Particle Swarm Optimisation and Statistical Clustering for Feature Selection

Mitchell C. Lane, Bing Xue, Ivy Liu, and Mengjie Zhang

Victoria University of Wellington, P.O. Box 600, Wellington 6140, New Zealand
{Mitchell C. Lane,Bing.Xue,Mengjie.Zhang}@ecs.vuw.ac.nz,
Ivy.Liu@msor.vuw.ac.nz

Abstract. Feature selection is an important issue in classification, but it is a difficult task due to the large search space and feature interaction. Statistical clustering methods, which consider feature interaction, group features into different feature clusters. This paper investigates the use of statistical clustering information in particle swarm optimisation (PSO) for feature selection. Two PSO based feature selection algorithms are proposed to select a feature subset based on the statistical clustering information. The new algorithms are examined and compared with a greedy forward feature selection algorithm on seven benchmark datasets. The results show that the two algorithms can select a much smaller number of features and achieve similar or better classification performance than using all features. One of the new algorithms that introduces more stochasticity achieves the best results and outperforms all other methods, especially on the datasets with a relatively large number of features.

Keywords: Feature selection, Particle swarm optimisation, Statistical clustering.

1 Introduction

A machine learning technique (e.g. a classification algorithm) often suffers from the problem of high dimensionality. Feature selection aims to select a small subset of relevant features to reduce the dimensionality, maintain or increase the classification performance and simplify the learned classifiers [1].

Feature selection is a difficult task due mainly to the large search space and the feature interaction problem [2]. Most of the existing methods suffer from the problem of stagnation in local optima. Particle swarm optimisation (PSO) [3,4] is an arguable global search technique, which has been successfully applied to many areas, including feature selection [5,6]. In PSO, a candidate solution is represented by a particle in the swarm. Particles fly in the search space to find the optimal solutions by updating the velocity and position of each particle. In binary PSO (BPSO) [7], each particle is encoded as a binary string (i.e. "1" and "0"). The velocity value represents the probability of the corresponding dimension in the position taking value "1". The detailed description of BPSO is not presented here due to the page limit and it can be seen in [7].

Many statistical measures have been applied to form the evaluation function in a feature selection algorithm [1,8]. However, all of them are used in filter approaches, which can not achieve as good classification performance as wrapper approaches [1]. This paper uses a new statistical clustering method [9,10] that groups relatively homogeneous

S. Cranefield and A. Nayak (Eds.): AI 2013, LNAI 8272, pp. 214–220, 2013.

features together based on a statistical model. The method considers all features simultaneously and takes the feature interaction into account. Features in the same cluster are similar and they are dissimilar to features in other clusters. Since the feature interaction is an important issue in feature selection, the statistical feature interaction information found by the clustering method can be used to develop a good feature selection algorithm. However, this has seldom been investigated.

1.1 Goals

The overall goal of this paper is to investigate the use of statistical clustering information in PSO for feature selection. To achieve this goal, a statistical clustering method as a preprocessing step is performed on part of the training set to group features to different clusters. A simple greedy forward search (GFFS) is developed to select one feature from each cluster and then two new PSO based algorithms are proposed to search for a better combination of features from each cluster. Specifically, we will investigate:

- whether the simple GFFS can effectively use the clustering information to select a small number of features and achieve similar or even better classification accuracy than using all features,
- whether PSO with the clustering information produced by statistical clustering can achieve better performance than GFFS, and
- whether the introduction of a greater amount of stochasticity to the above new PSO based algorithm can further improve the classification accuracy.

2 Proposed Feature Selection Approaches

In this work, we use a newly developed clustering method proposed by Pledger and Arnold [9] and Matechou et. al. [10], which is not described here due to the page limit. The clustering method is performed as a preprocessing step on a small number of training instances to cluster features into different groups. Features in the same cluster are considered as similar features. Selecting multiple features from the same cluster may bring redundancy. Features from different clusters are more likely to be complementary to each other, which can increase the classification performance. Therefore, we first develop a simple greedy forward selection algorithm to select a single feature from each cluster to investigate whether the selected features can obtain similar or better classification performance than using all features. We then propose two BPSO based feature selection algorithms to search for a better feature subset.

2.1 Greedy Forward Feature Selection (GFFS)

GFFS is proposed based on the idea of sequential forward selection, where features are sequentially added to the feature subset, but the key part of GFFS is the use of the statistical clustering information.

GFFS starts with an empty feature set S and features are sequentially added into S according to the classification performance. Each individual feature is first used for classification on the training set. Features are then ranked according to the classification accuracy. The highest ranked feature that has the best classification performance is added to the feature subset S and other features in the same cluster are removed. For the remaining features, the feature combined with which S can achieve better classification performance than with others is added to S. The other features in the same cluster are removed. This procedure is repeated until all clusters have been visited and only one feature is selected from each cluster. The number of features selected by GFFS is the number of feature clusters, which is much smaller than the total number of features.

2.2 BPSO for Feature Selection Based on Maximum Probability (PSOMP)

Traditionally, when using PSO for feature selection, features are selected from the whole feature set [5,6]. In order to select one feature from each cluster, we first develop a new BPSO based feature selection algorithm named (PSOMP), where features are selected from each cluster according to the maximum probability calculated by BPSO.

When using PSO for feature selection, each feature corresponds to one dimension in the position and velocity. "1" in the position means the corresponding feature is selected and "0" otherwise. BPSO may select more than one feature from each cluster. Therefore, PSOMP is proposed to select a single feature from each cluster. When using clustering information, a cluster of features correspond to a number of dimensions. Selecting one feature from each cluster means only one of these dimensions in the position can be updated to "1". To achieve this, the maximum probability mechanism is developed in PSOMP, where the motivation is that the velocity in BPSO represents the probability of the corresponding dimension taking value "1" [7]. In terms of feature selection, the velocity represents the probability of a feature being selected, i.e. the feature with the highest velocity has the maximum probability to be selected. Therefore, PSOMP updates the position value of only one feature (with the highest velocity) to "1", and updates all the other position values in the same cluster to "0".

Algorithm 1 shows the pseudo-code of PSOMP. The classification performance of the selected features is used to form the fitness function in PSOMP. The number of features selected equals to the number of feature clusters.

2.3 PSOMP with Tournament Feature Selection (PSOTFS)

PSOMP is based solely upon the probability of each feature, which allows PSOMP to select a single feature from each cluster, but may result in the quick (premature) convergence of the swarm. To resolve this problem, a tournament feature selection operator is introduced to PSOMP to develop a new algorithm named PSOTFS.

The goal of using the tournament selection operator is to introduce some stochasticity to the swarm to ensure the diversity of the population. Note that the tournament selection operator is not applied to the individual particles in PSO, but to the features in the same cluster to select a sub-group of features. The tournament selection operator is applied before the position updating procedure in Algorithm 1 (after Line 10). It randomly selects a sub-group of features from a feature cluster. Then the maximum probability mechanism (in Line 11) is applied on the selected sub-group (instead of on the whole cluster in PSOMP) to find the feature with the highest probability in the

Algorithm 1: Pseudo-code of PSOMP

```
1  begin
2      initialise position x and velocity v of each particle,
3      random select one feature from each cluster;
4      while Maximum Iterations has been not met do
5          evaluate the classification performance of the selected features;
6          update pbest and gbest of each particle;
7          for i=1 to Swarm Size do
8              for d=1 to Dimensionality do
9                  update v_i ;                                   /* Update velocity */
10             for C=1 to Clusters Size do
11                 find the dimension (LD) with the largest velocity in the cluster C ;   /* feature with
                   the highest probability */
12                 update the position value in dimension LD to 1 ;          /* Update position */
13                 update other dimensions(features) in C to 0 ;             /* Update position */
14     calculate the training and testing classification performance of the selected features;
15     return gbest, the training and testing classification performance.
```

Table 1. Datasets

Dataset	# Features	# Instances	# Classes	# Clusters
Australian Credit Approval (Aus.)	14	690	2	7
Vehicle	18	846	4	5
German	24	1000	2	10
World Breast Cancer Diagnostic (WBCD)	30	569	2	8
Lung Cancer	56	32	3	7
Sonar	60	208	2	10
Musk Version 1 (Musk1)	166	476	2	12

sub-group. The position value of this feature is updated to "1" and that of all other features in the same cluster are updated to "0". Algorithm 1 can show the pseudo-code of PSOTFS by adding the tournament selection after Line 10 and replacing the cluster in Line 11 with the sub-group selected by the tournament selection.

3 Experimental Design

Seven benchmark datasets (Table 1) were chosen from the UCI machine learning repository [11] to test the performance of the proposed algorithms, GFFS, PSOMP and PSOTFS. The number of clusters obtained from the statistical clustering method is listed in the last column of Table 1. The instances in each dataset are split randomly into a training set (70%) and a test set (30%). K-Nearest Neighbour (KNN) with K=5 is used to evaluate the classification performance of the selected features.

The parameters of PSOMP and PSOTFS are set as follows: $w = 0.7298$, $c_1 = c_2 = 1.49618$, population size is 30, the maximum number of iterations is 100 and the fully connected topology is used. These values are chosen based on the common settings in the literature [4]. The size of the tournament feature selection in PSOTFS is half of the number of features in the cluster. On each dataset, GFFS obtained an unique solution because it is a deterministic algorithm. PSOMP and PSOTFS have been conducted for 50 independent runs on each dataset. Student's T-tests (Z-tests) are performed to compare their classification performances, where the significance level was selected as 0.05.

Table 2. Experimental Results

Dataset	Method	NO. of Features	Accuracy				Dataset	Method	NO. of Features	Accuracy			
			Ave (Best)	Std	T1	T2				Ave (Best)	Std	T1	T2
	All	14	70.05					All	18	83.86			
	GFFS	7	70.53		+			GFFS	5	84.84		+	
Aus.	PSOMP	7	73.43 (73.43)	0	+	+	Vehicle	PSOMP	5	84.41 (84.84)	8.1E-3	+	-
	PSOTFS	7	73.43 (73.43)	0	+	+		PSOTFS	5	84.84 (84.84)	1E-015	+	=
	All	24	68.33					All	30	92.98			
	GFFS	10	68.67		+			GFFS	8	89.47		-	
German	PSOMP	10	69.67 (72.00)	0.0047	+	+	WBCD	PSOMP	8	93.91 (94.74)	0.00574	+	+
	PSOTFS	10	69.67 (69.67)	1E-015	+	+		PSOTFS	8	92.98 (92.98)	0	=	+
	All	56	70					All	60	76.19			
	GFFS	7	90		+			GFFS	10	76.19		=	
Lung	PSOMP	7	80.2 (90.00)	0.0424	+	-	Sonar	PSOMP	10	75.65 (82.54)	0.0322	=	=
	PSOTFS	7	80.8 (90.00)	0.0337	+	-		PSOTFS	10	76.29 (85.71)	0.0337	=	=
	All	166	83.92										
	GFFS	12	79.02		-								
Musk1	PSOMP	12	80.8 (86.01)	0.0233	-	+							
	PSOTFS	12	81.62 (87.41)	0.0266	-	+							

4 Results and Discussions

Experimental results are shown in Table 2, where "T1" represents the results of the T-Test between the classification performance of each new algorithm and that of using all features. "T2" represents the results of the T-Test between the classification accuracy achieved by GFFS and that of PSOMP or PSOTFS.

Note that since each algorithm is only allowed to select a single feature from each cluster, the number of features selected by all the three algorithms is the same as the number of feature clusters. Therefore, each algorithm selected a significantly smaller number of features than the total number of features in the dataset.

Results of GFFS. According to Table 2, on five of the seven datasets, GFFS maintained or improved the classification performance by using only the selected small number of features. On the WBCD and Musk1 datasets, although the classification performance of GFFS is slightly decreased, the number of features is significantly reduced. The results suggest that this simple greedy forward selection algorithm can utilise the information provided by the clustering method to effectively reduce the number of features and achieve similar or higher classification performance than using all features.

Results of PSOMP. Table 2 shows that PSOMP achieved significantly higher classification accuracy than using all features on five of the seven datasets and similar performance on one dataset. Although on Musk1, the average classification performance of PSOMP is slightly (3%) lower than using all features, PSOMP removed around 92% of the original features, which considerably reduced the classification time and dimensionality. Meanwhile, the best classification performance of PSOMP is 2% higher than using all features. The results suggest that PSOMP using the statistical information to guide the search of BPSO can successfully address feature selection problems.

PSOMP discovered feature subsets with significantly better or similar classification performance to GFFS in most cases. The results suggest that PSOMP using BPSO as the search technique can better search the solution space to obtain better results than GFFS. Meanwhile, rather than obtaining a single solution by GFFS, PSOMP can generate multiple results, which has a higher probability to achieve better performance.

Results of PSOTFS. According to Table 2, PSOTFS selected a significantly smaller number of features and achieved similar or significantly higher accuracy than using all features on six of the seven datasets. Only on the Musk1 dataset, the average classification accuracy of PSOTFS is around 2% lower than using all features, but its best accuracy is around 4% higher and it selected only around 7% of the original features. PSOTFS achieved similar or higher accuracy than GFFS on six of the seven datasets. Compared with PSOMP, PSOTFS achieved similar performance to PSOMP on five of the seven datasets, where the number of features is relatively small. On the Sonar and Musk1 datasets with a slightly larger number of features, PSOTFS outperformed PSOMP in terms of both the average and the best classification performance.

The results suggest that a greater amount of stochasticity in PSOTFS maintains the swarm diversity to avoid premature convergence. Therefore, PSOTFS achieved higher classification accuracy than PSOMP in most cases, especially on the datasets with a larger number of features and the solution space is more complex.

5 Conclusions

The goal of this paper was to investigate the use of statistical clustering methods in PSO for feature selection. The goal was successfully achieved by developing two new PSO based feature selection approaches, PSOMP and PSOTFS, to select a single feature from each cluster. The proposed algorithms are compared with a simple greedy forward feature selection algorithm (GFFS) on seven datasets. The experiments show that by using the statistical clustering information, GFFS selected a small number of features and achieved better classification performance than using all features. The basic PSOMP outperformed GFFS in most cases and PSOTFS achieved better classification performance than PSOMP because of the introduction of stochasticity to the swarm.

This study is a preliminary work of successfully using statistical clustering in feature selection, which motivates us to further investigate this research topic, such as using PSO to select multiple or zero features from each cluster to further improve the performance and using statistical clustering information for feature construction.

References

1. Dash, M., Liu, H.: Feature selection for classification. Intelligent Data Analysis 1, 131–156 (1997)
2. Guyon, I., Elisseeff, A.: An introduction to variable and feature selection. The Journal of Machine Learning Research 3, 1157–1182 (2003)
3. Kennedy, J., Eberhart, R.: Particle swarm optimization. IEEE International Conference on Neural Networks 4, 1942–1948 (1995)
4. Shi, Y., Eberhart, R.: A modified particle swarm optimizer. In: IEEE International Conference on Evolutionary Computation (CEC1998), pp. 69–73 (1998)
5. Xue, B., Zhang, M., Browne, W.: Particle swarm optimization for feature selection in classification: A multi-objective approach. IEEE Transactions on Systems, Man, and Cybernetics, Part B: Cybernetics (2012), doi:10.1109/TSMCB.2012.2227469

6. Wang, X., Yang, J., Teng, X., Xia, W.: Feature selection based on rough sets and particle swarm optimization. Pattern Recognition Letters 28, 459–471 (2007)
7. Kennedy, J., Eberhart, R.: A discrete binary version of the particle swarm algorithm. In: IEEE International Conference on Systems, Man, and Cybernetics, Computational Cybernetics and Simulation, vol. 5, pp. 4104–4108 (1997)
8. Bach, F.R., Jordan, M.I.: A probabilistic interpretation of canonical correlation analysis. Technical report (2005)
9. Pledger, S., Arnold, R.: Multivariate methods using mixtures: correspondence analysis, scaling and pattern detection. Computational Statistics and Data Analysis (2013), `http://dx.doi.org/10.1016/j.csda.2013.05.013`
10. Matechou, E., Liu, I., Pledger, S., Arnold, R.: Biclustering models for ordinal data. In: Presentation at the NZ Statistical Assn. Annual Conference. University of Auckland (2011)
11. Bache, K., Lichman, M.: UCI Machine Learning Repository (2013)

Evaluating the Seeding Genetic Algorithm

Ben Meadows[1], Patricia J. Riddle[1],
Cameron Skinner[2], and Michael W. Barley[1]

[1] Department of Computer Science, University of Auckland, NZ
[2] Amazon Fulfillment Technologies, Seattle, WA

Abstract. In this paper, we present new experimental results supporting the Seeding Genetic Algorithm (SGA). We evaluate the algorithm's performance with various parameterisations, making comparisons to the Canonical Genetic Algorithm (CGA), and use these as guidelines as we establish reasonable parameters for the seeding algorithm. We present experimental results confirming aspects of the theoretical basis, such as the exclusion of the deleterious mutation operator from the new algorithm, and report results on GA-difficult problems which demonstrate the SGA's ability to overcome local optima and systematic deception.

Keywords: genetic algorithm, evolutionary algorithm, seeding genetic algorithm, seeding operator.

1 Introduction

The field of evolutionary computation is well-established, but still poorly understood. Evolutionary algorithms' behaviour, particularly the low-level interaction of their constituent parts, is often counterintuitive. For example, it has been demonstrated [1] that a naive selection scheme can lead to reduced performance of the evolutionary algorithm through extinction of the fundamental *building blocks.* that appear in the solution. While the *crossover* operator is capable of propagating and recombining building blocks, its capacity for recombination is often stunted in practice by low building block diversity. Conversely, operators that are good at discovery (e.g., mutation, random search) tend to be much more disruptive of existing building blocks. We claim that it should be possible to (a) explicitly segregate building block discovery and combination phases and optimise them independently, and (b) adjust a selection scheme to greatly reduce building block extinction without unduly reducing performance.

Skinner and Riddle [2,3] have addressed these claims with work on a *seeding operator* that significantly outperforms mutation on the task of building block discovery, reduces extinction events, and is able to navigate local optima. The primary purpose of this paper is to report the results of further experiments with this new operator in an exploration and delimitation of the abilities of an algorithm using it: the *Seeding Genetic Algorithm* (SGA). This paper has been shortened for publication; an expanded version is available online. [4] In the remainder of this paper we report on a series of experiments examining the SGA's behaviour on four problems, establishing effective and generalisable parameters for the algorithm, and demonstrating the seeding operator's flexibility.

S. Cranefield and A. Nayak (Eds.): AI 2013, LNAI 8272, pp. 221–227, 2013.

2 The Seeding Genetic Algorithm

Perhaps the core problem for the GA is maintaining population diversity. Numerous techniques have been demonstrated to combat early convergence (e.g., crowding [5]). However, few provide a strong guarantee that they indefinitely prevent premature convergence. The SGA makes such a guarantee. [2,3] Skinner's analysis [2,3] elicited a number of key concepts, including: (a) the positive correlation between number of bits per block and importance of high discovery rate; (b) the distinctive 'rapid combination' and 'slow completion' phases of the Canonical Genetic Algorithm (CGA); (c) The generally poor discovery-to-destruction ratios of mutation and crossover;[1] (d) The capacity of random search to outperform common genetic operators; and (e) Discovery in the CGA is largely due to crossover, rather than mutation. These motivated the creation of the Seeding Genetic Algorithm, which does away with mutation and divides the functionality of the genetic algorithm into selection, discovery and combination.

The algorithm's basic functionality is as folllows. In a single initial stage, it samples the search space with random search, selecting candidates of above-average fitness to create a 'seed pool' containing a number of the lowest-level building blocks in the problem. It then runs as a GA, but during the crossover step candidate parents from the seed pool may be inserted into the main population via seeding. Besides serving the role of discovery operator, seeding has two additional effects. When a particular building block becomes extinct in the main population the seed pool retains a copy that can then be incorporated into the population via recombination. Furthermore, those parts of seed pool individuals' genomes that do not contribute to fitness will be copied into the main population along with the building blocks, boosting diversity and preventing early convergence. For further details, see the long form of this paper. [4]

3 Experimental Results

We will report on a series of experiments that examine how the SGA behaves on a number of well-known fixed-length problems. We choose problems with known building block structure, on the basis that the GA's power comes from its ability to combine different building blocks. In particular we are interested in the ability of the algorithm to discover new building blocks and to select pairs of parents that have some hope of producing highly fit offspring.

We report on three problem sets. The classic *Royal Road* problem[2] was designed as a GA-easy problem, but found to be unusually difficult. It lacks deception, and has discrete building blocks and a single solution. The more difficult recursive *Hierarchical If-And-Only-If* [7] problem[3] has multiple global optima.

---

[1] Mutation can only discover new blocks where local search would be useful, and may destroy more than 99% of the blocks it discovers.

[2] Our implementation of this problem, *RR*, following Mitchell and Forrest [6], is a 64-bit Royal Road consisting of eight adjacent, coherent blocks of eight bits each.

[3] We implement a Hierarchical If-And-Only-If problem, $HIFF$, of length 64.

HIFF has 254 blocks of six different lengths, with massive interdependence, a massively vertiginous fitness landscape,[4] and two possible maximally different solutions. The *Deceptive Trap* problem[5] is designed to foil optimisation algorithms such as gradient descent; Deceptive Trap does so at the building block level. In this deceptive problem, [8] each block decreases in fitness as it comes closer to its complete form. Incremental improvements for a block lead the algorithm towards a local maxima. A complete inversion is required to go from the trapped state to the maximum for that block. Any hill-climbing type search will lead directly away from the global maximum.

In the following sections, we use the *bootstrap test* [9] for statistical significance. This usage is motivated by our having censored data (not all runs of the algorithm complete). The test draws many "bootstrap samples" from the combined results of two algorithms A and B to establish whether the difference in their means is significantly different from the 0 predicted by the null hypothesis. We set $p < 0.05$ to determine statistical significance. Extra numerical results and descriptions not given here are available online. [4]

Seeding Probability against Mutation Rate. We first apply an experimental framework to the question of whether there is any point to retaining mutation alongside seeding; the theoretical work of Skinner determined mutation to be ineffectual as a discovery operator, but did not test this. We ran tests on RR with 2-point crossover, a presample size of 1000, and a seed pool size of 50. We took the mean of 100 runs each time, testing various combinations of parameters, including seeding probabilities varying from 0.1 to 0.3, 1/64 or 1/1024 mutation rates, and five forms of tournament selection, rank selection and fitness proportional selection. For lower seeding probabilities, the SGA always significantly outperformed the CGA, and mutation reduced its performance. The 'best' level of mutation appeared to be at least partly dependent on the seed probability, suggesting there is some interference between the operators. We discard the mutation operator for the remainder of our tests.

Seeding with Random Individuals. We wish to test Skinner's [3] theory-based statement that the usefulness of seeding lies in introducing 'superior' genetic material (building blocks) – as opposed to the benefit it provides by preventing early convergence. To this end, we compare the CGA to a form of SGA that seeds with randomly generated individuals.[6] The test was run on RR with 5-tournament and 2-point crossover, for five different seed probabilities. In every comparable instance, the CGA outperformed the random-seeding SGA. We conclude that the seeding operator is not just enforcing diversity, but introducing the building blocks vital to the algorithm.

---

[4] Every (n)-value local or global optimum has two smaller, less-fit ($n-1$)-value fitness peaks equidistant from it in solution space.

[5] Our implementation, DT, is a 60-bit string consisting of 12 traps of length 5 each.

[6] Either mutation or seeding could technically introduce new blocks, but the chance is remote; we are thus effectively comparing the power of the mutation operator in the CGA and the power of the seeding operator in the SGA to prevent early convergence.

Table 1. Mean generations for the SGA to complete the RR problem; comparing presample size (rows) and seed probabilities (columns). The best value is given in **bold**. Values worse than those for the CGA are given in *italics*. Numbers presented here are normalised for the cost of seed pool generation.

	0.1	**0.15**	**0.2**	**0.25**	**0.3**
500	*363.7*	*315.7*	*323.4*	*342.2*	*506.1*
1000	221.7	181.8	169.7	196.7	*394.8*
2000	126.2	99.1	91.5	107.1	235.9
5000	122.9	98.5	95.1	96.4	229.9
10000	123.3	94.9	**88.2**	101.4	199.3
20000	118.6	99.9	97.9	103.8	189.6
30000	124.6	106.6	107.4	111.4	177.3
40000	128.6	112.5	111.8	118.5	165.5
50000	135.5	122.4	119.9	125.4	167.2
100000	189.6	185.8	182.9	188.2	206.3
125000	225.2	220.4	220.8	222.9	238.6
150000	*265.2*	*258.9*	*258.4*	*260.9*	*275.6*

We now turn to investigation of the various parameters of the SGA. We will *normalise* results of the SGA to account for the computational cost of seed pool generation,[7] by assuming that the computational bottleneck is the number of fitness evaluations performed.[8] [3]

Presample Size and Seed Probability. We first varied the seeding probability against the presample size. Skinner [3] used a presample size of 1000, and chose the probability $\frac{1}{3}$ so that approximately half of the crossover operations would involve replacement of at least one parent. We conducted experiments varying the size of the presample from 500 to 750000. We tested seeding probabilities in the range 0.1 to 0.35, using typical parameters, including 5-tournament selection, 2-point crossover, and seed pool size $n = 50$.

The results for RR with 2-point crossover are given in Table 1. The CGA completed in a mean 256.6 generations. The SGA was able to reduce this by as much as 233 generations (a 90% reduction, or 66% after normalisation). While a seed probability near 0.2 is best, the seeding algorithm significantly outperforms the CGA with seed probabilities between 0.1 and 0.3. The algorithm encounters diminishing returns with presample sizes above 150,000. The SGA performed comparably with different variations and parameterisations of the problem.

The CGA did not succeed on HIFF: one in 1000 runs succeeded with 2-point crossover; the "weaker" selection operators (rank selection and 2-tournament) were more able to escape local optima and performed slightly better. The SGA solved HIFF fairly easily, but even with a presample of 500,000, at least 10% of the runs did not complete. We therefore ran a set of experiments of 100 runs each, increasing the maximum number of generations from 1000 to 20,000, using the higher seed probabilities earlier experiments indicated. According to Table 2

[7] Without normalisation, an arbitrarily large presample will always be 'best' in the sense that at some point, the presample will actually include a solution string.

[8] The canonical GA makes N fitness calculations per generation, depending on various factors; a presample size of N is then equivalent to running one extra generation of the CGA, and so a presample of size X has an additional cost of X/N.

Table 2. Tests of the SGA on deceptive problems: HIFF (left) and DT (right). Each value is a mean over 100 runs up to 20,000 generations. Rows are different presample sizes; columns are different seed probabilities. The best value for each problem is given in **bold**. Numbers presented here are normalised for the cost of seed pool generation.

	0.2	0.25	0.3	0.35
100	8016.4	3932.5	7794.6	N/A
500	5029.7	2860.9	6794.4	N/A
1000	6521.5	2149.4	6301.2	10152.0
5000	5781.4	2261.7	4318.8	8800.1
10000	4239.2	1774.0	3087.8	9409.9
50000	3309.3	1291.0	2019.4	6071.1
100000	3505.3	1480.9	1109.4	4404.2
200000	1734.1	1126.4	1257.7	2981.6
300000	2845.7	1467.7	**1084.5**	2365.3
400000	3397.8	1394.7	1099.4	2379.8
500000	2505.6	1336.9	1101.9	2232.5

	0.1	0.15	0.2
100	3182.8	4279.3	4684.1
500	1458.9	1623.0	1566.1
1000	1063.0	1251.1	966.7
5000	585.0	563.0	556.2
10000	513.6	386.3	458.0
25000	401.9	394.8	331.5
50000	372.0	**282.8**	307.4
75000	355.6	332.9	325.5
100000	371.0	360.1	356.5
150000	492.3	397.6	417.5
200000	491.4	479.5	481.1

(left), the CGA remained inadequate on HIFF, but the SGA solved it rapidly (after normalisation) with a seed probability[9] of 0.3 and presample size around 300,000. Even a relatively small presample size is still useful: all runs found a solution for seed probabilities up to 0.3 with presamples of size ≥10,000.

The CGA *cannot* solve DT: it did not succeed once in a total of over 15,000 tests. Remarkably, the SGA solved DT even more easily than HIFF. With a large presample of 500,000, the SGA is sufficiently powerful to solve the DT problem every time, but for smaller presamples, not all runs completed. Again, we ran additional tests to 20,000 generations. The SGA then consistently found a solution for all tested seed probabilities when the presample was at least 25,000. Table 2 (right) suggests that the SGA had a 'spread' of generally good performance for presamples ranging from tens of thousands to hundreds of thousands. Even when a seed pool of 50 was drawn from a presample of 100, the SGA was sufficiently robust to solve DT more than two-thirds of the time.

Presample Size and Seed Pool Size. To find the best seed pool size for given presample sizes, we varied the size of the seed pool from $n = 5$ to 1000, fixing the seed probability at a previously discovered reasonable level. Given that Section 3 showed that diversity maintenance is not the seeding operator's most important contribution, we expected using a smaller proportion of the presample (i.e., seed quality) to be at least as important as sheer presample size (i.e., seed diversity).

We ran experiments on RR with a seed probability of 0.2. Our initial seed pool size of 50, following Skinner, [3] was a good choice (95.1 normalised generations for presample 5000); 100 may have been slightly better (78.0 normalised generations), but not to a statistically significant level. The viable seed pool size increases with the presample size, but the optimal size is fairly consistent. The SGA was able to improve on the performance of the CGA (256.6 generations) with as small a seed pool as 25. On the HIFF problem, using the previously-determined 'good' seed probability of 0.3, we found that the SGA performs well

[9] The algorithm becomes steadily more effective with higher seed probabilities before dropping off suddenly after 0.3. Interestingly, this value represents a ~50% chance that at least one parent in a reproduction event will be replaced by a seed individual.

with a seed pool size around 25, finishing in a mean normalised 401.7 generations at a presample size of 50,000. Diminishing returns appeared before the presample reached 100,000 (much lower for larger seed pools). We ran the same experiments on the DT problem, using the 'good' seed probability 0.175. We found a trend towards a larger seed pool size than usual; with increasing presample size, the first seed pool size on which all runs completed was 1000; the second was 500. 98.8 - 99.9% of runs completed for seed pool sizes 50 and 100 and presample size 50,000, in a mean ∼305 generations.

4 Concluding Remarks

We confirmed the SGA'S efficacy compared to the CGA. A continual supply of all building blocks combined with a suitable selection scheme leads to a significant reduction in number of fitness evaluations required. We have demonstrated an 80% or greater reduction in the difficulty of problems the CGA struggles with, and the ability to solve problems the CGA cannot, including those with extreme local optima and block-level deception. We have run the SGA over 1,200,000 times to obtain an initial set of 'reasonable' parameters.[10]

Many extensions to the CGA have addressed the failings we overcame here. The utility of the seeding operator lies in its ability to prevent early convergence *and* overcome local optima *and* solve deceptive problems.[11] Much future work is possible, including exploring the relationship between selection pressure and seed probability, refinement of the selection operator and replacement strategy[12] to take advantage of the SGA's strengths, and testing the algorithm on real-world domains. For further comments, see our extended online paper. [4]

References

1. Forrest, S., Mitchell, M.: Relative building-block fitness and the building-block hypothesis. In: Whitley, L. (ed.) Foundations of Genetic Algorithms, pp. 109–126 (1993)
2. Skinner, C., Riddle, P.: Random search can outperform mutation. In: IEEE Congress on Evolutionary Computation, CEC 2007 (2007)
3. Skinner, C.: On the discovery, selection and combination of building blocks in evolutionary algorithms. PhD thesis, Citeseer (2009)
4. Meadows, B., Riddle, P., Skinner, C., Barley, M.: Evaluating the seeding genetic algorithm (2013), `http://www.cs.auckland.ac.nz/~pat/AI2013-long.pdf`

[10] For real-world domains, 'best' parameters will be unknown, but we note that default parameters of 2-point crossover, 0.9 crossover rate, 0.2 seed probability, a seed pool of 50 and a presample of 100,000 surpassed the CGA on all of our problems.

[11] The fraction of high-quality individuals in the next generation can be calculated in terms of their parents, making the rate *fixed* over a run. This gives the SGA an advantage over other GA extensions which only *delay* early convergence.

[12] A replacement scheme that varied the seed probability throughout the run might deal better with the increasing homogenisation of fitter individuals.

5. De Jong, K.A.: Analysis of the behavior of a class of genetic adaptive systems. PhD thesis, University of Michigan Ann Arbor, MI (1975)
6. Mitchell, M., Forrest, S.: B. 2.7. 5: Fitness landscapes: Royal road functions. Handbook of evolutionary computation (1997)
7. Watson, R.A., Pollack, J.B.: Recombination without respect: Schema combination and disruption in genetic algorithm crossover. In: Proceedings of Genetic and Evolutionary Computation Conference (GECCO), pp. 112–119 (2000)
8. Goldberg, D.E.: Simple genetic algorithms and the minimal, deceptive problem. Genetic Algorithms and Simulated Annealing 74 (1987)
9. Cohen, P., Kim, J.: A bootstrap test for comparing performance of programs when data are censored, and comparisons to Etzioni's test. Technical report, University of Massachusetts (1993)

A Constructive Artificial Chemistry to Explore Open-Ended Evolution

Thomas J. Young and Kourosh Neshatian

University of Canterbury, Christchurch, New Zealand
thomas.young@pg.canterbury.ac.nz

Abstract. We introduce a simple Artificial Chemistry to provide an open-ended representation for the exploration of artificial evolution. The chemistry includes an energy model based on the conservation of total kinetic and potential energy, and a constructive reaction model where possible reactions are discovered "on-the-fly". The implementation is built on an existing open-source cheminformatics toolkit for performance and has a feature-set that prioritises the needs of Artificial Life over fidelity to real-world chemistry, unlike many existing artificial chemistries.

Keywords: Emergence, Artificial Chemistry, Constructive, Energy.

1 Introduction

We are interested in the factors that enable "true" open-ended evolution in an artificial system. Previous work has emphasized that simply creating the conditions for ongoing evolution is unlikely to result in outcomes that surprise or intrigue [3,10].

In this paper we concentrate on the open representation component of true open-ended evolution: the mechanism for describing the targets of the evolutionary component. In addition to the basic function of description, any representation must be capable of unrestricted extension, be understandable by humans, and be computationally efficient. Previous work (e.g., [7]) has supported our belief that these requirements can be met by an Artificial Chemistry (see [5] for an introduction).

2 An Artificial Chemistry for Open-Ended Evolution

Following Dittrich et al. [5], we describe our Artificial Chemistry implementation by (S, R, A): a set of molecules S, a set of rules R describing the possible transformations of the molecules, and a reactor algorithm A to select and order reactions.

In general, for systems where S represents a set of discrete molecules, there are two main strategies for reaction selection [7], sometimes combined hierarchically: spatial and aspatial. In a spatial strategy, molecules have position and velocity, and reactions occur when two molecules collide. Modelling the motion of a large

S. Cranefield and A. Nayak (Eds.): AI 2013, LNAI 8272, pp. 228–233, 2013.

population of molecules can however be computationally expensive. By contrast, in an aspatial strategy, molecules are assumed to occupy a well-mixed container. Our broader motivation is to identify a set of factors that may lead to open-ended evolution. With this in mind, an aspatial approach has several advantages: first, it is computationally simpler than a spatial chemistry. Second, it is likely to be more robust in respect to parameter selection as it removes the need for models of molecular movement and spatial proximity.

With regard to our specific chemistry, to the best of our knowledge the artificial chemistry described by Ducharme et al. [6] is the only other work similar in both goals and approach to our own. The approach taken is to model the energy changes associated with reactions. The chemistry however is spatial; atoms are arranged on a 2-dimensional grid and have velocity. When two atoms pass within a particular distance, they interact. The possible types of interactions are pre-specified, with the type chosen being driven by the atomic composition and energies of the interacting atoms. Reactions are therefore between atoms rather than molecules. Although computational costs are not reported, it seems plausible that the calculation of intersections on a spatial grid will be expensive for large molecular populations. Another cost comes from the required rearrangement of molecules post-reaction into energy-efficient configurations. We are interested in establishing whether, for the purposes of modelling open-ended evolution, these extra capabilities with their associated performance costs are actually required.

2.1 Molecules (*S*)

Our model is based on RDKit [9], open-source software for cheminformatics. RDKit provides a number of useful capabilities including format conversions to and from SMILES [4] and graphical forms of molecules; standard sanity checks for molecular structure, and molecular manipulations, but most importantly, is well-tested and optimised for performance.

Molecules are modelled as an extension of standard RDKit *Mol* objects, constructed from RDKit *Atoms* connected with *Bonds*. Standard Lewis dot structures built on the inherited atomic properties are used to identify possible bonds, and a formal charge model is used to record the charge changes associated with modifications to the molecular structure caused by reactions.

2.2 Reaction Rules (*R*)

A constructive artificial chemistry [8] is one where new components may be generated through the action of other components, and where those new components may themselves take part in new types of reactions, and so on. This appears fundamental to an open-ended representation. In our model, all reactions emerge solely from the properties of the reacting molecules. For each reaction between two molecules we generate a list of reaction alternatives by enumerating all possible bond additions, bond subtractions, and changes in bond type between the reactants. For example, the reactants H_2 and O_2 generate three reaction alternatives: breaking of the H-H bond, breaking of the O=O double bond, and a transformation of the O=O double bond to a single bond. The reactants H^+ and OH^- give two alternative reactions: breaking of the O-H bond (giving $H+H^++O^-$) and formation of a single bond between H^+ and O to give H_2O. The Reactor Algorithm selects one of these reaction alternatives by choosing from a distribution of reaction alternatives weighted by associated energy changes (see sec. 2.3.)

Our energy model enforces conservation of mass so reactions can be represented solely by bond changes in RDKit. This follows the approach taken in graph-based chemistries such as GGL/ToyChem [1,2] where reactions are modelled as a series of changes to graph edges, or bonds, only.

2.3 Reactor Algorithm (A)

Our reactor algorithm is an implementation of a well-stirred reaction container where every reactant has equal probability of participating in the next reaction. Reactions are modelled as head-on elastic collisions between two reactants, chosen at random from the population, with changes to kinetic energy equalling the increase or decrease in molecular potential energy associated with the creation, destruction or change of order of bonds. Creation of a bond results in a reduction of molecular potential energy and an increase to kinetic energy; destruction results in the reverse. A change in bond type is modelled as the sum of a bond creation and of a bond destruction. Total energy in the system is always constant, and equal to the sum of the initial kinetic energy of all molecules plus the sum of their potential energies.

The magnitude of the change in potential energy, measured in arbitrary energy units, is taken for simplicity from a table of approximate real-world chemical bond energies for each combination of atoms and bond type. For example, the creation of an H-H bond releases 104.2 units; the breaking of a C=O double bond takes 185 energy units.

We select a reaction to fire from the possible alternatives for the reactants by probabilistic selection from the alternatives biased towards options that release rather than consume energy. Fig. 1 shows an example of the shift in products that occurs as a result of this weighting as the overall quantity of energy in the system is changed.

3 Experimental Evaluation

The reaction model and energy model leads to the following expectations:

1. Given two reactants, changing the reaction energy should result in different sets of reaction products.
2. Molecular quantities reach equilibrium—that is, the set of interacting molecules is constant, with fluctuations expected in quantities. We expect molecular concentrations to stabilize at non-extreme values (equilibrium rather than driven to an extreme) after some transition period from the initial conditions.

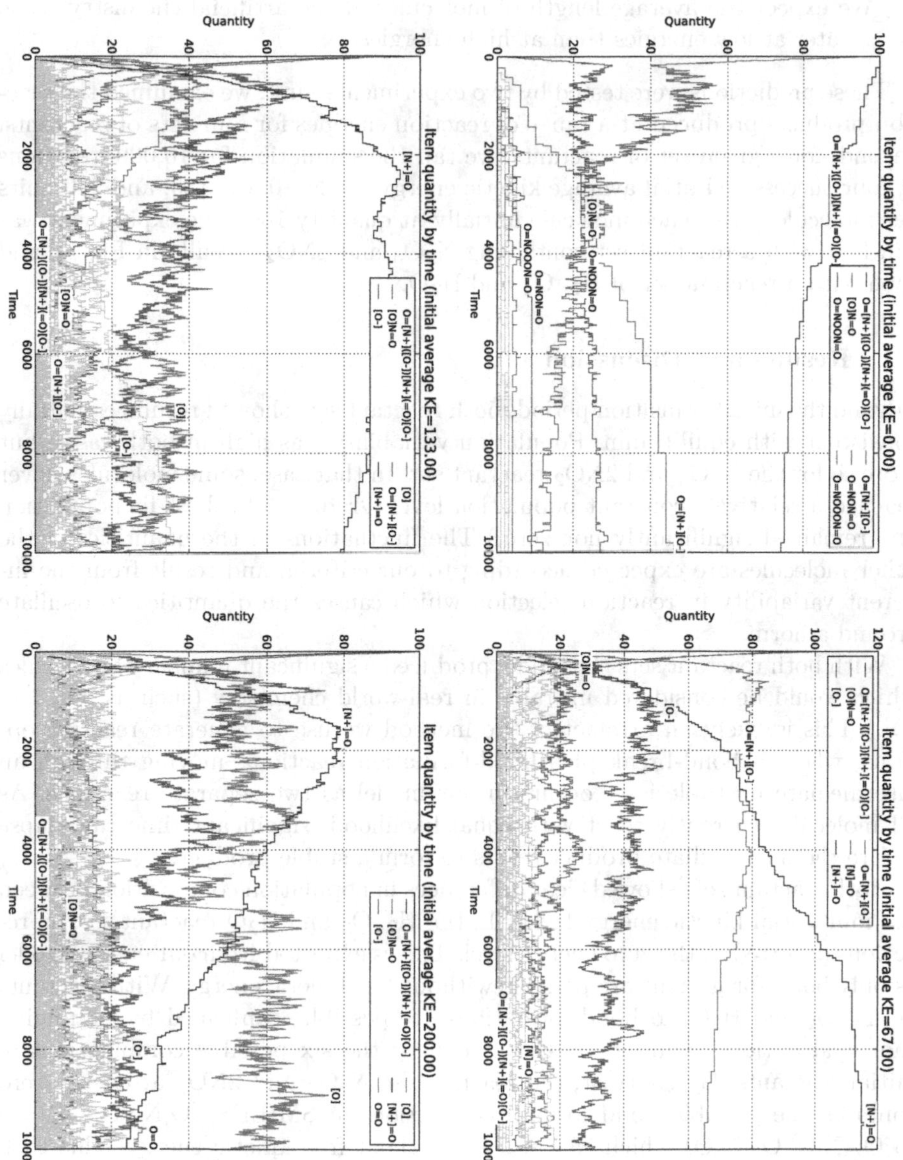

Fig. 1. Molecular quantities over time for initial population of N_2O_4 and $2NO_2$ with initial average KE ranging from 0 to 200 units (only molecules with significant quantities are labelled; remainder appear as light-grey lines). Molecules represented in SMILES notation.

3. The equilibrium point depends on the energy of the system. Our energy model preferentially forms bonds at low energies, and breaks bonds at high. We expect the average length of molecules in the artificial chemistry to be greater at low energies than at high energies.

These predictions were tested by two experiments: first, we examined the reaction products produced at a range of reaction energies for four sets of reactants. Second, for a given set of reactants, we ran the simulation for 10,000 iterations at four successive initial average kinetic energy levels—0, 67, 133, and 200 units per molecule—with each molecule initially at quantity 100. The experiment was run first with a reactant set containing N_2O_4 and $2NO_2$ (results in Fig.1), and then with a reactant set of H_2, O_2 and H_2O.

3.1 Results and Discussion

Beyond the initial transition period, both reactant sets showed results essentially consistent with equilibrium. Population variability was high in both cases, but more so for the N_2O_4 and $2NO_2$ reactant set. In that case, some molecules never reached a relatively constant population level (gradient of a best-fit population line remained significantly non-zero.) The fluctuations in the quantities of the other molecules are expected according to our criteria, and result from the inherent variability in reaction selection which causes the quantities to oscillate around a norm.

With both reactant sets the model produced a significant number of molecules which would be considered unstable in real-world chemistry (such as [O-] and [O].) This is likely an artefact of the method we use to generate reaction options, where a bond-break plus bond-formation reaction—moving through an intermediate unstable ion—occurs in our model as two separate reactions. As all molecules currently react with equal likelihood, significant time can elapse before the intermediate product reacts to form a stable product.

Both reactant sets showed clear differences in population composition between the four initial kinetic energy levels. In the H_2, O_2 and H_2O reactant set, no reactions occurred at the zero energy level. This is expected from our energy model as only bond-formations are possible without free kinetic energy. With reactants of H_2, O_2 and H_2O no bond formations are possible, confirmed by examining the bond options returned by the model for the six possible combinations of initial reactants. By contrast, the reactant set N_2O_4 and $2NO_2$ at energy zero contains one possible bond formation reaction (in SMILES, [O]N=O.[O]N=O to O=N[O][O]N=O) which can proceed without free kinetic energy. This then releases a product which can also react, and so on, thus explaining the different results between the reaction sets.

4 Conclusions

Open-ended evolution is, as far as we know, an emergent phenomenon, and so an open-ended system is more likely to be discovered than designed. Our approach therefore will be to identify a set of plausible factors that may lead to emergent evolution; to systematically search through the set of chemistries based on varying values of those factors; and to use a reasonable set of measures, such as the number and types of reaction cycles, to identify promising chemistries from our search. This is quite different from other works which postulate a particular design at the outset.

Our initial artificial chemistry appears to be at least compatible with the requirements for the future exploration of open-ended evolution. The model is simpler than comparable alternatives, and the energy and reaction models produce results consistent with our predictions for the system's behaviour (with the exception of achieving equilibrium with the N_2O_4 and $2NO_2$ reactant set). An aspatial approach does however come with restrictions. Most obviously, as there is no concept of proximity in the chemistry, there can be no boundaries or membranes or even basic distinctions between *inside* and *outside*. This is critical in biology but it is unclear if this is equally important in non-biological systems. We expect that experimental comparison between the aspatial and spatial approaches in the course of our exploratory experiments will help to clarify this.

References

1. Benkö, G., Flamm, C., Stadler, P.F.: A graph-based toy model of chemistry. Journal of Chemical Information and Computer Sciences 43(4), 1085–1093 (2003), `http://pubs.acs.org/doi/abs/10.1021/ci0200570` pMID: 12870897
2. Benkö, G., Flamm, C., Stadler, P.F.: The toychem package: A computational toolkit implementing a realistic artificial chemistry model (2005), `http://www.tbi.univie.ac.at/~xtof/ToyChem/`
3. Channon, A.: Unbounded evolutionary dynamics in a system of agents that actively process and transform their environment. Genetic Programming and Evolvable Machines 7(3), 253–281 (2006), doi:10.1007/s10710-006-9009-3
4. Daylight Chemical Information Systems, I.: Daylight theory manual (2011), `http://www.daylight.com/dayhtml/doc/theory/index.html`
5. Dittrich, P., Ziegler, J., Banzhaf, W.: Artificial chemistries-a review. Artificial Life 7(3), 225–275 (2001), `http://www.mitpressjournals.org/doi/abs/10.1162/106454601753238636`
6. Ducharme, V., Egli, R., Legault, C.Y.: Energy-based artificial chemistry simulator. In: Adami, C., Bryson, D.M., Ofria, C., Pennock, R.T. (eds.) Proceedings of the Thirteenth International Conference on the Simulation and Synthesis of Living Systems (Artificial Life 13), pp. 449–456 (2012)
7. Faulconbridge, A.: RBN-World: sub-symbolic artificial chemistry for artificial life. Ph.D. thesis, University of York (2011)
8. Fontana, W., Wagner, G.P., Buss, L.W.: Beyond digital naturalism. Artificial Life 1(2), 211–227 (1994)
9. Landrum, G.: Rdkit: Open-source cheminformatics (2013), `http://www.rdkit.org`
10. Maley, C.: Four steps toward open-ended evolution. In: GECCO-99: Proceedings of the Genetic and Evolutionary Computation Conference, pp. 1336–1343. Morgan Kaufmann (1999)

Game Description Language Compiler Construction*

Jakub Kowalski and Marek Szykuła

Institute of Computer Science, University of Wrocław, Poland
{kot,msz}@ii.uni.wroc.pl

Abstract. We describe a multilevel algorithm compiling a general game description in GDL into an optimized reasoner in a low level language. The aim of the reasoner is to efficiently compute game states and perform simulations of the game. This is essential for many General Game Playing systems, especially if they use simulation-based approaches. Our compiler produces a faster reasoner than similar approaches used so far. The compiler is implemented as a part of the player Dumalion. Although we concentrate on compiling GDL, the developed methods can be applied to similar Prolog-like languages in order to speed up computations.

Keywords: General Game Playing, Game Description Language, Compiler Construction.

1 Introduction

The aim of General Game Playing (GGP) is to develop a system that can play variety of games with previously unknown rules. Unlike standard artificial game playing, where designing an agent requires special knowledge about the game, in GGP the key is to create an universal algorithm performing well in different situations and environments. As such, General Game Playing was identified as a new Grand Challenge of Artificial Intelligence and from 2005 the annual AAAI GGP Competition is taking place to foster and monitor progress in this research area [5]. Because of its universal domain, GGP combines multiple disciplines [19] from searching, planning, learning [1,4,12,15,17] to evolutionary algorithms, distributed algorithms and compiler construction [11,14,16,20].

In many General Game Playing systems it is crucial to have an efficient reasoning algorithm performing simulations of the game. More computations means a larger game tree traversed, more gained knowledge, deeper search or more simulations in Monte Carlo algorithms. In response to these needs, we developed our compiler. Because of used optimizations, it produces very effective reasoners which can compute game states faster than other so far known approaches.

The paper is organized as follows. Section 2 provides necessary background and describes current state of the art. Step by step details of our construction are presented in Section 3. Section 4 contains overview of experimental results. We conclude in Section 5.

* This research was supported in part by Polish MNiSW grant IP2012 052272.

S. Cranefield and A. Nayak (Eds.): AI 2013, LNAI 8272, pp. 234–245, 2013.

2 Game Description Language

To develop a general game playing system, there is a need for a standard to encode game rules in a formal way. For the sake of World Wide GGP Competition, Game Description Language (GDL) [5,10] is used. This first-order logic language based on Datalog has enough expression power to describe all finite, turn-based, deterministic games with full information, simultaneous moves and a fixed number of players. By "finite" we mean that the set of possible game states, and the set of actions (moves) which players can choose in each state should be finite. Also every match should end after a finite number of turns. Players perform actions simultaneously, which means that in each turn all players select their moves, without knowing the decision of the others. Sequential games can be simulated by explicit adding some `noop` move for the players which should normally wait. Another strong restrictions are that no game element can be random and all players should have the full information about the game state. Extension of the GDL called GDL-II [18] removes these limits, but this leads to a more complicated system where general playing is even harder.

Syntactically GDL is very similar to Prolog. It is purely axiomatic, so there is no arithmetic or other complex game concepts (like pieces or boards) included, every such thing must be explicitly stated in the code. GDL is rule based which means that gaining information about a game state is equivalent to applying rules and extending the set of holding (true) facts. As example, in listing 1.1, we show some rules of the game *Goldrush* from Dresden GGP Server [6].

Listing 1.1. Part of the *Goldrush* game GDL code

```
(role Green) (role Red)
(init (OnMap Green 1 1)) (init (OnMap Red 7 7))
(init (OnMap Obstacle 1 6)) (init (OnMap Obstacle 2 4)) ...
(init (OnMap (Gold 2) 1 7)) (init (OnMap (Gold 1) 3 6)) ...
(init (OnMap (Item Blaster 3) 7 4)) ...
(init (OnMap (Item Stoneplacer 3) 1 4)) ...
(<= (legal ?r (Move ?nx ?y))
    (role ?r) (true (OnMap ?r ?x ?y)) (InBoard ?nx))
    (or (+ ?x 1 ?nx) (- ?x 1 ?nx))
(<= (next (OnMap ?r ?x ?y))
    (role ?r) (does ?r (Move ?x ?y)))
(+ 0 0 0) (+ 1 0 1) (+ 2 0 2) (+ 3 0 3) ...
(<= (- ?x ?y ?z) (+ ?y ?z ?x))
(InBoard 1) (InBoard 2) (InBoard 3) ... (InBoard 7)
```

Predicates that are arguments of `init`, `true` and `next` can be considered as a minimal set of predicates enough to restore all information about the state, which we will call as the *base predicates*. This means that the full game state (the *view* of a state) is a set of facts closed under application on the *base facts* and the next state is computed based on the previous full state and the players actions.

This leads us to the notion of the reasoner. This is an essential part of every player, allowing it to shift the state based on information from the game controller. During competition, the game controller sends to a player only moves made by all players, so computing the next state, legal moves and so forth should be made at the player's side. In other words the reasoner is an implementation of the game loop (Fig. 1) described by game rules.

2.1 Reasoner Implementations

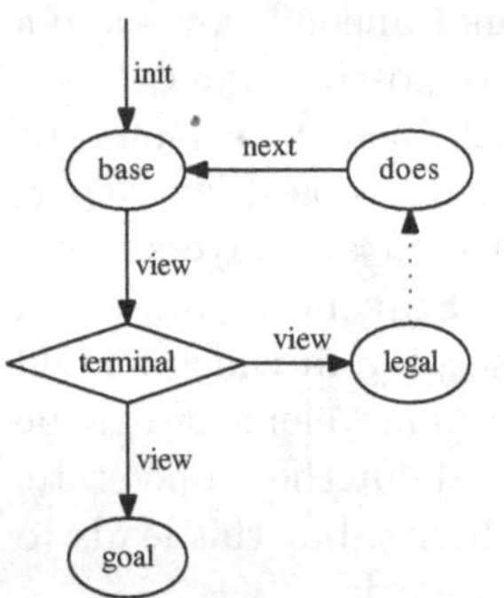

Fig. 1. Game loop: the *view* of a state is computed based on the current *base* facts. Then either game ends or the program should wait for actions of the players and then compute a next *base*.

An efficient reasoner implementation, while not connected with playing algorithms, is one of the most important part of a general game player. This applies to currently dominating simulation-based approach [3], but also some knowledge based ([7,8,15]) players make benefits from performing game simulations to tune their heuristic evaluation functions [1]. The problem of playing general games is so computationally difficult, that most of players use distributed architecture to support parallel computations on many machines [11,13]. Complementary approach to parallelism is speeding up the process of reasoning itself. Most common implementations of the reasoner are based on prolog engines built into another programming language. The benefit from this approach is that it requires only syntactic modifications of GDL, so it results in simplicity of implementation. But using a very general inference engine to compute rules of strictly given form is a drawback causing lack of computations speed.

The most obvious method of avoiding this is compiling GDL to other language. Straightforward rewriting GDL code to C++ language, done in a top-down manner was described in [20] and Java version of this approach can be found in [13]. On the other hand, the method of forward chaining GDL to OCaml compilation with some optimizations was proposed in detail in [14]. Other approaches to make reasoning faster contains usage of propositional networks [2] and instantiating games to use binary decision diagrams [9].

In the following section we will describe in details our method of constructing GDL compiler. We use combination of a few ideas such us careful ordering of computations, optimizing control flow and designing dedicated structures to perform queries. All of these, result in a significant improvement in efficiency compared to the methods used before.

3 Compilation

Our compiler takes as an input the game rules written in GDL and outputs a structure called *compilation plan*, decoding the computation strategy for the reasoner. The plan can be then translated to some efficient low-level programming language like C/C++. In this section we describe all major steps of constructing the plan. Our general aim is to achieve better efficiency of computing game states by the resulted reasoner. Looking to a process of playing GDL game, it can be considered as an usage of a database. The predicates are containers with some facts and we have to perform reading (queries) and writing (insertions) to these containers. Our method uses such techniques as flattening domains, optimizing data structures for containers, reordering operations, which are mostly apart from the target language.

3.1 Calculating Domains and Flattening

The first thing we need to do, after parsing GDL to some abstract tree structure, is to compute domains of the predicates' arguments. Let P be a predicate. Because predicates in GDL can be nested, every occurrence of P form a tree of its arguments. In that case we can describe such occurrence as a function from vectors encoding positions in tree to arguments symbols, so e.g. position of `3` in `OnMap (Item Blaster 3) 7 4` can be described as $\langle 0, 2 \rangle$ (positions at every tree level are enumerated from 0). We want to calculate domain as a function which takes a pair of a predicate P and a tree position $\tilde{p}$ (possible for P), and returns a set of symbols that can occur in this position. Such domains are in fact a supersets of the real predicates' domains and also lose information about dependencies between arguments. However this is enough for the further calculations.

The method proposed in [9] requires computing set of dependencies where $(P, \tilde{p}) \rhd (Q, \tilde{q})$ ($\rhd$ is "depends on" relation) if and only if there exists a rule with P in the head and Q in the body, where at the positions $\tilde{p}$ and $\tilde{q}$ respectively, the same variable occurs. This means that every symbol in domain of $(Q, \tilde{q})$ should be also in domain of $(P, \tilde{p})$. In this approach calculating domains means resolving dependencies by extending appropriate domains until a fixpoint is obtained.

We improved this method to handle nested predicates and compute smaller domains. In our case extending domains include also domains of every subtree of variable occurrence. If $(P, \tilde{p}) \rhd (Q, \tilde{q})$ then for every $\tilde{q}''$ which has $\tilde{q}$ as a prefix (so $\tilde{q}'' = \tilde{q} + \tilde{q}'$ for some position vector $\tilde{q}'$), also $(P, \tilde{p} + \tilde{q}') \rhd (Q, \tilde{q}'')$. These dependencies must be dynamically computed because during the algorithm new predicates positions can be found.

Instead of $\rhd$ we use relation $\rhd_R$ where $(P, \tilde{p}) \rhd_R (Q, \tilde{q})$ if dependency is created by rule R in CNF form (note that every game described in GDL can be easily converted do CNF). Let $\odot$ be the operator of domain conjunction defined as $(\odot\, d_1 \ldots d_n)\ v = d_1(v) \cap \ldots \cap d_n(v)$. For every rule R we create set $\pi^R_{(P,\tilde{p})}$ containing every $(Q, \tilde{q})$ such that $(P, \tilde{p}) \rhd_R (Q, \tilde{q})$ holds. Then for every $\pi^R_{(P,\tilde{p})}$ we extend domain of $(P, \tilde{p})$ by $\odot\, d_i$ for $d_i \in \pi^R_{(P,\tilde{p})}$. This simulates conjunction which takes place in GDL rules and prevent domains from containing symbols unused in practice. The procedure loops for every pair $(P, \tilde{p})$ and finishes when a fixpoint is found.

To illustrate this algorithm, consider a subset of game rules shown in Listing 1.1. Calculating domains based on this example appoints the following domain of predicate `OnMap`:

$(\texttt{OnMap}, \langle 0\rangle) \rightarrow \{\texttt{Green,Red,Obstacle}\}$

$(\texttt{OnMap}, \langle 0,1\rangle) \rightarrow \{\texttt{Gold,Item}\}$

$(\texttt{OnMap}, \langle 0,2\rangle) \rightarrow \{\texttt{1,2,Blaster,Stoneplacer}\}$

$(\texttt{OnMap}, \langle 0,3\rangle) \rightarrow \{\texttt{3}\}$

$(\texttt{OnMap}, \langle 1\rangle) \rightarrow \{\texttt{1,2,3,4,5,6,7}\}$

$(\texttt{OnMap}, \langle 2\rangle) \rightarrow \{\texttt{1,2,3,4,5,6,7}\}$

Because of arguments nesting, the number of leaves in parse tree can vary for one predicate. But to effectively perform queries, we need to have predicates without nesting and with a fixed arity. To achieve this, we developed a notion of *flattened* predicate and algorithms to convert standard (nested) predicate to flattened form and to perform reversed conversion.

The arity of a flattened predicate is the number of leaves in the widest of arguments assignments found in domain calculating phase. Tighter occurrences of the predicate are then stretched using special non-GDL symbol `#nil`, and each variable occurrence is extended by introducing new variables with added suffixes to avoid ambiguity. From now on, each mentioned predicate is flattened. Conversion from flattened predicate to its standard GDL form is necessary when the player needs to send a move to the game controller, having a flattened move given by the reasoner. As we made proper algorithms, we can stand:

Lemma 1. *For every occurrence of a valid GDL predicate with a fixed domain, there exist its unique flattened form. There exists an algorithm that converts these forms.*

A small part of flattened *Goldrush* game is shown in Listing 1.2 as an example. As it shows at line 10, it can create rules with unbound variables, but the values of these variables are explicitly set to `#nil` during further calculations.

Listing 1.2. Flattened GDL code

```
2  (init (OnMap Green #nil #nil 1 1)) ...
3  (init (OnMap Obstacle #nil 1 6)) ...
4  (init (OnMap Gold 2 #nil 1 7)) ...
5  (init (OnMap Item Blaster 3 7 4)) ...
6  (init (OnMap Item Stoneplacer 3 1 4)) ...

10 (<= (next (OnMap ?r0 ?r1 ?r2 ?x0 ?y0))
11       ( role ?r0 ) ( does ?r0 ( Move ?x0 ?y0 #nil ) ) )
```

3.2 Predicates Dependency Graph and Layering

Let say that a predicate P depends on Q if there exists a game rule R such that P is the head of R and the body of R contains Q. We consider the *dependency graph*, which is a directed graph representing the dependency relation of predicates.

After the complete dependency graph is built, it is split up to subgraphs representing each of the game *phases*, depending on what we are going to compute. The phases are: *init*, *term*, *goal*, *legal* and *next* and they correspond to the solid arrows from the game loop visualization (Fig. 1), where *term*, *goal* and *legal* belong to *view*. In such a way, the dependency graph gives us information about the predicates usage.

We can get rid of all predicates that are not needed to compute any of `legal`, `terminal`, `goal`, `next`. *Constant* predicates can be fully precomputed during the initialization phase and they stay unchanged during the rest of the game. These and base predicates belong to the *init* phase. The predicates reachable in the reversed dependency graph from `terminal`, `goal`, `legal` and `next` belong to corresponding phases respectively. There is one exception: predicates reachable from both `goal` and `legal` are put in the *term* phase. We note that it is not necessary required to compute the `goal` predicate while the state is non-terminal. This loses possible information about scores in non-terminal states, but it saves computation time and in Monte Carlo approach simulations go to the end anyway, so checking the `goal` values in non-terminals can be avoided.

All proper GDL games must be stratified, which means that for all predicates P and Q if P depends on `not` Q then P must be in a higher stratum, and all facts of a lower stratum should be deducted before deducting the upper stratum starts (which is always possible). This mechanism allows to treat GDL deduction as continuously adding facts to a database, without worrying of withdrawing them if computations are made in the proper order. In a top-down approach right computation order is for free, but in a bottom-up ordering is more flexible and can lead to better efficiency.

Despite stratification as a result of negations placement, we consider *layering*. This is a more general and a more complex approach based on dependency graphs. Each layer corresponds to a set of strongly connected components of the dependency graph. There are two types of layers:

Acyclic layer is a set of predicates such that there is no path in the dependency graph between any two predicates from this set. This means that, if only all the lower layers are computed, all the rules with these predicates in the head can be computed simultaneously and only once.

Cyclic layer is a set of predicates that are reachable from any other from this set (by using at least one edge). In this case the number of rules applications to deduct these predicates is unknown, and computations must take place until a fixpoint is reached (no new fact is added after an iteration).

Partitioning of the dependency graph to layers should be done in a way, that acyclic layers should be as large as possible, and cyclic layers as small as possible (which reduces number of computations). Currently we create the layers incrementally from the nodes without ingoing edges (so the first layer contains all "leaves" of graph). If there is a choice which layer cyclic or acyclic should be considered as a lower, the lower (first to compute) goes cyclic one.

3.3 Defining the Rules Computation Order

Mapping from predicates to layers does not make ordering of rule computation unambiguous. Consider a rule R and let L_h^R be the layer where the head of the rule belongs, and let $L_{b\ \max}^R$ be the maximal (the highest) layer of predicates in body of R. This means that R must be computed before layer $L_h^R + 1$ and after

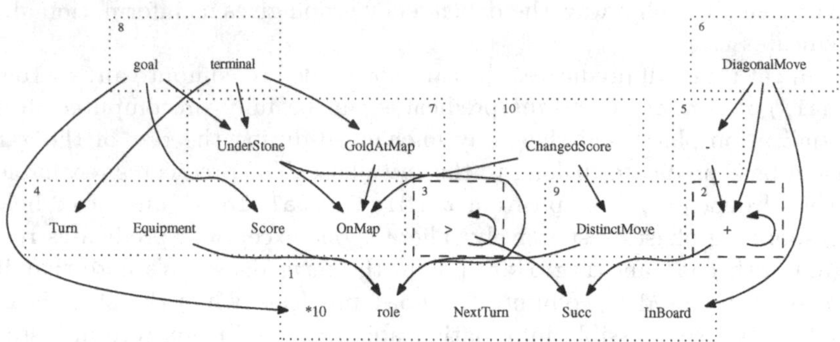

Fig. 2. Dependency graph of the game *Goldrush* (predicates `legal` and `next` are omitted due to visibility) with cyclic layers marked with dashed border

layer $L^R_{b\,\max}$, so the rule can be placed in any layer between these values. An exceptional situation is when $L^R_h = L^R_{b\,\max}$, which happens only for some rules from cyclic layers. In this case rule placing is unambiguous.

Intelligent rule placement can lead to some speed improvement in two main cases. First, when the predicate in the head of a rule is from a cyclic layer but the body is not ($L^R_h < L^R_{b\,\max}$) then the rule can be computed cheaper, because it is computed only once within a lower (assuming non cyclic) layer. Sometimes even special layers for such rules can be created. The second reason to move rules is that in some layers in admissible range there are rules similar in construction and some sharing computations between them can be done.

3.4 Filter Trees

With computed order of the rule computations we can produce the final plan. In such a plan the symbols and predicates get their unique id, and each predicate has bound information about its (flattened) domain and container type. To appoint exact ordering of game state computations, structures called *filter trees* are created.

Filter trees contains nodes, which can have child nodes. The whole computation process is just traversing the tree and performing actions according to the types of nodes. During computation a set of local variables and a set of containers are maintained. In the root the sets are empty. A variable (similarly container) defined in a node has scope bounded to the children nodes. A variable defined in a parent node is bound in the children and cannot change its value. The nodes have the following types:

- *Sequence* A node with many children which should be executed in the order.
- *Query* Queries the specified container for a given subset of facts. There can be either symbols or bound and unbound variables. For each fact in the container which matches the query, define unbound variables by setting

the values to match the fact and go into the child node. For optimization purposes a query can have additional explicit domain filter.
- *Accept* Inserts a fact to the specified container. All of the variables used in an insertion must be bound. If a new fact is added, a special *repeat* flag is set to inform a cyclic layer to be repeated.
- *Repeat* Repeats computation of its subtree until the *repeat* flag is unset. The flag is checked and cleared each time between iterations.
- *If* It has three children. The child called *test* is executed first until it is finished or a special *Return* node is reached. If *Return* was reached the child *true* is executed, otherwise *false*.

A GDL rule can be simply transformed to a filter tree without any significant modifications. A conjunction is simply nested children, an alternative can be decoded as *Sequence* and negated terms can be put in *If* filter with *Accept* in *false* subtree. Distinct between two variables is converted to a *Query* with unspecified container but with a list of distinct variables, while distinct with a variable and a symbol is just *Query* with reduced variable's domain.

A careful way of constructing filter trees can reduce much of computations. At first, if the same query occurs in two rules in the same layer it can be shared. Because every nested query can potentially cut out variable domains, the right order of nested queries can also improve efficiency. The last main optimization takes place when all variables from the heads of the rules are already defined in some query. Since the added fact is fully defined it remains only to check if the rest of queries can be satisfied. This can be realized by putting into *If* the rest nested queries, so a single positive pass through them is sufficient to immediately return and insert a fact.

We observe that only the values of variables which can reach *Insert* or *Return* node are necessary to be considered. We can restrict the domain in advance in queries defining these variables, instead of filtering them by nested queries.

We create five filter trees, one for each of the phases. An example filter tree for *next* phase of the game *blocker* [6] is presented in Fig. 3. Creating the filter trees finishes our construction, allowing generation of the final code.

3.5 Data Structures for Containers

Queries can have very different shapes. They consist of a predicate and a fixed number of arguments, depending on the predicate's arity. The arguments can be constants or variables. The simplest are those asking about existence of a particular fact like `cell 1 3 b`. More complicated are queries mixing both constants and variables, including bound variables like in `cell ?x ?x ?t`. Efficiency of performing queries depends on the data structure used to implement the container for a given predicate. An elementary analysis can be used to estimate query and insert costs for various data structures. Although more deep optimizations can take care of the proportion between queries and insertions, and occurrences of query shapes.

In our compiler we use a few different data structures. We describe them here and perform a simple efficiency analysis. Consider a container and assume that

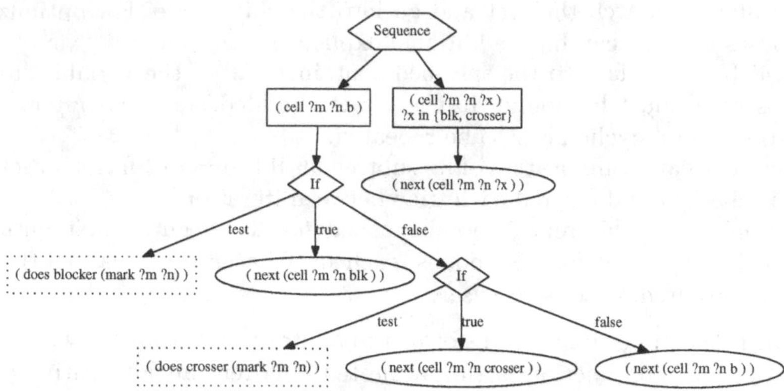

Fig. 3. *Next* filter tree of the game *blocker*. Box nodes represent *Queries* while ellipses are *Accept* filters. Every *test* branch of *If* ends with unmarked *Return* node.

d is the arity of facts in the container. Then let $c_1, \dots, c_d$ be the sizes of the domains of the arguments, that is the i-th argument can take one of c_i possible values. Thus the container can hold at most $C = \prod_{i=1}^{d} c_i$ facts. Assume that n is the number of currently stored facts in the container. We assume that d is a fixed constant and comparing any two values (symbols) takes $O(1)$ time.

Querying a subset of facts in the container takes time at least $\Omega(s)$, where s is the size of the queried subset, because of further processing these facts. For simplicity we do not consider the cases with duplicated unbound variables in a single query, which are also rare cases. Thus an optimal data structure would take $O(s)$ time performing a query, and $O(1)$ time performing an insertion. A special careful should be taken for querying for a particular fact, because such an operation is often required during insertions to prevent storing duplicated facts in the container.

One of the simplest data structures is a standard dynamically-sized *vector*. Inserting to a vector takes $O(1)$ time, but querying for a particular fact or a subset of facts can take $O(n)$ time in the worst case. In opposition to the vector there is a complete *lookup array* with the fixed size C. Each allowed fact has a fixed position in the array storing a flag indicating if the fact is in the container. Insertion to an array, as well as querying for a particular fact, takes $O(1)$ time. However querying for a larger subset of facts can take $O(C)$ time, depending on the number of possible facts matching the query. Thus vectors are better for larger domains and smaller number of stored facts, and arrays conversely.

Hash and tree sets are quite efficient for insertions and querying for a particular fact, which take $O(1)$ time in a hash set and $O(\log n)$ in a balanced tree set with lexicographically ordering of facts. However querying for a subset of facts may take $O(n)$ in a hash set, and as well in a tree set if the first argument is an unbound variable.

A *trie* is a more complex tree-like structure. Levels correspond to the arguments. At the level h, each node contains d_h pointers to nodes at the height

$h+1$ (except the last). These correspond to all d_h allowed values of the h-th argument. A fact is encoded by a path from the root to a leaf. A pointer is *null* if there are no facts with the corresponding value. A trie grows as more facts are added. An insertion takes $O(\sum_{i=1}^{d} c_i)$ time in the worst case (empty trie), because we must create a node at each level. Querying a particular fact takes $O(1)$ time. Efficiency of a trie in querying for a specified subset depends on the order of the arguments. A query can cost from $O(s)$ time when the constant arguments are at the beginning, up to $O(n)$ time when they are at the end. We consider tries as an universal balanced structure, since it performs quite well in most cases.

With an assumption that we have a constant number of query shapes, we developed two nearly optimal *composed structures*. Composed structure consist of a set of other structures made especially for efficient maintaining of different query shapes. The first such structure is the *trie-composed* structure based on tries, and the second is the *tree-composed* structure based on balanced tree sets. It seems that the *trie-composed* structure is better, because queries usually occur more frequently than insertions, and it generally has a lower constant factor.

Lemma 2. *The* trie-composed *structure takes $O(s)$ time for a query and $O(c_1 + \ldots + c_d)$ time for an insertion.*

Lemma 3. *The* tree-composed *structure takes $O(s + \log n)$ time for a query and $O(\log n)$ time for an insertion.*

3.6 Final Code Generation

We have implemented GDL compilation to C++. The compiled reasoner is just a module allowing to maintain game states. In particular the phases are functions initializing the reasoner and computing the *init state*, computing a *view state* given a *base state*, or computing a next *base state* given a *base state*, *view state* and moves of the players. It also provides an interface for answering if a state is terminal, getting the goal or the legal moves.

Each node of the filter tree is directly inlined in the code. In this way many technical optimizations are possible by using the context, since for example, each query can have different set of domains for the variables and we can perform explicit iteration through them.

4 Experimental Results

We have implemented the compiler as a Java program producing a reasoner in C++ as output. Our benchmark results are presented in Table 1. The resulted reasoners were compiled by g++. The main program performed uniformly random simulations of the games. Comparative reasoners uses *ECLiPSe Prolog* system. The benchmarks were done on Intel(R) Core(TM) i7-3610QM 2.3GHz with 8GB of RAM.

Table 1. The numbers of performed random simulations and visited game states per second for different games

Game	The reasoner of Dumalion			Prolog	
	Compilation	Simulations	States	Simulations	States
Tic-Tac-Toe	0.736 s	331,647	2,860,888	2,076	15,829
Blocker	0.645 s	194,628	1,729,674	1,020	8,049
Connect Four	0.857 s	15,092	353,283	287	6,424
Breakthrough	1.074 s	3,086	200,901	55	3,553
Checkers	10.482 s	211	21,291	12	1,186
Skirmish	7.810 s	71	7,114	5	518

Although differences between computation speed between a simple Prolog engine and the optimized and compiled code are outstanding as expected, an interesting observation is that the improvement factor is less for more complicated games (such a tendency is also visible in benchmarks from [13,14]). In other hand, while the improvement factor is smaller, the numbers of performed simulations and visited states were increased several time, and this is especially crucial for very difficult games when computing states is hard and even small speed up can give a big advantage.

Because of hardware differences and chosen method of benchmarking it is hard to make a straightforward comparison between our compiler and other approaches described in [13,14,20]. But after recalculating all the results to a common base simulations over a second, a roughly comparison shows that the reasoner of Dumalion can compute simulations from 2 to about 10 times faster (depending on the game) than the compiling methods described so far, and from 10 to 160 times faster than a standard Prolog engine.

5 Conclusions and Future Work

Using a compiler generator to create reasoners requires far more work than running a Prolog engine on syntactically changed GDL code, but the benefit in computation speed is significant. We mention here a few of inconveniences in our method. At first the produced reasoner must be compiled into a native code. This can take quite long if the game is complicated. The second problem is that we lose all the structure information, for example we cannot ask about a specified predicate defined in the original GDL, since it is possible that the corresponding container does not exist at all due to optimizations. Another drawback is that the process of compilation itself make the whole GGP system more complicated and harder to handle, especially if it should support parallelism.

There are many other ways of further optimizations of the plan. They include introducing new temporary containers, reordering of arguments, more careful selection of container types, reordering of queries and splitting them. As the future work, we have plan to construct GGP architecture with the aim of efficient maintain compiled code in a scalable, parallel system.

References

1. Clune, J.: Heuristic Evaluation Functions for General Game Playing. In: AAAI, pp. 1134–1139. AAAI Press (2007)
2. Cox, E., Schkufza, E., Madsen, R., Genesereth, M.: Factoring General Games using Propositional Automata. In: Proceedings of the IJCAI Workshop on General Game Playing, GIGA 2009 (2009)
3. Finnsson, H., Bjornsson, Y.: Simulation-based Approach to General Game Playing. In: AAAI. AAAI Press (2008)
4. Finnsson, H., Bjornsson, Y.: CadiaPlayer: Search-Control Techniques. KI 25(1), 9–16 (2011)
5. Genesereth, M., Love, N., Pell, B.: General game playing: Overview of the AAAI competition. AI Magazine 26, 62–72 (2005)
6. Gunther, M., Schiffel, S.: Dresden General Game Playing Server, **http://ggpserver.general-game-playing.de**
7. Haufe, S., Michulke, D., Schiffel, S., Thielscher, M.: Knowledge-Based General Game Playing. KI 25(1), 25–33 (2011)
8. Haufe, S., Thielscher, M.: Pushing the Envelope: General Game Players Prove Theorems. In: Li, J. (ed.) AI 2010. LNCS, vol. 6464, pp. 1–10. Springer, Heidelberg (2010)
9. Kissmann, P., Edelkamp, S.: Instantiating General Games Using Prolog or Dependency Graphs. In: Dillmann, R., Beyerer, J., Hanebeck, U.D., Schultz, T. (eds.) KI 2010. LNCS, vol. 6359, pp. 255–262. Springer, Heidelberg (2010)
10. Love, N., Hinrichs, T., Haley, D., Schkufza, E., Genesereth, M.: General Game Playing: Game Description Language Specification. Technical report, Stanford Logic Group (2008)
11. Mehat, J., Cazenave, T.: A Parallel General Game Player. KI 25(1), 43–47 (2011)
12. Michulke, D., Thielscher, M.: Neural Networks for State Evaluation in General Game Playing. In: Buntine, W., Grobelnik, M., Mladenić, D., Shawe-Taylor, J. (eds.) ECML PKDD 2009, Part II. LNCS, vol. 5782, pp. 95–110. Springer, Heidelberg (2009)
13. Möller, M., Schneider, M., Wegner, M., Schaub, T.: Centurio, a General Game Player: Parallel, Java- and ASP-based. KI 25(1), 17–24 (2011)
14. Saffidine, A., Cazenave, T.: A Forward Chaining Based Game Description Language Compiler. In: IJCAI Workshop on General Intelligence in Game-Playing Agents (GIGA 2011), pp. 69–75 (2011)
15. Schiffel, S., Thielscher, M.: Fluxplayer: A Successful General Game Player. In: Proceedings of the 22nd AAAI Conference on Artificial Intelligence (AAAI 2007), pp. 1191–1196. AAAI Press (2007)
16. Sharma, S., Kobti, Z., Goodwin, S.: General Game Playing with Ants. In: Li, X., et al. (eds.) SEAL 2008. LNCS, vol. 5361, pp. 381–390. Springer, Heidelberg (2008)
17. Sharma, S., Kobti, Z., Goodwin, S.: Knowledge Generation for Improving Simulations in UCT for General Game Playing. In: Wobcke, W., Zhang, M. (eds.) AI 2008. LNCS (LNAI), vol. 5360, pp. 49–55. Springer, Heidelberg (2008)
18. Thielscher, M.: A General Game Description Language for Incomplete Information Games. In: Proceedings of the AAAI Conference on Artificial Intelligence, pp. 994–999. AAAI Press (2010)
19. Thielscher, M.: General Game Playing in AI Research and Education. In: Bach, J., Edelkamp, S. (eds.) KI 2011. LNCS, vol. 7006, pp. 26–37. Springer, Heidelberg (2011)
20. Waugh, K.: Faster State Manipulation in General Games using Generated Code. In: IJCAI Workshop on General Game Playing, GIGA 2009 (2009)

Model Checking for Reasoning about Incomplete Information Games

Xiaowei Huang[1], Ji Ruan[2], and Michael Thielscher[1]

[1] University of New South Wales, Australia
{xiaoweih,mit}@cse.unsw.edu.au
[2] Auckland University of Technology, New Zealand
jiruan@aut.ac.nz

Abstract. GDL-II is a logic-based knowledge representation formalism used in general game playing to describe the rules of arbitrary games, in particular those with incomplete information. In this paper, we use *model checking* to automatically verify that games specified in GDL-II satisfy desirable temporal and knowledge conditions. We present a systematic translation of GDL-II to a model checking language, prove the translation to be correct, and demonstrate the feasibility of applying model checking tools for GDL-II games by four case studies.

1 Introduction

The general game description language GDL, as the input language for general game-playing systems [7], has recently been extended to GDL-II to incorporate games with nondeterministic actions and where players have incomplete/imperfect information [20]. However, not all GDL-II descriptions correspond to games, let alone meaningful and non-trivial games. Genesereth *et al.* [7] list a few properties that are necessary for well-formed GDL games, including guaranteed termination and the requirement that all players have at least one legal move in non-terminal states. The introduction of incomplete information raises new questions, e.g., can players *always know* their legal moves in non-terminal states or *know* their goal values in terminal states?

Temporal logics have been applied to the verification of computer programs, and more broadly computer systems [13,3]. The programs are in certain states at each instant, and the correctness of the programs can be expressed as temporal specifications. A good example is the temporal logic formula "$AG\,\neg deadlock$" meaning *the program can never enter a deadlock state*. Epistemic logics, on the other hand, are formalisms for reasoning about knowledge and beliefs. Their application in verification was originally motivated by the need to reason about communication protocols. One is typically interested in what knowledge different parties to a protocol have before, during and after a run (i.e., an execution sequence) of the protocol. Fagin *et al.* [4] give a comprehensive study on epistemic logic for multi-agent interactions.

Ruan and Thielscher [16] have shown that the situation at any stage of a game in GDL-II can be characterized by a multi-agent epistemic (i.e., S5-) model. Yet, this result only provides a static characterization of what players know (and don't know) at a certain stage. Our paper extends this recent analysis with a temporal dimension, and also provides a practical method for verifying temporal and epistemic properties using a model checker MCK [5]. We present a systematic translation from GDL-II

S. Cranefield and A. Nayak (Eds.): AI 2013, LNAI 8272, pp. 246–258, 2013.

into equivalent specifications in the model specification language of MCK. Verifying a property φ for a game description G is then equivalent to checking whether φ holds for the translation $\mathtt{trs}(G)$. The latter can be automatically checked in MCK.

The paper is organized as follows. Section 2 introduces GDL-II and MCK. Section 3 presents the translation along with possible optimizations and a proof of its correctness. Experimental results for four case studies are given in Section 4. The paper concludes with a discussion of related work and directions for further research.

2 Background

Game Description Language GDL-II. A complete game description consists of the names of (one or more) players, a specification of the initial position, the legal moves and how they affect the position and the players' knowledge thereof, and the terminating and winning criteria. The emphasis of game description languages is on *high-level, declarative game rules* that are easy to understand and maintain. Background knowledge is not required—a set of rules is all a player needs to know to be able to play a hitherto unknown game. Meanwhile, GDL and its successor GDL-II have a precise semantics and are fully machine-processable.

The GDL-II rules in Fig. 1 formalize a simple but famous game called *Monty Hall*, where a car prize is hidden behind one of three doors and where a candidate is given two chances to pick a door. Highlighted are the pre-defined *keywords* of GDL-II. The intuition behind the rules is as follows. Line 1 introduces the players' names (the game host is modelled by the pre-defined role called `random`). Line 2 defines the four features that comprise the initial game state. The possible moves are specified by the rules for `legal`: in step 1, the `random` player must decide where to hide the car (line 3) and, simultaneously, the candidate chooses a door (line 7); in step 2, `random` opens a door that is not the one that holds the car nor the chosen one (lines 4–5); finally, the candidate can either stick to their earlier choice (noop) or switch to the other, yet unopened door (line 9 and 10, respectively). The candidate's only percept throughout the game is to see the door opened by the host (line 14) and where the car is after step 3 (line 15). The remaining rules specify the state update (rules for `next`), the conditions for the game to end (rule for `terminal`), and the payoff for the player depending on whether they got the door right in the end (rules for `goal`).

GDL-II is suitable for describing synchronous n-player games with randomness and imperfect information. Valid game descriptions must satisfy certain syntactic restrictions, which ensure that all necessary inferences "$\vdash$" in Definition 1 below are finite and decidable; see [12] for details. In the following, we assume the reader to be familiar with basic notions and notations of logic programming, as can be found in e.g. [11].

A state transition system can be obtained from a valid GDL-II game description by using the notion of the *stable models* of logic programs with negation [6]. The syntactic restrictions in GDL-II ensure that all logic programs we consider have a *unique* and *finite* stable model [12,20]. Hence, the state transition system for GDL-II has a finite set of players, finite states, and finitely many legal moves in each state. By $G \vdash p$ we denote that ground atom p is contained in the unique stable model, denoted as $\text{SM}(G)$, for a stratified set of clauses G. In the following definition of the game semantics for GDL-II, *states* are identified with the set of ground atoms that are true in them.

```
role(candidate). role(random).
init(closed(1)). init(closed(2)). init(closed(3)). init(step(1)).
legal(random,hide_car(?d))  <= true(step(1)), true(closed(?d)).
legal(random,open_door(?d)) <= true(step(2)), true(closed(?d)),
                               not true(car(?d)), not true(chosen(?d)).
legal(random,noop)          <= true(step(3)).
legal(candidate,choose(?d)) <= true(step(1)), true(closed(?d)).
legal(candidate,noop)       <= true(step(2)).
legal(candidate,noop)       <= true(step(3)).
legal(candidate,switch)     <= true(step(3)).
next(car(?d))  <= does(random,hide_car(?d)).
...
next(step(4))  <= true(step(3)).
sees(candidate,?d) <= does(random,open_door(?d)).
sees(candidate,?d) <= true(step(3)), true(car(?d)).
terminal   <= true(step(4)).
goal(candidate,100) <= true(chosen(?d)), true(car(?d)).
goal(candidate,  0) <= true(chosen(?d)), not true(car(?d)).
```

Fig. 1. G_{MH} - a GDL-II description of the Monty Hall game adapted from [21]

Definition 1. *[20] Let G be a valid GDL-II description. The state transition system* $(R, s_0, \tau, l, u, \mathcal{I}, \Omega)$ *of G is given by*

- *roles* $R = \{i \mid \mathtt{role}(i) \in \mathrm{SM}(G)\}$*;*
- *initial position* $s_0 = \mathrm{SM}(G \cup \{\mathtt{true}(f) \mid \mathtt{init}(f) \in \mathrm{SM}(G)\})$*;*
- *terminal positions* $\tau = \{s \mid \mathtt{terminal} \in s\}$*;*
- *legal moves* $l = \{(i, a, s) \mid \mathtt{legal}(i, a) \in s\}$*;*
- *state update function* $u(M, s) = \mathrm{SM}(G \cup \{\mathtt{true}(f) \mid \mathtt{next}(f) \in \mathrm{SM}(G \cup s \cup M)\})$*, for all* joint *legal moves M (i.e., where each role in R takes one legal move);*
- *information relation* $\mathcal{I} = \{(i, M, s, p) \mid \mathtt{sees}(i, p) \in \mathrm{SM}(G \cup s \cup M)\}$*;*
- *goal relation* $\Omega = \{(i, n, s) \mid \mathtt{goal}(i, n) \in s\}$.

Note that a state s contains all ground atoms that are true in the state, which includes the "fluent atoms" $\mathtt{true}(f)$ in, respectively, $\{\mathtt{true}(f) \mid \mathtt{init}(f) \in \mathrm{SM}(G)\}$ (for the initial state) and $\{\mathtt{true}(f) \mid \mathtt{next}(f) \in \mathrm{SM}(G \cup s \cup M)\}$ (for the successor state of s and M), and all other atoms that can be derived from G *and* these fluent atoms.

Different runs of a game can be described by *developments*, which are sequences of states and moves by each player up to a certain round. A player *cannot distinguish* two developments if the player has made the same moves and perceptions in both of them.

Definition 2. *[20] Let* $(R, s_0, \tau, l, u, \mathcal{I}, \Omega)$ *be the state transition system of a GDL-II description G, then a* development δ *is a finite sequence*

$$\langle s_0, M_1, s_1, \ldots, s_{d-1}, M_d, s_d \rangle$$

such that for all $k \in \{1, \ldots, d\}$ $(d \geq 0)$*,* M_k *is a joint move and* $s_k = u(M_k, s_{k-1})$.

A terminal development is a development such that the last state is a terminal state, i.e., $s_d \in \tau$*. The* length *of a development* δ*, denoted as len*(δ)*, is the number of states in* δ*. By* $M(i)$ *we denote agent i's move in the joint move* M*. Let* $\delta|_k$ *be the prefix of* δ *up to length* $k \leq len(\delta)$.

A player $i \in R \backslash \{\mathtt{random}\}$ cannot distinguish *two developments* $\delta = \langle s_0, M_1, s_1, \ldots \rangle$ *and* $\delta' = \langle s_0, M'_1, s'_1 \ldots \rangle$ *(written as* $\delta \sim_i \delta'$*) iff len*$(\delta) =$ *len*(δ') *and for any* $1 \leq k \leq len(\delta) - 1$: $M_k(i) = M'_k(i)$*, and* $\{p \mid (i, M_k, s_{k-1}, p) \in \mathcal{I}\} = \{p \mid (i, M'_k, s'_{k-1}, p) \in \mathcal{I}\}$.

Model Checker MCK. In this paper, we will use MCK (for: "Model Checking Knowledge"), which is a model checker for temporal and knowledge specifications [5]. The overall setup of MCK supposes a number of agents acting in an environment. This is modelled by an *interpreted system*, formally defined below, where agents perform actions according to protocols. Actions and the environment may only be partially observable at each instant in time. In MCK, different approaches to the temporal and epistemic interaction and development are implemented. Knowledge may be based on current observations only, on current observations and clock value, or on the history of all observations and clock value. The last corresponds to *synchronous perfect recall* and is used in this paper. In the temporal dimension, the specification formulas may describe the evolution of the system along a single computation, i.e., use linear time temporal logic; or they may describe the branching structure of all possible computations, i.e., use branching time or computation tree logic. We give the basic syntax of Computation Tree Logic of Knowledge (CTL*K$_n$).

Definition 3. *The language of CTL*K$_n$ (with respect to a set of atomic propositions Φ), is given by the following grammar:*

$$\varphi ::= p \mid \neg\varphi \mid \varphi \vee \psi \mid A\varphi \mid X\varphi \mid \varphi\mathcal{U}\psi \mid K_i\varphi.$$

The other logic constants and connectives $\top, \bot, \vee, \rightarrow$ are defined as usual. In addition, $F\varphi$ (read: finally, φ) is defined as $\top\mathcal{U}\,\varphi$, and $G\varphi$ (read: globally, φ) as $\neg F\neg\varphi$.

The semantics of the logic can be given using *interpreted systems* [4]. Let S be a set, which we call the set of environment states, and Φ be the set of atomic propositions. A *run* over environment states S is a function $r : \mathbf{N} \rightarrow S \times L_1 \times \ldots \times L_n$, where each L_i is called the set of *local states of agent* i. These local states are used to concretely represent the information on the basis of which agent i computes its knowledge. Given run r, agent i, and time m, we write $r_i(m)$ for the $(i+1)$-th component (in L_i) of $r(m)$, and $r_e(m)$ for the first component (in S). An *interpreted system* over environment states S is a tuple $\mathcal{IS} = (\mathcal{R}, \pi)$, where $\mathcal{R}$ is a set of runs over environment states S, and $\pi : \mathcal{R} \times \mathbf{N} \rightarrow \mathcal{P}(\Phi)$ is an interpretation function. A *point* of $\mathcal{IS}$ is a pair (r, m) where $r \in \mathcal{R}$ and $m \in \mathbf{N}$.

Definition 4. *Let $\mathcal{IS}$ be an interpreted system, (r, m) be a point of $\mathcal{IS}$, and φ be a CTL*K$_n$ formula.* Semantic entailment $\models$ *is defined inductively as follows:*

- $\mathcal{IS}, (r, m) \models p$ *iff* $p \in \pi(r, m)$;
- *the propositional connectives* $\neg, \wedge$ *are defined as usual;*
- $\mathcal{IS}, (r, m) \models A\varphi$ *iff* $\forall r' \in \mathcal{R}$ *with* $r'(k) = r(k)$ *and* $\forall k \in [0..m]$, *we have* $\mathcal{IS}, (r', m) \models \varphi$;
- $\mathcal{IS}, (r, m) \models X\varphi$ *iff* $\mathcal{IS}, (r, m + 1) \models \varphi$;
- $\mathcal{IS}, (r, m) \models \varphi\mathcal{U}\,\psi$ *iff* $\exists m' \geq m$ *s. t.* $\mathcal{IS}, (r, m') \models \psi$ *and* $\mathcal{IS}, (r, k) \models \varphi$ *for all* $k \in [m..m')$;
- $\mathcal{IS}, (r, m) \models K_i\varphi$ *iff* $\forall (r', m')$ *with* $r_i(m) = r'_i(m')$, *we have* $\mathcal{IS}, (r', m') \models \varphi$.

Syntax of MCK Input Language. An MCK description consists of an environment and one or more agents. An environment model represents how states of the environment are affected by the actions of the agents. A protocol describes how an agent selects an action under a certain environment.

Formally, an *environment model* is a tuple $\mathcal{M}_e = (Agt, Acts, Var_e, Init_e, Prog_e)$ where Agt is a set of agents, $Acts$ is a set of actions available to the agents, Var_e is a set of environment variables, $Init_e$ is an initial condition, in the form of a boolean formula over Var_e, and $Prog_e$ is a standard program for the environment e to be defined below.

Let $ActVar(\mathcal{M}_e) = \{i.a \mid i \in Agt,\ a \in Acts\}$ be a set of *action variables* generated for each model $\mathcal{M}_e$. An atomic statement in $Prog_e$ is of the form $x := expr$, where $x \in Var_e$ and $expr$ is an expression over $Var_e \cup ActVar(\mathcal{M}_e)$.

A *protocol for agent* i in $\mathcal{M}_e$ is a tuple $Prot_i = (PVar_i, OVar_i, Acts_i, Prog_i)$, where $PVar_i \subseteq Var_e$ is a set of *parameter variables*, $OVar_i \subseteq PVar_i$ is a set of *observable variables*, $Acts_i \subseteq Acts$, and $Prog_i$ is a standard program. An atomic statement in $Prog_i$ is either of the form $x := expr$, or of the form $\ll a \gg$ with $a \in Acts_i$.

A *standard program* over a set Var of variables and a set A of atomic statements is either the terminated program ϵ or a sequence P of the form $stat_1 ; \ldots ; stat_m$, where each $stat_k$ is a simple statement and ';' denotes sequential composition.

Simple statement $stat_k$ can be *atomic statements* in A; or *nondeterministic branching statements* of the form: if $g_1 \rightarrow a_1 [] \ldots [] g_m \rightarrow a_m$ fi; or *nondeterministic iteration statements* of the form: do $g_1 \rightarrow a_1 [] \ldots [] g_m \rightarrow a_m$ od, where each a_k is an atomic statement in A and each *guard* g_k is a boolean expressions over Var.

Each atomic statement a_k can be executed only if its corresponding guard g_k holds in the current state. If several guards hold simultaneously, one of the corresponding actions is selected nondeterministically. The last guard g_m can be "$otherwise$", which is shorthand for $\neg g_1 \wedge \cdots \wedge \neg g_{m-1}$. An *if*-statement executes once but a *do*-statement can be repeatedly executed.

Semantics of MCK Input Language. Based on a set of agents running protocols in the context of a given environment, we can define an interpreted system as follows.

Definition 5. *A* system model $\mathcal{S}$ *is a pair* $(\mathcal{M}_e, Prot)$ *where* $\mathcal{M}_e = (Agt, Acts, Var_e, Init_e, Prog_e)$ *and* $Prot$ *a joint protocol with* $Prot_i = (PVar_i, OVar_i, Acts_i, Prog_i)$ *for all* $i \in Agt$.

Let a state with respect to $\mathcal{S}$ *be an assignment* s *over the set of variables* Var_e*. A* transition model *over* $\mathcal{S}$ *is* $\mathcal{M}(\mathcal{S}) = (S, I, \{O_i\}_{i \in Agt}, \rightarrow, V)$*, where* S *is the set of states of* $\mathcal{S}$*;* I *is the set of initial states* s *such that* $s \models Init_e$*;* $O_i(s) = s \upharpoonright OVar_i$ *is the partial assignment given on the observable variables of agent* i*,* $\rightarrow$ *is a transition relation on* $S \times S$*;*[1] *and a valuation function* V *is given by: for any boolean variable* x*,* $x \in V(s)$ *iff* $s(x) = true$*.* [2]

An infinite sequence of states $s_0 s_1 \ldots$ *is an* initialized computation *of* $\mathcal{M}(\mathcal{S})$ *if* $s_0 \in I$*,* $s_k \in S$ *and* $s_k \rightarrow s_{k+1}$ *for all* $k \geq 0$*. An* **interpreted system over** $\mathcal{S}$ *is* $\mathcal{IS}(\mathcal{S}) = (\mathcal{R}, \pi)$*, where* $\mathcal{R}$ *is the set of runs such that each run* r *corresponds to an initialized computation* $s_0 s_1 \ldots$ *with* $r_e(m) = s_m$*, and* $r_i(m) = O_i(s_0) O_i(s_1) \ldots O_i(s_m)$*; and* $\pi(r, m) = V(s_m)$*.*

[1] More precisely, $s \rightarrow s'$ if s' is obtained by executing the parallel program $Prog_e \,||_{i \in Agt}\, Prog_i$ on s; see [14] for details.

[2] For simplicity, we assume x to be boolean; this can be easily extended to enumerated type variables: Suppose x is a variable with type $\{e_1, \ldots, e_m\}$, then use m booleans $x.e_1, \ldots, x.e_m$ such that $x.e_k \in V(s)$ iff $s(x) = e_k$.

3 Translation from GDL-II to MCK

Our main contribution in this paper is a systematic translation from a GDL-II description G into an MCK description $\mathtt{trs}(G)$. The translation is provably correct in that the game model derived from G using the semantics of GDL-II satisfies the exact same formulas as the model that is derived from $\mathtt{trs}(G)$ using the semantics of MCK. This will be formally proved later in this section. We use the GDL-II description of the Monty Hall game from Fig. 1, denoted as G_{MH}, to illustrate the whole process. The translation trs can be divided into the following steps.

Preprocessing. The first step is to obtain a variable-free (i.e., ground) version of the game description G. We can compute the domains, or rather supersets thereof, for all predicates and functions of G by generating a domain dependency graph from the rules of the game description, following [19]. The nodes of the graph are the arguments of functions and predicates in game description, and there is an edge between two nodes whenever there is a variable in a rule of the game description that occurs in both arguments. Connected components in the graph share a (super-)domain. E.g., lines 2–3 in G_{MH} give us the domain graph as above, from which it can be seen that the arguments of both closed() and hide_car() range over $\{1, 2, 3\}$.

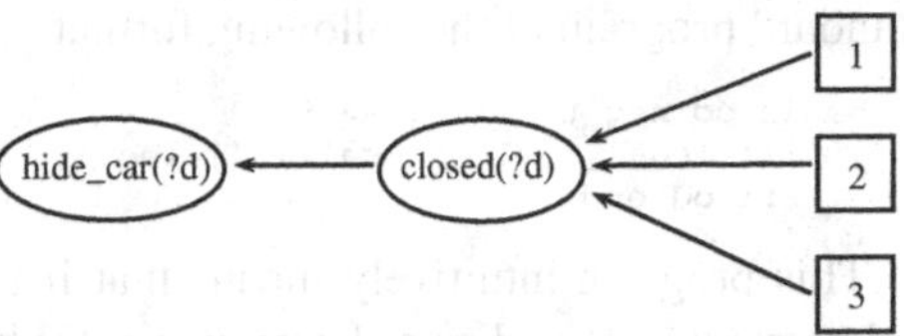

Once we have computed the domains, we instantiate all the variables in G to obtain all ground atoms, e.g., true(closed(1)), legal(random, hide_car(1)), etc. Our following translation operates on an equivalent variable-free version of G, which for convenience we still refer to as G.

Deriving Environment Variables. This step derives all the environment variables Var_e. Let AT be the set of ground atoms in G. Define the following subsets of AT according to the keywords: $\mathtt{AT_t} = \{\mathtt{h} \in \mathtt{AT} \mid \mathtt{h} = \mathtt{true(p)}\}$, $\mathtt{AT_n} = \{\mathtt{h} \in \mathtt{AT} \mid \mathtt{h} = \mathtt{next(p)}\}$, $\mathtt{AT_d} = \{\mathtt{h} \in \mathtt{AT} \mid \mathtt{h} = \mathtt{does(i,a)}\}$, $\mathtt{AT_i} = \{\mathtt{h} \in \mathtt{AT} \mid \mathtt{h} = \mathtt{init(p)}\}$, $\mathtt{AT_s} = \{\mathtt{h} \in \mathtt{AT} \mid \mathtt{h} = \mathtt{sees(r,p)}\}$, and $\mathtt{AT_l} = \{\mathtt{h} \in \mathtt{AT} \mid \mathtt{h} = \mathtt{legal(r,p)}\}$. Let p be obtained by replacing '(' and ',' with '_' and by removing ')' in a ground atom p. Define t as follows:

- $t(\mathtt{init(p)}) = p$, $t(\mathtt{true(p)}) = p\_old$ and $t(\mathtt{next(p)}) = p$;
- $t(\mathtt{does(i,a)}) = did\_i$;
- $t(\mathtt{p}) = p$ for all $\mathtt{p} \in \mathtt{AT} \setminus (\mathtt{AT_i} \cup \mathtt{AT_t} \cup \mathtt{AT_n} \cup \mathtt{AT_d})$.

Note that the ground atoms with keywords legal, terminal, goal are all in $\mathtt{AT} \setminus (\mathtt{AT_i} \cup \mathtt{AT_t} \cup \mathtt{AT_n} \cup \mathtt{AT_d})$. As an example, $t(\mathtt{sees(i,a)}) = sees\_i\_a$ and $t(\mathtt{legal(i,a)}) = legal\_i\_a$. The set of environment variable Var_e is then $\{t(\mathtt{p}) \mid \mathtt{p} \in \mathtt{AT}\}$. For convenience, we denote $t(A)$ as $\{t(x) \mid x \in A\}$.

The type of each variable $did\_i \in t(\mathtt{AT_d})$ is the set of legal moves of agent i plus two additional moves, INIT and STOP, that do not appear in G, i.e., $\{\mathtt{a} \mid \mathtt{legal(i,a)} \in \mathtt{AT}\} \cup \{\mathtt{INIT}, \mathtt{STOP}\}$. The type of variables in $Var_e \setminus t(\mathtt{AT_d})$ is Bool.

Initial Condition. This step specifies the environment initial condition $Init_e$, which is an assignment over Var_e. By using the semantics of G and $\mathtt{AT_i}$, we first compute the initial state s_0 (see Definition 1). Then for any $\mathtt{p} \in \mathtt{AT_i}$, we add boolean expression

"$t(\mathtt{p}) == true$" to $Init_e$ as a conjunct; and for all $did\_i \in t(\mathtt{AT_d})$, we add "$did\_i ==$ INIT". For the rest, add "$t(\mathtt{p}) == true$" if $\mathtt{p} \in s_0$, and "$t(\mathtt{p}) == false$" if $\mathtt{p} \notin s_0$.

Agent Protocols. This step specifies the agents and their protocols. The names of the agents are read off the `role()` facts. Let $Prot_i = (PVar_i, OVar_i, Acts_i, Prog_i)$ be the protocol of agent i, such that $PVar_i = Var_e$, $OVar_i = \{sees\_i\_p \mid sees\_i\_p \in t(\mathtt{AT_s})\} \cup \{did\_i\}$ includes all the variables representing i's percept and i's move, and $Acts_i = \{\mathtt{a} \mid \mathtt{legal(i,a)} \in G\}$ includes all the legal moves of agent i. Note that $Acts_i$ does not include the two special moves in the protocol. The last component $Prog_i$ is a standard program of the following format:

```
begin do neg terminal ->
    if legal_i_a1 -> <<a1>> [] legal_i_a2 -> <<a2>> [] ...
    fi od end
```

This program intuitively means that if the current state is not terminal, then a legal move is selected non-determinstically by i. The statements between do $\cdots$ od are executed repeatedly. The variables inside <<>> represent moves.

State Transition. This step specifies the environment program $Prog_e$. Each environment variable is updated in correspondence with the rules in G. The main task is *to translate these rules into MCK statements in a correct order.* In GDL-II, the order of the rules does not matter as the stable model semantics [12,20] always gives the same unique model, but MCK uses the imperative programming style in which the order of the statements does matter; e.g., executing "$x := 0; x := 1;$" results in a different state than "$x := 1; x := 0;$". To take care of the order, we separate the program $Prog_e$ into three parts.

The first part updates the variables in $t(\mathtt{AT_d})$ using the following template (for i):

```
if i.a1    -> did_i := a1 [] i.a2    -> did_i := a2 []
     ...   otherwise -> did_i := STOP
fi;
```

The second part of $Prog_e$ updates the variables in $t(\mathtt{AT_t})$ and $t(\mathtt{AT_n} \cup \mathtt{AT_s})$. For all $p\_old \in t(\mathtt{AT_t})$, an atomic statement of the form $p\_old := p$ is added to ensure that the value of p is remembered before it is updated. For any atom $h \in t(\mathtt{AT_n} \cup \mathtt{AT_s})$, suppose $h = t(\mathtt{h})$ and $Rules(\mathtt{h})$ is the set of rules in G with head h:

$$r_1 : \mathtt{h} \Leftarrow b_{11}, \cdots, b_{1j}$$
$$\cdots \quad \cdots$$
$$r_k : \mathtt{h} \Leftarrow b_{k1}, \cdots, b_{kj}$$

where b_{xy} is a literal over AT. Define a translation tt as follows:

- $tt(\mathtt{does(i,a)}) = did\_i == a$;
- $tt(\mathtt{not\ x}) = neg\ tt(\mathtt{x})$; and other cases are same as t.

The translation of $Rules(\mathtt{h})$ has the following form:

$$h := (tt(b_{11}) \wedge \cdots \wedge tt(b_{1j})) \vee \cdots \vee (tt(b_{k1}) \wedge \cdots \wedge tt(b_{kj}))$$

This simplifies to $h := true$ if one of the bodies is empty. Essentially, this is a form of the standard Clark Completion [2], which captures the idea that h will be false in the next state unless there is a rule to make it true. The statements with $t(\mathtt{AT_t})$ should be given before those with $t(\mathtt{AT_n} \cup \mathtt{AT_s})$.

The third part deals the variables in $t(\mathtt{AT} \setminus (\mathtt{AT_t} \cup \mathtt{AT_n} \cup \mathtt{AT_s} \cup \mathtt{AT_d} \cup \mathtt{AT_i}))$. Pick such an atom h and take $Rules(\mathtt{h})$. The literals in the body of these rules are translated differently from the last case, as h refers to the current instead of the next state. Define a new translation tt' as follows:

- $tt'(\mathtt{true(p)}) = p$ and all other cases are identical to tt.

The translation of $Rules(\mathtt{h})$ is similar to the above by replacing tt by tt'. The statements in the third part are ordered according to the dependency graph. If $\mathtt{h}'$ depends on h, then the statement of $tt'(\mathtt{h})$ must appear before that of $tt'(\mathtt{h}')$. The fact that GDL rules are stratified ensures that a desirable order can always be found.

Optimizations. The above translation can be further optimized to make the model checking more efficient by reducing the number of variables.

(1) Using definitions. The variables in $t(\mathtt{AT} \setminus (\mathtt{AT_t} \cup \mathtt{AT_n} \cup \mathtt{AT_d} \cup \mathtt{AT_i}))$ (refer to the third part of the State Transition step) can be represented as definitions to save memory space for variables. The assignment statement $h := expr$ is swapped with definition *define* $h = expr$. MCK replaces h using the boolean expression $expr$ during its preprocessing stage, so h does not occupy memory during the main stage.

(2) Removing static atoms. We distinguish three special kinds of atoms in GDL-II: those (a) appearing in the rules with empty bodies, (b) never appearing in the heads of rules, (c) only appearing in the rules with (a) and (b). Under the GDL-II semantics, atoms in (a) are always true, those in (b) are always false, and those in (c) do not change their value during gameplay. Therefore we can replace them universally with their truth values. E.g., consider the following rules:

```
succ(1,2). succ(2,3).
next(step(2)) <= true(step(1)), succ(1,2).
next(step(3)) <= true(step(2)), succ(2,3).
```

Both succ(1, 2), succ(2, 3) are always true, so we replace them using their truth values. Then we can further simplify this by removing the "*true*" conjuncts universally (and by removing the rules with a "*false*" conjunct in the body):

```
next(step(2)) <= true(step(1)).     next(step(3)) <= true(step(2)).
```

(3) Converting booleans to typed variables. The atoms in $\mathtt{AT} \setminus \mathtt{AT_d}$ are translated to booleans in our non-optimized translation. There often are sets of booleans B such that at each state exactly one of them is true. We can then convert the booleans in B into one single variable v_B with the type $\{b_1, \ldots, b_{|B|}\}$, where $|B|$ is the size of B. This results in a logarithmic space reduction on B: $2^{|B|}$ is reduced to $|B|$. Reusing the example just discussed, we can create a variable v_{step} with type $\{1, 2, 3\}$.

Translation Correctness

The above completes the translation from G to $\mathsf{trs}(G)$. As our main theoretical result, we show as follows that our translation is correct: first the game model derived from a GDL-II description G is proved to be isomorphic to the interpreted system that is derived from its translation $\mathsf{trs}(G)$, then a CTL*K$_n$ formula is shown to have an equivalent interpretation (i.e., the same truth value) over these two models.

We first extend the concept of finite developments in Definition 2 to infinite ones.

Definition 6 (Infinite Developments and GDL-II Models). *Let $\langle R, s_0, t, l, u, \mathcal{I}, g\rangle$ be the state transition system of a game description G, and $\delta = \langle s_0, M_1, s_1, \ldots, M_d, s_d\rangle$ a finite terminal development of G, then an* infinite extension *of δ is an infinite sequence $\langle s_0, M_1, s_1, \ldots, M_d, s_d, M_{d+1}, s_{d+1}, \ldots\rangle$ such that M_{d+k} is the joint move where all players take a special move* STOP *and $s_{d+k} = s_d$ for all $k \geq 1$.*

Given a GDL-II description G, the game model $\mathsf{GM}(G)$ *is a tuple $(D, \{\sim_i | i \in Agt\})$, where D is the set of infinite developments δ such that either δ is an infinite development without terminal states, or δ is an infinite extension of a finite terminal development; and $\sim_i$ is agent i's indistinguishability relation defined on the finite prefixes of $\delta|_k$ as in Definition 2.*

For a given δ, let $\delta(k)$ denote the k-th state s_k; $\delta(k)^M$ the k-th joint move M_k; and (δ, k) the pair (M_k, s_k).

Definition 7 (Isomorphism). *Let* $\mathsf{GM} = (D, \{\sim_i | i \in Agt\})$ *be a game model and $\mathcal{IS} = (\mathcal{R}, \pi)$ an interpreted system.* GM *is* isomorphic *to $\mathcal{IS}$ if there is a bijection* w *between the ground atoms of* GM *and the atomic propositions of $\mathcal{IS}$, and a bijection* z *between D and $\mathcal{R}$ satisfying the following:* $\mathsf{z}(\delta) = r$ *iff for any ground atom* p: $\mathsf{p} \in \delta(k)$ *iff* $\mathsf{w}(\mathsf{p}) \in \pi(r, k)$, *and* $\mathtt{does}(i, a) \in \delta(k)^M$ *iff did_i == a is true in (r, k).*

Intuitively, z associates a point (δ, k) in a development to a point (r, k) in a run such that they coincide in the interpretation of basic and move variables. The following proposition is the first step in showing the correctness of our translation.

Proposition 1. *Given a GDL-II description G, let* trs *be the translation from GDL-II to MCK, then the game model* $\mathsf{GM}(G)$ *is isomorphic to the interpreted system* $\mathcal{IS}(\mathsf{trs}(G))$.

For the technical details of the proof we must refer to [17].

Let w be a bijection from the set of ground atoms of G to the set of atomic propositions of CTL*K$_n$ and w^{-1} be its inverse. The semantics of CTL*K$_n$ over GDL-II Game Models can be given as relation $\mathsf{GM}(G), (\delta, m) \models \varphi$ in analogy to the semantics of CTL*K$_n$ over interpreted systems; e.g., $\mathsf{GM}(G), (\delta, m) \models p$ iff $\mathsf{w}^{-1}(p) \in \delta(m)$, and $\mathsf{GM}(G), (\delta, m) \models K_i\varphi$ iff for all states (δ', m') of $\mathsf{GM}(G)$ that satisfy $\delta|_m \sim_j \delta'|_{m'}$ we have $\mathsf{GM}(G), (\delta', m') \models \varphi$.

The following proposition then shows that checking φ against a game model of G is equivalent to checking φ against the interpreted system of $\mathsf{trs}(G)$.

Proposition 2. *Given a GDL-II description G, let* trs *be the translation from GDL-II to MCK; φ a CTL*K$_n$ formula over the set of atomic propositions in* $\mathsf{trs}(G)$; *and* w, z *the bijections from the isomorphism between* $\mathsf{GM}(G)$ *and* $\mathcal{IS}(\mathsf{trs}(G))$ *then:*

$$\mathsf{GM}(G), (\delta, m) \models \varphi \textit{ iff } \mathcal{IS}(\mathsf{trs}(G)), (\mathsf{z}(\delta), m) \models \varphi.$$

This follows from Proposition 1 by an induction on the structure of φ and completes the proof of our main result.

Our optimization techniques do not affect the isomorphism. So we can follow a similar argument as Proposition 1 and 2 to show that the optimized translation is correct.

4 Experimental Results

We present experimental results on four GDL-II games from the repository at *general-game-playing.de*: Monty Hall (MH), Krieg-TicTacToe (KTTT), Transit, and Meier. Same games were also used in Haufe and Thielscher [9]. MCK (v1.0.0) runs on Intel 3.3 GHz CPU and 8GB RAM with GNU Linux 2.6.32.

Temporal and epistemic specifications. The temporal logic formulas can be used to specify the *objective* aspects of a game. The following three properties represent the basic requirements from [7]. (Let $Legal_i$ and $Goal_i$ be the set of legal moves and goals of i respectively.)

$$AF\ terminal \tag{1}$$

$$AG(\neg terminal \rightarrow \bigwedge_{i\in Agt} \bigvee_{p\in Legal_i} p) \tag{2}$$

$$\bigwedge_{i\in Agt} \neg AG\neg goal\_i\_100 \tag{3}$$

Property (1) says that the game always terminates. Property (2) expresses *playability*: at every non-terminal state, each player has a legal move. Property (3) expresses *fairness*: every player has a chance to win, i.e., to eventually achieve the maximal goal value 100. These properties apply both to GDL and GDL-II games. The next three properties concern the *subjective* views of the players under incomplete-information situations, hence are specific to GDL-II games.

$$\bigwedge_{i\in Agt} G(terminal \rightarrow K_i terminal) \tag{4}$$

$$\bigwedge_{i\in Agt} G\ (\neg terminal \rightarrow \bigwedge_{p\in Legal_i} (K_i p \vee K_i \neg p)) \tag{5}$$

$$\bigwedge_{i\in Agt} G\ (terminal \rightarrow \bigwedge_{p\in Goal_i} (K_i p \vee K_i \neg p)) \tag{6}$$

Property (4) says that once the game has terminated, all players know this. Property (5) says that any player always knows its legal moves in non-terminal states; and property (6) says that in a terminal state, all players know their outcome.

φ	MH	KTTT	Transit	Meier	Meier′	φ	MH	KTTT	Transit	Meier	Meier′
(1)	0.47	1864.81	12.17	6.41	8079.52	**(4)**	0.60	22847.06	14.91	7.00	NA
(2)	0.48	3528.14	7.54	9.75	13192.91	**(5)**	0.56	22643.12	14.39	23.28	NA
(3)	0.67	303.04	11.02	17.06	15056.29	**(6)**	0.43	5498.03	45.15	11.01	NA

The table above shows the runtimes (in seconds) on five translations. The first four translations use all three optimization techniques on the four games. The last translation, Meier′, is partially optimized with the third technique applied only for the variables in $t(\mathtt{AT_s})$. As a consequence, Meier′ uses 126 booleans for what in the fully optimized Meier is represented by 4 enumerated type variables of a size equivalent to about 22 booleans, i.e., the state space of Meier is only $(1/2)^{104}$ of the state space of Meier′. The time is measured in seconds and "NA" indicates that MCK did not return a result after

10 hours. A comparison of the two translations of Meier shows that our optimization can be very effective. Somehow surprisingly, the result shows that the game Meier is not well-formed as it does not satisfy property (1). The last three properties were also checked by Haufe and Thielscher [9], but for Transit, their approach could not prove or disprove these properties; in contrast, our approach obtains the results fully. Note that although we only show the experiment results for four games, our approach is not a specialised solution for these four games only. It is general enough to deal with all GDL and GDL-II games.

5 Related Work and Further Research

There are a few papers on reasoning about games in GDL and GDL-II. Haufe *et al.* [8] use Answer Set Programming for verifying temporal invariance properties against a given game description by structural induction. Haufe and Thielscher [9] extend [8] to deal with epistemic properties for GDL-II. Their approach is restricted to positive-knowledge formulas unlike ours, which can handle more expressive epistemic and temporal formulas.

Ruan *et al.* [15] provide a reasoning mechanism for strategic and temporal properties but restricted to the original GDL for complete information games. Ruan and Thielscher [16] examine the epistemic logic behind GDL-II and in particular show that the situation at any stage of a game can be characterized by a multi-agent epistemic (i.e., S5-) model. Ruan and Thielscher [18] provide both semantic and syntactic characterizations of GDL-II descriptions in terms of a strategic and epistemic logic, and show the equivalence of these two characterizations. The current paper does not handle strategies but is able to provide practical results by using a model checker.

Kissmann and Edelkamp [10] instantiate GDL descriptions and utilise BDDs to construct a symbolic search algorithm to solve single- and two-player turn-taking games with complete information. This is related to our work in the sense that we also do an instantiation of GDL descriptions and uses the BDD-based symbolic model checking algorithms of MCK to verify properties. But our approach is more general and in particular handles games with incomplete information.

Other existing work is related to our paper in that they too deal with declarative languages. Chang and Jackson [1] show the possibility of embedding declarative relations and expressive relational operators into a standard CTL symbolic model checker. Whaley *et al.* [22] propose to use Datalog (which GDL is based upon) with Binary Decision Diagrams (BDDs) for program analysis.

We conclude by pointing out some directions for further research. Firstly our results suggest that the optimization we have applied allows us to verify some formulas quickly, but it is still difficult to deal with a game like Blind TicTacToe. However a hand-made version of this game (with more abstraction) in MCK does suggest that MCK has no problem to cope with the number of reachable states in this game. So the question is, what other optimization techniques can we find for the translation? Secondly, we would like to investigate how to make MCK language more expressive by allowing declarative relations such as shown in [1]. Our current translation maps GDL-II to MCK's input, and MCK internally encodes that into BDDs for symbolic checking. So a more direct map from GDL-II to BDDs may result in a significant efficiency gain. Thirdly, we want to explore the use of bounded model checking as MCK has implemented some model checking algorithms for this.

Acknowledgements. We thank the anonymous reviewers for their helpful comments. This research was supported by the Australian Research Council (ARC; project DP 120102023). Michael Thielscher is the recipient of an ARC Future Fellowship (FT 0991348). He is also affiliated with the University of Western Sydney. Most of the work was completed while Ji Ruan was working at the University of New South Wales.

References

1. Chang, F.S.H., Jackson, D.: Symbolic model checking of declarative relational models. In: Osterweil, L.J., Rombach, H.D., Soffa, M.L. (eds.) ICSE, pp. 312–320. ACM (2006)
2. Clark, K.L.: Negation as Failure. In: Gallaire, H., Minker, J. (eds.) Logic and Data Bases, pp. 292–322. Plenum Press, New York (1978)
3. Clarke, E.M., Emerson, E.A.: Design and synthesis of synchronization skeletons using branching time temporal logic. In: Kozen, D. (ed.) Logic of Programs 1981. LNCS, vol. 131, pp. 52–71. Springer, Heidelberg (1982)
4. Fagin, R., Halpern, J.Y., Moses, Y., Vardi, M.Y.: Reasoning About Knowledge. The MIT Press, Cambridge (1995)
5. Gammie, P., van der Meyden, R.: MCK: Model checking the logic of knowledge. In: Alur, R., Peled, D.A. (eds.) CAV 2004. LNCS, vol. 3114, pp. 479–483. Springer, Heidelberg (2004)
6. Gelfond, M., Lifschitz, V.: The stable model semantics for logic programming. In: Kowalski, R., Bowen, K. (eds.) Proceedings of IJCSLP, pp. 1070–1080. MIT Press, Seattle (1988)
7. Genesereth, M., Love, N., Pell, B.: General game playing: Overview of the AAAI competition. AI Magazine 26(2), 62–72 (2005)
8. Haufe, S., Schiffel, S., Thielscher, M.: Automated verification of state sequence invariants in general game playing. Artificial Intelligence Journal 187-188, 1–30 (2012)
9. Haufe, S., Thielscher, M.: Automated verification of epistemic properties for general game playing. In: Proceedings of KR (2012)
10. Kissmann, P., Edelkamp, S.: Gamer, a general game playing agent. KI 25(1), 49–52 (2011)
11. Lloyd, J.: Foundations of Logic Programming, 2nd edn. Springer (1987)
12. Love, N., Hinrichs, T., Haley, D., Schkufza, E., Genesereth, M.: General Game Playing: Game Description Language Specification. Tech. Rep. LG–2006–01, Stanford (2006)
13. Manna, Z., Pnueli, A.: The Temporal Logic of Reactive and Concurrent Systems. Springer, Berlin (1992)
14. van der Meyden, R., Gammie, P., Baukus, K., Lee, J., Luo, C., Huang, X.: User manual for mck 1.0.0. Tech. rep., University of New South Wales (2012)
15. Ruan, J., van der Hoek, W., Wooldridge, M.: Verification of games in the game description language. Journal Logic and Computation 19(6), 1127–1156 (2009)
16. Ruan, J., Thielscher, M.: The epistemic logic behind the game description language. In: Proceedings of AAAI, San Francisco, pp. 840–845 (2011)
17. Ruan, J., Thielscher, M.: Model checking games in GDL-II: the technical report. Tech. Rep. CSE-TR-201219, University of New South Wales (2012)
18. Ruan, J., Thielscher, M.: Strategic and epistemic reasoning for the game description language GDL-II. In: Proceedings of ECAI, Montpellier, pp. 696–701 (2012)

19. Schiffel, S., Thielscher, M.: Fluxplayer: A successful general game player. In: Proceedings of AAAI, pp. 1191–1196. AAAI Press (2007)
20. Thielscher, M.: A general game description language for incomplete information games. In: Proceedings of AAAI, pp. 994–999 (2010)
21. Thielscher, M.: The general game playing description language is universal. In: Proceedings of IJCAI, Barcelona, pp. 1107–1112 (2011)
22. Whaley, J., Avots, D., Carbin, M., Lam, M.S.: Using datalog with binary decision diagrams for program analysis. In: Yi, K. (ed.) APLAS 2005. LNCS, vol. 3780, pp. 97–118. Springer, Heidelberg (2005)

Neuroevolution for Micromanagement in the Real-Time Strategy Game Starcraft: Brood War

Jacky Shunjie Zhen and Ian Watson

Department of Computer Science, University of Auckland
szhe024@aucklanduni.ac.nz, ian@cs.auckland.ac.nz

Abstract. *Real-Time Strategy (RTS) games have become an attractive domain for AI research in recent years, due to their dynamic, multi-agent and multi-objective environments. Micromanagement, a core component of many RTS games, involves the control of multiple agents to accomplish goals that require fast, real time assessment and reaction. In this paper, we present the application and evaluation of a Neuroevolution technique in evolving micromanagement agents for the RTS game Starcraft: Brood War (SC:BW). The NeuroEvolution of Augmented Topologies (NEAT) algorithm, both in its standard form and its real-time variant (rtNEAT) is comparatively evaluated in micromanagement tasks. Preliminary results suggest the general viability of these techniques in comparison to traditional, non-adaptive AI. Further analysis of each algorithm identified differences in task performance and learning rate.*

Keywords: Real-Time Strategy Games, Neuroevolution, Evolutionary Computation.

1 Introduction

It was predicted more than a decade ago, that interactive computer games would emerge as an ideal platform for Artificial Intelligence research [1]. Due to their increasingly complex and realistic simulations, video games have become fine approximations of real world environments. AI techniques can be developed and evaluated in a cost effective and contained manner, before being applied to more complicated real world problems [1,2]. The popularity of video games as an entertainment medium has resulted in a consistently growing, multi-billion dollar software industry [3]. This in turn is a driver of video game technology and research, of which AI is a vital component [4].

The popularity and ease of access to videogame hardware and software has increased the accessibility of computing power and simulation environments for AI research. On the other hand, the contribution of AI research to commercial game development has been lacking in recent years [5]. This has resulted in high dependency on deterministic and non-adaptive AI techniques in commercial games that limit their realism, replayability and challenge [6].

Real Time Strategy (RTS) games are a genre of video games that provide unique challenges to AI research [7]. Characteristic of the genre is a real-time,

S. Cranefield and A. Nayak (Eds.): AI 2013, LNAI 8272, pp. 259–270, 2013.

stochastic environment, with multiple objectives and enormous action and state space. These features require AI agents with multiple levels of abstraction and reasoning, fast reaction and expert game knowledge. An example of the genre is Starcraft: BroodWar (SC:BW)[1], an RTS game that is a popular game environment for AI research. Using a third party plugin called the Brood War API (BWAPI)[2], it is possible to create complex AI agents to play matches of SC:BW.

In this paper, we aim to evaluate the effectiveness of Neuroevolution (NE) techniques in developing learning agents for playing SC:BW. The goal is to contribute to the development of a complete AI system capable of learning and executing human expert level strategy in SC:BW. In particular we focus on 'micromanagement', a crucial level of abstraction in the RTS domain, handling the fast combat component of the overall game. NE applies evolutionary algorithms to train artificial neural networks that are known to be effective approximators of complex, non-linear functions. Meanwhile, RTS games have a large state and action space that is suitable for neural networks. Furthermore, research on the NEAT algorithm [8] has shown the effectiveness of NE in reinforcement learning tasks, of which SC:BW has been successfully modelled [9,10].

We first implemented a micromanagement agent for SC:BW that uses NEAT and a real time variant rtNEAT for learning behavior. Next, the viability of the agent was evaluated against the existing SC:BW AI in multiple experiments. Finally, we analyzed the difference in performance between standard NEAT and the real-time variant, both in the rate of learning and in match performance. The rest of the paper is structured as follows: we provide a survey of related work around SC:BW AI and the NEAT algorithm. Next, we describe an overview of the implementation of our AI agent and the usage of NEAT, followed by the evaluation of the agent and a discussion of results. Finally, we give concluding remarks and highlight areas of future research.

2 Related Work

In many RTS games, the game strategy can be roughly divided into two levels of abstraction. Macromanagement is the level that is concerned with high level strategic decision making such as resource planning and opponent modeling. Techniques dealing with macromanagement must choose and adapt sequences of strategic actions to meet goals of varied hierarchy. Example of techniques applied to this domain include Case Based Reasoning[11] and Goal Driven Autonomy[12]. Micromanagement is the level concerned with direct combat tactics and unit control. Traditionally, micromanagement is accomplished via static AI techniques such as scripts based on simple metrics [13]. More complicated techniques in the literature include Reinforcement Learning (RL) and Evolutionary Algorithm approaches.

RL combined with neural networks has been applied to SC:BW micromanagement [10]. Agent learning was accomplished using the online Sarsa RL algorithm,

[1] Starcraft: Brood War: `http://us.blizzard.com/en-us/games/sc/`

[2] Brood War API: `http://code.google.com/p/bwapi/`

with neural-networks to approximate the state-action value function. Results showed a significant winning advantage against standard Starcraft AI, but required thousands of training rounds and are limited in the type and number of units represented. In [9], a comparative evaluation of RL techniques applied to SC:BW was presented. Four variants of RL algorithms were applied to a specific micromanagement task, involving a long ranged unit with high mobility against numerous melee (close ranged) enemies. Evaluations identified strengths and weaknesses of the different algorithms, and showed a high win rate against the default SC:BW AI. However, the results are derived from a very limited scenario, and the author acknowledges it is only the first part of a larger RL based SC:BW agent.

Work that is most related to ours is from [14], in which rtNEAT was applied to SC:BW micromanagement. Units were controlled by separate neural networks, specifying actions to take in real time. The network typology is evolved over generations of 12 vs 12 unit combat. Evaluations showed a significant win rate within 300 training generations and also claimed the rtNEAT algorithm allowed fast, real-time strategy adaptation. A limitation of the study is the use of a custom SC:BW map that replenishes unit numbers with up to 100 unit reserves. This is an unrealistic depiction of real SC:BW combat where units are not replenished immediately to replace dead units. Furthermore, real-time fitness improvement occurs only over unit combat time that is minimal in a full SC:BW match. However, the work showcased the potential of applying NEAT based algorithms to SC:BW micromanagement that our work analyzes further.

2.1 Neuroevolution and NEAT

Neuroevolution (NE) has shown effectiveness compared to standard RL, in problems with continuous and high-dimensional state spaces [8]. Traditional NE worked on pre-defined topologies, and searched over the space of connection weights. Topology and Weight Evolving Artificial Neural Networks (TWEANNs) attempts to also evolve the topology of the network, and has the potential to improve training speed and accuracy of solutions [15]. Furthermore, it reduces the uncertainty and effort of deciding on network topology by researchers [16].

However, TWEANN techniques face numerous challenges, such as complications with network structure in crossover and problems with genetic encoding. Work by [16] developed the NEAT algorithm to address these challenges. The classic NEAT algorithm was expanded to a real-time variant[17], in which evolution occured over a real time environment. The rtNEAT algorithm was demonstrated in a game called NERO, where agents are evolved and adapted in real time to tackle changing objectives. Regular NEAT has been successfully applied to RTS by [18], where neural networks are evolved to become AI players in an ensemble process. In [6] both NEAT and rtNEAT were used to automatically balance the challenge of the AI player in an RTS game. A novel challenge metric coupled with a fitness function guided the evolution of neural networks. The AI was continuously evolved to converge to the same challenge level as the human player, thus creating a more balanced gaming experience.

3 Implementation

We use the BWAPI open source framework for creating and executing AI modules. It exposes functionality to retrieve information about the game state and to issue commands to game units. Units are encapsulated as BWNEAT units, with an accompanying neural network for decision making. The SC:BW game state is updated once per frame, i.e. every 56 milliseconds on normal game speed. BWAPI triggers an event on every game state update, allowing AI code to react. During this event, each BWNEAT unit feeds internal and external percepts from the AI Module as inputs to its neural network, and interprets the network output as the next action to be performed. The NEAT Manager module is responsible for instantiating the BWNEAT Units and is an interface to the NEAT and rtNEAT algorithms. It receives evaluated fitness from each unit, performs NEAT evolution and reassigns neural networks to BWNEAT units. Fig. 1 summarizes the agent architecture.

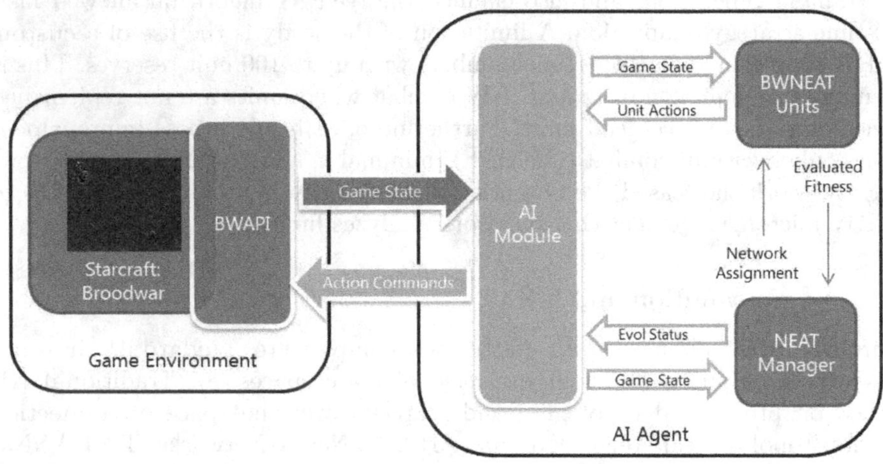

Fig. 1. An overview of the agent architecture

3.1 Neural Network Architecture

The basic neural network architecture is fully connected and feed forward, with randomized starting weights (Fig. 2). It begins with 0 hidden nodes and gradually allows nodes and connections to be added via the NEAT and rtNEAT algorithms. The inputs were chosen as important percepts to induce learning behavior that would allow a unit to inflict as much damage to the enemy units, while taking as little damage as possible. For example, when the unit's weapon is on cooldown (a pause between consecutive attacks), it should do its best to avoid damage. Another aspect is the range of unit weapons. If for example, the unit has a longer weapon range than the enemy units, it is possible to perform a hit-and-run strategy. These percepts are based on domain knowledge of SC:BW and is common in many RTS games.

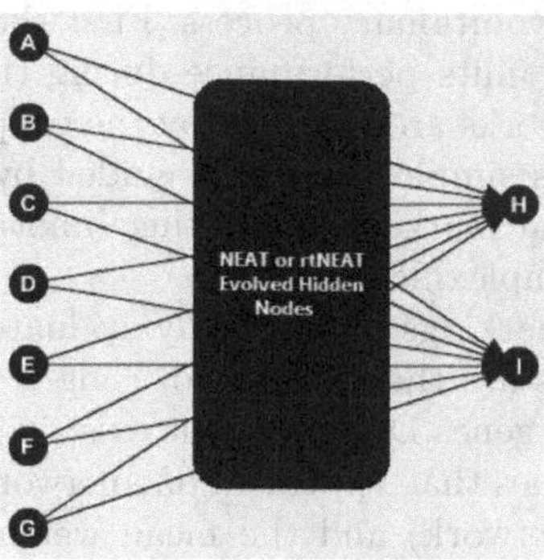

Fig. 2. Initial network architecture. Nodes A to G (*Bias, WeaponCooldown, RemainingHealth, WeaponRange, EnemyWeaponRange, NumAlliesInRange and NumEnemiesInRange*) denote a mixture of agent internal and external percepts as input to the network, while nodes H and I (*Fight and Retreat*) denote the outputs as two possible unit actions.

The output of the neural network corresponds to two unit actions: fight or retreat. The action with the largest corresponding output is chosen by the unit. If the fight action is taken, the unit executes a simple routine that targets the enemy unit with the lowest hit points (health) within its weapon range. The retreat action makes the unit move a small distance away from enemies and obstacles, via a weighted vector. These actions are based on similar implementations in [9] and [10], as they are simple to implement, but complicated enough to produce sophisticated behavior when performed in varying sequences.

3.2 Fitness Function

The NEAT and rtNEAT algorithms are guided by a fitness metric. In the context of SC:BW unit micromanagement, the fitness should reflect the performance of an individual unit. For both NEAT and RTNEAT, we define the fitness F_i for a unit $_i$ as:

$$F_i = \frac{TDD_i - HPL_i}{IHP_i} + 1 \tag{1}$$

The function takes in the total damage dealt by the unit (TDD), its hit point loss (HPL) accrued over the match and its initial hit point (IHP). In theory, the fitness is only upperbound by the total hit points of enemy units. However in practise, the average fitness of each unit falls under $[0, 2]$ where at its lowest the unit has produced no damage and dies, and at its highest value it has dealt twice as much damage than it has taken.

3.3 NEAT Evolution

Training via the classic NEAT algorithm occurs over generations of SC:BW matches. After a match, regardless of win or loss, the population of neural networks go through an evolutionary process. First the fitness of each network is evaluated based on the units performance during the match. Next, some of the worst performing networks are replaced by the offspring of some of the best performing networks. This simple process is guided by three principles: tracking evolution via historical markers, protecting innovation via speciation and minimizing search via 'complexification' [8].

NEAT uses historical markings to efficiently evaluate similarity between network topologies. Networks are then speciated using a similarity metric formed via the number of disjoint genes D (genes that exist in one network and not the other), excess genes E (genes that appear in one network later in evolution than any genes on the other network) and the mean weight difference of matching genes W[8]:

$$S = \frac{c_1 E}{N} + \frac{c_2 D}{N} + c_3 W \tag{2}$$

$c1$, $c2$ and $c3$ are adjustible weighting coefficients, and E and D are normalized by dividing N, the number of genes in the larger network. Networks are grouped into species via this similarity metric, and a compatibility threshold that can be modified to specify species bounds. A network shares its evaluated fitness with other members of its species, in order to encourage diversification of solutions and prevent single species dominance. NEAT adjusts each network's fitness based on its similarity metric against all other organisms in the population. The number of offsprings spawned by a species after each generation is based on the proportion of its average species fitness to the total of all average species fitness [8].

3.4 rtNEAT Evolution

The real-time variant of the NEAT algorithm is designed specifically to operate in a continuous, real-time domain. In particular when adapted to video games, the performance of AI agents is able to improve gradually as the game is played, without abrupt changes over a whole generation of evolution. In the context of SC:BW, rtNEAT applies evaluation and replacement on game units every n ticks of game time. The number of game ticks between replacement is an important factor that affects evolution. If new organisms are replaced too quickly, then they cannot be evaluated accurately and new innovations may be needlessly thrown away. A law of eligibility is formed by [17], stating the number of ticks between replacements, with respect to the fraction of the population that is too young to be replaced I, the minimum time alive m and the population size P:

$$n = \frac{m}{|P|I} \tag{3}$$

In our experiments we empirically define m as 300 game frames, to offset the delay between the start of a match and the first enemy encounter. We follow [17] in defining 50% of the population as eligible for replacement, and $p = 12$ the number of units which is constant. This gives us $n = 50$, the number of games frames between replacement in rtNEAT experiments. In the next section we describe in more detail, evaluations that incorporate these algorithms and principles to analyse the effectiveness of NEAT and rtNEAT for SC:BW micromanagement.

4 Experimentation and Results

We devised experiments to gauge the effectiveness of NEAT and rtNEAT evolved micromanagement agents against the standard SC:BW AI. A variety of unit setups were used in order to simulate different micromanagement scenarios. From these experiments, we saw very high fluctuation and variation in the fitness and win rate of agents over generations. In order to adjust for these fluctuations, we ran experiments to find the number of generations taken for each algorithm to converge to a suitable solution, when evolution is halted upon finding a potential candidate.

4.1 Experiment Setup

SC:BW units vary on attributes such as race, weapon and armour type. In order to keep the experimental variables constant and to avoid an explosion of unit type permutations, we based our experimental setup on [14] that compared 4 unit type variations (melee vs. melee, ranged vs. ranged, melee vs. ranged, ranged vs. melee). The number of units is kept at a constant 12 vs. 12, which is the maximum selectable number of units for a human controlled squad. The scenario used throughout experimentation is a flat map, based on those used in the AIIDE 2010 Starcraft micromanagement tournament[3].

4.2 Evolutionary Process Experiment

We first compared the performance of NEAT and rtNEAT algorithms on each of the 4 unit matchup variations, over 300 generations of evolution. Each matchup is repeated 25 times to reduce randomness in network starting weights. The average unit fitness and the match outcome is recorded over 300 generations, and averaged over the 25 runs.

Match Variations	NEAT WR	NEAT SD	NEAT CL 95%	rtNEAT WR	rtNEAT SD	rtNEAT CL 95%
Range vs. Range	60.39%	9.86%	1.12%	72.35%	11.53%	1.31%
Range vs. Melee	97.59%	7.95%	0.90%	69.48%	10.44%	1.19%
Melee vs. Melee	23.80%	8.26%	0.94%	49.73%	11.63%	1.32%
Melee vs. Range	58.89%	10.79%	1.23%	47.87%	10.26%	1.17%

Fig. 3. Summary statistics for our first experiment. Mean win rate (WR) over 300 generations, standard deviation (SD) and the 95% confidence level (CL) is shown.

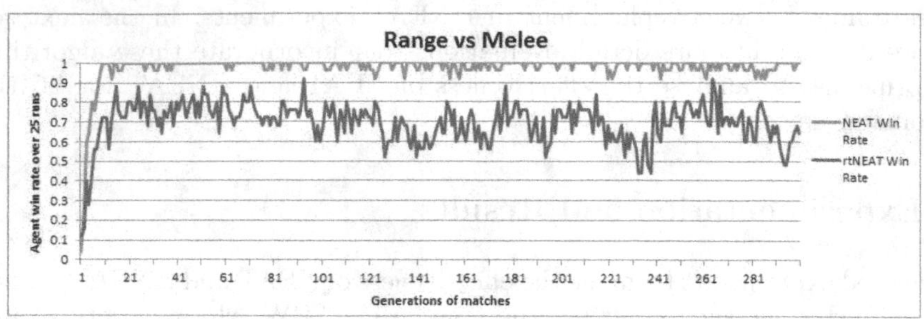

Fig. 4. Plot of the best performing match up (range vs. melee) average win rate

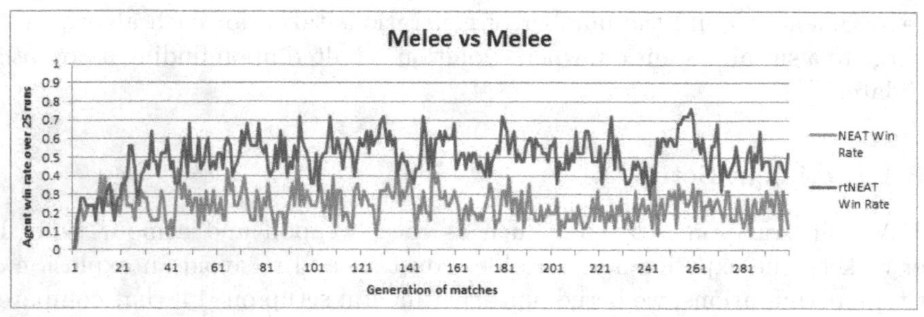

Fig. 5. Plot of the worst performing match up (melee v.s melee) average win rate

The results suggest that there is no single algorithm dominating all match variations (Fig. 3). Mean win rate is higher for NEAT on range vs. melee (mean 97.59%, SD 7.95%) and melee vs. range (58.89% mean, 10.79% SD), while rt-NEAT is higher on range vs. range (60.39% mean, 9.86% SD) and melee vs melee (49.73% mean, 11.63% SD). If we consider a win rate higher than 50% to indicate better than baseline performance against the built-in SC:BW AI, then both NEAT and rtNEAT show effectiveness on range vs. range and range vs. melee battles. NEAT is also effective in melee vs. range (58.89% mean ± 1.23% at 95% confidence level). From examining the agent behavior, the effectiveness of controlling ranged units can be attributed to having learnt the hit-and-run micromanagement strategies. It is interesting to note the generally poor performance of both algorithms when controlling melee units. We discuss in more detail the evolved behavior of the agents contributing to the performance in Section 5.

Fig. 4 and Fig. 5 show plots of some of the best and worst performing match up variations. The average fitness plot is not shown, but is highly correlated to the average win rate. From these plots, we see a trend of initial poor performance and a quick convergence to some local optima. This is typical of evolutionary algorithms where the initial starting solutions are randomized and are

[3] AIIDE 2010 micromanagement tournament: `eis.ucsc.edu/starcrafttournament1`

not expected to perform well. On all variations the first convergence to a local optimal occurs between the 10th to 20th generation. There is significant fluctuation of win rate throughout generations that is further illustrated by high variance shown by their standard deviations. In general, the standard deviation is higher on rtNEAT runs, suggesting greater variation in evolutionary success over generations than standard NEAT. This may be due to the nature of the rtNEAT algorithm, in introducing real time change. Both algorithms are capable of producing high performing solutions at various generations, but are also quick to introduce mutations weakening the solutions. This is partly due to the nature of the experiment where we allow evolution to continue even after achieving a winning solution.

4.3 Generational Convergence Experiment

By defining a success criteria, we can halt the evolution of both the NEAT and rtNEAT algorithms once an acceptable solution is achieved. We defined a successful solution to be an agent that achieves 10 consecutive wins (the probability an agent with 50% win rate can win 10 consecutive games is $< 0.1\%$). This is a strict criteria as an agent may still be high performing even though it loses 1 game out of 10 (e.g. due to the stochastic nature of the game state). However, we use this to simplify the running of the experiment and to show that it is possible to robustly generate agents of this level of performance. We keep all other variables the same as in our previous experiment, except that the evolution terminates when a solution reaches 10 wins, or after 1000 generations. When a solution achieves a win, evolution is halted until the agent either achieves 10 wins, or a loss is encountered, where upon evolution continues. After a successful solution is achieved, or if no solution is found after 1000 generations, the experiment is reset to an initial population with randomized weights. For each algorithm and each matchup, we stopped the experiment at 60 runs and analyzed the results.

Match Variations	NEAT MNG	NEAT SD	NEAT CL 95%	rtNEAT MNG	rtNEAT SD	rtNEAT CL 95%
Range vs. Range	116.03	124.39	32.13	18.33	12.52	3.24
Range vs. Melee	4.15	1.76	0.46	19.95	26.03	6.73
Melee vs. Melee	42.52	53.15	13.73	26.78	18.30	4.73
Melee vs. Range	9.60	5.90	1.52	22.07	16.00	4.13

Fig. 6. Summary statistics for generational convergence experiment. Mean number of generations (MNG) taken to produce an acceptable solution is shown. Standard deviation (SD) and the 95% confidence level (CL) for the mean was also calculated.

In all experiments, an acceptable solution was found before 1000 generations, with most converging under 100 generations (Fig. 6). There was high variability in the number of generations required to arrive at an acceptable solution, evident by the high standard deviation in some match ups (e.g. 124.39 SD and 116.03 Mean generations for NEAT range vs range). The mean number of generations taken between NEAT and rtNEAT is comparable to the average win rate performance of the previous experiment: NEAT converges faster for range vs. melee and melee vs. range match ups, while rtNEAT is faster for range vs. range and melee vs. melee. The range in generations taken between different matchups is higher for NEAT (4.15 mean for range.vs melee and 116.03 mean for range vs. range) than for rtNEAT (18.33 mean for range vs. range and 26.78 mean for melee vs. melee). This suggests the performance of rtNEAT is more stable under different unit variations.

Overall, the experiment showed that both algorithms were capable of generating effective solutions for micromanagement against the default SC:BW AI. However, it was necessary to establish an acceptance criteria for which to halt evolution and to preserve winning behavior. In the next section, we discuss further implications of the experimental results.

5 Discussion

In the first experiment, the fluctuation of fitness and success rate of solutions can be due to a number of reasons. Firstly, it suggests that any structural innovations introduced were making significant differences in the performance of the neural networks. This is probably due to the simplicity of the network design, where only 2 outputs exist, such that any structural change may affect the action selected. A simple neural network allows faster convergence by reducing the search space of initial nodes and weights. But it also means it is faster to diverge from the local optima. On top of this, the stochastic nature of the game environment can result in the same solution having varied success over different runs.

This also explains the general poor performance of both algorithms on melee match ups in the first experiment. Melee units do best in direct attack as they lack the weapon range to perform hit-and-run maneuvers. Any innovation introduced to make melee units run will immediately reduce the success rate of the solution. In the second experiment, the algorithms have no problem generating a solution for melee match ups, when no new innovations were introduced after a solution begins to do well. Another factor is an interesting behavior exhibited by the units over generations of evolution: some units are evolved to retreat when enemies are first found, but come back to fight after allied units are engaged in combat. These units tend to generate more fitness than those directly attacking from the beginning, as they do not receive as much damage over time. However, as the population begins to favour this behavior, there is a breaking point in which no units will stay to fight, leading to a match loss and a return to evolution favouring units that do not retreat. This cycle is highly correlated with the fluctuation of the win rate over generations.

Interestingly, the first experiment showed that rtNEAT produced higher variation in the success of solutions than NEAT over time. However in the second experiment, the average number of generations for an acceptable solution was less varied across different match ups than NEAT. This suggests real-time evolution can be quicker in introducing changes that reduce fitness, but also allow a more robust convergence to a solution regardless of unit variation (variability in state and solution space). This is intuitive, since rtNEAT should be faster in reacting to changes in the environment in real time, than regular NEAT evolution between generations.

It is possible to complicate the initial neural network architecture, by incorporating more percepts as input nodes and providing finer grain output decisions. For example, the inputs can incorporate a deeper ontology of unit quality and type variations (armour, weapon and ability types etc) and more precise directional and distance data. Instead of fight or retreat actions, the decisions can be to move at specific angles for specific distances, and to explicitly decide which units to attack. Enemy target selection is itself complicated enough to be a separate learning task, perhaps requiring the optimization of a separate neural network that takes into consideration enormous unit type variations and the location of units. More complicated neural network designs allow for agents with more complicated behaviors that are able to perform well under a higher variety of conditions. The disadvantage is a greater number of dimensions to search and optimize for, resulting in slower training time.

There are limitations to the evaluation methodology to be addressed. For example, while the experiments show that the technique is able to learn to defeat the standard SC AI, the results do not extend to human opponents. However, testing against the standard SC AI is a baseline measure used in much of the related work, particularly for micromanagement. It is difficult to evaluate against human players, due to the number of games required to be played, and the lack of an objective human skill measure for the micromanagement task (current measures exist only for full SC games). It is possible to evaluate against other micromanagement AI, but there is a lack of a standardized evaluation methodology to do so.

6 Conclusions

Our evaluations confirmed the viability of NEAT and rtNEAT algorithms in evolving agents for various SC:BW micromanagement scenarios. When the algorithms are allowed to run non-stop, the win rate of agents against the default SC:BW AI fluctuates highly over generations. However, when evolution is halted upon reaching an acceptable level of performance, both algorithms are able to consistently generate winning agents, with most under 100 generations. Each algorithm differs in the variability of performance over different unit matchups. Factors contributing to the difference in performance include the complexity of the network starting topology and the variation in unit types. There is room to explore network complexity further, and a need to establish standardized evaluation methods for micromanagement agent evaluations. More work is needed to adapt these techniques for commercial RTS game deployment, but results here have shown promising performance in a learning AI capable of defeating scripted AI under short training time.

References

1. Laird, J., VanLent, M.: Human-level AI's killer application: Interactive computer games. AI Magazine 22(2), 15–26 (2001)
2. Buro, M.: Call for AI research in RTS games. In: Proceedings of the AAAI 2004 Workshop on Challenges in Game AI, pp. 2–4 (2004)
3. Siwek, S.E.: Video Games in the 21st Century. Technical report. Entertainment Software Association (2010)
4. Yildirim, S., Stene, S.B.: A survey on the need and use of ai in game agents. In: Proceedings of the 2008 Spring Simulation Multiconference, pp. 124–131 (2008)
5. Mehta, M., Ontañón, S., Amundsen, T., Ram, A.: Authoring behaviors for games using learning from demonstration. In: Workshop on Case-Based Reasoning for Computer Games, ICCBR (2009)
6. Olesen, J.K., Yannakakis, G.N., Hallam, J.: Real-time challenge balance in an RTS game using rtNEAT. In: 2008 IEEE Symposium on Computational Intelligence and Games, pp. 87–94 (2008)
7. Buro, M., Furtak, T.M.: RTS games and real-time AI research. In: Proceedings of the Behavior Representation in Modeling and Simulation Conference, pp. 63–70 (2004)
8. Stanley, K.O., Miikkulainen, R.: Efficient Evolution of Neural Network Topologies. In: Proceedings of the 2002 Congress on Evolutionary Computation (CEC 2002). IEEE (2002)
9. Wender, S., Watson, I.: Applying reinforcement learning to small scale combat in the real-time strategy game StarCraft:Broodwar. In: Computational Intelligence and Games (CIG), pp. 402–408 (2012)
10. Shantia, A., Begue, E., Wiering, M.: Connectionist reinforcement learning for intelligent unit micro management in starcraft. In: The 2011 International Joint Conference on Neural Networks (IJCNN), pp. 1794–1801 (2011)
11. Cadena, P., Garrido, L.: Fuzzy Case-Based Reasoning for Managing Strategic and Tactical Reasoning in StarCraft. In: Batyrshin, I., Sidorov, G. (eds.) MICAI 2011, Part I. LNCS, vol. 7094, pp. 113–124. Springer, Heidelberg (2011)
12. Weber, B., Mateas, M., Jhala, A.: Applying goal-driven autonomy to StarCraft. In: Artificial Intelligence and Interactive Digital Entertainment, AIIDE 2010 (2010)
13. Davis, I.L.: Strategies for strategy game AI. In: Proceedings of the AAAI Spring Symposium on Artificial Intelligence and Computer Games, pp. 24–27 (1999)
14. Gabriel, I., Negru, V., Zaharie, D.: Neuroevolution based multi-agent system for micromanagement in real-time strategy games. In: Proceedings of the Fifth Balkan Conference in Informatics - BCI 2012, p. 32 (2012)
15. Yao, X.: Evolving artificial neural networks. Proceedings of the IEEE 87, 1423–1447 (1999)
16. Stanley, K.O., Miikkulainen, R.: Evolving neural networks through augmenting topologies. Evol. Comput. 10(2), 99–127 (2002)
17. Stanley, K.O.: Evolving neural network agents in the NERO video game. In: Proceedings of the IEEE 2005 Symposium on Computational Intelligence and Games, pp. 182–189 (2005)
18. Jang, S.H., Yoon, J.W., Cho, S.B.: Optimal strategy selection of non-player character on real time strategy game using a speciated evolutionary algorithm. In: Proceedings of the 5th International Conference on Computational Intelligence and Games, pp. 75–79 (2009)

Towards General Game-Playing Robots: Models, Architecture and Game Controller

David Rajaratnam and Michael Thielscher

The University of New South Wales, Sydney, NSW 2052, Australia
{daver,mit}@cse.unsw.edu.au

Abstract. General Game Playing aims at AI systems that can understand the rules of new games and learn to play them effectively without human intervention. Our paper takes the first step towards *general game-playing robots*, which extend this capability to AI systems that play games in the real world. We develop a formal model for general games in physical environments and provide a systems architecture that allows the embedding of existing general game players as the "brain" and suitable robotic systems as the "body" of a general game-playing robot. We also report on an initial robot prototype that can understand the rules of arbitrary games and learns to play them in a fixed physical game environment.

1 Introduction

General game playing is the attempt to create a new generation of AI systems that can understand the rules of new games and then learn to play these games without human intervention [5]. Unlike specialised systems such as the chess program Deep Blue, a general game player cannot rely on algorithms that have been designed in advance for specific games. Rather, it requires a form of general intelligence that enables the player to autonomously adapt to new and possibly radically different problems. General game-playing systems therefore are a quintessential example of a new generation of software that end users can customise for their own specific tasks and special needs.

This paper makes the first step towards *general game-playing robots*, which we define to be autonomous systems that can understand descriptions of new games and learn to play them in a physical game environment. Ultimately, a truly general game-playing robot must be able not only to accommodate new game rules but also adapt to unknown game environments. However, as a first step we aim at robots that learn to play new games within a fixed physical world. These robots should accept arbitrary game descriptions for such environments and then play these games effectively by manipulating the actual objects, e.g. pieces, of the real game.

In seeking to build a novel system such as a general game-playing robot, it is necessary to provide some justification for why such a system is of interest. We briefly highlight three disparate reasons. Firstly, interaction with a game-playing robot requires both mental and physical activity from a human opponent and could therefore be of interest in the area of rehabilitation and elderly care [3]. Secondly, games have a wide variety of physical requirements, and consequently offer a vast array of well-defined and controlled environments in which to benchmark techniques for robot recognition

S. Cranefield and A. Nayak (Eds.): AI 2013, LNAI 8272, pp. 271–276, 2013.

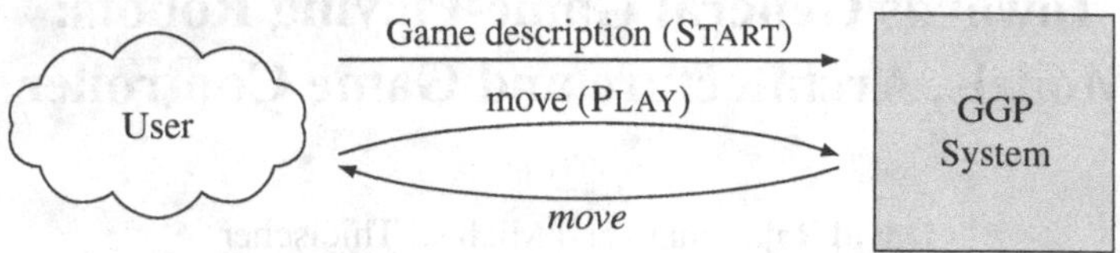

Fig. 1. Functionality of a general game player

and manipulation. Finally, this research has applications beyond games. Many tasks have game-like properties. For example, a domestic robot fetching an item has to adapt to changes in the environment; the item may not be where the robot expected, or the operator may change locations. Viewing such a task as a game can provide a natural framework for controlling the robot's behaviour.

Our specific contribution to general game-playing robots in this paper is three-fold. First, we develop a formal model to identify essential requirements for game descriptions when translating gameplay moves from a purely virtual to a physical environment (Section 3). Second, we provide a systems architecture that integrates research in General Game Playing with research in Robotics (Section 4). This architecture is generic in that it allows the embedding of both existing general game players and suitable robotic systems. Third, to evaluate the feasibility of our framework, we have built a first general game-playing robot and tested it on different games of variable difficulty played on a physical chess-like board with moving pieces (Section 5).

2 Background

General Game Playing. The annual AAAI GGP Competition [5] defines a general game player as a system that can understand the rules of any n-player game given in the general Game Description Language (GDL) and is able to play those games effectively. The functionality is illustrated in Fig. 1: A player must first accept any communicated game description (START message). After a given time period, and without human intervention, the player then makes his opening move, accepts legal moves by other players (PLAY message) and continues to play until the game terminates.

Since the first AAAI competition in 2005, General Game Playing has evolved into a thriving AI research area. Established methods include Monte Carlo tree search [2], the automatic generation of evaluation functions [4] and knowledge acquisition [7]. Several complete general game-playing systems also have been described [12,2].

Robotics. Building a robot that can recognise and manipulate objects in a physical environment is a complex and challenging problem. While basic robot kinematics is well understood, the ability to recognise and manipulate arbitrary objects in a complex environment remains an active area of research [9]. However, dealing with a predetermined set of rigid objects, such as common household items or game boards and pieces, is more amenable to known techniques and technologies [10,6]. Consequently, the research in domestic robotics provides an excellent platform for the development of general game playing robots that can play a wide variety of games.

3 GDL Descriptions for Physical Game Environments

Game descriptions must satisfy certain basic requirements to ensure that the game is effectively playable; for example, players should always have at least one legal move in nonterminal positions [5]. In bringing gameplay from mere virtual into physical environments, general game-playing robots add a new class of desirable properties that concern the manifestation of the game rules in the real world. Notably, a good GDL description requires all moves deemed legal by the rules to be actually physically possible. In this section we develop a framework for mathematically describing the application of game descriptions to physical environments that allows us to formalise such properties.

Environment Models. Consider the robotic game environment shown in Fig. 2(a). It features a 4×4 chess-like board with an additional row of 4 marked positions on the right. Tin cans are the only type of objects and can be moved between the marked positions (but cannot be stacked). To formally analyse the use of physical environments like this to play a game, we model them by *state machines* (Σ, S, s_0, δ), where

- Σ denotes the *actions* that can be performed in the environment;
- S are the *states* the environment can be in, including the special *failure* $\in S$;
- $s_0 \in S$ is the *starting state*;
- $\delta : S \times \Sigma \to S$ is the *transition function*.

Example. The following is a model for our cans-on-a-chessboard environment.

- We want to allow players two actions: doing nothing and moving a can. Formally, $\Sigma = \{\mathtt{noop}\} \cup \{\mathtt{move}(u, v, x, y)\colon\ u, x \in \{\mathtt{a}, \mathtt{b}, \mathtt{c}, \mathtt{d}, \mathtt{x}\};\ v, y \in \{\mathtt{1}, \mathtt{2}, \mathtt{3}, \mathtt{4}\}\}$.
- Any location can either be empty or house a can. Hence, the environment can be in any of 2^{20} different states plus *failure* . Fig. 2(b) illustrates two example states.
- Any configurations of cans can be set up as the starting state s_0.
- For the transition function δ, action $\mathtt{noop}$ has no effect, hence $\delta(s, \mathtt{noop}) = s$. Action $\mathtt{move}(u, v, x, y)$ maps any state with a can at (u, v) and no can at (x, y) to the same state except that now there is a can at (x, y) and none at (u, v). If no can is at (u, v) or there already is one at (x, y) , then $\delta(s, \mathtt{move}(u, v, x, y)) =$ *failure*. E.g., $s_2 = \delta(\mathtt{move}(\mathtt{d}, 4, \mathtt{b}, 2), \delta(\mathtt{move}(\mathtt{x}, 2, \mathtt{a}, 1), s_1))$ where s_1 and s_2 respectively denote the top and bottom states depicted in Fig. 2(b).

Projecting Games onto Physical Environments. When we use a physical environment to play a game, the real objects become representatives of entities in the abstract game. A pawn in chess, for example, is typically manifested by an actual wooden piece of a certain shape and colour. But any other physical object, including a tin can, can serve the same purpose. Conversely, any game environment like our 4×4 board with cans can be interpreted in countless ways as physical manifestation of a game. Thus our example board can not only be used for all kinds of mini chess-like games (cf. top of Fig. 2(c)) but also to play, say, the single-player 8-PUZZLE (cf. bottom of Fig. 2(c)). For this the cans represent numbered tiles that need to be brought in the right order.

The manifestation of a game in a physical environment can be mathematically captured by *projecting* the positions from the abstract game onto actual states of the game

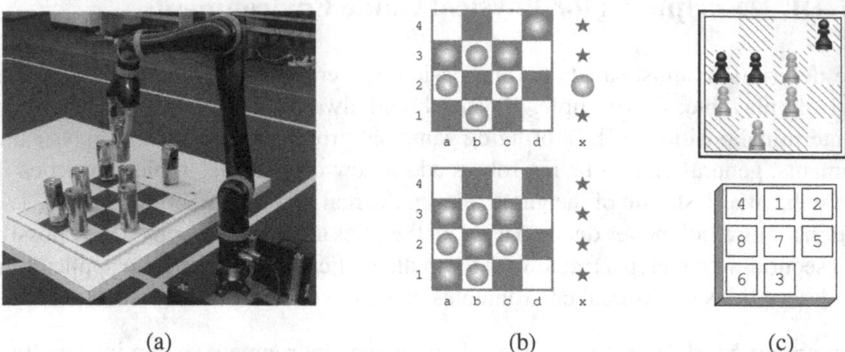

Fig. 2. Game environment with robot, and games projected onto this environment

environment. Formally, if Positions is the set of all possible positions in the game, then a *projection function* is of the form $\pi : \texttt{Positions} \rightarrow S \setminus \{failure\}$, where S is the set of states in the environment model as above.

Example. Fig. 2(c) shows two positions from two different games, a mini-variant of Breakthrough and the 8-puzzle. We can view the states depicted to the left of each position (Fig. 2(b)) as their projection onto our example physical game environment, in which the extra row can be used to park captured pieces. For the sake of simplicity, the robot in our environment is not required to distinguish between different types of cans. Hence, it is only through the projection function that the robotic player knows whether a can stands for a white or a black pawn, say. The reader should note that a similar feature is found in many games humans play; for example, the pieces on a chessboard alone not telling us whose move it is or which side still has castling rights.

Requirements for Game Descriptions for Physical Environments. The standard semantics for GDL [13] defines precisely how to interpret a given set of rules as a game. In particular, every valid game description determines the following.

- The set Positions of all possible positions in the game.
- The initial position, $\texttt{Init} \in \texttt{Positions}$.
- The set $\texttt{LegalMoves}(p)$ of legal moves in each position p.
- The function $\texttt{Update}(p, m)$ to compute the new position after move m in p.

All of these elements of a general, abstract game can be related to gameplay in a physical game environment via the projection function that maps game positions onto physical states. Specifically, we can formalise fundamental requirements concerning the *playability* of an arbitrary game in a given environment (Σ, S, s_0, δ) as follows.

1. All legal moves must correspond to actions in the physical environment. Formally, $\forall p \in \texttt{Positions}. \forall m \in \texttt{LegalMoves}(p). m \in \Sigma$.
2. Initial position and initial physical state correspond. Formally, $\pi[\texttt{Init}] = s_0$.
3. All legal position updates correspond to a possible physical state update. Formally, $\forall p \in \texttt{Positions}. \forall m \in \texttt{LegalMoves}(p). \pi[\texttt{Update}(p, m)] = \delta(\pi[p], m)$.

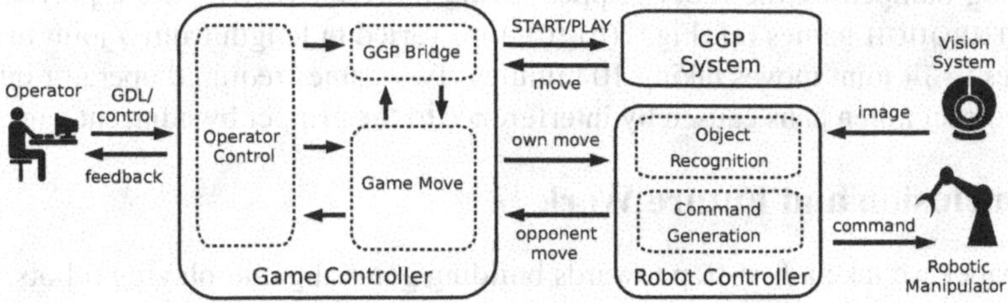

Fig. 3. Systems architecture

4 Systems Architecture

In the following, we describe a general systems architecture for general game-playing robots (Fig. 3). Its defining feature is to allow for the seamless integration of any existing general game player as the "brain" and any suitable robotic system as the "body."

The **Robot Controller** serves as the low-level executor of the robot. The required functionality is, (1) to process sensor data in order to recognise moves by detecting changes in the environment, and (2) to command the manipulator to execute any instance of the defined actions in the environment. Performing a single action may require a complex series of operations; for example, the sequence of operations required to move a can from one location to another. The robot controller needs to monitor the execution of such an action and, if possible, recover from failures whenever they occur.

The **Game Controller** provides the link between the low-level executor and the high-level decision making. It accepts any new GDL description sent through the interface and transmits these game rules to the general game-playing system in the form of a standard START message. During gameplay, the game controller passes on any move decided upon by the high-level controller to the low-level controller, and it converts opponent moves reported by the robot controller into standard PLAY messages for the general game player. The latter involves validating the legality of all opponent moves.

The systems architecture allows for plugging in as the high-level control module any **General Game-Playing System** that follows the standard specification [5].

5 Prototype

We constructed a prototype system using a Kinova Jaco arm, pictured in Fig. 2(a), and the ROS robotic software [11]. The highly successful CadiaPlayer [2] was used as the underlying GGP reasoner. While the basic system architecture has been implemented as described in Sections 4, the prototype simulates the vision module through operator input. For this reason the robot relies on the pieces being in the correct game position.

We ran tests with three games, each repeated 10 times. Each round of the short single-player 4-QUEENS game lasted approximately 2.5 minutes. At no point was operator intervention required. 8-PUZZLE, a single-player game with a cluttered physical environment (cf. Fig. 2(b), bottom), lasted approximately 16-17 minutes with 30 moves. Two out of 10 games required operator intervention to pick up a can that had fallen after being bumped by the robot gripper. Being more interactive, the 2-player MINI-BREAKTHROUGH games (cf. Fig. 2(b)-(c), top) varied in length from 5 joint moves in 4 minutes to 14 joint moves taking 10 minutes. Two games required operator intervention to correct fallen cans caused by interference to the gripper by adjacent cans.

6 Conclusion and Future Work

In this paper we take a first step towards building general game-playing robots; robots that can understand and play arbitrary games in a physical environment. The system was realised by drawing together two topical yet disparate areas of artificial intelligence research: general game playing and robotics. A prototype system was built and experiments performed to validate the general architecture and feasibility of the project.

Beyond the necessary task of implementing the object recognition system, there are a number of directions for future work. Firstly, given some GDL game it should be possible to automatically construct the rules of a meta-game that incorporates recovery from known error states. The recovery behaviour would then simply be a result of the robot playing the larger meta-game. A second broad area for future work is that of learning. Instead of having to describe new games in GDL, it would be beneficial if the robot were able to learn how to play a game by being shown how by a user, as in [8,1].

References

1. Barbu, A., Narayanaswamy, S., Siskind, J.: Learning physically-instantiated game play through visual observation. In: Proc. of ICRA, pp. 1879–1886. IEEE Press (2010)
2. Björnsson, Y., Finnsson, H.: CADIAPLAYER: A simulation-based general game player. IEEE Transactions on Computational Intelligence and AI in Games 1(1), 4–15 (2009)
3. Broekens, J., Heerink, M., Rosendal, H.: Assistive social robots in elderly care: a review. Gerontechnology 8(2) (2009)
4. Clune, J.: Heuristic evaluation functions for general game playing. In: Proc. of AAAI, pp. 1134–1139 (2007)
5. Genesereth, M., Love, N., Pell, B.: General game playing: Overview of the AAAI competition. AI Magazine 26(2), 62–72 (2005)
6. Goldfeder, C., Ciocarlie, M.T., Dang, H., Allen, P.K.: The columbia grasp database. In: Proc. of ICRA, pp. 1710–1716. IEEE Press (2009)
7. Haufe, S., Schiffel, S., Thielscher, M.: Automated verification of state sequence invariants in general game playing. Artificial Intelligence 187-188, 1–30 (2012)
8. Kaiser, Ł.: Learning games from videos guided by descriptive complexity. In: Proc. of AAAI, pp. 963–969 (2012)
9. Kemp, C.C., Edsinger, A., Torres-Jara, E.: Challenges for robot manipulation in human environments. IEEE Robotics & Automation Magazine 14(1), 20–29 (2007)
10. Lai, K., Bo, L., Ren, X., Fox, D.: A large-scale hierarchical multi-view rgb-d object dataset. In: Proc. of ICRA, pp. 1817–1824. IEEE Press (2011)
11. Quigley, M., Conley, K., Gerkey, B., Faust, J., Foote, T., Leibs, J., Wheeler, R., Ng, A.: ROS: an open-source robot operating system. In: ICRA Workshop on Open Source Software (2009)
12. Schiffel, S., Thielscher, M.: Fluxplayer: A successful general game player. In: Proc. of AAAI, pp. 1191–1196 (2007)
13. Schiffel, S., Thielscher, M.: A Multiagent Semantics for the Game Description Language. In: Filipe, J., Fred, A., Sharp, B. (eds.) ICAART 2009. CCIS, vol. 67, pp. 44–55. Springer, Heidelberg (2010)

Bisimulation for Single-Agent Plausibility Models

Mikkel Birkegaard Andersen[1], Thomas Bolander[1], Hans van Ditmarsch[2], and Martin Holm Jensen[1]

[1] DTU Compute, Technical University of Denmark
{mibi,tobo,mhje}@dtu.dk
[2] LORIA, CNRS, Université de Lorraine
hans.van-ditmarsch@loria.fr

Abstract. Epistemic plausibility models are Kripke models agents use to reason about the knowledge and beliefs of themselves and each other. Restricting ourselves to the single-agent case, we determine when such models are indistinguishable in the logical language containing conditional belief, i.e., we define a proper notion of bisimulation, and prove that bisimulation corresponds to logical equivalence on image-finite models. We relate our results to other epistemic notions, such as safe belief and degrees of belief. Our results imply that there are only finitely many non-bisimilar single-agent epistemic plausibility models on a finite set of propositions. This gives decidability for single-agent epistemic plausibility planning.

1 Introduction

A typical approach in belief revision involves preferential orders to express degrees of belief and knowledge [10,13]. This goes back to the 'systems of spheres' in [11,9]. Dynamic doxastic logic was proposed and investigated in [14] in order to provide a link between the (non-modal logical) belief revision and modal logics with explicit knowledge and belief operators. A similar approach was pursued in belief revision in dynamic epistemic logic [3,19,17,5,20], that continues to develop strongly [7,18]. We focus on the proper notion of structural equivalence on (static) models encoding knowledge and belief simultaneously. A prior investigation into that is [8], which we relate our results to at the end of the paper. Our motivation is to find suitable structural notions to reduce the complexity of planning problems. Such plans are sequences of actions, such as iterated belief revision. It is the dynamics of knowledge and belief that, after all, motivates our research.

The semantics of belief depend on the structural properties of models. To relate the structural properties of models to a logical language we need a notion of structural similarity, known as bisimulation. A bisimulation relation relates a modal operator to an accessibility relation. Epistemic plausibility models do not have an accessibility relation as such but a plausibility relation. This induces a set of accessibility relations: the *most plausible* states are the *accessible* states for the modal belief operator; and the *plausible* states are the *accessible* states for

S. Cranefield and A. Nayak (Eds.): AI 2013, LNAI 8272, pp. 277–288, 2013.

the modal knowledge operator. But it contains much more information: to each modal operator of conditional belief (or of degree of belief) one can associate a possibly distinct accessibility relation. This begs the question how one should represent the bisimulation conditions succinctly. Can this be done by reference to the plausibility relation directly, instead of by reference to these, possibly many, induced accessibility relations? It is now rather interesting to observe that relative to the modal operations of knowledge and belief the plausibility relation is already in some way too rich.

Example 1. The (single-agent) epistemic plausibility model on the left in Figure 1 consists of three worlds w_1, w_2, and w_3. p is only false in w_2, and $w_1 < w_2 < w_3$[1]: the agent finds it most plausible that p is true, less plausible that p is false, and even less plausible that p is true. As p is true in the most plausible world, the agent believes p. If we go to slightly less plausible, the agent is already uncertain about the value of p, she only knows trivialities such as $p \vee \neg p$. The world w_3 does not make the agent even more uncertain. We therefore can discard that other world where p is true. This is the model in the middle in Figure 1. It is bisimilar to the model on the left! Therefore, and that is the important observation: *having one world more or less plausible than another world in a plausibility model does not mean that in any model with the same logical content we should find a matching pair of worlds.* This is evidenced in the figure: on the left w_3 is less plausible than w_2, but in the middle no world is less plausible than v_2; there is no match.

Now consider retaining w_3 and making it as plausible state as w_1. This gives the plausibility model on the right in Figure 1, where u_1 and u_3 are equiplausible (equally plausible), written $u_1 \simeq u_3$. This model is bisimilar to both the left and the middle model. But the right and middle one share the property that more or less plausible in one, is more or less plausible in the other: now there is a match. This makes for another important observation: *we can reshuffle the plausibilities such that models with the same logical content preserve the plausibility order.*

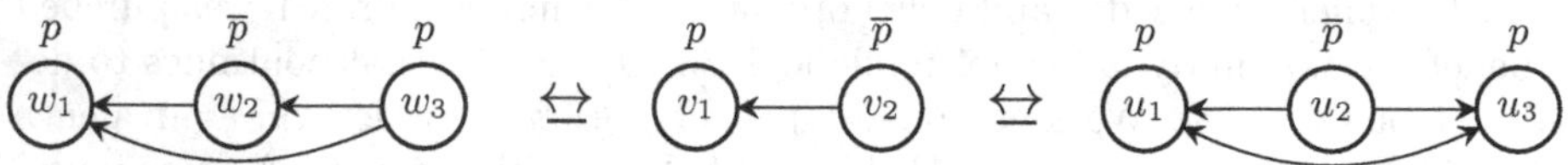

Fig. 1. All three models are bisimilar. The models in the middle and on the right are normal, the model on the left is not normal. An arrow $w_1 \leftarrow w_2$ corresponds to $w_1 \leq w_2$. Reflexive edges are omitted. $\overline{p}$ means that p does not hold.

In Section 2 we define the epistemic doxastic logic, the epistemic plausibility models on which it is interpreted, the suitable notion of bisimulation, and demonstrate the adequacy of this notion via a correspondence between modal equivalence and bisimilarity. The final sections 3, 4, and 5 respectively translate our results to degrees of belief and safe belief, discuss the problematic generalization to the multi-agent case, and demonstrate the relevance of our results for epistemic planning.

[1] If $s < t$, we have $s \leq t$ and $t \not\leq s$.

2 Single-Agent Plausibility Models and Bisimulation

2.1 Language, Structures, and Semantics

Definition 1 (Epistemic doxastic language). *For any countable set of propositional symbols P, we define the epistemic-doxastic language $\mathcal{L}_P$ by:*

$$\varphi ::= p \mid \neg\varphi \mid \varphi \wedge \varphi \mid K\varphi \mid B^{\varphi}\varphi$$

where $p \in P$, K is the epistemic modality (knowledge) and B^{φ} the conditional doxastic modality (conditional belief). We use the usual abbreviations for the other boolean connectives as well as for $\top$ and $\bot$, and the abbreviation B for $B^{\top}$. The dual of K is denoted $\widehat{K}$, and the dual of B^{φ} is denoted $\widehat{B}^{\varphi}$.

We consider epistemic plausibility models as in [5]. A *well-preorder* on a set S is a reflexive and transitive relation $\leq$ on S such that every non-empty subset has minimal elements. The set of *minimal elements* of a subset T of S is given by:

$$Min_{\leq}T = \{s \in T \mid s \leq s' \text{ for all } s' \in T\}.$$

This is a non-standard notion of minimality, taken from [5]. Usually a minimal element of a set is an element that is not greater than any other element. On total preorders the two notions of minimality coincide. In fact, using the definition of minimality above, any well-preorder is total: For any pair of worlds s, t, $Min_{\leq}\{s, t\}$ is non-empty, and therefore $s \leq t$ or $t \leq s$.[2] These well-preorders are the *plausibility relations* (or *plausibility orderings*), expressing that a world is considered at least as plausible as another. This encodes the doxastic content of a model.

We can define such epistemic plausibility models with the plausibility relation as a primitive and with the epistemic relation as a derived notion. Alternatively, we can assume both as primitive relations, but require that more plausible means (epistemically) possible. We chose the latter.

Definition 2 (Epistemic plausibility model). *An* epistemic plausibility model *(or simply* plausibility model*) on a set of propositional symbols P is a tuple $\mathcal{M} = (W, \leq, \sim, V)$, where*

- *W is a set of* worlds, *called the* domain.
- *$\leq$ is a well-preorder on W, called the* plausibility relation.
- *$\sim$ is an equivalence relation on W called the* epistemic relation. *We require, for all $w, v \in W$, that $w \leq v$ implies $w \sim v$.*
- *$V : W \to 2^P$ is a* valuation.

For $w \in W$ we name $(\mathcal{M}, w)$ a pointed epistemic plausibility model, *and refer to w as the* actual world *of $(\mathcal{M}, w)$.*

---

[2] A well-preorder is not the same as a well-founded preorder; e.g., $y \leq x$, $z \leq x$ is a well-founded preorder, but not a well-preorder, as z and y are incomparable. Well-founded preorders are not necessarily total.

As we require that $\leq$-comparable worlds are indistinguishable, totality of $\leq$ gives that $\sim$ is the universal relation $W \times W$.

Definition 3 (Satisfaction Relation). *Let $\mathcal{M} = (W, \leq, \sim, V)$ be a plausibility model on P. The satisfaction relation is given by, for $w \in W$, $p \in P$, $\varphi, \varphi' \in \mathcal{L}_P$,*

$$\begin{array}{ll} \mathcal{M}, w \models p & \textit{iff } p \in V(w) \\ \mathcal{M}, w \models \neg\varphi & \textit{iff not } \mathcal{M}, w \models \varphi \\ \mathcal{M}, w \models \varphi \wedge \varphi' & \textit{iff } \mathcal{M}, w \models \varphi \textit{ and } \mathcal{M}, w \models \varphi' \\ \mathcal{M}, w \models K\varphi & \textit{iff } \mathcal{M}, v \models \varphi \textit{ for all } v \sim w \\ \mathcal{M}, w \models B^{\psi}\varphi & \textit{iff } \mathcal{M}, v \models \varphi \textit{ for all } v \in Min_{\leq}[\![\psi]\!]_{\mathcal{M}}, \end{array}$$

where $[\![\psi]\!]_{\mathcal{M}} := \{w \in W \mid \mathcal{M}, w \models \psi\}$. We write $\mathcal{M} \models \varphi$ to mean $\mathcal{M}, w \models \varphi$ for all $w \in W$. Further, $\models \varphi$ (φ is valid) means that $\mathcal{M} \models \varphi$ for all models $\mathcal{M}$, and $\Phi \models \varphi$ (φ is a logical consequence of the set of formulas Φ) stands for: for all $\mathcal{M}$ and $w \in \mathcal{M}$, if $\mathcal{M}, w \models \psi$ for all $\psi \in \Phi$, then $\mathcal{M}, w \models \varphi$.[3]

Example 2. Consider again the the models in Figure 1. The model on the left is of the form $\mathcal{M} = (W, \leq, \sim, V)$ with $W = \{w_1, w_2, w_3\}$ and $\leq$ defined by: $w_1 \leq w_2$, $w_2 \leq w_3$, $w_1 \leq w_3$ (plus the reflexive edges). The valuation V of the model on the left maps w_1 into $\{p\}$, w_2 into $\emptyset$ and w_3 into $\{p\}$. In all three models of the figure, the formula $Bp \wedge \neg Kp$ holds, that is, p is believed but not known.

2.2 Normal Epistemic Plausibility Models and Bisimulation

The examples and proposal of Section 1 are captured by the definition of bisimulation that follows after these preliminaries. First, given a plausibility model $\mathcal{M} = (W, \sim, \leq, V)$ consider an equivalence relation on worlds defined as follows:

$$w \approx w' \quad \text{iff} \quad V(w) = V(w').$$

The $\approx$-equivalence class of a world is defined as usual as $[w]_{\approx} = \{w' \in W \mid w' \approx w\}$. Next, the ordering $\leq$ on worlds in W can be lifted to an ordering between sets of worlds $W', W'' \subseteq W$ in the following way:

$$W' \leq W'' \quad \text{iff} \quad w' \leq w'' \text{ for all } (w', w'') \in W' \times W''.$$

Finally, the lifted ordering leads us to a formalization of normal models of Example 1.

Definition 4 (Normal Plausibility Relation). *Given a plausibility model $\mathcal{M} = (W, \leq, \sim, V)$, the* normal plausibility relation *on $\mathcal{M}$ is the relation on W defined by:*

$$w \preceq w' \quad \textit{iff} \quad Min_{\leq}[w]_{\approx} \leq Min_{\leq}[w']_{\approx}.$$

$\mathcal{M}$ is called normal *if $\preceq = \leq$. The* normalisation *of $\mathcal{M} = (W, \leq, \sim, V)$ is $\mathcal{M}' = (W, \preceq, \sim, V)$. As for $<$, we write $w \prec w'$ for $w \preceq w'$ and $w' \not\preceq w$.*

[3] For an axiomatization of this logic see e.g. [16].

Note that if $u, v \in Min_{\leq} W'$ for some set W' then, by definition of $Min_{\leq}$, both $u \leq v$ and $v \leq u$. Hence, the condition $Min_{\leq}[w]_{\approx} \leq Min_{\leq}[w']_{\approx}$ above is equivalent to the existence *some* minimal element of $[w]_{\approx}$ being $\leq$-smaller than *some* minimal element of $[w']_{\approx}$.

Lemma 1. *Let w and w' be two worlds in the normal model $\mathcal{M} = (W, \preceq, \sim, V)$. If w and w' have the same valuation, they are equiplausible.*

Proof. As $w \approx w'$, we have $[w]_{\approx} = [w']_{\approx}$, and thus $Min_{\preceq}[w]_{\approx} = Min_{\preceq}[w']_{\approx}$. By Definition 4 we $w \preceq w'$ and $w' \preceq w$, which is equivalent to $w \simeq w'$.

Example 3. Take another look at the models of Figure 1 (for reference, we name them $\mathcal{M}_1$, $\mathcal{M}_2$ and $\mathcal{M}_3$). We want models $\mathcal{M}_1$ and $\mathcal{M}_2$ to be bisimilar via the relation $\mathfrak{R}$ given by $\mathfrak{R} = \{(w_1, v_1), (w_3, v_1), (w_2, v_2)\}$ (see Section 1). Usually, in a bisimulation, every modal operator has corresponding back and forth requirements. For our logic of conditional belief there is an infinity of modal operators, as there is an infinity of of conditional formulas. (Having *only* unconditional belief $B\varphi$ defined as $B^{\top}\varphi$ is not enough, see Example 4.) Instead, we define our bisimulation indirectly by way of the plausibility relation. Example 1 showed that we cannot match 'more plausible' in $\mathcal{M}_1$ with 'more plausible' in $\mathcal{M}_2$ using simply $\leq$. With $\leq$ as seen in $\mathcal{M}_3$ (the normalization of $\mathcal{M}_1$) where $\leq = \preceq$, we can.

Definition 5 (Bisimulation). *Let plausibility models $\mathcal{M} = (W, \leq, \sim, V)$ and $\mathcal{M}' = (W', \leq', \sim', V')$ be given. Let $\preceq, \preceq'$ be the respective derived normal plausibility relations. A non-empty relation $\mathfrak{R} \subseteq W \times W'$ is a* bisimulation *between $\mathcal{M}$ and $\mathcal{M}'$ if for all $(w, w') \in \mathfrak{R}$:*

[atoms] $V(w) = V'(w')$.
[forth$_{\preceq}$] *If $v \in W$ and $v \preceq w$, there is a $v' \in W'$ s.t. $v' \preceq' w'$ and $(v, v') \in \mathfrak{R}$.*
[back$_{\preceq}$] *If $v' \in W'$ and $v' \preceq' w'$, there is a $v \in W$ s.t. $v \preceq w$ and $(v, v') \in \mathfrak{R}$.*
[forth$_{\sim}$] *If $v \in W$ and $w \sim v$, there is a $v' \in W'$ s.t. $w' \sim' v'$ and $(v, v') \in \mathfrak{R}$.*
[back$_{\sim}$] *If $v' \in W'$ and $w' \sim' v'$, there is a $v \in W$ s.t. $w \sim v$ and $(v, v') \in \mathfrak{R}$.*

A total bisimulation *between $\mathcal{M}$ and $\mathcal{M}'$ is a bisimulation with domain W and codomain W'. For a bisimulation between pointed models $(\mathcal{M}, w)$ and $(\mathcal{M}', w')$ it is required that $(w, w') \in \mathfrak{R}$. If a bisimulation between $(\mathcal{M}, w)$ and $(\mathcal{M}', w')$ exists, the two models are called* bisimilar *and we write $(\mathcal{M}, w) \leftrightarroweq (\mathcal{M}', w')$. Two worlds w, w' of a model $\mathcal{M}$ are called* bisimilar *if there exists a bisimulation $\mathfrak{R}$ between $\mathcal{M}$ and itself with $(w, w') \in \mathfrak{R}$.*

This definition gives us the bisimulation put forth in Example 3. As $\sim$ is the universal relation on W, [forth$_{\sim}$] and [back$_{\sim}$] enforce that all bisimulations are total.

If $\sim$ was not a primitive, we could instead have conditions [up-forth$_{\preceq}$] and [up-back$_{\preceq}$] (that consider less plausible v and v'), in place of [forth$_{\sim}$] and [back$_{\sim}$]. This would define the same bisimulations.

2.3 Correspondence between Bisimilarity and Modal Equivalence

In the following we prove that bisimilarity implies modal equivalence and vice versa. This shows that our notion of bisimulation is proper for the language and models at hand. First we define modal equivalence.

Definition 6 (Modal equivalence). *Given are models $\mathcal{M} = (W, \leq, \sim, V)$ and $\mathcal{M}' = (W', \leq', \sim', V')$ on P with $w \in W$ and $w' \in W'$. We say that $(\mathcal{M}, w)$ and $(\mathcal{M}', w')$ are* modally equivalent *iff for all $\varphi \in \mathcal{L}_P$, $\mathcal{M}, w \models \varphi$ iff $\mathcal{M}', w' \models \varphi$. In this case we write $(\mathcal{M}, w) \equiv (\mathcal{M}', w')$.*

Lemma 2. *If two worlds of a model are $\approx$-equivalent, they are bisimilar.*

Proof. Assume worlds w and w' of a model $\mathcal{M} = (W, \leq, \sim, V)$ have the same valuation. Let $\mathfrak{R}$ be the relation that relates each world of $\mathcal{M}$ to itself and additionally relates w to w'. We want to show that $\mathfrak{R}$ is a bisimulation. This amounts to showing [atoms], [forth$_\preceq$], [back$_\preceq$], [forth$_\sim$] and [back$_\sim$] for the pair $(w, w') \in \mathfrak{R}$. [atoms] holds trivially since $w \approx w'$. [forth$_\sim$] and [back$_\sim$] also hold trivially, by choice of $\mathfrak{R}$. For [forth$_\preceq$], assume $v \in W$ and $v \preceq w$. We need to find a $v' \in W$ such that $v' \preceq w'$ and $(v, v') \in \mathfrak{R}$. Letting $v' = v$, it suffices to prove $v \preceq w'$. Since $w \approx w'$ this is immediate: $v \preceq w$ iff $Min_{\leq}[v]_{\approx} \leq Min_{\leq}[w]_{\approx}$ iff (because $w \approx w'$) $Min_{\leq}[v]_{\approx} \leq Min_{\leq}[w']_{\approx}$ iff $v \preceq w'$. [back$_\preceq$] is proved similarly.

Proposition 1. *Bisimilarity implies modal equivalence.*

Proof. We will prove that for all formulas $\varphi \in \mathcal{L}_P$, if $\mathfrak{R}$ is a bisimulation between pointed models $(\mathcal{M}, w)$ and $(\mathcal{M}', w')$ then $\mathcal{M}, w \models \varphi$ iff $\mathcal{M}', w' \models \varphi$. The proof is by induction on the structure of φ. The base case is when φ is propositional. Then the required follows immediately from [atoms], using that $(w, w') \in \mathfrak{R}$. For the induction step, we have the following cases of φ: $\neg\psi, \psi \wedge \gamma, K\psi, B^\gamma\psi$. We skip the first three, fairly standard cases and show only $B^\gamma\psi$.

Let $\mathfrak{R}$ be a bisimulation between $(\mathcal{M}, w)$ and $(\mathcal{M}', w')$ with $\mathcal{M} = (W, \leq, \sim, V)$ and $\mathcal{M} = (W', \leq', \sim', V')$. We only prove $\mathcal{M}, w \models B^\gamma\psi \Rightarrow \mathcal{M}', w' \models B^\gamma\psi$, the other direction being proved symmetrically. So assume $\mathcal{M}, w \models B^\gamma\psi$, that is, $\mathcal{M}, v \models \psi$ for all $v \in Min_{\leq}[\![\gamma]\!]_{\mathcal{M}}$. We need to prove $\mathcal{M}', v' \models \psi$ for all $v' \in Min_{\leq'}[\![\gamma]\!]_{\mathcal{M}'}$. So let $v' \in Min_{\leq'}[\![\gamma]\!]_{\mathcal{M}'}$. Choose $x \in Min_{\leq}\{u \in W \mid u \approx z \text{ and } (z, v') \in \mathfrak{R} \text{ for some } z \in W\}$. Let $y \in [\![\gamma]\!]_{\mathcal{M}}$ be chosen arbitrarily, and choose y' with $(y, y') \in \mathfrak{R}$ (recall that any bisimulation is total). The induction hypothesis implies $\mathcal{M}', y' \models \gamma$. Let $y'' \approx y'$ be chosen arbitrarily. Lemma 2 implies the existence of a bisimulation $\mathfrak{R}'$ between $(\mathcal{M}', y'')$ and $(\mathcal{M}', y')$. Since $\mathcal{M}', y' \models \gamma$, the induction hypothesis gives us $\mathcal{M}', y'' \models \gamma$, that is, $y'' \in [\![\gamma]\!]_{\mathcal{M}'}$. Since v' was chosen $\leq'$-minimal in $[\![\gamma]\!]_{\mathcal{M}'}$, we must have $v' \leq' y''$. Since y'' was chosen arbitrarily with $y'' \approx y'$, we get $v' \leq' Min_{\leq'}[y']_{\approx}$. We can now conclude $Min_{\leq'}[v']_{\approx} \leq' v' \leq' Min_{\leq'}[y']_{\approx}$, and hence $v' \preceq y'$.

By [back$_\preceq$] there is a v such that $(v, v') \in \mathfrak{R}$ and $v \preceq y$. By choice of x, $x \leq Min_{\leq}[v]_{\approx}$. Since $v \preceq y$ we now get: $x \leq Min_{\leq}[v]_{\approx} \leq Min_{\leq}[y]_{\approx} \leq y$. Since y was chosen arbitrarily in $[\![\gamma]\!]_{\mathcal{M}}$, we can conclude:

$$x \leq u \text{ for all } u \in [\![\gamma]\!]_{\mathcal{M}}. \tag{1}$$

By choice of x, there is a $z \approx x$ with $(z, v') \in \mathfrak{R}$. From $z \approx x$, Lemma 2 implies the existence of a bisimulation $\mathfrak{R}''$ between $(\mathcal{M}, x)$ and $(\mathcal{M}, z)$. Since $\mathfrak{R}''$ is a bisimulation between $(\mathcal{M}, x)$ and $(\mathcal{M}, z)$, and $\mathfrak{R}$ is a bisimulation between $(\mathcal{M}, z)$ and $(\mathcal{M}', v')$, the composition $\mathfrak{R}'' \circ \mathfrak{R}$ must be a bisimulation between $(\mathcal{M}, x)$ and $(\mathcal{M}', v')$. Applying the induction hypothesis to the bisimulation $\mathfrak{R}'' \circ \mathfrak{R}$, we can from $v' \in [\![\gamma]\!]_{\mathcal{M}'}$ conclude $x \in [\![\gamma]\!]_{\mathcal{M}}$. Combining this with (1), we get $x \in Min_{\leq}[\![\gamma]\!]_{\mathcal{M}}$. By original assumption this implies $\mathcal{M}, x \models \psi$. Applying again the induction hypothesis to the bisimulation $\mathfrak{R}'' \circ \mathfrak{R}$, this gives us $\mathcal{M}, v' \models \psi$, as required, thereby concluding the proof.

We proceed now to the converse, that modal equivalence with regard to $\mathcal{L}_P$ implies bisimulation, though first taking a short detour motivating the need for conditional belief.

Example 4. The normal plausibility models $(\mathcal{M}_1, w_1)$ and $(\mathcal{M}_2, v_1)$ of Figure 2 are modally equivalent for the language with only unconditional belief. We can show this by first demonstrating that $\mathcal{M}_1$ and $\mathcal{M}_2$ have the same modal description Φ (a modal description Φ of a model $\mathcal{M}$ is a set of formulas such that $\Phi \models \psi$ iff $\mathcal{M} \models \psi$). We observe that the description of both models is

$$B(p_1 \wedge \neg p_2 \wedge \neg p_3) \wedge K((p_1 \wedge \neg p_2 \wedge \neg p_3) \vee (\neg p_1 \wedge p_2 \wedge \neg p_3) \vee (\neg p_1 \wedge \neg p_2 \wedge p_3))$$

To see why, note that w_1 and v_1 are both the only minimal worlds in their respective models, so belief in (description of the valuation) $p_1 \wedge \neg p_2 \wedge \neg p_3$ will be the same. Further, in both models all three constituent worlds are epistemically possible, so K cannot distinguish either between the models (the disjunction sums up the three different valuations). We then note that, as both w_1 and v_1 satisfy $p_1 \wedge \neg p_2 \wedge \neg p_3$, $(\mathcal{M}_1, w_1)$ and $(\mathcal{M}_2, v_1)$ of Figure 2 must be modally equivalent: any boolean formula must be a consequence of $p_1 \wedge \neg p_2 \wedge \neg p_3$, whereas any belief or knowledge formula evaluated in the points of these models must be a model validity that is a consequence from the model description Φ.

On the other hand, $(\mathcal{M}_1, w_1)$ and $(\mathcal{M}_2, v_1)$ are not bisimilar. Pairs in the bisimulation must have matching valuations, so the only option is the relation $\{(w_1, v_1), (w_2, v_2), (w_3, v_3)\}$. But this does neither satisfy [forth$_{\preceq}$] nor [back$_{\preceq}$].

We do not want that these models are modally equivalent in, for example, a *dynamic* epistemic language. Consider an agent learning $\neg p_1$ from a public announcement. This deletes w_1 and v_1 from their respective models. After this announcement in $\mathcal{M}_1$, the agent believes p_2. In $\mathcal{M}_2$ this is not the case. Here the agent will believe p_3. With conditional belief we can capture this distinction already in the static language ($\mathcal{M}_1 \models B^{\neg p_1} p_2$, while $\mathcal{M}_2 \not\models B^{\neg p_1} p_2$).

Fig. 2. The models $\mathcal{M}_1$ and $\mathcal{M}_2$ of Example 4. For visual clarity, we leave out false propositional variables.

Definition 7 (Δ). *Let two worlds w, w' of a model $\mathcal{M} = (W, \leq, \sim, V)$ on P be given where $V(w) \neq V(w')$. If there is a $p \in V(w) - V(w')$, then let $\delta_{w,w'}$ be such a p; otherwise, let $\delta_{w,w'} = \neg q$ for some $q \in V(w') - V(w)$. Any such choice of $\delta_{w,w'}$ for a given pair w, w' is called a* propositional difference *between w and w'. If instead $V(w) = V(w')$, let $\delta_{w,w'} = \top$. Finally, let $\Delta_w = \bigwedge_{w' \prec w} \delta_{w,w'}$ be the conjunction of some propositional difference between w and each world strictly more $\preceq$-plausible than w (the empty conjunction when no such world exist).*

Continuing Example 4, we can choose $\Delta_{w_2} = \neg p_1$. We then have that $\widehat{B}^{\Delta_{w_2}} p_2$ distinguishes $\mathcal{M}_1$ and $\mathcal{M}_2$ by evaluating belief on worlds no more plausible than w_2 and v_2 respectively. However, choosing $\Delta_{w_2} = p_2$ would not distinguish, so we add an additional disjunct for w_3. Regardless of which propositional differences are used in Δ_{w_2} and Δ_{w_3}, $\widehat{B}^{\Delta_{w_2} \vee \Delta_{w_3}} p_2$ distinguishes the models. This is, of course, not sufficient for constructing distinguishing formulas in the general case, but for our purposes of proving Proposition 2 it is enough.

Lemma 3. *Let w and w' be worlds of the model $\mathcal{M} = (W, \leq, \sim, V)$ s.t. $w' \preceq w$, and φ a formula of $\mathcal{L}_P$, s.t. $\mathcal{M}, w' \models \varphi$. Then $\mathcal{M}, w \models \widehat{B}^{\Delta_w \vee \Delta_{w'}} \varphi$.*

Proof. In the following we abbreviate $\Delta_w \vee \Delta_{w'}$ by $\Delta_{w,w'}$. We need to show that $\exists u \in Min_{\leq}[\![\Delta_{w,w'}]\!]_{\mathcal{M}}$, s.t. $\mathcal{M}, u \models \varphi$. By construction of $\Delta_{w,w'}$, we have that for all $s \in [\![\Delta_{w,w'}]\!]_{\mathcal{M}}$, either $s \approx w$, $s \approx w'$ or ($w \preceq s$ and $w' \preceq s$). By choice of w and w', we have $w' \preceq w$, meaning that $\exists w'' \in Min_{\leq}[\![\Delta_{w,w'}]\!]_{\mathcal{M}}$ such that $w' \approx w''$. Lemma 2 then says that w' and w'' are bisimilar, and Proposition 1 that they are modally equivalent. Thus $\mathcal{M}, w'' \models \varphi$. This is the u we are looking for, giving $\mathcal{M}, w \models \widehat{B}^{\Delta_{w,w'}} \varphi$.

Proposition 2. *On the class of image-finite models, modal equivalence implies bisimilarity.*

Proof. Let $\mathcal{M} = (W, \leq, \sim, V)$ and $\mathcal{M}' = (W', \leq', \sim', V')$ be two image-finite, plausibility models on P, and define $\mathfrak{R} \subseteq W \times W'$, such that $(w, w') \in \mathfrak{R}$ iff $(\mathcal{M}, w) \equiv (\mathcal{M}', w')$. We show that $\mathfrak{R}$ is in fact a bisimulation of the kind defined in Definition 5. Showing that $\mathfrak{R}$ satisfies [atoms] is trivial. We skip the, less trivial, [forth$_\sim$], and [back$_\sim$] and show the considerably more complicated case of [forth$_\preceq$] ([back$_\preceq$] is similar) as follows: Assume $(\mathcal{M}, w) \equiv (\mathcal{M}', w')$, $v \in W$ and $v \preceq w$ and show that assuming that for all $v' \in W'$, $v' \preceq w'$ implies $(\mathcal{M}, v) \not\equiv (\mathcal{M}', v')$, leads to a contradiction. This then gives $(\mathcal{M}, v) \equiv (\mathcal{M}', v')$ and therefore $(v, v') \in \mathfrak{R}$.

Let $S' = \{v' \mid v' \preceq w'\} = \{v'_1, \ldots v'_n\}$ be the successors of w'. This set is finite, due to image-finiteness of the model. If v and no successor of w' is modally equivalent, there exists formulae $\varphi^{v'_i}$, such that $\mathcal{M}, v \models \varphi^{v'_i}$ and $\mathcal{M}', v'_i \not\models \varphi^{v'_i}$. Therefore, $\mathcal{M}, v \models \varphi^{v'_1} \wedge \cdots \wedge \varphi^{v'_n}$. For notational ease, let $\Phi = \varphi^{v'_1} \wedge \cdots \wedge \varphi^{v'_n}$.

With $\mathcal{M}, v \models \Phi$, Lemma 3 gives $\mathcal{M}, w \models \widehat{B}^{\Delta_{w,v}} \Phi$ ($\Delta_{w,v}$ is finite due to image-finiteness of the models). Now, $\mathcal{M}', w' \models \widehat{B}^{\Delta_{w,v}} \Phi$ (which we must have due to modal equivalence) iff there exists a $u' \in Min_{\leq}[\![\Delta_{w,v}]\!]_{\mathcal{M}'}$ such that $\mathcal{M}', u' \models \Phi$. By construction of Φ, no world v'_i exists such that $v'_i \preceq w'$ and $\mathcal{M}', v'_i \models \Phi$, so we must have $w' \prec u'$. There are two cases for (the weakest requirements for) this u' to be minimal. Either (i) $u' \leq w'$ or (ii) $w' < u'$ and $w' \notin [\![\Delta_{w,v}]\!]_{\mathcal{M}'}$. If (i) is the case, we must have a world w'', with $w'' \approx w'$ and $w'' < u'$, or we couldn't have $w' \prec u'$. But $w'' < u'$ means that u' cannot be minimal unless $w' \notin [\![\Delta_{w,v}]\!]_{\mathcal{M}'}$, because otherwise $w'' \in [\![\Delta_{w,v}]\!]_{\mathcal{M}'}$. So, for (i) and (ii) both, we must have $w' \notin [\![\Delta_{w,v}]\!]_{\mathcal{M}'}$. This yields $\mathcal{M}', w' \models \neg\Delta_{w,v}$. But as $\mathcal{M}, w \models \Delta_{w,v}$, we get the sought after contradiction of $(\mathcal{M}, w) \equiv (\mathcal{M}', w')$.

3 Degrees of Belief and Safe Belief

In this section we sketch some further results that can be obtained for our single-agent setting of the logic of knowledge and conditional belief. Apart from *conditional belief*, other familiar epistemic notions in the philosophical logical and artificial intelligence community are *safe belief* [16] and *degrees of belief* [10,15]. Our results generalize fairly straightforwardly to such other notions. An agent has *safe belief* in formula φ iff it will continue to believe φ no matter what *true* information conditions its belief.[4]

Definition 8 (Safe belief). *We extend the inductive language definition with a clause $\Box\varphi$ for safe belief in φ. The semantics are $\mathcal{M}, w \models \Box\varphi$ iff ($\mathcal{M}, w \models B^{\psi}\varphi$ for all ψ such that $\mathcal{M}, w \models \psi$).*

Degrees of belief are a quantitative alternative to conditional belief. The zeroth degree of belief $B^0\varphi$ is defeasible belief $B\varphi$ as already defined. For $\mathcal{M}, w \models B^1\varphi$ to hold φ should be true in (*i*) all minimal worlds accessible from w; but additionally, (ii) if you take away those from the equivalence class, in all worlds that are now minimal. If we do this with the normal plausibility relation we get what we want (otherwise, we run into the same problems as before — our treatment is not compatible with e.g. Spohn's approach [15], that allows 'gaps' (layers without worlds) in between different degrees of belief).

$$
\begin{array}{lll}
Min^0_{\preceq}[w]_{\sim} & := Min_{\preceq}([w]_{\sim}) & \\
Min^{n+1}_{\preceq}[w]_{\sim} & := Min^n_{\preceq}[w]_{\sim} & \text{if } Min^n_{\preceq}([w]_{\sim}) = [w]_{\sim} \\
Min^{n+1}_{\preceq}[w]_{\sim} & := Min^n_{\preceq}[w]_{\sim} \cup Min_{\preceq}([w]_{\sim} \setminus Min^n_{\preceq}[w]_{\sim}) & \text{otherwise}
\end{array}
$$

We now can define the logic of knowledge and degrees of belief.

Definition 9 (Degrees of belief). *We replace the clause for conditional belief in the inductive language definition by a clause $B^n\varphi$ for belief in φ to degree n, for $n \in \mathbb{N}$. The semantics are*

$$\mathcal{M}, w \models B^n\varphi \textit{ iff for all } v \in Min^n_{\preceq}([w]_{\sim}) : \mathcal{M}, v \models \varphi$$

In an extended version of this paper we are confident that we will prove that the logics of conditional belief and knowledge, of degrees of belief and knowledge, and both with the addition of safe belief are all expressively equivalent.

[4] This definition is conditional to modally definable subsets, unlike [5,16] where it is on any subset. In that case safe belief is not bisimulation invariant and increases the expressivity of the logic.

4 Multi-agent Epistemic Doxastic Logic

For a finite set A of agents and a set of propositional symbols P the *multi-agent epistemic-doxastic language* $\mathcal{L}_{P,A}$ is

$$\varphi ::= p \mid \neg\varphi \mid \varphi \wedge \varphi \mid K_a\varphi \mid B_a^{\varphi}\varphi,$$

where $p \in P$ and $a \in A$. *Epistemic plausibility models* are generalized similarly, we now have plausibility relations $\leq_a$ and epistemic relations $\sim_a$ for each agent a. For each agent the domain is partitioned into (possibly) various equivalence classes, such that each class is a well-preorder. The single-agent results do not simply transfer to the multi-agent stage. We give an example.

Example 5. Consider Figure 3. The solid arrows represent the plausibilities for agent a and the dashed arrow for agent b. In our example, the partition for a is $\{w_0\}, \{w_1, w_2, w_3\}$, whereas the partition for b is $\{w_0, w_1\}, \{w_2\}, \{w_3\}$. Unlike before, the two p-states are not bisimilar, because in the state w_1 agent b is uncertain about the value of p but defeasibly believes p (there is a less plausible alternative w_0, whereas in state w_3 agent b knows (and believes) that p. In both worlds, of course, agent a still believes that p, but a distinguishing formula between the two is now, for example, $\neg K_b p \wedge B_a p$, true in w_1 but false in w_3.

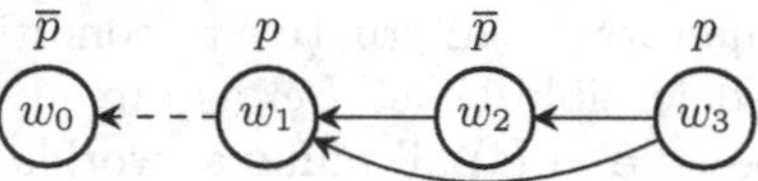

Fig. 3. A plausibility model wherein the two p worlds are not bisimilar, because they have different higher-order belief properties

It will be clear from Example 5 that we cannot, for each agent, derive a normal plausibility relation $\preceq_a$ from a given plausibility relation $\leq_a$ by identifying worlds with the same valuation: $w \approx_a w'$ iff $V(w) = V(w')$ and $w \sim_a w'$ does not work (worlds w_1 and w_3 in Example 3 satisfy different formulas). Some strengthening guarantees that bisimilarity still implies modal equivalence. An example is, using the above $\approx_a$:

$$\begin{array}{lll} w \approx w' & \text{iff} & (\text{for all agents } a : w \approx_a w') \\ w \preceq_a w' & \text{iff} & (Min_{\leq_a}[w]_{\approx} \leq_a Min_{\leq_a}[w']_{\approx}) \end{array}$$

Unfortunately we do not get that modal equivalence then implies bisimilarity. The strongest possible approach is of course to require that [$w \approx w'$ iff (w, w') is a pair in the bisimulation relation]. This works, but it is is rather self-defeating. In due time we hope to find a proper generalisation in between these two extremes.

5 Planning

In planning an agent is tasked with finding a course of action (i.e. a plan) that achieves a given goal. A planning problem implicitly represents a state-transition system, where transitions are induced by actions. Exploring this state-space is a common method for reasoning about and synthesising plans. A growing community investigates planning in dynamic epistemic logic [6,12,4,1], and using the framework presented here we can in similar fashion consider planning with doxastic attitudes. To this end we identify states with plausibility models, and the goal with a formula of the epistemic doxastic language. Further we can describe the dynamics of actions by using e.g. hard announcements or soft announcements [17], or yet more expressive notions such as event models [5].

With the state-space consisting of plausibility models, model theoretic results become pivotal to the development of planning algorithms. In general, we cannot require even single-agent plausibility models (even on a finite set of propositional symbols) to be finite. Also, normal plausibility models need not be finite — obvious, as the 'normalising' procedure in which we replace $\leq$ by $\preceq$ does not change the domain. Our definition of bisimulation has a crucial property in this regard: By Lemma 2 the bisimulation contraction of a model will contain no two worlds with the same valuation, hence any bisimulation minimal model on a finite set of propositions is finite. Moreover, two bisimulation minimal models are bisimilar exactly when they are isomorphic, and it follows that are only finitely many distinct bisimulation minimal epistemic plausibility models. With the reasonable assumption that actions preserve bisimilarity (this is the case for the types of actions mentioned above), our investigations on the proper notion of bisimulation therefore allow us to employ a smaller class of models in planning. This is a chief motivation for our work here, and an immediate consequence is that determining whether there exists a plan for a plausibility planning problem is *decidable* (see [2]).

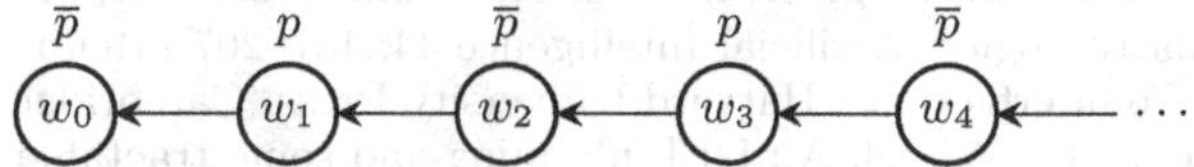

Fig. 4. Uncontractable chain of p and $\neg p$-worlds

It is remarkable that the approach of [8] to defining bisimulation for epistemic plausibility models does not yield decidability of planning problems, not even for single-agent models defined on a single proposition. It has, for instance, that the model in Figure 4 consisting of an infinite 'directed chain' of alternating p and $\neg p$ worlds (a copy of the natural numbers axis) is bisimulation minimal. In our approach the bisimulation minimal model would be the middle one of Figure 1, regardless of the number of worlds. Though [8] also shows that bisimilarity implies modal equivalence and vice versa (for image finite models), this is not inconsistent with our results here. Another difference between our approach and [8] lies in the semantics of safe belief. There, safe belief is relative to any subset (see also Footnote 4). For a 'directed chain' model, the safe belief semantics of [8] permits counting the number of p and $\neg p$ worlds. Such more expressive semantics naturally come at a cost, namely having no finite bound on the size of minimal single-agent models.

Acknowledgements. We thank the reviewers of the AI 13 conference for their comments. Hans van Ditmarsch is also affiliated to IMSc (Institute of Mathematical Sciences), Chennai, as research associate. He acknowledges support from European Research Council grant EPS 313360.

References

1. Andersen, M.B., Bolander, T., Jensen, M.H.: Conditional epistemic planning. In: del Cerro, L.F., Herzig, A., Mengin, J. (eds.) JELIA 2012. LNCS, vol. 7519, pp. 94–106. Springer, Heidelberg (2012)
2. Andersen, M.B., Bolander, T., Jensen, M.H.: Don't plan for the unexpected: Planning based on plausibility models. Logique et Analyse (to appear, 2014)
3. Aucher, G.: A combined system for update logic and belief revision. In: Barley, M.W., Kasabov, N. (eds.) PRIMA 2004. LNCS (LNAI), vol. 3371, pp. 1–17. Springer, Heidelberg (2005)
4. Aucher, G.: DEL-sequents for regression and epistemic planning. Journal of Applied Non-Classical Logics 22(4), 337–367 (2012)
5. Baltag, A., Smets, S.: A qualitative theory of dynamic interactive belief revision. In: Proc. of 7th LOFT. Texts in Logic and Games, vol. 3, pp. 13–60. Amsterdam University Press (2008)
6. Bolander, T., Andersen, M.B.: Epistemic planning for single and multi-agent systems. Journal of Applied Non-classical Logics 21(1), 9–34 (2011)
7. Britz, K., Varzinczak, I.: Defeasible modalities. In: Proc. of the 14th TARK (2013)
8. Demey, L.: Some remarks on the model theory of epistemic plausibility models. Journal of Applied Non-Classical Logics 21(3-4), 375–395 (2011)
9. Grove, A.: Two modellings for theory change. Journal of Philosophical Logic 17, 157–170 (1988)
10. Kraus, S., Lehmann, D., Magidor, M.: Nonmonotonic reasoning, preferential models and cumulative logics. Artificial Intelligence 44, 167–207 (1990)
11. Lewis, D.K.: Counterfactuals. Harvard University Press, Cambridge (1973)
12. Löwe, B., Pacuit, E., Witzel, A.: DEL planning and some tractable cases. In: van Ditmarsch, H., Lang, J., Ju, S. (eds.) LORI 2011. LNCS, vol. 6953, pp. 179–192. Springer, Heidelberg (2011)
13. Meyer, T.A., Labuschagne, W.A., Heidema, J.: Refined epistemic entrenchment. Journal of Logic, Language, and Information 9, 237–259 (2000)
14. Segerberg, K.: Irrevocable belief revision in dynamic doxastic logic. Notre Dame Journal of Formal Logic 39(3), 287–306 (1998)
15. Spohn, W.: Ordinal conditional functions: a dynamic theory of epistemic states. In: Harper, W.L., Skyrms, B. (eds.) Causation in Decision, Belief Change, and Statistics, vol. II, pp. 105–134 (1988)
16. Stalnaker, R.: Knowledge, belief and counterfactual reasoning in games. Economics and Philosophy 12, 133–163 (1996)
17. van Benthem, J.: Dynamic logic of belief revision. Journal of Applied Non-Classical Logics 17(2), 129–155 (2007)
18. van Benthem, J.: Logical Dynamics of Information and Interaction. Cambridge University Press (2011)
19. van Ditmarsch, H.: Prolegomena to dynamic logic for belief revision. Synthese (Knowledge, Rationality & Action) 147, 229–275 (2005)
20. van Ditmarsch, H., Labuschagne, W.A.: My beliefs about your beliefs – a case study in theory of mind and epistemic logic. Synthese 155, 191–209 (2007)

An Efficient Tableau for Linear Time Temporal Logic

Ji Bian, Tim French, and Mark Reynolds

The University of Western Australia,
Crawley, WA 6009, Australia

Abstract. Practical reasoning aids for dense-time temporal logics are not at all common despite a range of potential applications from verification of concurrent systems to AI. There have been recent suggestions that the temporal mosaic idea can provide implementable tableau-style decision procedures for various linear time temporal logics beyond the standard discrete natural numbers model of time. In this paper we extend the established idea of mosaic tableaux by introducing a novel abstract methodology of partiality which allows a partial mosaic to represent many mosaics. This can significantly reduce the running time of building a tableau. We present partial mosaics, partial mosaic-based tableau and algorithms for building the tableau.

Keywords: dense-time temporal logic, partial mosaic, tableau.

1 Introduction

The development [5,1,11] of temporal logics for reasoning applications using various dense and general linear models of time has progressed for decades alongside the more well-known logics for discrete time [10]. Applications range from engineering of reactive hardware [1] to continuous time multi-agent systems [8]. Being able to efficiently automate reasoning in continuous systems is an important aspect of building autonomous hybrid systems [15].

A basic syntax underlying this work is $L(\mathcal{U}, \mathcal{S})$ with natural-language style Until and Since connectives originally introduced in [7]. We denote by US/LIN the logic of $L(\mathcal{U}, \mathcal{S})$ over the class of all linear flows of time. Other logics for more specific flows of time such as the reals or dense-time can be obtained by a semantic restriction of the same language [11].

It was only recently [12] that deciding satisfiability of US/LIN formulas was shown to be in PSPACE (just like PLTL [14]). However, the formulation of reasoning tools for US/LIN has been a much longer process in the making: axioms [4], decidability [5], complexity [12], and the outline of a new tableau technique [13] based roughly on the idea of *mosaic* games [11].

In this paper we continue the work with making this proposed tableau technique considerably more efficient and practical. The tableau is used to decide the satisfiability (or equally the validity) of any given formula in US/LIN. The mosaics we use are the description of small pieces of a model: they describe which

S. Cranefield and A. Nayak (Eds.): AI 2013, LNAI 8272, pp. 289–300, 2013.

formulas hold at a pair of points from a model and which hold everywhere in between. Other sorts of mosaics have been used to develop reasoning techniques for other temporal and modal logics [9,6]. We decide whether a finite set of such small pieces is sufficient to describe a model of a given formula.

This paper makes a big step towards building more practical tableau systems by proposing a new theoretical object, the partial mosaic, which represents a whole range of related mosaics, in one finite syntactic object. By supporting dealing with a whole set of similar mosaics at once, we can speed up large parts of the reasoning task by significant amounts.

We start with describing partial mosaics. In this part, we will see that partial mosaics, which each represents several mosaics, can reduce searching times when building a tableau tree. Next we show the equivalence of the existence of a certain saturated set of partial mosaics to the existence of a linear structure witnessing the mosaics and introduce a partial mosaic-based tableau. In the last part, we show this new mosaic technique makes a measurable step towards practicality, by running experiments comparing the partial mosaic approach with the older full mosaic approach.

2 The Logic

Fix a countable set $\mathbf{L}$ of propositional atoms. *Frames* $(T, <)$, or *flows of time*, will be irreflexive linear orders. *Structures* $\mathbf{T} = (T, <, h)$ will have a frame $(T, <)$ and a *valuation* h for the atoms i.e. for each atom $p \in \mathbf{L}$, $h(p) \subseteq T$. The idea is that if $t \in h(p)$ then the proposition p is true at time t.

The language $L(\mathcal{U}, \mathcal{S})$ is is determined by the 2-place temporal connectives $\mathcal{U}$ (Until) and $\mathcal{S}$ (Since) along with classical $\neg$ and $\wedge$. The well-formed formulas of the language are built up recursively from atoms and $\top$ (for truth) using those connectives. That is, for formulas α and β, $\neg\alpha$, $\alpha \wedge \beta$, $\mathcal{U}(\alpha, \beta)$ and $\mathcal{S}(\alpha, \beta)$ are included in the language.

Formulas are evaluated at points in structures $\mathbf{T} = (T, <, h)$. We write $\mathbf{T}, x \models \alpha$ when α is true at the point $x \in T$:

$\mathbf{T}, x \models p$ iff $x \in h(p)$, for p atomic;

$\mathbf{T}, x \models \mathcal{U}(\alpha, \beta)$ iff there is $y > x$ in T such that $\mathbf{T}, y \models \alpha$ and for all $z \in T$ such that $x < z < y$ we have $\mathbf{T}, z \models \beta$; and

$\mathbf{T}, x \models \mathcal{S}(\alpha, \beta)$ iff there is $y < x$ in T such that $\mathbf{T}, y \models \alpha$ and for all $z \in T$ such that $y < z < x$ we have $\mathbf{T}, z \models \beta$.

Abbreviations include the usual classical and temporal ones such as $F\alpha = \mathcal{U}(\alpha, \top)$ and $G\alpha = \neg F(\neg\alpha)$ and their mirror images P and H. The resulting logic is refered to as US/Lin.

As a simple example of the problems we aim to solve, we can give a dense-time point-based temporal language and a synthesis algorithm to solve planning problems.

The states are represented as propositions: g, p, q, r, and actions are represented by propositions that may hold over extended intervals: a, b, c

Suppose that we want to achieve goal g from starting state s, under the following conditions:

In order to achieve g you need to do a from state p and also (in parallel) do b from state s;
To get to p you need to do c from s but as a red herring you can also achieve p from q by doing a;
There is no action that ends in q.

Here are a list of formulas to capture this, and some other basic assumptions (such as the uniqueness of the start state s):

$$G(g \to (\mathcal{S}(p,a) \wedge \mathcal{S}(s,b))) \qquad G(p \to (\mathcal{S}(s,c) \vee \mathcal{S}(q,a)) \qquad G(\neg q)$$
$$H(\neg g \wedge \neg p \wedge \neg q \wedge \neg r) \qquad H(\neg a \wedge \neg b \wedge \neg c) \qquad H\neg s \wedge s \wedge G\neg s$$

The synthesis task is to find a model of Fg along with the conjunction of the clauses above. In this case, there is a model and it tells us a plan to achieve the goal.

3 Partial Mosaics for $\mathcal{U}$ and $\mathcal{S}$

To decide the satisfiability of a formula, we construct a complete structure for the formula using mosaics as building blocks. Mosaics correspond to sets of labelled structures and building a tableau of mosaics is proved equivalent to building a whole model in [13]. However, the huge numbers of mosaics often make exhaustive searches infeasible in practice, especially for unsatisfiable formulas. Results for satisfiable formulas are also often slow or not feasible.

Here we introduce partial mosaics where each partial mosaic represents a set of mosaics. This representation reduces the search space for the tableau. Partial mosaics can be used to represent structures at different levels of abstraction. We will see the representation of models and abstraction of partial mosaics in the following subsection. To build a whole model, the set of partial mosaics has be free of *defects* that necessitates inclusion of some others partial mosaics. A *full decomposition* is the resulting sequence of partial mosaics after all defects have been cured. The existence of partial mosaic-based tableau proves to be equivalent to a structure via a *saturated set.*

3.1 A Mosaic

Mosaics are concerned with a finite set of formulas:

Definition 1. *For each formula ϕ, define the* closure *of ϕ to be* $\text{Cl}\phi = \{\psi, \neg\psi \mid \psi \leq \phi\}$ *where $\chi \leq \psi$ means that χ is a subformula of ψ.*

We can think of $\text{Cl}\phi$ as being closed under negation: treat $\neg\neg\alpha$ as if it was α.

A mosaic is a description of a small piece of a model. Each mosaic consists of three parts: a set of formulas held at a earlier point in a model, a set of formulas held at a final point in a model, and a set of formulas held by all points in a between. Any mosaic m' is defined to a triple $S(m'), C(m'), E(m')$ in [13] where both $S(m')$ and $E(m')$ are sets of formula. The sets $S(m')$ and $E(m')$ are restricted to be propositionally consistent and maximal (relative to a closure set) so they describe the set of formulas true at two points in a structure, and the set $C(m')$ is all formulas that are true at all points in between. (Note that $C(m')$ may not be maximal , since formulas may be true at some points and false at others, and may not be consistent, because there may be no points between the start and the end of the mosaic. For a formal description, see [13]). We can see that for a formula consisting on n symbols, there could be as many as 2^{3n} mosaics, making exhaustive search very costly. A partial mosaic generalizes many mosaics allowing us to search large regions of the mosaic state space simultaneously.

3.2 A Partial Mosaic

We will define a partial mosaic to be a triple $(S(m), C(m), E(m))$ of sets of formulas as well. The intuition is that this corresponds to two points from a structure: $S(m)$ is the start set of formulas (from $\mathrm{Cl}\phi$) true at the starting point, $E(m)$ is the end set of formulas true at the final point, and $C(m)$ is the cover set of formulas which are held or possibly held by all points strictly in between. A partial mosaic m, like a full mosaic, consists of three sets of formulas $(S(m), C(m), E(m))$: the *start set*, $S(m)$, contains of formulas that must be true at the start of the mosaic; the *end set*, $E(m)$, contains formulas that must be true at the end of the mosaic; and the *cover set*, $C(m)$, describes formula that must be true in the middle of the model. These sets are not maximal, so for example, we may find that neither p nor $\neg p$ is in the start set. In this case the partial mosaic matches full mosaics that have p in the start set, and full mosaics that have $\neg p$ in the start set. For the cover set, we are interested in whether a formula is true over an interval, so we introduce **S5** modalities, $\Box$ and $\Diamond$, to generalize these scenarios.

Definition 2. *The* partial closure *of ϕ is the set* $P\mathrm{Cl}\phi = \{\Box\alpha, \Diamond\alpha | \alpha \in \mathrm{Cl}\phi\}$

Intuitively, if $\Box\alpha$ is in the cover set, then α is true everywhere between the start and the end of the mosaic, and if $\Diamond\alpha$ is in the cover set, the α is true somewhere between the start and the end of the mosaic. It is important that these sets of formulas are consistent.

Definition 3. *Suppose $\phi \in L(\mathcal{U}, \mathcal{S})$ and $S \subseteq \mathrm{Cl}\phi$. Say S is* propositionally consistent *(PC) iff there is no substitution instance of a tautology of classical propositional logic of the form $\neg(\alpha_1 \wedge ... \wedge \alpha_n)$ with each $\alpha_i \in S$.*
Say S is maximally propositionally consistent *(MPC) iff S is maximal in being a subset of $\mathrm{Cl}\phi$ which is PC.*
Suppose $S \subseteq P\mathrm{Cl}\phi$. Then S is partially propositionally consistent *if for every $\Diamond\alpha \in S$, $\{\alpha\} \cup \{\beta | \Box\beta \in S\}$ is propositionally consistent.*

Note, that it is possible that there is no formula $\Diamond\alpha \in C(m)$, in which case it is possible that m describes a structure with no points between the start and the end. We are now able to formally define partial mosaics.

Definition 4. *Suppose ϕ is from $L(U,S)$. A ϕ-*partial mosaic *is a triple* m. $S(m) \subseteq \mathrm{Cl}\phi$, $E(m) \subseteq \mathrm{Cl}\phi$ *and* $C(m) \subseteq \mathrm{PCl}\phi$ *such that:*

C0 $S(m)$ *and* $E(m)$ *are PC, and* $C(m)$ *is partially propositionally consistent.*
C1 $\Diamond\alpha \in C(m)$ *and* $\Box\beta \in C(m)$ *imply* $\mathcal{U}(\alpha,\beta) \in S(m)$.
C2 $\Box\beta \in C(m)$ *and* $\alpha \in E(m)$ *imply* $\mathcal{U}(\alpha,\beta) \in S(m)$.
C3 $\Box\beta \in C(m)$, $\beta \in E(m)$ *and* $\mathcal{U}(\alpha,\beta) \in E(m)$ *imply* $\mathcal{U}(\alpha,\beta) \in S(m)$.
C4 $\mathcal{U}(\alpha,\beta) \in S(m)$ *implies either:* $\Diamond\alpha \in C(m)$;
or $\Box\beta \in C(m)$ *and* $\alpha \in E(m)$;
or $\Box\beta \in C(m)$, $\beta \in E(m)$ *and* $\mathcal{U}(\alpha,\beta) \in E(m)$.
C5 $\Diamond\neg\beta \in C(m)$ *and* $\Box\neg\alpha \in C(m)$ *imply* $\neg\mathcal{U}(\alpha,\beta) \in S(m)$.
C6 $\Box\neg\alpha \in C(m)$, $\neg\beta \in E(m)$ *and* $\neg\alpha \in E(m)$ *imply* $\neg\mathcal{U}(\alpha,\beta) \in S(m)$.
C7 $\Box\neg\alpha \in C(m)$, $\neg\alpha \in E(m)$ *and* $\neg\mathcal{U}(\alpha,\beta) \in E(m)$ *imply* $\neg\mathcal{U}(\alpha,\beta) \in S(m)$.
C8 $\neg\mathcal{U}(\alpha,\beta) \in S(m)$ *implies either:* $\Diamond\neg\beta \in C(m)$;
or $\Box\neg\alpha \in C(m)$, $\neg\beta \in E(m)$ *and* $\neg\alpha \in E(m)$;
or $\Box\neg\alpha \in C(m)$, $\neg\alpha \in E(m)$ *and* $\neg\mathcal{U}(\alpha,\beta) \in E(m)$.
C9- C16 mirrors of C1 to C8.

We now confine our attention to the semantics of a partial mosaic. The semantics of a partial mosaic is a set of some fragments of structures it represents.

Definition 5. *If* $\mathbf{T} = (T,<,h)$ *is a structure and* $\phi \in L(\mathcal{U},\mathcal{S})$ *then for some* $x < z$ *from* T *we say that a partial mosaic* m *represents* $\mathbf{T}$ *from* x *to* z *iff:*

1. $S(m) \subseteq \{\alpha \in \mathrm{Cl}\phi | \mathbf{T}, x \models \alpha\}$;
2. $C(m) \cap \mathrm{Cl}\phi \subseteq \{\beta \in \mathrm{Cl}\phi |$ *for all* $y \in T$, *if* $x < y < z$ *then* $\mathbf{T}, y \models \beta\}$, *and*
3. $E(m) \subseteq \{\gamma \in \mathrm{Cl}\phi | \mathbf{T}, z \models \gamma\}$

A partial mosaic m *is satisfiable if and only if there is some structure* $\mathbf{T} = (T,<,h)$ *and some* $x, z \in T$ *such that* m *represents* $\mathbf{T}$ *from* x *to* z.

With these semantics we can now see the syntactic criteria of Definition 4. If $\mathcal{U}(\alpha,\beta) \in S(m)$ it is either satisfied before the end of the mosaic (C1), at the end of the mosaic (C2), or after the end of the mosaic (C3), and one of these cases must hold (C4). Likewise, if $\neg\mathcal{U}(\alpha,\beta) \in S(m)$ then $\mathcal{U}(\alpha,\beta)$ must be invalidated before the end of the mosaic (C5), at the end of the mosaic (C6), or after the end of the mosaic (C7), and one of these conditions must hold (C8). Similar constraints can be given for $\mathcal{S}(\alpha,\beta)$ which we omit due to space reasons.

3.3 Mosaics vs Partial Mosaics

A partial mosaic can refer to a higher level of abstraction than a full mosaic and so it can serve to express a whole range of related mosaics. The intuition is that the $S(m)$ and $E(m)$ of a mosaic that are maximally propositionally consistent are less abstract than $S(m)$ and $E(m)$ of a partial mosaic that are only propositionally consistent; and $C(m)$ of a mosaic that is subset of $\mathrm{Cl}\phi$ can be captured by a $C(m)$ of a partial mosaic that is partially propositionally consistent. Given any full mosaic m, the corresponding partial mosaic is $\overline{m} = (S(m), \{\Box\alpha | \alpha \in C(m)\}, E(m))$. However, as partial mosaics do not need to have maximal propositional consistent sets for the start and end of a mosaic, they can represent structures at varying levels of abstraction. We first demonstrate different levels of the abstraction in $C(m)$ by introducing an ordering over partial mosaic.

Definition 6. *For ϕ-partial mosaics m_1 and m_2, we say $m_1 \subseteq m_2$ iff $S(m_1) \subseteq S(m_2)$, $C(m_1) \subseteq C(m_2)$ and $E(m_1) \subseteq E(m_2)$, and we say that m_1 is more* general *than m_2*

If m_1 is more general than m_2 then m_1 will typically represent more models than m_2.

Example: a partial mosaic for $\mathcal{U}(p, q)$ is

$$(\{q\}, \{\Diamond\neg p, \Diamond q\}, \{\}).$$

this partial mosaic can serve to express another partial mosaic and a mosaic as follows:

$$(\{q, \mathcal{U}(p, q)\}, \{\Diamond\neg p, \Diamond q, \Diamond\mathcal{U}(p, q)\}, \{\neg p, \neg q, \neg\mathcal{U}(p, q)\})$$
$$(\{p, q, \mathcal{U}(p, q)\}, \{\Diamond\neg p, \Diamond q, \Box q\}, \{p, \neg q, \neg\mathcal{U}(p, q)\})$$

As partial mosaics represent many full mosaics, if we are able to show that a partial mosaic does not correspond to a linear temporal structure, then we have ruled out a large number of full mosaics, which now do not have to be explored.

3.4 Defects

When building a model we may find that a certain set of required partial mosaics necessitates the inclusion of another set of partial mosaics. This includes satisfying formulas in the form of $\mathcal{U}(\alpha, \beta)$ in the start set, $\mathcal{S}(\alpha, \beta)$ in the end set, and formulas that are required to be true at a point between the start and the end. For example, if we have $\mathcal{U}(\alpha, \beta)$ holding at $x < z$, it possibly requires point y with $x < y < z$ witnessing α such that α true at y and β true everywhere between x and y. Another example is if we have $\neg\beta$ not true everywhere between x and z then we need to find a point y in between x and z witnessing β. Here a *defect* is introduced as a not yet satisfied formula, and we have to *cure* it.

Here we introduce 5 types of defects:

Definition 7. *A* defect *in a partial mosaic m is one or more of the followings:*

1. *a formula $\mathcal{U}(\alpha, \beta) \in S(m)$ with either*
   1.1 $\Box\beta \notin C(m)$,
   1.2 $\alpha \notin E(m)$ and $\beta \notin E(m)$, or
   1.3 $\alpha \notin E(m)$ and $\mathcal{U}(\alpha, \beta) \notin E(m)$;
2. *a formula $\mathcal{S}(\alpha, \beta) \in E(m)$ with either*
   2.1 $\Box\beta \notin C(m)$,
   2.2 $\alpha \notin S(m)$ and $\beta \notin S(m)$, or

2.3 $\alpha \notin S(m)$ *and* $S(\alpha, \beta) \notin S(m)$;
3. *a formula* $\beta \in \mathrm{Cl}\phi$ *with* $\Diamond\beta \in C(m)$; *or*
4. *a formula* $\mathcal{U}(\alpha, \beta) \in E(m)$;
5. *a formula* $\mathcal{S}(\alpha, \beta) \in S(m)$.

A sequence of partial mosaics is usually used to cure defects, and we have to string them together in the linear order. The following definition allows us to relate sequences of mosaics to a single mosaic.

Definition 8. *Suppose* $C(m_1)$, $C(m_2)$, *and* $C(m_3)$ *are partially propositionally consistent. We introduce* partial intersection $\cap_*$ *as:*

$$C(m_1) \cap_* C(m_2) = \{\Box\alpha, \Diamond\beta \mid \Box\alpha \in C(m_1) \cap C(m_2),\ \Diamond\beta \in C(m_1) \cup C(m_2)\}.$$

Definition 9. *We say that* ϕ*-partial mosaics* m_1 *and* m_2 *are* composable *iff* $E(m_1) = S(m_2)$. *In that case, their* composition *is* m , *in which* $S(m) = S(m_1)$, $E(m) = E(m_2)$, *and*

$$C(m) = ((C(m_1) \cap_* C(m_2)) - \{\Box\alpha | \alpha \notin E(m_1)) \cup \{\Diamond\alpha | \alpha \in E(m_1)\}$$

The idea of composition between two partial mosaics can be easily extended to a sequence of composable partial mosaics, by successive compositions.

Definition 10. *We say a sequence of composable partial mosaics is a* decomposition *of a partial mosaic* m_1 *iff its composition* m_2 *has* $m_1 \subseteq m_2$.

A *full decomposition* of is intended to provides witnesses to the cure of every type 1, 2 and 3 defect in the partial mosaic.

Definition 11. *The decomposition of* $< m_1, m_2, ..., m_n >$ *of partial mosaic* m *is* full *iff the following three conditions all hold:*

1. *for all* $\mathcal{U}(\alpha, \beta) \in S(m)$ *we have either*
   1.1. $\Box\beta \in C(m)$ *and either* ($\beta \in E(m)$ *and* $\mathcal{U}(\alpha, \beta) \in E(m)$) *or* $\alpha \in E(m)$,
   1.2. or there is some i *with* $1 \leq i < n$ *such that* $\alpha \in E(m_i)$, $\Box\beta \in C(m_j)$ *(all* $j \leq i$*) and* $\beta \in E(m_j)$ *(all* $j < i$*);*
2. *the mirror image of 1.; and*
3. *for each* $\beta \in \mathrm{Cl}\phi$, *we have* $\Diamond\beta \in C(m)$; *and there is* i *such that* $1 \leq i < n$ *and* $\beta \in E(m_i)$.

1.2 aims at curing type 1 defect in m and α is witnessed by the end of m_i. Similarly for the mirror image for type 2 defect; 3 is for type 3 defect: the end of m_i witnesses β. Similar constructions can be defined to cure type 4 and 5 defects, by appending mosaics to the start of end of a mosaic. We will omit the details due to space considerations.

3.5 Satisfiability Formulas and Partial Mosaics

We say a formula ϕ is satisfiable iff there a point in a structure $\mathbf{T} = (T, <, h)$ such that ϕ is true at that point. Here we check the satisfiability of a formula ϕ in $L(U, S)$ with satisfiable partial mosaics. Specifically, if we check the satisfiability of a ϕ-partial mosaic m with either $\phi \in S(m)$ or $\phi \in E(m)$, then ϕ is satisfiable.

The main idea is similar to situation in [13] with mosaic, but a partial mosaic is more general than a mosaic as shown by the following lemma.

Lemma 1. *A ϕ-partial mosaic m is satisfiable iff there is at least a satisfiable ϕ-mosaic m' such that $m \subseteq m'$.*

Clearly, if a ϕ-partial mosaic m is not satisfiable then any mosaic m' with $m \subseteq m'$ is also not satisfiable. Checking the satisfiability of partial mosaics rather than mosaics can greatly reduce the search space. A proof of Lemma 1 can be found in [2].

In the procedure of proving US/LIN satisfiability, we are to guess a partial mosaic $(S(m), C(m), E(m))$ for ϕ and then check that $(S(m), C(m), E(m))$ is satisfiable. Thus we now focus on deciding whether a partial mosaic is satisfiable.

We introduce a technique: *a saturated set of partial mosaics* (SSSPM) as seen in earlier work [9].

Definition 12. *A* full special expansion *for a standard partial mosaic m is any finite, composing sequence of partial mosaics $< m_1, ..., m_i, ..., m_j, ..., m_n >$ with m_i and $m_j (1 \leq i \leq j \leq n)$ such that*
1. $S(m_i) = S(m)$, and $E(m_j) = E(m)$,
2. m_1, m_2, ..., m_{i-1} is a left expansion of m,
3. m_i, ..., m_j is a full decomposition of m,
4. m_{j+1}, ..., m_n is a right expansion of m.

A SSSPM is a set of partial mosaics with each one in the set having a full special expansion containing only partial mosaics from the set. It also summarises the organization of partial mosaics in it, and show cures for any defects in any partial mosaic in the set.

4 Tableau of Partial Mosaic

The ultimate aim is to determine whether a given formula ϕ in $L(U, S)$ is satisfiable in US/LIN or not. Deciding validity is simply equivalent to determining satisfiability of the negation. We can show a formula ϕ is satisfiable if and only if there is a satisfiable for ϕ-partial mosaic, and a mosaic is satisfiable if and only if it has a SSSPM. To show that a SSSPM exists, we use a tableau method to progressively cure defects in mosaics.

The tableau is a tree where each node is labelled by a partial mosaic, and each node has a finite number of children (which are ordered). The labels on the nodes of the children must form a full decomposition of the label on the node of the parent. Nodes whose labels do not have defects do not require children, and nor do nodes whose label is equal to the label of an ancestor. If a finite tableau exists with a mosaic m at the root, then m is satisfiable. A complete description can be found in [9].

Rather than the depth-first search approach used in [9], the partial mosaic lend themselves to an iterative deepening approach, where a finer level of abstraction is only explored if it is required to cure a defect.

We are able to show that the satisfiability of a formula, is equivalent to the existence of a saturated set of partial mosaics (SSSPM) containing a ϕ-partial mosaic, and that such a SSSPM exists if and only if there is a successful tableau with m at its root.

Lemma 2. *A ϕ-partial mosaic m has a successful tableau iff there exists a saturated set of partial mosaics(SSSPM) for ϕ with m in it,*

Theorem 1. *Suppose that ϕ is a formula of $L(U, S)$. Then ϕ-partial mosaic m is satisfiable iff there a successful tableau with m as its root*

For full proofs of Lemma 2 and Theorem 1, see [2].

When using tableaux to decide US/LIN, we have to see that a partial mosaic represents model with more than two points. However some formulas are only satisfiable in one point models such as $\neg\mathcal{U}(\top,\top) \wedge \neg\mathcal{S}(\top,\top)$. We introduce a singleton tableau that is PC set of subformulas of ϕ with no formulas of the form $\mathcal{U}(\alpha, \beta)$ or $\mathcal{S}(\alpha, \beta)$. Then tableau is able to solve the satisfiability for all formulas

The strategy for constructing a successful tableau is to search through every possible tableau for all partial mosaics for ϕ. The strategy is similar in constructing a mosaics-based tableau [13]. However, rather than expanding the labels of new nodes via only full expansions of mosaics, here we not only expand the labels of new children nodes via full expansions of partial mosaics but also we sometimes have to expand the leaf nodes of neighbouring branches with a single and more specific child node. This is because a partial mosaic may not contain a critical formula, nor its negation.

4.1 Examples

Here are some examples. Tableaux shown in diagrams as trees grow down from the root. Left-expanding children are indicated by l, right by r. If left-expanding or right-expanding children don't exist, we use n short for null to represent it. Successful branches are indicated by $leaf$.

See the example in Figure 1 a successful tableau for a ϕ-mosaic containing $\phi = \mathcal{U}(q,r) \wedge \mathcal{S}(p,r)$. Here, there is no type 3 for the root partial mosaic. However, it has both type 4 defect $\mathcal{U}(q,r)$ in the end set and type 5 defect $\mathcal{S}(p,r)$ in the start set. So, it has only left-expanding children and right-expanding children.

See the example in Figure 2 for an incomplete tableau for a ϕ-mosaic containing $\phi = p \wedge FPH\neg p$. The last two non-null-branches end successfully. There is no way, however, of completing the first non-null branch from m_1 to m_2 as there is no way of curing the type 3 defect $\neg H \neg p$ in m_2. Note that although the branch from m_1 to m_2 is doomed to failure, the formulas ϕ is satisfiable, and an alternative satisfying tableaux can be found.

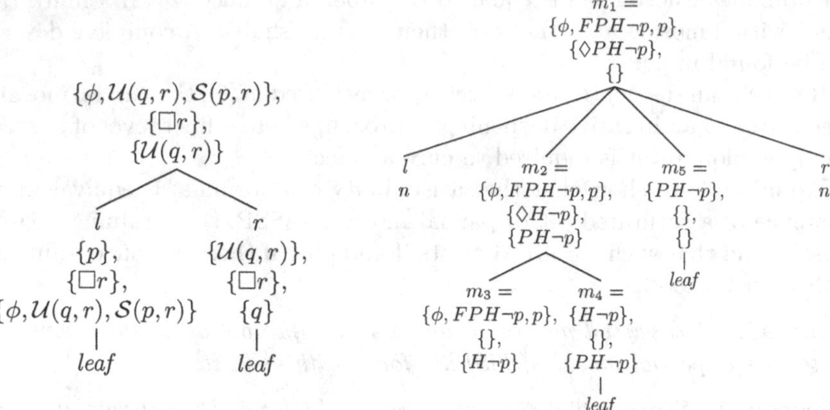

Fig. 1. Tableau for $\phi = \mathcal{U}(q,r) \wedge \mathcal{S}(p,r)$

Fig. 2. Partial tableau for $\phi = p \wedge FPH\neg p$

5 Experimental Results

A comprehensive set of benchmarks has been collected for discrete linear time, and some application benchmark are directly the specification of actual systems including traffic-light controlling system and mutual exclusion protocol. Despite the significant accumulation, there are no such systematic application benchmark formulas for US/LIN or dense time. Here we try to introduce some benchmark formulas to compare the two reasoners in Table 1: The formulas 2, 3 and 10 are three of six Burgess-Xu axioms, and all of them are valid in US/LIN; we create 16 and 18 that intend to violate another two axioms and they are unsatisfiable in US/LIN (both of them involve exhaustive search for mosaics/partial mosaics and trigger exponential behaviour in both reasoners); 12, 13 and 14 are a series of benchmarks to specify various mutual exclusions protocols, especially 14 is for general linear time; 19 is to test the performances of reasoners on dense time only; 9 comprises many eventualities to be checked, and even some eventualities are contradicted; 5 requires us to check eventualities for Until and also for Since; 17 and 20 are random unsatisfiable formulas; and 1 and 4 are random satisfiable formulas.

We present timing and space results in Table 1. Column Sat? indicates whether it is satisfiable; Run contains the time taken by full mosaics (M) and partial mosaics (PM). The value U indicates exceeding 3 hours but without any results. The column $Nodes$ is the number of nodes explored. We see that there is great improvement of performance for most case no matter what formulas they are(either satisfiable or unsatisfiable formulas). Some performance improvement is attributed to not having to check a large number of root nodes. See formula $r \wedge \mathcal{U}(p \wedge \neg\mathcal{S}(r,s)), s)$. We have to check 213,330 mosaics as the root, while we only check 5 partial mosaics as the root. The decreased difficulty of decomposing a node also contributes to the speedup. Intuitively, more defects requires longer expansion to cure and more running time. Mosaics easily introduce much more defects than partial mosaics, especially type 3 defect. This is mainly attributed to their different cover set. See formula $\neg(\mathcal{U}(p,q) \rightarrow \mathcal{U}(p, q \wedge \mathcal{U}(p,q)))$. Some mosaics even have up to 12 type 3 defects including some roots, while a partial mosaic only at most 4. Type 3 defects are expensive to cure. Even if we cure type 3 defects in one level, the curing witnesses may possibly have type 1 or 2 defects in the next level. This is the reason why it even takes several minutes to decompose a mosaic for $r \wedge \mathcal{U}(p \wedge \neg\mathcal{S}(r,s)), s)$ and $\neg(\mathcal{U}(p,q) \rightarrow \mathcal{U}(p, q \wedge \mathcal{U}(p,q)))$. There is another interesting result that tableau of partial mosaics has shorter full decomposition in comparison with that of full mosaic tableaux. This generally implies less nodes to decompose that improves the running time. This is the reason for the improvement in $p \wedge G(p \rightarrow Fp) \wedge G(\mathcal{U}(\top, \bot)) \wedge FG\neg p$ and $G(p \rightarrow G\neg q) \wedge G(q \rightarrow G\neg p) \wedge FG(\neg p \wedge \neg q)$. An online implementation of this solver is provided [3].

Table 1. Run-time and memory usage of partial mosaic and full mosaic tableaux. A full table and detailed results is available in [2].

Num	Formula	Sat?	Run(s)		Nodes	
			M	PM	M	PM
1	$G(\mathcal{U}(\neg p, p)) \wedge \neg F(\mathcal{U}(\neg p, \bot)) \wedge \neg F(\mathcal{U}(p, \bot))$	Y	888	< 1	52	8
2	$G(p \rightarrow q) \rightarrow (\mathcal{U}(r,p) \rightarrow \mathcal{U}(r,q))$	Y	U	1	$> 2^{20}$	7
3	$(p \wedge \mathcal{U}(q,r)) \rightarrow \mathcal{U}((q \wedge \mathcal{S}(p,r)), r)$	Y	U	4	$> 2^{20}$	5
4	$G(\mathcal{U}(F\neg p, p)) \wedge G(\neg p \rightarrow G\neg p)$	Y	61	< 1	47	4
5	$G((\mathcal{S}(p, \neg p) \wedge \mathcal{U}(p, \neg p)) \vee (\mathcal{S}(\neg p, p) \wedge \mathcal{U}(\neg p, p)))$	Y	U	< 1	$> 2^{20}$	3
9	$G(p \rightarrow \mathcal{U}(q, 0) \wedge Fq \rightarrow FFq \wedge Fq \wedge Fq \rightarrow p)$	Y	U	1	$> 2^{10}$	2
10	$G(p \rightarrow q) \rightarrow (\mathcal{U}(p,r) \rightarrow \mathcal{U}(q,r))$	Y	U	< 1	$> 2^{10}$	2
12	$Fp \wedge G(p \rightarrow Fp) \wedge G(\mathcal{U}(\top, \bot)) \wedge FG\neg p$	Y	24	< 1	13	3
13	$G(p \rightarrow G\neg q) \wedge G(q \rightarrow G\neg p) \wedge FG(\neg p \wedge \neg q)$	Y	U	< 1	$> 2^{10}$	4
14	$p \wedge (p \rightarrow Fp) \wedge G(p \rightarrow Fp) \wedge G(\mathcal{U}(p, \neg p)) \wedge F\neg p$	Y	14	< 1	7	2
16	$r \wedge \mathcal{U}(p \wedge \neg\mathcal{S}(r,s)), s)$	N	U	< 1	$> 2^{20}$	5
17	$\mathcal{U}(\mathcal{U}(p,q), \mathcal{U}(q,p)) \wedge G\neg p$	N	U	< 1	$> 2^{20}$	9
18	$\neg(\mathcal{U}(p,q) \rightarrow \mathcal{U}(p, q \wedge \mathcal{U}(p,q)))$	N	U	< 1	$> 2^{20}$	3
19	$\neg\mathcal{U}(\neg p, p) \wedge \neg\mathcal{U}(p, \neg p) \wedge G\neg\mathcal{U}(\top, \bot)$	Y	189	30	102	21
20	$FPp \wedge \neg Fp \wedge \neg p \wedge \neg Pp$	N	9	< 1	2	2

6 Conclusion and Future Work

We have been able to formulate the novel idea of a partial mosaic, representing a whole set of more or less similar mosaics. We have presented a partial mosaic-based tableau for US/LIN and shown that it is a sound and complete reasoning system via saturated sets of partial mosaics. We have also shown that the partial mosaic tableau system is a big step towards a practical and efficient reasoning system for US/LIN. Experiments have shown significant speedups in comparison with the existing tableau of mosaics.

Particular directions to pursue in future work include finding good heuristics for the choice of partial mosaics to expand nodes or enhancing the level of abstraction for partial mosaics.

References

1. Alur, R., Dill, D.L.: A theory of timed automata. Theoretical Computer Science 126, 183–235 (1994)
2. Bian, J.: Efficient tableaux for temporal logic. Doctoral thesis, in preparation (2013), `http://www.csse.uwa.edu.au/~jibian/thesis/thesis.pdf`
3. Bian, J.: Linear time tableaux for partial mosaic, Online solver (2013), `http://www.csse.uwa.edu.au/~jibian/partialmosaictab/`
4. Burgess, J.: Axioms for tense logic I: 'Since' and 'Until'. Notre Dame J. Formal Logic 23(2), 367–374 (1994)
5. Burgess, J., Gurevich, Y.: The decision problem for linear temporal logic. Notre Dame J. Formal Logic 26(2), 115–128 (1985)
6. Hirsch, R., Hodkinson, I., Marx, M., Mikulas, S., Reynolds, M.: Mosaics and step-by-step. Remarks on "A modal logic of relations". In: Orlowska, E. (ed.) Logic at Work: Essays Dedicated to the Memory of Helen Rasiowa. STUDFUZZ, vol. 24, pp. 158–167. Springer, Heidelberg (1999)
7. Kamp, H.: Tense logic and the theory of linear order. PhD thesis, University of California (1968)
8. Kesten, Y., Manna, Z., Pnueli, A.: Temporal verification of simulation and refinement. In: de Bakker, J.W., de Roever, W.-P., Rozenberg, G. (eds.) REX 1993. LNCS, vol. 803, pp. 273–346. Springer, Heidelberg (1994)
9. Marx, M., Mikulas, S., Reynolds, M.: The mosaic method for temporal logics. In: Dyckhoff, R. (ed.) TABLEAUX 2000. LNCS (LNAI), vol. 1847, pp. 324–340. Springer, Heidelberg (2000)
10. Pnueli, A.: The temporal logic of programs. In: Proceedings of the Eighteenth Symposium on Foundations of Computer Science, pp. 46–57. Springer (1977)
11. Reynolds, M.: Dense time reasoning via mosaics. In: TIME 2009: Proceedings of the 2009 16th International Symposium on Temporal Representation and Reasoning, pp. 3–10. IEEE Computer Society (2009)
12. Reynolds, M.: The complexity of temporal logics over linear time. Journal of Studies in Logic 3, 19–50 (2010)
13. Reynolds, M.: A tableau for until and since over linear time. In: Proc. of 18th International Symposium on Temporal Representation and Reasoning (TIME 2011), Lubeck, Germany. IEEE Computer Society Press (September 2011)
14. Sistla, A., Clarke, E.: Complexity of propositional linear temporal logics. J. ACM 32, 733–749 (1985)
15. Yin, Z., Tambe, M.: Continuous time planning for multiagent teams with temporal constraints. In: Proceedings of the 22nd International Joint Conference on Artificial Intelligence, IJCAI 2011, pp. 465–471 (2011)

Updates and Uncertainty in CP-Nets

Cristina Cornelio[1], Judy Goldsmith[2], Nicholas Mattei[3], Francesca Rossi[1], and K. Brent Venable[4]

[1] University of Padova, Italy
{cornelio,frossi}@math.unipd.it
[2] University of Kentucky, USA
goldsmit@cs.uky.edu
[3] NICTA and UNSW, Australia
nicholas.mattei@nicta.com.au
[4] Tulane University and IHMC, USA
kvenabl@tulane.edu

Abstract. In this paper we present a two-fold generalization of conditional preference networks (CP-nets) that incorporates uncertainty. CP-nets are a formal tool to model qualitative conditional statements (cp-statements) about preferences over a set of objects. They are inherently static structures, both in their ability to capture dependencies between objects and in their expression of preferences over features of a particular object. Moreover, CP-nets do not provide the ability to express uncertainty over the preference statements. We present and study a generalization of CP-nets which supports changes and allows for encoding uncertainty, expressed in probabilistic terms, over the structure of the dependency links and over the individual preference relations.

Keywords: Preferences, Graphical Models, Probabilistic Reasoning, CP-nets.

1 Introduction

CP-nets are used to model conditional information about preferences [2]. Preferences play a key role in automated decision making [9] and there is some experimental evidence suggesting qualitative preferences are more accurate than quantitative preferences elicited from individuals in uncertain information settings [19]. CP-nets are compact, arguably quite natural, intuitive in many circumstances, and widely used in many applications in computer science such as recommendation engines [8].

Real life scenarios are often dynamic. A user can change his mind over time or the system under consideration can change its laws. Preferences may change over time. Thus, we need a structure that can respond to change through updates, without the need to completely rebuild the structure. Additionally, we often meet situations characterized by some form of uncertainty. We may be uncertain about our preferences or on what features our preferences depend. In order to model this, we need a structure that includes probabilistic information. The need for encoding uncertain, qualitative information has seen some work in the recommendation engine area [7,16] and is a motivating example.

Consider a household of two people and their Netflix account. The recommendation engine only observes what movies are actually watched, what time they are watched,

S. Cranefield and A. Nayak (Eds.): AI 2013, LNAI 8272, pp. 301–312, 2013.

and their final rating. There are two people in this house and let us say that one prefers drama movies to action movies while the other has the opposite preference. When making a recommendation about what type of movie to watch, the engine may have several solid facts. Comedies may always be watched in the evening, so we can put a deterministic, causal link between time of day and type of movie. However, we cannot observe which user is sitting in front of the television at a given time. There is strong evidence from the behavioral social sciences showing that adding uncertainty to preference frameworks may be a way to reconcile transitivity when eliciting input from users [17], among other nice properties [12]. Using this idea, we add a probabilistic dependency between our belief about who is in front of the television and what we should recommend. We may want to update the probability associated with this belief based on the browsing or other real-time observable habits of the user. To do this we need a updateable and changeable structure that allows us to encode uncertainty.

We propose and study the complexity of reasoning with *PCP-nets*, for Probabilistic CP-nets, which allow for uncertainty and online modification of the dependency structure and preferences. PCP-nets provide a way to express probabilities over dependency links and probability distributions over preference orderings in conditional preference statements. Given a PCP-net, we show how to find the most probable optimal outcome. Additionally, since a PCP-net defines a probability distribution over a set of CP-nets, we also show how to find the most probable induced CP-net.

2 Background and Related Work

Probabilistic reasoning has received a lot of attention in Computer Science [8] and other areas [12]. Elicitation and modeling of preferences has also been considered in probabilistic domains such as POMDPs [3]. Recently, another generalization of CP-nets to include probabilities was introduced by Bigot et al. [1]. The model proposed by Bigot et al. restricts probabilities to be defined on orderings. We allow for probabilities on edges but, as we will show, this is a somewhat redundant specification that is useful for elicitation. Moreover, Bigot et al. focus on optimization and dominance testing in the special tractable case of tree-structured networks, we base our algorithmic approach on a more general connection with Bayesian networks. Reconciling these two models is an important direction for future work.

2.1 CP-Nets

CP-nets are a graphical model for compactly representing conditional and qualitative preference relations [2]. They exploit conditional preferential independence by decomposing an agent's preferences via the *ceteris paribus* (cp) assumption (all other things being equal). CP-nets bear some similarity to Bayesian networks (see 2.2). Both use directed graphs where each node stands for a domain variable, and assume a set of features $F = \{X_1, \dots, X_n\}$ with finite domains $\mathscr{D}(X_1), \dots, \mathscr{D}(X_n)$. For each feature X_i, each user specifies a set of *parent* features $Pa(X_i)$ that can affect her preferences over the values of X_i. This defines a *dependency graph* in which each node X_i has $Pa(X_i)$ as its immediate predecessors. Given this structural information, the user explicitly specifies her preference over the values of X_i for *each complete assignment* on $Pa(X_i)$. This preference is a total or partial order over $\mathscr{D}(X)$ [2].

Note that the number of complete assignments over a set of variables is exponential in the size of the set. Throughout this paper, we assume there is an implicit constant that specifies the maximum number of parent features, $|Pa(X)|$, that any feature may have. With this restriction, and an implicit bound on $|\mathscr{D}(X)|$, we can and do treat the size of the conditional preference representation for any X as a constant.

An *acyclic* CP-net is one in which the dependency graph is acyclic. A CP-net need not be acyclic. For example, my preference for the entree may depend on the choice of the main course, and my preference for the main course may depend on the choice of the entree. However in this paper we focus on acyclic CP-nets.

The semantics of CP-nets depends on the notion of a *worsening flip*. A worsening flip is a change in the value of a variable to a value which is less preferred by the cp-statement for that variable. We say that one outcome α is *better* than another outcome β (written $\alpha > \beta$) if and only if there is a chain of worsening flips from α to β. This definition induces a preorder over the outcomes.

In general, finding optimal outcomes and testing for optimality in this ordering is NP-hard. However, in acyclic CP-nets, there is only one optimal outcome and this can be found in as many steps as the number of features via a *sweep forward procedure* [2]. We sweep through the CP-net, following the arrows in the dependency graph and assigning at each step the most preferred value in the preference table. Each step in the sweep forward procedure is exponential in the number of parents of the current feature, and there are as many steps as features. In this paper we assume the number of parents is bounded, so this algorithm takes time polynomial in the size of the CP-net.

Determining if one outcome is better than another according to this ordering (called a dominance query) is NP-hard even for acyclic CP-nets [6,10]. Whilst tractable special cases exist, there are also acyclic CP-nets in which there are exponentially long chains of worsening flips between two outcomes.

2.2 Bayesian Networks

Bayesian networks (BNs) allow for a compact representation of uncertain knowledge and for a rigorous way of reasoning with this knowledge [15]. A BN is a directed graph where each node corresponds to a random variable; the set of nodes is denoted by V; a set of directed edges connects pairs of nodes (if there is an edge from node X to node Y, X is said to be a *parent* of Y); the graph has no directed cycles and hence is a directed acyclic graph (DAG); each node X_i has a conditional probability distribution $\mathbb{P}(X_i|Parents(X_i))$ that quantifies the effect of the parents on the node. If the nodes are discrete variables, each X_i has a *conditional probability table (CPT)* that contains the conditional probability distribution, $\mathbb{P}(X_i|Parents(X_i))$. Each CPT row must therefore have probabilities that sum to 1.

Inference in a BN corresponds to calculating $\mathbb{P}(X|E)$ where both X and E are sets of variables of the BN, or to finding the most probable assignment for X given E. The variables in E are called *evidence*.

There are three standard inference tasks in BNs: *belief updating*, which is finding the probability of a variable or set of variables, possibly given evidence; finding the *most probable explanation (MPE)*, that is, the most probable assignment for all the variables given evidence; and finding the *maximum a-posteriori hypothesis (MAP)*, where we are interested in a subset of m variables $A_1, \cdots, A_m$ and we want to compute the most probable assignment of $\{A_1, \cdots, A_m\}$ by summing over the values of all combinations of $V \setminus \{A_1, \cdots, A_m\} \cup E$, where E is a (possibly empty) set of evidence variables.

The inference tasks are computationally hard. However, they can be solved in polynomial time if we impose some restrictions on the topology of the BNs such as bounding the induced width [4, 5]. Given an ordering of the variables of a BN, these algorithms have a number of steps linear in the number of variables, and each step is exponential in the number of variables preceding the current one in the ordering and connected to it in the BN graph. The largest of these numbers is the induced width of the graph of the BN. Different variable orderings give steps with different complexity. Finding a good variable ordering is a difficult problem. If we assume the induced width is bounded, the overall algorithm is polynomial, and if $|Pa(X)|$ is bounded by a constant, then the induced width is also bounded.

3 Probabilistic CP-Nets

We define a generalization of traditional CP-nets with probabilities on individual cp-statements as well as on the dependency structure.We assume that the probabilities expressed over the dependency structure is consistent with the probabilities expressed over the variable orderings themselves. A model defined in this way allows us to use algorithms and techniques from BNs to efficiently compute outputs for common queries when the size of the dependency graph is bounded.

A PCP-net (for Probabilistic CP-net) is a CP-net where: **(1)** each dependency link is associated with a probability of existence consistent with the given variable ordering; and **(2)** for each feature A, instead of giving a preference ordering over the domain of A, we give a probability distribution over the set of all preference orderings for A.

More precisely, given a feature A in a PCP-net, its *PCP-table* is a table associating each combination of the values of the parent features of A to a probability distribution over the set of total orderings over the domain of A.

Probabilities expressed on the dependency links and the corresponding PCP-tables are not independent. If we consider all the possible ways in which we can obtain a CP-table from PCP-table by choosing specific orderings we see that we can divide the CP-tables into two classes: those representing a "true" dependency and those representing independence of the child feature. Each induced CP-table is associated to the joint probability of the orderings it contains. The probability of activation or non-activation of a dependency *must* coincide with the sum of probabilities associated to the CP-tables where the dependency is activated or not activated. Otherwise, the probability of the dependency and the probability of the ordering are not reconcilable and the structure itself expresses an impossible relationship.

Example 1. Consider the PCP-net $\mathscr{C}$ shown with two features, A and B, with domains $\mathscr{D}_A = \{a_1, a_2\}$ and $\mathscr{D}_B = \{b_1, b_2\}$. The preferences on B depend on the assignment to A with probability p. Given the probability assignment to the orderings of B given A we have that $p = q_1 \cdot (1 - q_2) + (1 - q_1) \cdot q_2$.

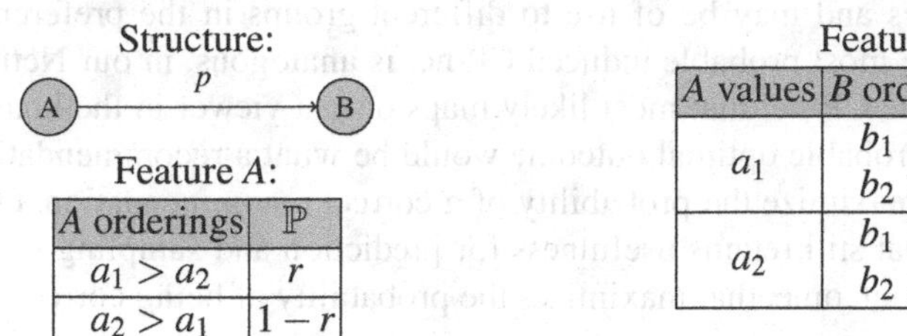

Feature A:

A orderings	$\mathbb{P}$
$a_1 > a_2$	r
$a_2 > a_1$	$1 - r$

Feature B:

A values	B orderings	$\mathbb{P}$
a_1	$b_1 > b_2$	q_1
	$b_2 > b_1$	$1 - q_1$
a_2	$b_1 > b_2$	q_2
	$b_2 > b_1$	$1 - q_2$

The induced CP-net with probability $\mathbb{P} = (1 - r) \cdot (1 - q_1) \cdot q_2$ is shown below.

Structure: A → B

Feature A:

A orderings
$a_2 > a_1$

Feature B:

A values	B orderings
a_1	$b_2 > b_1$
a_2	$b_1 > b_2$

Given a PCP-net $\mathscr{C}$, a *CP-net induced by* $\mathscr{C}$ has the same features and domains as $\mathscr{C}$. The dependency edges of the induced CP-net are a subset of the edges in the PCP-net which must contain all edges with probability 1. CP-nets induced by the same PCP-net may, therefore, have different dependency graphs. Moreover, the CP-tables are generated accordingly for the chosen edges. For each independent feature, one ordering over its domain (i.e., a row in its PCP-table) is selected. Similarly, for dependent features, an ordering is selected for each combination of the values of parent features. Each induced CP-net has an associated probability obtained from the PCP-net by taking the product of the probabilities of the deterministic orderings chosen in the CP-net.

One may note that the probabilities on edges are redundant whenever the probabilities in the PCP-tables are completely specified. However, we have chosen the presented formalism as it may be useful for elicitation purposes. Consider a settings where we are attempting to determine the strength of a relationship between two variables, such as the relationship between time of day and type of movie desired. It may be easier for people to describe this relationship directly rather than express the underlying joint probability distribution as humans are generally poor at estimating and working with probability directly [20]. Using this elicitation method we could then assume some underlying distribution for the variable ordering (skewed one way or another based on evidence). We leave an exploration of this topic for future work and focus on the base case, where PCP-nets are consistent, for the current work.

Since we have a probability distribution on the set of all induced CP-nets, it is important to be able to find the *most probable induced CP-net*. We are also interested in finding the *most probable optimal outcome*. Given a PCP-net and an outcome (that is, a value for each feature), the probability of such an outcome being optimal corresponds to the sum of the probabilities of the CP-nets that have that outcome as optimal.

4 Reasoning with PCP-Nets

Given a PCP-net we study mainly two tasks: finding the most probable induced CP-net and finding the most probable optimal outcome. These two reasoning tasks have slightly different semantics and may be of use to different groups in the preference reasoning community. The most probable induced CP-net is analogous, in our Netflix example from earlier, to the CP-net that most likely maps onto a viewer in the household. Whereas, the most probable optimal outcome would be what a recommendation engine should suggest to maximize the probability of a correct recommendation. One is an aggregated model, that still retains usefulness for prediction and sampling while the other is an aggregated outcome, that maximizes the probability of being correct.

4.1 The Most Probable Induced CP-Net

We reduce the problem of finding the most probable induced CP-net to that of finding an assignment with maximal joint probability of an appropriately defined BN.

Given a PCP-net $\mathscr{C}$, we define the BN called *general network*, or *G-net($\mathscr{C}$)*, associated with $\mathscr{C}$, as follows. We create a variable for each independent feature A of the PCP-net, with domain equal to the set of all possible total orderings over the domain of A. The probability distribution over the orderings is given by the PCP-table of A. For each dependent feature B of the PCP-net, we add as many variables to the G-net as there are combinations of value assignments to the parents. Each of these variables B_1 to B_n will have the same domain: the set of total orderings over the domain of B.

Consider the PCP-net with two features, $\mathscr{C}$, from Example 1 whose corresponding G-net is shown below. The variables have domains $\mathscr{D}_A = \{a_1 > a_2, a_2 > a_1\}$, $\mathscr{D}_{B_{a_1}} = \{b_1 > b_2, b_2 > b_1\}$, and $\mathscr{D}_{B_{a_2}} = \{b_1 > b_2, b_2 > b_1\}$.

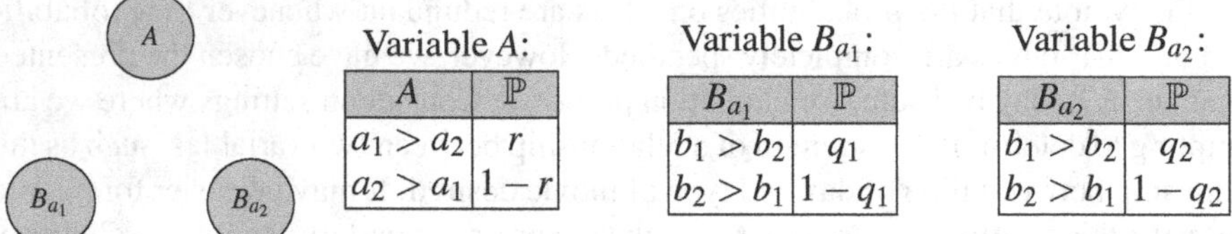

Variable A:

A	$\mathbb{P}$
$a_1 > a_2$	r
$a_2 > a_1$	$1-r$

Variable B_{a_1}:

B_{a_1}	$\mathbb{P}$
$b_1 > b_2$	q_1
$b_2 > b_1$	$1-q_1$

Variable B_{a_2}:

B_{a_2}	$\mathbb{P}$
$b_1 > b_2$	q_2
$b_2 > b_1$	$1-q_2$

Theorem 1. *Given a PCP-net $\mathscr{C}$ and the corresponding G-net N, there is a one-to-one correspondence between the assignments of N and the induced CP-nets of $\mathscr{C}$.*

Theorem 2. *Given a PCP-net $\mathscr{C}$, the probability of realizing one of its induced CP-nets $\mathscr{C}_i$, is the joint probability of the corresponding assignment in the G-net for $\mathscr{C}$.*

Proof. There is a one-to-one correspondence between rows in the PCP-tables and nodes in the G-net. Additionally, choosing a particular ordering in a PCP-net row corresponds to an assignment to a variable in the G-net. □

Theorem 3. *The probabilities over the induced CP-nets of a certain PCP-net form a probability distribution.*

Proof. The probability defined in Theorem 2 is computed as a product of non-negative factors, thus it is non-negative. Moreover, the sum of the probabilities of all the CP-nets in the set of the induced CP-nets is equal to 1, because there's a 1-1 correspondence between the assignments of the G-net with positive probability and the induced CP-nets, and the sum of the probabilities of all the assignments of a BN is equal to 1. □

Theorem 4. *Given a PCP-net $\mathscr{C}$ and its induced CP-nets, the most probable of the induced CP-nets is the variable assignment with maximal joint probability in the G-net for $\mathscr{C}$.*

4.2 The Most Probable Optimal Outcome

The most probable optimal outcome is the outcome that occurs with the greatest probability as the optimal in the set of induced CP-nets. The probability that an outcome o is optimal corresponds to the sum of the probabilities of the CP-nets that have o as the optimal outcome. Observe that the most probable optimal outcome may not be the optimal outcome of the most probable CP-net. Consider a PCP-net with only one feature A with domain $\mathscr{D}_A = \{a_1, a_2, a_3\}$ and let $a_1 > a_2 > a_3 = 0.3$, $a_1 > a_3 > a_2 = 0.3$, and $a_3 > a_2 > a_1 = 0.4$. The most probable CP-net is the one corresponding to the third ordering and it has the optimal outcome a_3. The other CP-nets have a_1 as optimal, so $\mathbb{P}(a_1) = 0.6$ and $\mathbb{P}(a_3) = 0.4$. The most probable optimal outcome is therefore a_1 but the optimal outcome of the most probable CP-net is a_3.

To find the most probable optimal outcome, we cannot find the most probable induced CP-net by the G-net procedure described above and then find its optimal outcome; we must make use of another BN which we call the *optimal network*.

Given a PCP-net $\mathscr{C}$, the *optimal network (Opt-net)* for $\mathscr{C}$ is a BN with the same dependencies graph as $\mathscr{C}$. Thus, the Opt-net has a variable for each of the PCP-net's features. The domains of the variables in the Opt-net are the values of the corresponding features that are ranked first in at least one ordering with non-zero probability. The conditional probability tables of the Opt-net are obtained from the corresponding PCP-tables as follows: for each assignment of the parent variables, we consider the corresponding probability distribution over the values of the dependent variable defined in the PCP-table. The probability of a value for the dependent variable is the sum of the probabilities of the orderings that have that particular value as most preferred according to that distribution. Notice that our construction applies even when there are cyclic dependences in the corresponding PCP-net.

Example 2. Consider the PCP-net $\mathscr{C}$ with three features A, B and C with domains $\mathscr{D}_A = \{a_1, a_2\}$, $\mathscr{D}_B = \{b_1, b_2\}$ and $\mathscr{D}_C = \{c_1, c_2, c_3\}$.The Opt-net has the same dependency graph as $\mathscr{C}$, with three variables A, B and C with domains: $\mathscr{D}_A = \{a_1, a_2\}$, $\mathscr{D}_B = \{b_1, b_2\}$ and $\mathscr{D}_C = \{c_1, c_2\}$, and two edges AC and BC. The domain of variable C in the Opt-net does not contain value c_3 because it never appears as most preferred in any ordering. Therefore, the Opt-net has a table for entry $a_1 b_2$ where c_1 appears with probability 0.2 and c_2 appears with probability 0.8.

Structure:

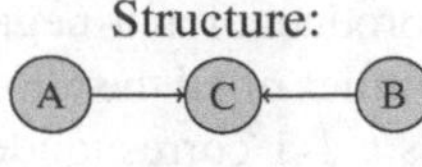

Variable A:

ordering for A	$\mathbb{P}$
$a_1 > a_2$	0.8
$a_2 > a_1$	0.2

Variable B:

ordering for B	$\mathbb{P}$
$b_1 > b_2$	0.7
$b_2 > b_1$	0.3

Variable C:

A & B	ordering for C	$\mathbb{P}$
$a_1\ b_1$	$c_1 > c_2 > c_3$	0.3
	$c_2 > c_1 > c_3$	0.7
$a_1\ b_2$	$c_1 > c_2 > c_3$	0.2
	$c_2 > c_1 > c_3$	0.4
	$c_2 > c_3 > c_1$	0.4
$a_2\ b_1$	$c_1 > c_3 > c_2$	0.4
	$c_2 > c_3 > c_1$	0.6
$a_2\ b_2$	$c_1 > c_2 > c_3$	0.1
	$c_2 > c_1 > c_3$	0.9

Theorem 5. *Given a PCP-net $\mathscr{C}$ and its Opt-net, there is a one-to-one correspondence between the assignments (with non-zero probability) of the Opt-net and the outcomes that are optimal in at least one induced CP-net of $\mathscr{C}$.*

Theorem 6. *Given a PCP-net $\mathscr{C}$, the probability that an outcome is optimal is the joint probability of the corresponding assignment in the optimal network. If no such corresponding assignment exists, then the probability of being optimal is 0.*

Proof. By construction, the set of assignments of the Opt-net of $\mathscr{C}$ is a subset of those of $\mathscr{C}$. By the definition of the Opt-net, if an assignment of $\mathscr{C}$ is not an assignment of the Opt-net, then it cannot be optimal in any induced CP-net.

Let us now focus on the assignments of $\mathscr{C}$ that have a corresponding assignment in the Opt-net. Let $x = (x_1, x_2, ..., x_n)$ be one of these assignments. We denote by $P_{opt}(x)$ the joint probability of x, $\mathbb{P}(X_1 = x_1, ..., X_n = x_n)$ in the Opt-net. We recall that the probability that x is optimal in the PCP-net is the sum of the probabilities of the induced CP-nets that have assignment x as optimal. We call this probability $P_{cp}(x)$. We must prove that $P_{opt}(x) = P_{cp}(x)$.

Let us consider A_x, the set of induced CP-nets that have x as their optimal assignment; giving $P_{cp}(x) = \sum_{\mathscr{C} \in A_x} \mathbb{P}(\mathscr{C})$. When we compute the optimal value for a CP-net, we sweep forward, starting from the independent features, assigning features their most preferred value. This means that only one subset of the rows of the CP-tables is considered when computing the optimal outcome. We can thus split a CP-net $\mathscr{C}$ into two parts, one affecting the choice of the optimal outcome (denoted with $\mathscr{C}_*$) and one not involved in it (denoted with $\mathscr{C}_{-*}$). If we consider the probability that that CP-net is induced by the PCP-net, we see that these two parts are independent. Thus we have $P_{cp}(x) = \sum_{\mathscr{C} \in A_x} \mathbb{P}(\mathscr{C}) = \sum_{\mathscr{C} \in A_x} \mathbb{P}(\mathscr{C}_*)\mathbb{P}(\mathscr{C}_{-*})$.

Regarding $\mathscr{C}_*$, observe that the optimal outcome x can be produced in many different ways, as there can be many different orderings that produce the same result. For example the orderings $a_1 > a_2 > a_3$ and $a_1 > a_3 > a_4$ produce both the optimal value a_1 for variable X_1. So we can do a disjoint partition of the set A_x into k subsets $A_{x_1}, ..., A_{x_k}$ for some k.

Two CP-nets $\mathscr{C}$ and $\mathscr{D}$ that belong to the same A_{x_i} are equal in the part that actively affects the choice of the optimal value and different in the other parts: $\mathscr{C}_* = \mathscr{D}_*$ and $\mathscr{C}_{-*} \neq \mathscr{D}_{-*}$.

Let $\mathscr{C}_*^i$ be the part that is equal for all the members of A_{x_i}. The probability becomes: $P_{cp}(x) = \sum_{i=1}^{k} \mathbb{P}(\mathscr{C}_*^i) \sum_{\mathscr{C} \in A_{x_i}} \mathbb{P}(\mathscr{C}_{-*})$. We note that $\sum_{\mathscr{C} \in A_{x_i}} \mathbb{P}(\mathscr{C}_{-*}) = 1 \; \forall i = 1,...,k$, since we are summing the probability of all possible cases regarding $\mathscr{C}_{-*}$. Thus the probability becomes $P_{cp}(x) = \sum_{i=1}^{k} \mathbb{P}(\mathscr{C}_*^i)$. However, we have $\mathbb{P}(X_1 = x_1,...,X_n = x_n) = \sum_{i=1}^{k} \mathbb{P}(\mathscr{C}_*^i)$ and, thus, $P_{cp}(x) = P_{opt}(x)$. since we built the rows of the probability tables for the variables $X_1,...,X_n$ by summing the probability of the orderings that have the same head. This is the same as summing the probabilities over the subset A_{x_i}. □

Theorem 7. *To find the most probable optimal outcome for a PCP-net $\mathscr{C}$, it is sufficient to compute the assignment with the maximal joint probability of its optimal network.*

5 PCP-Nets and Induced CP-Nets

A PCP-net defines a probability distribution over a set of induced CP-nets. However, this step is not always reversible: below we show that, given a probability distribution over a set of CP-nets, all with the same features and domains, there may be no PCP-net such that the given CP-nets are its induced CP-nets. However, the function that maps a PCP-net to its set of induced CP-nets is injective. Therefore, if there is a PCP-net which induces a set of CP-nets, we can find it quickly. This observation may be an interesting starting point for future work. We may be able to use CP-nets elicited from individuals to generate a PCP-net with which to "hot start" and create highly probable configurations for a recommendation system that is responsible for suggesting configurations for products to new customers [7, 16].

Theorem 8. *Given a probability distribution over a set of CP-nets (even if they have the same dependency graph), there may exist no PCP-net inducing it.*

Proof. Consider the following four CP-nets ($\mathscr{C}_1$, $\mathscr{C}_2$, $\mathscr{C}_3$ and $\mathscr{C}_4$) defined on the same variables: A and B. The two features have domains $\mathscr{D}_A = \{a_1, a_2\}$ and $\mathscr{D}_B = \{b_1, b_2\}$. The probability distribution on the four CP-nets is defined as follows:

- $\mathscr{C}_1$ has probability $\mathbb{P}(\mathscr{C}_1) = 0.3$ and CP-tables:

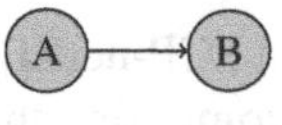

ordering for A
$a_1 > a_2$

A	ordering for B
a_1	$b_1 > b_2$
a_2	$b_2 > b_1$

- $\mathscr{C}_2$ has probability $\mathbb{P}(\mathscr{C}_2) = 0.2$ and CP-tables:

ordering for A
$a_1 > a_2$

ordering for B
$b_1 > b_2$

- $\mathscr{C}_3$ has probability $\mathbb{P}(\mathscr{C}_3) = 0.1$ and CP-tables:

ordering for A
$a_1 > a_2$

ordering for B
$b_2 > b_1$

- $\mathscr{C}_4$ has probability $\mathbb{P}(\mathscr{C}_4) = 0.4$ and CP-tables:

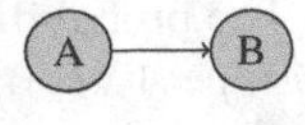

ordering for A
$a_1 > a_2$

A	ordering for B
a_1	$b_2 > b_1$
a_2	$b_1 > b_2$

If $\mathscr{C}_1, \mathscr{C}_2, \mathscr{C}_3$ and $\mathscr{C}_4$ were all the induced CP-nets of a PCP-net, this PCP-net would have the dependency graph (on the features A and B with the relationship having probability k of occurring):

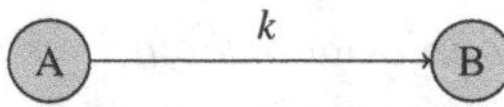

and the following PCP-tables:

ordering for A
$a_1 > a_2$

A	ordering for B	probability
a_1	$b_1 > b_2$	p_1
	$b_2 > b_1$	p_2
a_2	$b_1 > b_2$	p_3
	$b_2 > b_1$	p_4

where the values p_1, p_2, p_3 and p_4 need to be solutions of the following system of equations:

$$\begin{cases} p_1 p_3 = 0.2 \\ p_2 p_3 = 0.4 \\ p_1 p_4 = 0.3 \\ p_2 p_4 = 0.1 \end{cases} \quad \begin{cases} 0 \leq p_1 \leq 1 \\ 0 \leq p_2 \leq 1 \\ 0 \leq p_3 \leq 1 \\ 0 \leq p_4 \leq 1 \end{cases} \quad \begin{cases} p_1 + p_2 = 1 \\ p_3 + p_4 = 1 \end{cases}$$

However, such a system has no solution. □

Theorem 9. *Given a probability distribution over a set of CP-nets, we can compute a PCP-net to fit this distribution, if it exists.*

6 Updating Probabilistic CP-Nets

We now turn our attention to modifications to the structure of a PCP-net. These changes can be implemented in an efficient way and their effects on computing the most probable optimal outcome and the most probable induced CP-net are minimal, in terms of complexity. Modifying the structure of the PCP-net is similar to entering evidence in a BN framework. By changing an arc or setting an ordering for a variable we can fix parts of the probability distribution and compute the outcomes of the resulting structure.

To add or delete a dependency or feature we just update the respective probability tables. This may involve deleting redundancy when we delete a feature. Due to the independence assumptions, we can modify probabilities over ordering and features at a local level, with no need to recompute the entire structure when new information is added.

When we modify a PCP-net $\mathscr{C}$ we also need to modify the probability tables in the associated G-net. This can change the most probable induced CP-net and therefore we need to recompute the outcome of the G-net.

To add or delete a dependency or feature, independent or dependent, we need to add or delete (or both) a number of nodes in the G-net which is exponential in the maximum number of parents which we assume to be bounded. The same can be said with respect to updating a probability table with either evidence or changing the distribution.

When we modify a PCP-net $\mathscr{C}$, the changes affect its Opt-net. Consider the dependency of feature B on feature A. When we add or delete this dependency, or when we change its probability, we only need to recompute the probability table of B in the Opt-net. When computing the most probable optimal outcome, we note that, in the worst case, we must recompute the whole maximal joint probability of the Opt-net. The same can be said when we delete a feature, as this amounts to the deletion of a set of dependencies, or when we modify the probability distribution over the orderings on B for a specific assignment to all of its parents. When we add a feature A to $\mathscr{C}$, we must add the corresponding node in the Opt-net and generate the corresponding probability table. This new node is independent. Thus, revising the current most probable optimal outcome is easy: the new optimal is the current one extended with the optimal value of the new feature.

7 Conclusions and Future Work

We have defined and shown how to reason with a generalized version of CP-nets, called PCP-nets, which can model probabilistic uncertainty and be updated without recomputing their entire structure. We have studied how to reason with these new structures in terms of optimality. PCP-nets can be seen as a way to bring together BNs and CP-nets, thus allowing to model preference and probability information in one unified structure.

We plan to study dominance queries and optimality tests in PCP-nets, as well as to study appropriate eliciting methods for both preferences and probabilities. Bigot et al. [1] have begun this line of inquiry on their model and show that, for PCP-nets that have a tree structure, dominance testing is tractable. We would also like to further explore, as Bigot et al., how our results related to the notion of local Condorcet winners in CP-net aggregation [21] as well as other issues in CP-net aggregation such as bribery [13, 14] and joint decision making [11]. Additionally, we have made several assumptions to bound the reasoning complexity of PCP-nets; we would like to relax these bounds or obtain results about approximability when these assumptions are lifted. We also plan to consider the use of PCP-nets in a multi-agent setting, where classical CP-nets have already been considered [18]. In this setting, PCP-nets can be used to represent probabilistic information on the preferences of a population.

Acknowledgments. This work is supported by the National Science Foundation, under grants CCF-1049360 and IIS-1107011. Any opinions, findings, and conclusions or recommendations expressed in this material are those of the authors and do not necessarily reflect the views of the National Science Foundation.

NICTA is funded by the Australian Government through the Department of Broadband, Communications and the Digital Economy and the Australian Research Council (ARC) through the ICT Centre of Excellence program. This research is also funded by AOARD grant 124056.

References

1. Bigot, D., Fargier, H., Mengin, J., Zanuttini, B.: Probabilistic conditional preference networks. In: Proc. 29th Conf. on Uncertainty in Artificial Intelligence, UAI (2013)
2. Boutilier, C., Brafman, R., Domshlak, C., Hoos, H., Poole, D.: CP-nets: A tool for representing and reasoning with conditional ceteris paribus preference statements. Journal of Artificial Intelligence Research 21, 135–191 (2004)
3. Boutilier, C.: A POMDP formulation of preference elicitation problems. In: Proc. 18th AAAI Conference on Artificial Intelligence, pp. 239–246 (2002)
4. D'Ambrosio, B.: Inference in Bayesian Networks. AI Magazine 20(2), 21 (1999)
5. Dechter, R.: Bucket elimination: A unifying framework for reasoning. Artificial Intelligence 113(1-2), 41–85 (1999)
6. Domshlak, C., Brafman, R.: CP-nets: Reasoning and consistency testing. In: Proc. 8th Intl. Conf. on Principles and Knowledge Representation and Reasoning, KRR (2002)
7. Faltings, B., Torrens, M., Pu, P.: Solution generation with qualitative models of preferences. Computational Intelligence 20(2), 246–263 (2004)
8. Fürnkranz, J., Hüllermeier, E.: Preference Learning: An Introduction. Springer (2010)
9. Goldsmith, J., Junker, U.: Preference handling for artificial intelligence. AI Magazine 29(4) (2009)
10. Goldsmith, J., Lang, J., Truszczynski, M., Wilson, N.: The computational complexity of dominance and consistency in CP-nets. Journal of Artificial Intelligence Research 33(1), 403–432 (2008)
11. Maran, A., Maudet, N., Pini, M.S., Rossi, F., Venable, K.B.: A framework for aggregating influenced CP-nets and its resistance to bribery. In: Proc. 27th AAAI Conference on Artificial Intelligence (2013)
12. Marden, J.I.: Analyzing and Modeling Rank Data. CRC Press (1995)
13. Mattei, N., Pini, M.S., Rossi, F., Venable, K.B.: Bribery in voting over combinatorial domains is easy. In: Proc. 11th Intl. Joint Conf. on Autonomous Agents and Multiagent Systems, AAMAS (2012)
14. Mattei, N., Pini, M.S., Rossi, F., Venable, K.B.: Bribery in voting with CP-nets. Annals of Mathematics and Artificial Intelligence (2013)
15. Pearl, J.: Probabilistic Reasoning in Intelligent Systems: Networks of Plausible Inference. Morgan Kaufmann (1988)
16. Price, R., Messinger, P.R.: Optimal recommendation sets: Covering uncertainty over user preferences. In: Proc. 20th AAAI Conference on Artificial Intelligence, pp. 541–548 (2005)
17. Regenwetter, M., Dana, J., Davis-Stober, C.P.: Transitivity of preferences. Psychological Review 118(1) (2011)
18. Rossi, F., Venable, K., Walsh, T.: mCP nets: representing and reasoning with preferences of multiple agents. In: Proc. 19th AAAI Conference on Artificial Intelligence, pp. 729–734 (2004)
19. Roth, A.E., Kagel, J.H.: The handbook of experimental economics, vol. 1. Princeton University Press, Princeton (1995)
20. Tversky, A., Kahneman, D.: Judgement under uncertainty: Heuristics and biases. Science 185, 1124–1131 (1974)
21. Xia, L., Conitzer, V., Lang, J.: Voting on multiattribute domains with cyclic preferential dependencies. In: Proc. 23rd AAAI Conference on Artificial Intelligence, pp. 202–207 (2008)

Some Complexity Results for Distance-Based Judgment Aggregation

Wojciech Jamroga[1] and Marija Slavkovik[2]

[1] University of Luxembourg, Luxembourg
wojtek.jamroga@uni.lu
[2] University of Bergen, Norway
marija.slavkovik@infomedia.uib.no

Abstract. Judgment aggregation is a social choice method for aggregating information on logically related issues. In distance-based judgment aggregation, the collective opinion is sought as a compromise between information sources that satisfies several structural properties. It would seem that the standard conditions on distance and aggregation functions are strong enough to guarantee existence of feasible procedures. In this paper, we show that it is not the case, though the problem becomes easier under some additional assumptions.

1 Introduction

It is often convenient to ascribe information-related stances (such as judgments, opinions, beliefs, etc.) to collectives of agents. The need for modeling collective opinions can be either external or internal. External agents may ascribe opinions to institutions and groups in order to simplify their model of the world and reason about it. Agents inside the group may need to reach consensus about issues of interest, and in particular to obtain collective decisions that will lead to consistent collective action. In this paper, we focus on one of the formal frameworks that try to explain how collective judgments are formed from individual stances, namely *judgment aggregation theory* [33].

Several formal theories within artificial intelligence have tried to explain how collective judgments arise from individual judgments. Epistemic logic [41] proposes to aggregate agents' views by aggregating the underlying models, i.e., indistinguishability relations over different valuations of atomic sentences. By different operations on epistemic relations we obtain different notions of group knowledge: mutual knowledge, common knowledge, distributed knowledge etc. [24]. On the other hand, Dempster-Shafer theory [9, 39] shows how probabilistic beliefs can be merged into a single collective belief. However, epistemic logic requires *complete* individual views, that is, everybody's opinions about *every* conceivable state of the world must be given as input. Dempster rule of combination admits incomplete models, but may yield logically inconsistent judgments, i.e., ones that violate logical interdependencies between issues (even if the input consists of consistent individual judgments). Thus, both theories make assumptions that turn out too strong for most cases of practical reasoning.

S. Cranefield and A. Nayak (Eds.): AI 2013, LNAI 8272, pp. 313–325, 2013.

Another formal framework is provided by social choice theory. It develops and analyses (on a more abstract level) methods for reaching group decisions through aggregating individual stances. For instance, voting rules used in political elections are social choice methods. Deriving collective opinions from a partial representation of individual judgments on a set of mutually dependent issues may be obtained by a method of similar kind. More precisely, the problem of aggregating binary valuations assigned to each element of a set of logically related elements into a consistent set of valuations is studied by *judgment aggregation theory* [33].

Distance-based judgment aggregation [34, 38] comprises the largest class of judgment aggregation rules. Inspired by belief merging rules, the idea is to define the collective opinion as a "well-behaved compromise" among the individual opinions of the group members. That is, we assume that each member is willing to give up some of their judgments as long as the resulting aggregate judgment set does not stray too far from their individual ones. Distance-based aggregation rules are supposed to satisfy a number of structural constraints (see Section 3 for details) to make sure their output is indeed "well-behaved" in the mathematical sense. It seems – at least at the first glance – that the constraints should lead to computationally well-behaved procedures. In this paper, we show that it is not necessarily true.

Why is computational complexity important for aggregating judgments? Essentially, judgment aggregation provides an intuitive representation for decision problems in collective reasoning. In this context, its computational complexity is crucial. More specifically, judgment aggregation rules are *procedures* that determine the collective view based on individual inputs. The procedure is only useful if it returns the result in reasonable time. This is perhaps not that crucial in case of a jury consisting of 10 members and deliberating over 5 connected issues. Consider, however, a team of 100 robots reaching a collective decision based on the input from 400 sensors with different (but overlapping) range, or 500 stakeholders trying to agree on a company agenda. Scalability of the procedure becomes clearly of utmost importance.

The paper is structured as follows. In Section 3 we give the necessary definitions for a judgment aggregation problem and distance-based aggregation rules. In Section 4 we consider the problem of *verifying* whether a particular set of judgments can be selected as collective, for a collection of individual judgments, by a distance-based judgment aggregation rule. We also extend our results to aggregation of opinions expressed in multi-valued logics. In Section 5 we present our conclusions.

2 Related Work

Our paper fits in the area of *computational social choice* [6] which comprises interdisciplinary study of how computational analysis can be used to make social choice methods operational. Many contributions, e.g. [1, 7, 8, 23], have been made towards understanding the complexity-theoretic properties of voting rules.

In comparison, complexity-theoretic properties of judgment aggregation are not so well explored.

Complexity analysis of distance-based judgment aggregation has, to the best of our knowledge, been focused on analysis of *particular* aggregation rules. The following papers have addressed the complexity of judgment aggregation procedures: [2–4, 20–22]. Out of these works, [2–4] focus on the complexity and parameterized complexity of decision problems related to control and bribery in *quota judgment aggregation rules* which generalize issue-by-issue majority judgment aggregation [12].. [20] studies the complexity of deciding whether the so called *premise-based judgment aggregation rule* (a special type of quota rule [14]) can be applied to a given judgment aggregation problem. [21] investigates the complexity of deciding if a given judgment is selected by two alternative rules: the quota rule and the most "typical" distance-based aggregation rule that uses the sum of Hamming distances to compute the "score" for each judgment set. The work [22] gathers and deepens the results of [21] and [20].

Summarizing, complexity-theoretic properties of judgment aggregation are only partially explored. This applies especially to distance-based judgment aggregation, where the only existing studies refer to particular "natural" judgment aggregators, mainly based on the sum of Hamming distances. In contrast, *we take the opposite approach and explore the bounds of the framework*. That is, we investigate what kind of complexity can be expected from *arbitrary* distance-based aggregation rules.

Besides papers that explicitly refer to the complexity of judgment aggregation procedures, we must also mention works on complexity of distance-based belief merging [28] and especially distance-based preference aggregation [1, 18, 19].

Relation to Research on Preference Aggregation. The research on complexity in preference aggregation connects to the research on complexity of distance-based judgment aggregation through the result of [19] where it was shown that the Kemeny rule of voting coincides, for strict preference orders, with judgment aggregation based on the sum of Hamming distances. The complexity of the winner determination problem for the Kemeny preference aggregation rule, has been studied in [1] and [27], the latter proving it to be Θ_2^P complete.

It has been demonstrated that judgment aggregation is related to preference aggregation by showing when a preference aggregation problem can be translated to a judgment aggregation problem and vice versa [11, 26, 32]. Studies that formally establish the relationship between judgment aggregation rules and voting rules (or preference aggregation rules) on the general level are only now starting to be pursued [30], despite a number of discussions on the topic [13, 29, 35]. The general relationship between the complexity properties of preference aggregation rules and the complexity properties of the judgment aggregation rules that generalize them, is the next research step. We present some preliminary intuitions.

A judgment set can be used to characterize a strict preference order [11] by using a formula φ_b^a to represent that alternative a is preferred to alternative b. In complexity of preference aggregation, one is typically interested in the winner determination problem, that is, the problem of deciding whether an alternative is top ranked in at least one of preference orders produced by the preference aggregation rule. Considering only aggregation of strict preferences and following the analogy that a preference order is a judgment set, an alternative in preference aggregation corresponds to a judgment, and the winner determination problem can be interpreted as that of determining whether a particular judgment φ_b^a is a part of the collective judgment set produced by the judgment aggregation rule. The difficulty lies in the fact that a judgment aggregation rule can produce multiple collective judgment sets, some containing φ_b^a and some not. Therefore two different meaningful questions can be studied: (1) whether a judgment set as a whole can be selected as the collective opinion, corresponding to our definition of the winner set verification problem in Section 4, or (2) whether a given judgment is an element of all collective opinions, as in [22]. For preference aggregation, (1) corresponds to checking if a preference order is selected by the preference aggregation rule, while (2) is about determining whether a given alternative is highest ranked in all selected preference orders. Both decision problems are at least as hard as the problem of deciding whether an alternative is a winner of the election. Therefore we can expect decision problems in judgment aggregation to be no easier than their counterparts in preference aggregation.

Relations with Belief Merging. Judgment aggregation has been related with belief merging [38]. Both theories are concerned with aggregating sets of formulas, however the demands on the aggregation results are different. In judgment aggregation, the agenda limits the scope of issues whose consistent aggregated truth-value is of interest. In belief merging, the agenda does not exist. The interest focus in merging is on determining, not sets of formulas like in judgment aggregation, but the (closed under deduction) set of formulas that are logically entailed by the sets of formulas being merged. The computational complexity analysis in belief merging is concerned with the decision problem of whether one particular formula (judgment) is entailed by a given collection of belief sets [28].

3 Preliminaries

We first give a brief exposition of judgment aggregation and distance-based judgment aggregation.

3.1 Judgment Aggregation

Let $\mathcal{L}$ be a propositional language over a countable set of atomic propositions *Prop*, and let T be a set of truth values such that $1 \in T$ (i.e., it includes the value for "absolutely true"). Any $v : Prop \to T$ is called a propositional valuation; we denote the set of valuations as PV. Each $v \in PV$ extends to a valuation $val_v : \mathcal{L} \to T$ for all formulae of $\mathcal{L}$. In most of the paper we will assume that $\mathcal{L}$ is the language of classical propositional logic, $T = \{0, 1\}$, and val_v is defined by the classical Boolean semantics of negation, conjunction, etc.

Judgment aggregation can be defined as follows.[1] Let N be a finite set of *agents*, $\mathcal{A} \subseteq \mathcal{L}$ a finite *agenda* of issues, and $\mathcal{C} \subseteq \mathcal{L}$ a finite set of *admissibility constraints*. A *judgment set* is a consistent and admissible combination of opinions on issues from $\mathcal{A}$, that is, some $js : \mathcal{A} \to T$ for which there exists a valuation $v \in PV$ such that: (i) $val_v(\varphi) = js(\varphi)$ for every $\varphi \in \mathcal{A}$, and (ii) $val_v(\psi) = 1$ for every $\psi \in \mathcal{C}$. The set of all judgment sets is denoted by JS. Now, a *judgment profile* is a collection of judgment sets, one per agent, i.e., $jp : N \to JS$. With a slight abuse of notation, we will denote the set of all such profiles by $JS^{|N|}$. Note that we can conveniently represent judgment profiles as $|Agt| \times |\mathcal{A}|$ matrices of elements from T. Finally, a *judgment aggregation rule* $\nabla : JS^{|N|} \to \mathcal{P}(JS) \setminus \{\emptyset\}$ aggregates opinions from all the agents into a collective judgment set (or sets). We allow for more than one "winning" set to account for nondeterministic or inconclusive aggregation rules.

Example 1. *Consider 3 robots guarding a building, that have just observed a person. Each robot must assess whether the person is authorized to be there (proposition p_1), if it has malicious intent (p_2), and whether to classify the event as dangerous intrusion (p_3). Additionally, it is assumed that a non-authorized person with malicious intent implies intrusion: $\neg p_1 \wedge p_2 \to p_3$ (note that the converse does not have to hold). A possible judgment profile is shown in Figure 1. The figure also shows that the most "obvious" aggregation rule (majority) results in an inadmissible judgment set.*

	p_1	p_2	p_3
robot 1	1	1	0
robot 2	0	0	0
robot 3	0	1	1
majority	0	1	0

Fig. 1. Guarding robots. $N = \{1, 2, 3\}$, $\mathcal{A} = \{p_1, p_2, p_3\}$, $\mathcal{C} = \{\neg p_1 \wedge p_2 \to p_3\}$.

In case of binary (yes/no) judgments, this is equivalent to representing opinions as consistent and complete sets of propositional formulas. For example, the view of robot 1 in the Example 1 can be represented by the set $\{p_1, p_2, p_3\}$, the judgment set of robot 2 is $\{\neg p_1, \neg p_2, \neg p_3\}$, and for robot 3 it becomes $\{\neg p_1, p_2, p_3\}$. Issue-by-issue majority rule aggregates the sets into $\{\neg p_1, p_2, \neg p_3\}$ which is inconsistent with the constraint $\neg p_1 \wedge p_2 \to p_3$. Three-valued judgments can be modeled analogously by assuming that the third value is in place for p_i when neither p_i nor $\neg p_i$ occurs in the set (obviously, a set of judgments is then only required to be consistent but not necessarily complete). Representing judgments with more than 3 truth values by sets of formulas is not straightforward anymore.

There are two natural computational problems related to judgment aggregation: computing a "winning" judgment set and verifying that a judgment set is one of the winner sets. We look closer at the latter problem in Section 4.

3.2 Distance-Based Aggregation Rules

A distance-based aggregation rule [34, 38] looks for a collective opinion that does not stray too much from the individual judgments: Formally, such a rule

[1] Our definition of judgment aggregation combines features of logic-based aggregation [33] and algebraic aggregation [42]. It is easy to see that both formulations can be expressed in our notation.

is defined as $\nabla_{d,aggr}(jp) = \text{argmin}_{js \in JS} \{aggr(d(js, jp[1]), \ldots, d(js, jp[|N|]))\}$, where d is a *distance function* [10, p.3-4 and 45], and *aggr* an *aggregation function* [25, p.3], cf. the definitions below.

Definition 1. *An* algebraic aggregation *is a function* $aggr : (\mathbb{R}^+)^n \to \mathbb{R}^+$ *such that:* ***(minimality)*** $aggr(0^n) = 0$, *and* ***(non-decreasing)*** *if* $x \leq y$, *then* $aggr(x_1, \ldots, x, \ldots, x_n) \leq aggr(x_1, \ldots, y, \ldots, x_n)$.

Definition 2. *A* distance *over set* X *is a function* $d : X \times X \to \mathbb{R}^+ \cup \{0\}$ *such that:* ***(minimality)*** $d(x, y) = 0$ *iff* $x = y$, ***(symmetry)*** $d(x, y) = d(y, x)$, *and* ***(triangle inequality)*** $d(x, y) + d(y, z) \geq d(x, z)$.

Well known aggregators are: min, max, sum, and product. Well known distances are the Hamming distance $d_H(x, y) = \sum_{i=1}^{m} \delta_H(x[i], y[i])$, and the drastic distance $d_D(x, y) = \max_{i=1}^{m} \delta_H(x[i], y[i])$, while $\delta_H(x, y) = 0$ if $x = y$ and 1 otherwise.

In belief-merging, it is not required that d satisfies triangle inequality, d is a pseudo-distance, but the only two concrete distances used in belief merging, the Hamming and drastic distance, satisfy it. How necessary this property is in judgment aggregation, is not well studied, but since we do not know of d's that are not distances, we decided to use distances within the scope of this paper.

Example 2. *Consider the robots from Example 1, and let us use* d_H *as the distance and* $\sum$ *as the aggregator. Then, the winner sets are* $\{000, 011, 110\}$, *all with score (i.e., aggregate distance)* 3. *In other words, the agents cannot do better than to accept one of their individual opinions.*

4 Verification of Collective Opinions in Distance-Based Judgment Aggregation

In computational social choice various complexity-theoretic aspects of voting theory are studied, such as how difficult it is to find a winner of elections or how difficult it is to manipulate an election. There are two natural computational problems related to judgment aggregation: the function problem of computing a "winning" judgment set, and the decision problem of verifying that a given judgment set is one of the winner sets. We look closer at the latter.

4.1 Winner Set Verification

In judgment aggregation the "winner" of an aggregation is the resulting collective opinion, i.e., a set of judgments. Consequently one can consider complexity issues from the stance of a judgment on a particular issue, but also from the stance of an entire set of judgments. If one is concerned with particular judgments, then the interesting complexity-theoretic one-judgment question to study is: how complex is it to determine if a judgment value $t \in T$ was assigned to issue $a \in \mathcal{A}$. This stance is taken in the complexity analysis of [21]. A similar stance, of whether a given belief is included in the merging result of belief bases, is taken when studying the complexity-theoretic properties of belief merging [28]. We adopt a different approach, and look at the verification problem for a given complex opinion, i.e., a judgment set.

We begin by defining formally the problem of *winner set verification.* Then, we investigate the "absolute" complexity that one may face in distance-based aggregation. It turns out that the problem is undecidable in general. On the other hand, the problem becomes more feasible under some reasonable restrictions on the distance and algebraic aggregation functions. Finally, we determine the complexity of winner set verification for some natural aggregators.

The winner set verification problem for agenda $\mathcal{A}$, set of constraints $\mathcal{C}$, logic L and a rule $\nabla^{d,aggr}$, is defined as follows.

Definition 3. WINVER$_\nabla$ *is the decision problem defined as follows:*
Input: *Agents N, agenda $\mathcal{A}$, constraints $\mathcal{C}$, judgment profile $jp \in JS^{|N|}(\mathcal{A},\mathcal{C})$, and judgment set $js \in JS(\mathcal{A},\mathcal{C})$.*
Output: *true if $js \in \nabla(jp)$, else false.*

What is the complexity of WINVER? One could expect that, under the assumptions in Definitions 1 and 2, distance-based aggregation should behave reasonably in computational terms. Unfortunately, it is not the case.

4.2 Negative Results

Theorem 1. *There is a distance which is not Turing computable.*

Proof. We construct the *Turing distance* d_{TR} as follows. First, we assume a standard encoding of Turing machines in binary strings; we use $TM(X)$ to refer to the machine represented by the string of bits $X \in \{0,1\}^m$. We also assume by convention that strings starting with 0 or ending with 1 represent only machines that always halt (e.g., they can represent various TM's with only accepting states).

Let $halts(X) = 0$ if the $TM(X)$ halts, and 1 otherwise. Now, for any $js, js' \in \{0,1\}^m$, we take

$$d_{TR}(js,js') = d_D(js,js') + halts(h(js,js')),$$

where d_D is the drastic distance (i.e., $d_D(js,js') = 0$ if $js = js'$ and 1 otherwise), and $h(js,js') = (\delta_H(js[1],js'[1]),\ldots,\delta_H(js[m],js'[m]))$ is the Hamming sequence for (js,js'). In other words, we XOR the binary strings corresponding to js and js', interpret the resulting string as a TM, and set the distance to 0 or 1 depending on whether the TM halts or not. On top of that, we add 1 whenever js, js' are not exactly the same.

We check that d_{TR} is a distance:
1. $d_{TR}(js,js) = d_D(js,js) + halts(0^m) = 0$;
2. $d_{TR}(js,js') = 0 \Rightarrow d_D(js,js') = 0 \Rightarrow js = js'$;
3. $d_{TR}(js,js') = d_{TR}(js',js)$: straightforward;
4. Triangle inequality: the nontrivial case is $js \neq js' \neq js''$, then $d_{TR}(js,js') + d_{TR}(js',js'') \geq 2 \geq d_{TR}(js,js'')$.

For incomputability, we observe that $TM(X)$ halts iff $d_{TR}(X, 0^{|X|}) \leq 1$. Consider the following cases: (1) $X = 0^n$: $TM(0^n)$ halts and $d_{TR}(0^n, 0^n) = 0$; (2) $X \neq 0^n$ and $TM(X)$ halts: then, $d_{TR}(X, 0^{|X|}) = 1 + \mathit{halts}(X) = 1$; (3) $X \neq 0^n$ and $TM(X)$ does not halt: then, $d_{TR}(X, 0^{|X|}) = 1 + \mathit{halts}(X) = 2$. □

Theorem 2. *There is a distance and an aggregation function for which* WINVER *is undecidable.*

Proof. We construct a Turing reduction from the halting problem. Given is a representation $X \in \{0,1\}^m$ of a Turing machine (same assumptions as in Theorem 1, i.e., every X starting with 0 or ending with 1 represents a TM that halts). We take d_{TR} as the distance, and $aggr = \sum$. Let the agenda $\mathcal{A} = \{p_1, \dots, p_m\}$ consist of n unrelated atomic propositions, the set of constraints $\mathcal{C} = \emptyset$, and the judgment profile $jp = \{0^m, X\}$. Now, for $X = 1 \dots 0$ (the other cases of X trivially halt), we have that $TM(X)$ halts iff $js = 0^m, X$ are the only winner sets. To prove this, we first observe that: (i) there is no $Y \in \{0,1\}^m$ with the aggregate distance less than 1 (since the aggregate distance for Y is a sum of nonnegative elements that includes $d_D(Y, X) + d_D(Y, 0^m)$ and $X \neq 0^m$ by assumption); (ii) for all candidate judgment sets $Y \notin \{0^m, X\}$ the aggregate distance is at least 2 (by the analogous argument); (iii) for $Y = 1^m$ the aggregate distance is always exactly 2, the score being $d_D(1^m, 0^m) + \mathit{halts}(1^m) + d_D(1^m, X) + \mathit{halts}(\overline{X}) = 1 + 0 + 1 + 0$. $TM(1^m)$ halts because 1^m ends with 1, and $TM(\overline{X})$ halts because $\overline{X}$ begins with 0. Now we prove the equivalence:

⇒: Assume that $TM(X)$ halts. Then, the aggregate distance for X is 1, and the same for 0^m (because $d_{TR}(X, 0^m) = 1$ and $d_{TR}(X, X) = d_{TR}(0^m, 0^m) = 0$). Thus, by (i), $0^m, X$ must be winners, and by (ii) no other judgment set can be a winner.

⇐: Assume that $TM(X)$ does not halt. Then, the aggregate distance for X is 2, and likewise for 0^m (because $d_{TR}(X, 0^m) = 2$). By (ii), $0^m, X$ must be winners, but they are *not the only winners* – by (iii), 1^m must be a winner too.

We have proved that $TM(X)$ halts iff $js = 0^m, X$ are the only winner sets. Suppose now that deciding WINVER terminates in finite time. Then, the halting of $TM(X)$ could be verified by 2^m WINVER checks, i.e., also in finite time – which is a contradiction. □

Thus, the standard requirements on distance metrics and aggregation function are not sufficient to guarantee even decidability of the winner set verification problem. Of course, the judgment aggregation rule used in the proof of Theorem 2 is artificial and unlikely to be ever used in any practical context. Still, it shows that the framework allows – at least theoretically – for such ill-behaved rules. Note that the effect should be the same if the distance is based on solving any other undecidable problem. For example, it can be based on a solution to a certain game, and if the game assumes imperfect information and perfect recall of players then solving it is in general undecidable [37, 15]. Or, the distance can be defined in terms of resources needed by a group of agents to achieve a given task (for undecidability, cf. e.g. [5]). We believe that these two examples of hypothetical distance-based rules are not so far-fetched anymore.

Distance-based aggregation rules that are actually used have much better computational properties, as we demonstrate in Section 4.3. Yet, Theorem 2 is important because it shows the *bounds* of the framework: in principle, the complexity of related decision problems can be very bad. This means that, when trying a *new* variant of distance-based aggregation, one should be cautious, and carefully examine its computational characteristic beforehand.

4.3 Positive Results

We now prove that, under reasonable conditions, winner set verification sits in the first level of the polynomial hierarchy. We recall that $\mathbf{P}^{\mathbf{NP}[k]}$ is the class of problems solvable by a polynomial-time deterministic Turing machine asking at most k adaptive queries to an **NP** oracle. Clearly, $\mathbf{NP} \subseteq \mathbf{P}^{\mathbf{NP}[k]} \subseteq \mathbf{\Delta}_2^{\mathbf{P}} = \mathbf{P}^{\mathbf{NP}}$.

Theorem 3. *If aggr and d are computable in polynomial time then* WINVER *for* $\nabla_{d,aggr}$ *is in* $\mathbf{P}^{\mathbf{NP}[2]}$.

Proof. We prove the inclusion by showing Algorithm 1 for WINVER, which uses two oracles, given in Algorithms 2 and 3. Note that the js in the input of Algorithms 3 is always consistent.

Algorithm 1. Winver()

Input: js,jp,N,$\mathcal{A}$,$\mathcal{C}$,d,$aggr$
Output: **true** if js is a winner for jp under $aggr$, **false** otherwise
if *Consistent(js,$\mathcal{A}$,$\mathcal{C}$)* **and not** *ExistsBetter(js,jp,N,$\mathcal{A}$,$\mathcal{C}$,d,aggr)* **then**
  return true else return false

Algorithm 2. Oracle Consistent()

Input: js,$\mathcal{A}$,$\mathcal{C}$
Output: **true** if js is consistent for $\mathcal{A}$ and $\mathcal{C}$, **false** otherwise
guess a valuation $v \in PV$ for the atomic propositions in $\mathcal{A}$
if $val_v(\varphi) = js(\varphi)$ *for every* $\varphi \in \mathcal{A}$ **and** $val_v(\psi) = 1$ *for every* $\psi \in \mathcal{C}$ **then**
  return true else return false

Algorithm 3. Oracle ExistsBetter()

Input: js,jp,N,$\mathcal{A}$,$\mathcal{C}$,d,$aggr$
Output: **true** if there is a judgment set 'closer' to jp than js, **false** otherwise
guess $js' \in JS$
guess a valuation $v' \in PV$ for the atomic propositions in $\mathcal{A}$
if $val_{v'}(\varphi) = js'(\varphi)$ *for every* $\varphi \in \mathcal{A}$ **and** $val_v(\psi) = 1$ *for every* $\psi \in \mathcal{C}$
and$aggr\big(d(js', jp[1]), \ldots, d(js', jp[|N|])\big) < aggr\big(d(js, jp[1]), \ldots, d(js, jp[|N|])\big)$ **then**
  return true else return false

For combinations of most typical distances and aggregators, the following is a straightforward consequence.The problem of checking if a judgment is in at least one collective judgment set is already known to be **NP**-complete for $d = d_H, aggr = \sum$ [21].

Corollary 1. *If $aggr \in \{\min, \max, \sum, \prod\}$ and $d \in \{d_H, d_D\}$ then* WINVER *for $\nabla_{d,aggr}$ is in $\mathbf{P}^{\mathbf{NP}[2]}$.*

4.4 Aggregation of Non-binary Judgments

In this section, we briefly report that all the results from Sections 4.2 and 4.3 carry over to the case of judgments interpreted in a given k-valued logic.[2] In particular, we note that the algorithm in Section 4.3 depends neither on the set of truth values, nor on the way valuations of complex formulas derive from valuations of atomic propositions. Also, the Turing distance used in Section 4.2 is built on pointwise comparison of judgment sets that always results in a binary string. Thus, we can state the following.

Theorem 4. *For every $k \in \mathbb{N}$, there is a distance over $\{0, \ldots, k-1\}^m$ which is not Turing computable.*

Proof. Analogous to the proof of Theorem 1.

Theorem 5. *Let $k \in \mathbb{N}$, and $\mathcal{L}$ a k-valued logic constructed like in Section 3.1. Then, there is a distance and an aggregation function for judgment sets in $\mathcal{L}$ such that* WINVER *is undecidable.*

Proof. Analogous to the proof of Theorem 2.

Theorem 6. *If $aggr$ is an aggregation function over over $\{0, \ldots, k-1\}^n$, d is a distance metric over $\{0, \ldots, k-1\}^m$, and both $aggr$ and d are computable in polynomial time, then* WINVER *for $\nabla_{d,aggr}$ is in $\mathbf{P}^{\mathbf{NP}[2]}$.*

Proof. The claim is demonstrated by the same algorithm as in the proof of Theorem 3.

5 Conclusions

Complexity-theoretic properties of voting procedures are a frequent topic of study in computational social choice. In contrast, the complexity of judgment aggregation has drawn attention only recently. In this paper, we explore the complexity bounds of an important family of judgment aggregation rules, namely those based on minimization of aggregate distance. More precisely, we study the decision problem of verifying if a given judgment set can be selected as the

[2] We do not discuss motivation for using such judgments, and instead refer the interested reader e.g. to [36, 16, 31, 40, 17].

collective opinion. It turns out that feasibility of distance-based aggregation in general cannot be guaranteed, and should not be taken for granted. However, by assuming some requirements on the possible outcomes of the distance and aggregation functions, we can tame the complexity reasonably. We also show that the pattern of complexity does not change when the framework is extended to multi-valued judgments.

To our best knowledge, this paper is the first to analyze the complexity of verifying distance-based aggregate judgments on an abstract level. There are not many concrete judgment aggregation rules proposed in the literature; this aspect of the judgment aggregation theory has only now begun to be developed. Our results suggest that, when devising a new judgment aggregation rule, we should expect complexity traps, and carefully look for rules that are relatively efficient.

Acknowledgements. Wojciech Jamroga acknowledges the support of the FNR (National Research Fund) Luxembourg under project GALOT – INTER/DFG/12/06.

References

1. Bartholdi, J., Tovey, C.A., Trick, M.A.: Voting schemes for which it can be difficult to tell who won the election. Social Choice and Welfare 6, 157–165 (1989)
2. Baumeister, D., Erdélyi, G., Erdélyi, O.J., Rothe, J.: Control in judgment aggregation. In: Proceedings of the 6th European Starting AI Researcher Symposium, pp. 23–34 (2012)
3. Baumeister, D., Erdélyi, G., Erdélyi, O.J., Rothe, J.: Computational aspects of manipulation and control in judgment aggregation. In: Proceedings of the 3rd International Conference on Algorithmic Decision Theory, ADT 2013 (forthcomming, 2013)
4. Baumeister, D., Erdélyi, G., Rothe, J.: How hard is it to bribe the judges? A study of the complexity of bribery in judgment aggregation. In: Brafman, R., Roberts, F.S., Tsoukiàs, A. (eds.) ADT 2011. LNCS, vol. 6992, pp. 1–15. Springer, Heidelberg (2011)
5. Bulling, N., Farwer, B.: On the (un-)decidability of model checking resource-bounded agents. In: Proceedings of ECAI. Frontiers in Artificial Intelligence and Applications, vol. 215, pp. 567–572. IOS Press (2010)
6. Chevaleyre, Y., Endriss, U., Lang, J., Maudet, N.: A short introduction to computational social choice. In: van Leeuwen, J., Italiano, G.F., van der Hoek, W., Meinel, C., Sack, H., Plášil, F. (eds.) SOFSEM 2007. LNCS, vol. 4362, pp. 51–69. Springer, Heidelberg (2007)
7. Conitzer, V., Sandholm, T.: Complexity of manipulating elections with few candidates. In: Proceedings of AAAI 2002, pp. 314–319. American Association of Artificial Intelligence (2002)
8. Conitzer, V., Sandholm, T.: Communication complexity of common voting rules. In: Proceedings of the ACM Conference on Electronic Commerce, pp. 78–87 (2005)
9. Dempster, A.P.: Upper and lower probabilities induced by a multivalued mapping. The Annals of Mathematical Statistics 38(2) (1967)
10. Deza, M.M., Deza, E.: Encyclopedia of Distances. Springer (2009)

11. Dietrich, F.: A generalised model of judgment aggregation. Social Choice and Welfare 28(4), 529–565 (2007)
12. Dietrich, F.: Judgment aggregation by quota rules majority voting generalized. Journal of Theoretical Politics 19(4), 391–424 (2007)
13. Dietrich, F.: Scoring rules for judgment aggregation. Social Choice and Welfare (to appear, 2013)
14. Dietrich, F., Mongin, P.: The premiss-based approach to judgment aggregation. Journal of Economic Theory 145(2), 562–582 (2010)
15. Dima, C., Tiplea, F.L.: Model-checking atl under imperfect information and perfect recall semantics is undecidable. CoRR, abs/1102.4225 (2011)
16. Dokow, E., Holzman, R.: Aggregation of binary evaluations with abstentions. Journal of Economic Theory 145(2), 544–561 (2010)
17. Duddy, C., Piggins, A.: Many-valued judgment aggregation: Characterizing the possibility/impossibility boundary. Journal of Economic Theory 148(2), 793–805 (2013)
18. Eckert, D., Klamler, C., Mitlöhner, J., Schlötterer, C.: A distance-based comparison of basic voting rules. Central European Journal of Operations Research 14(4), 377–386 (2006)
19. Eckert, D., Mitlöhner, J.: Logical representation and merging of preference information. In: Proceedings of the IJCAI 2005 Multidisciplinary Workshop on Preference Handling (2005)
20. Endriss, U., Grandi, U., Porello, D.: Complexity of judgment aggregation: Safety of the agenda. In: Proceedings of AAMAS 2010, pp. 359–366 (2010)
21. Endriss, U., Grandi, U., Porello, D.: Complexity of winner determination and strategic manipulation in judgment aggregation. In: Proceedings of COMSOC 2010 (2010)
22. Endriss, U., Grandi, U., Porello, D.: Complexity of judgment aggregation. Journal of Artificial Intelligence Research 45, 481–514 (2012)
23. Escoffier, B., Lang, J., Öztürk, M.: Single-peaked consistency and its complexity. In: Proceedings of ECAI 2008, pp. 366–370. IOS Press (2008)
24. Fagin, R., Halpern, J.Y., Moses, Y., Vardi, M.Y.: Reasoning about Knowledge. MIT Press (1995)
25. Grabisch, M., Marichal, J.-L., Mesiar, R., Pap, E.: Aggregation Functions. Cambridge University Press (2009)
26. Grossi, D.: Correspondences in the theory of aggregation. In: Bonanno, G., Löwe, B., van der Hoek, W. (eds.) LOFT 2008. LNCS, vol. 6006, pp. 34–60. Springer, Heidelberg (2010)
27. Hemaspaandra, E., Spakowski, H., Vogel, J.: The complexity of kemeny elections. Theoretical Computer Science 349(3), 382–391 (2005)
28. Konieczny, S., Lang, J., Marquis, P.: Distance-based merging: A general framework and some complexity results. In: Proceedings of KR 2002, pp. 97–108 (2002)
29. Lang, J., Pigozzi, G., Slavkovik, M., van der Torre, L.: Judgment aggregation rules based on minimization. In: Proceedings of TARK, pp. 238–246 (2011)
30. Lang, J., Slavkovik, M.: Judgment aggregation rules and voting rules. In: Perny, P., Pirlot, M., Tsoukiás, A. (eds.) ADT 2013. LNCS, vol. 8176, pp. 230–243. Springer, Heidelberg (2013)
31. Li, N.: Decision paths in sequential non-binary judgment aggregation. Technical report, Universitat Autònoma de Barcelona (2010)
32. List, C., Pettit, P.: Aggregating sets of judgments: An impossibility result. Economics and Philosophy 18(01), 89–110 (2002)

33. List, C., Polak, B.: Introduction to judgment aggregation. Journal of Economic Theory 145(2), 441–466 (2010)
34. Miller, M.K., Osherson, D.: Methods for distance-based judgment aggregation. Social Choice and Welfare 32(4), 575–601 (2009)
35. Nehring, K., Pivato, M.: Incoherent majorities: The McGarvey problem in judgement aggregation. Discrete Applied Mathematics 159(15), 1488–1507 (2011)
36. Pauly, M., van Hees, M.: Logical constraints on judgement aggregation. Journal of Philosophical Logic 35(6), 569–585 (2006)
37. Peterson, G., Reif, J., Azhar, S.: Lower bounds for multiplayer noncooperative games of incomplete information. Computers and Mathematics with Applications 41(7), 957–992
38. Pigozzi, G.: Belief merging and the discursive dilemma: an argument-based account to paradoxes of judgment aggregation. Synthese 152(2), 285–298 (2006)
39. Shafer, G.: A Mathematical Theory of Evidence. Princeton University Press (1976)
40. Slavkovik, M., Jamroga, W.: Distance-based judgment aggregation of three-valued judgments with weights. In: Proceedings of the IJCAI Workshop on Social Choice and Artificial Intelligence
41. van der Hoek, W., Verbrugge, R.: Epistemic logic: A survey. Game Theory and Applications 8, 53–94 (2002)
42. Wilson, R.: On the theory of aggregation. Journal of Economic Theory 10(1), 89–99 (1975)

Supraclassical Consequence Relations

Tolerating Rare Counterexamples

Willem Labuschagne[1], Johannes Heidema[2], and Katarina Britz[3]

[1] University of Otago, Dunedin, New Zealand
willem@cs.otago.ac.nz
[2] University of South Africa, Pretoria, South Africa
[3] Centre for AI Research, CSIR Meraka Institute, Pretoria
University of KwaZulu-Natal, Durban, South Africa

Abstract. We explore a family of supraclassical consequence relations obtained by varying the criteria according to which counterexamples to classical entailment may be deemed tolerable. This provides a different perspective on the rational consequence relations of nonmonotonic logic, as well as introducing new kinds of entailment with a diversity of potential contextual applications.

Keywords: Nonmonotonic logic, Preference order, Induction, Abduction, Supraclassical consequence, Rational consequence, Dual preferential consequence, Correlative preferential consequence.

1 Introduction

Classical logic focuses on the relatively cautious consequence relation $\models$, which is used to represent inferences from premisses φ to consequences ψ that preserve truth: $\varphi \models \psi$ if and only if every model of φ is a model of ψ. Mathematical reasoning mainly employs classical consequence.

In contrast, nonmonotonic logic belongs to a tradition in logic that considers $\models$ merely a useful cognitive reference point among several consequence relations that may be of interest. As Johan van Benthem [12] puts it:

> "The idea that logic is about just one notion of 'logical consequence' is actually one very particular historical stance. It was absent in the work of the great pioneer Bernard Bolzano, who thought that logic should chart the many different consequence relations that we have, depending on the reasoning task at hand."

When defining alternative consequence relations, one may choose to go subclassical or to go supraclassical or both. To go subclassical is to disallow some inferences from premiss to consequence that would be legitimate according to $\models$. One may wish, for example, to introduce a constraint of pertinence between premiss and conclusion as in [2], which results in a stricter criterion than $\models$. The result is that some of the classically permitted inferences become illegitimate.

S. Cranefield and A. Nayak (Eds.): AI 2013, LNAI 8272, pp. 326–337, 2013.

Nonmonotonic logic chooses instead to go supraclassical, adopting a criterion less strict than $\models$ so as to accommodate various forms of common-sense reasoning in which agents compensate for limited information by using heuristic rules of thumb. Arguably, the most important thing about nonmonotonic logic is not its nonmonotonicity but its supraclassicality. Supraclassical consequence relations are ampliative extensions of $\models$, i.e. if $\mid\sim$ is supraclassical, then $\varphi \models \psi$ is a sufficient condition for $\varphi \mid\sim \psi$. Thus supraclassical consequence relations build on the pairs in $\models$ by adding new inference-pairs legitimised by the agent's heuristic information. It is these additional inference-pairs that may need to be retracted in the light of new evidence, giving rise to nonmonotonicity.

The endeavour to develop useful supraclassical logics was set forth in the landmark paper [8], in which the extra heuristic information needed for venturing beyond $\models$ was encoded in a preference order on states. Consequence relations satisfying different groups of postulates were described. One particular group of postulates (in some sense the strongest) characterised the rational consequence relations [9], which have since earned a privileged position in virtue of their close connection with AGM belief change [13]. However, focusing only on rational consequence relations would give nonmonotonic logic a rather monolithic face.

When we reflect on everyday human reasoning, it becomes apparent that this complex scene comprises diversified, even disparate, subfields, potentially exceeding Charles Sanders Peirce's division of reasoning into deduction, induction, and abduction [5,6]. Along one dimension the intentions and goals of the reasoning agents may differ: to induce a plausible prediction; to abduce a plausible cause; to test the plausibility of some purported entailment; to speculate (but not too wildly); etc. Along another dimension the heuristic information embodied in a specific preference order on states may be expressing a comparison relative to diverse attributes of those states: normality; typicality; likelihood; frequency of occurrence; resemblance to some real or ideal state; closeness in time according to the plan for some important event; place in some causal order in accord with relevant laws of nature; degree of compliance with some norm; etc.

We shall describe a family of supraclassical entailment relations among which are the rational consequence relations that constitute the industry standard for nonmonotonic logic. The family also includes hitherto unexamined relations that deserve scrutiny because of the very natural way in which they arise within our semantic framework. The product is an articulated range of consequence relations all sharing the preferential paradigm but varying in aptness for specific contexts.

The novelty of the contribution resides in the method by which different members of the family are generated. Whereas [8] arrives at different consequence relations by varying the type of order relation on states or the association between states and valuations, we shall take these to be fixed as if for rational consequence relations, and instead use the preference relation on states to control, in a nuanced way, the addition of new inference-pairs to $\models$.

2 Preliminaries

Henceforth L_A denotes a propositional language generated from the atomic sentences in A by the usual set of connectives $\{\neg, \wedge, \vee, \rightarrow, \leftrightarrow\}$, with $\top$ designating an arbitrary tautology. The language is equipped with a semantics in the manner familiar from nonmonotonic logic [8,9], so that the semantic structure comprises a set S of states, the set W_A of all valuations $w : A \longrightarrow \{0, 1\}$, a labelling function $\ell : S \longrightarrow W_A$, and a suitable order relation $\preccurlyeq$ on S. As is customary, $\preccurlyeq$ encodes heuristic information about preference, typicality, or likelihood that cannot in general be expressed propositionally in L_A (but see Section 8).

We shall deviate from [8] and [9] in three ways.

Firstly, we shall assume that A and S are finite. We consider this finiteness assumption to be justified both by an interest in everyday reasoning (as opposed to metamathematical applications) and as a technical simplification appropriate for an initial scrutiny of a new landscape. (A consequence of the finiteness assumption is that smoothness for $\preccurlyeq$ is automatic.)

Secondly, we shall assume that $\preccurlyeq$ is a total preorder on S (i.e. is reflexive on S, transitive, and connected in the sense that for all $s, t \in S$ at least one of $s \preccurlyeq t$ and $t \preccurlyeq s$ is the case). Total preorders provide a unified framework for both nonmonotonic logic and belief revision, and the strict modular partial orders used in [9] are just strict versions of total preorders.

Finally, we take states higher up in the order to be more preferred, more typical, or more likely to occur. For historical reasons the order is often inverted, as in [8] and [9], but we follow Shoham ([11], p.74) in respecting the common intuition that 'up' is 'more' when 'more' qualifies positive attributes such as normality, typicality, or likelihood rather than negative qualities such as abnormality, atypicality, or unlikelihood.

Satisfaction is defined as usual. For every sentence $\varphi \in L_A$, $Mod(\varphi)$ denotes the set of models of φ, i.e. the set of all $s \in S$ for which $\ell(s)$ renders φ true, and $Max(\varphi)$ the maximal models of φ with respect to $\preccurlyeq$.

For purposes of illustration we shall take $A = \{p, q\}$ with $S = W_A = \{11, 10, 01, 00\}$ where 10 denotes the state (i.e. valuation) in which p is true and q is false, and so forth. As an interpreted language, we may consider the system of interest to be either the Traffic System in which p abbreviates *The light for oncoming traffic is red* and q stands for *The oncoming car stops*, or the Light-Fan System in which p abbreviates *The light is on* and q stands for *The fan is on.*

3 From Counterexamples to Supraclassicality

Since $\varphi \models \psi$ if and only if $Mod(\varphi) \subseteq Mod(\psi)$, a strictly supraclassical consequence relation must have at least one pair (φ, ψ) for which $\varphi \not\models \psi$ and thus for which there is some model u of φ that fails to be a model of ψ. Think of state u as a *counterexample* to $\varphi \models \psi$, and thus in some sense a 'bad guy'. More generally, $Mod(\varphi \wedge \neg\psi)$ is the set of bad guys (i.e. counterexamples).

Not all bad guys are equally bad. The total preorder $\preccurlyeq$ on S stratifies the states into levels stacked from the least normal (most rare, most atypical) to the most normal (most likely, most typical). In this context some counterexample states may be relatively exceptional (less normal, less typical, or less likely than others). The guiding idea behind supraclassicality is to tolerate such not-so-very-bad guys, just as in life we commonly view those guilty of an uncharacteristic lapse of judgment less harshly than habitual criminals.

We wish to allow an inference-pair (φ, ψ) into the consequence relation as long as the counterexamples in $Mod(\varphi \wedge \neg\psi)$ are not maximally bad. The new insight of which we take advantage is that the worst counterexample states, those that are maximal, may be identified in four different ways. Maximality is relative not only to the ordering but also to the subset within which an element is considered to be maximal. Different supraclassical relations may be obtained by varying the subsets of S against which counterexamples are evaluated. These subsets must be supersets of $Mod(\varphi \wedge \neg\psi)$.

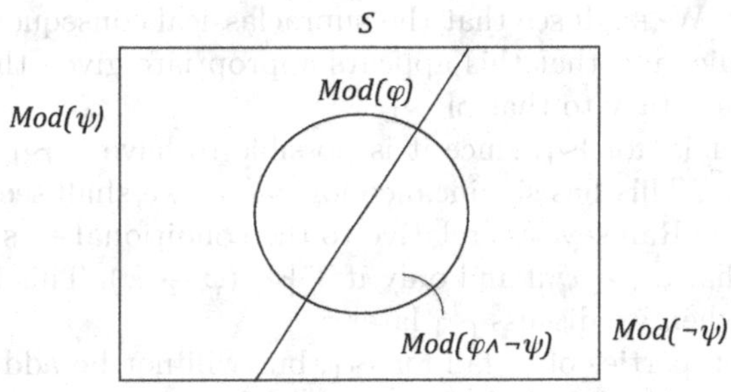

Fig. 1. Important subsets of S

On the face of it, there are four obvious supersets of $Mod(\varphi \wedge \neg\psi)$ within which counterexamples could be maximal:

1. $X_1 = Mod(\varphi)$
2. $X_2 = Mod(\neg\psi)$
3. $X_3 = Mod(\varphi) \cup Mod(\neg\psi)$
4. $X_4 = S$.

As a generic symbol for a supraclassical consequence relation we use $\mid\!\sim$ decorated by a subscript designating one of the four supersets of $Mod(\varphi \wedge \neg\psi)$. Thus we explore four supraclassical consequence relations, $\mid\!\sim_1$, $\mid\!\sim_2$, $\mid\!\sim_3$ and $\mid\!\sim_4$, which are all defined similarly:

$$\varphi \mid\!\sim_i \psi \text{ if and only if } Mod(\varphi \wedge \neg\psi) \cap Max(X_i) = \varnothing.$$

Relative to a fixed semantics, we may speak of 'the' relation $\mid\!\sim_i$ rather than 'a' relation $\mid\!\sim_i$.

4 Rational Consequence

Consider $\mid\sim_1$. In this case the set of counterexamples $Mod(\varphi \wedge \neg\psi)$ is viewed as a subset of $X_1 = Mod(\varphi)$, and thus $\varphi \mid\sim_1 \psi$ if and only if $Mod(\varphi \wedge \neg\psi) \cap Max(\varphi) = \varnothing$.

It follows that $\varphi \mid\sim_1 \psi$ if and only if every maximal model of φ is a model of ψ. In other words, $\mid\sim_1$ is the familiar rational consequence relation of nonmonotonic logic, defined as in [9].

A great deal is known about the properties of rational consequence relations and how they differ from those of $\models$. For purposes of comparison, we give a swift and selective recapitulation, beginning with some familiar properties of $\models$ that fail to hold for $\mid\sim_1$.

The relation $\mid\sim_1$ is famously *nonmonotonic*, i.e. there exist instances of $\mid\sim_1$ such that for some sentences $\varphi, \psi, \alpha \in L_A$ we have $\varphi \mid\sim_1 \psi$ but not $\alpha \wedge \varphi \mid\sim_1 \psi$. As hinted earlier, we should not ascribe overwhelming importance to nonmonotonicity. For some forms of reasoning, nonmonotonicity may be appropriate; for others, it may not. We shall see that the supraclassical consequence relation $\mid\sim_2$ is in fact monotonic, and that this appears appropriate given that its utility is in a sense complementary to that of $\mid\sim_1$.

Contraposition fails for $\mid\sim_1$, since it is possible to have $\varphi \mid\sim_1 \psi$ while failing to have $\neg\psi \mid\sim_1 \neg\varphi$. This has significance for $\mid\sim_2$, as we shall see.

Also, $\mid\sim_1$ fails the Ramsey test relative to the conditional $\rightarrow$ since it is not in general the case that $\varphi \mid\sim_1 \psi$ if and only if $\top \mid\sim_1 (\varphi \rightarrow \psi)$. This failure gains an interesting twist when we discuss $\mid\sim_4$ later.

Various other properties of $\models$ fail for $\mid\sim_1$, but will not be addressed here.

Turning to properties that do hold in general for $\mid\sim_1$, we may note the observance of a weaker form of monotonicity, rational monotonicity, which accounts for the *rational* in the name given to this type of consequence relation:

$$\text{if } \varphi \mid\sim_1 \psi \text{ and it is not the case that } \varphi \mid\sim_1 \neg\alpha, \text{ then } \alpha \wedge \varphi \mid\sim_1 \psi.$$

Although unqualified monotonicity fails for $\mid\sim_1$, the property called *right weakening*, which is a sort of dual of monotonicity obtained by replacing $\wedge$ by $\vee$ and left with right, does hold for $\mid\sim_1$:

$$\text{if } \varphi \mid\sim_1 \psi \text{ then } \varphi \mid\sim_1 \psi \vee \alpha.$$

We shall see the relevance of the duality between monotonicity and right weakening when we examine $\mid\sim_2$.

We may also make an algebraic observation. The reader will recall that in classical propositional logic, the equivalence classes of sentences form a Boolean algebra (hereafter referred to as the Lindenbaum-Tarski algebra) having $\models$ as the order relation, conjunction as meet, disjunction as join, the class $\top$ of tautologies as maximum, and the class $\bot$ of contradictions as minimum. The set of consequences of a premiss φ under $\models$ is a filter of the algebra, and in the other direction the set of premisses entailing a fixed consequence under $\models$ forms an ideal. This classical picture is neatly bisected by $\mid\sim_1$ and $\mid\sim_2$.

In respect of $\mid\!\sim_1$, we note that for a fixed premiss φ the set $\{\psi \mid \varphi \mid\!\sim_1 \psi\}$ is a filter in the Lindenbaum-Tarski algebra of propositions, just as in the classical case. In contrast, for a fixed conclusion ψ the set $\{\varphi \mid \varphi \mid\!\sim_1 \psi\}$ is not necessarily an ideal, because of nonmonotonicity.

Concluding our brief overview, the significance of the rational consequence relation $\mid\!\sim_1$ is that arguably it formalises inductive reasoning of the kind we might call *singular predictive inference* in order to distinguish it from other meanings of "induction", such as learning a default rule from a finite set of instances (generalisation). An example will serve to illustrate, but a deeper analysis may be found in [3] and [4], as well as in Section 7.

Example 1. Consider the language $L_{\{p,q\}}$ with a semantics consisting of $S = \{11, 10, 01, 00\}$ and the total preorder $\preccurlyeq$ that stratifies S into two levels, with 10 and 01 on the bottom and 11 and 00 at the top. Think of this as a knowledge representation language for the Traffic System. The preorder $\preccurlyeq$ depicts the heuristic that it is normal for the oncoming traffic to stop if their traffic light is red (the state 11) and it is normal for the oncoming traffic to continue driving without stopping if their traffic light is not red (the state 00). Suppose we are waiting at the intersection and observe that p is the case, i.e. that the light for oncoming cross traffic is red. We need to decide whether q is the case, i.e. whether the oncoming car will stop. Although $p \not\models q$, the reason we get to work every morning is that we are able to predict that the oncoming car will stop and so proceed fearlessly to cross the intersection ourselves. Our prediction is sanctioned by $\mid\!\sim_1$, since $p \mid\!\sim_1 q$. The prediction may of course turn out to be wrong, because $\mid\!\sim_1$ relies on uncertain heuristic information. In the case of a disastrous falsification of q, the disaster embodies a tragic, though presumably rare, counterexample to $p \models q$. Had the prediction instead been sanctioned by $\models$, we could have proceeded across the intersection secure in the knowledge that no drunk driver would run the red light. Sadly, everyday decision-making seldom enjoys the luxury of sufficient information to dispense with $\mid\!\sim_1$ and rely on $\models$.

5 Dual Preferential Consequence

Rational consequence in [9] (Section 3) corresponds to *modularity* of the preferential order, which is there taken to be a strict partial order on S, and in our exposition to *totality* of the preferential preorder. In [9] "preferential consequence" (and related terminology) does not connote modularity of the corresponding order. In this article, however, we use "preferential" throughout while always staying with the stipulation in Section 2 that $\preccurlyeq$ is a *total* preorder on S.

Consider $\mid\!\sim_2$. In this case the set of counterexamples $Mod(\varphi \wedge \neg\psi)$ is viewed as a subset of $X_2 = Mod(\neg\psi)$, and thus $\varphi \mid\!\sim_2 \psi$ if and only if $Mod(\varphi \wedge \neg\psi) \cap Max(\neg\psi) = \varnothing$.

By the definition, although it is possible that some model of φ may fail to satisfy ψ, that model is not to be a typical (i.e. maximal) model of $\neg\psi$ but instead is required to be somewhat atypical among the states that falsify ψ.

It is not hard to see that $\varphi \mid\!\sim_2 \psi$ if and only if $Mod(\varphi) \subseteq S \setminus Max(\neg\psi)$. Hence $\mid\!\sim_2$ is precisely the dual preferential consequence relation studied in [1].

The relation $\mid\!\sim_2$ is related to the rational consequence relation $\mid\!\sim_1$ in a manner that contraposition would have rendered trivial had this property held for $\mid\!\sim_1$:

$$\varphi \mid\!\sim_2 \psi \text{ if and only if } \neg\psi \mid\!\sim_1 \neg\varphi.$$

This intimate connection between $\mid\!\sim_1$ and $\mid\!\sim_2$ invites a quick comparison of features.

We first note that whereas $\mid\!\sim_1$ is nonmonotonic, monotonicity holds for $\mid\!\sim_2$:

$$\text{if } \varphi \mid\!\sim_2 \psi \text{ then } \alpha \wedge \varphi \mid\!\sim_2 \psi.$$

As previously observed, monotonicity is in some sense a dual of right weakening, and there is a general pattern of properties holding for $\mid\!\sim_2$ if they are the duals of properties holding for $\mid\!\sim_1$, as explained in [1]. For example, although right weakening fails to hold for $\mid\!\sim_2$, the weaker form of right weakening that is the dual of rational monotonicity does hold:

$$\text{if } \varphi \mid\!\sim_2 \psi \text{ and it is not the case that } \neg\alpha \mid\!\sim_2 \psi \text{ then } \varphi \mid\!\sim_2 \psi \vee \alpha.$$

Algebraically, for a fixed conclusion ψ the set $\{\varphi \mid \varphi \mid\!\sim_2 \psi\}$ is an ideal of the Lindenbaum-Tarski algebra, just as in the classical case, but for a fixed premiss φ the set of consequences $\{\psi \mid \varphi \mid\!\sim_2 \psi\}$ may fail to be a filter.

Finally, the significance of $\mid\!\sim_2$ is that it arguably formalises a kind of abductive reasoning, i.e. $\varphi \mid\!\sim_2 \psi$ can be interpreted as "φ partially explains ψ". An example will serve to illustrate, until Section 7.

Example 2. Consider again the language $L_{\{p,q\}}$ with a semantics consisting of $S = \{11, 10, 01, 00\}$ and the total preorder $\preccurlyeq$ that stratifies S into two levels, with 10 and 01 on the bottom and 11 and 00 at the top. Think of this as a knowledge representation language for the Light-Fan System, where p stands for "The light is on" and q for "The fan is on". Now suppose we observe that $\neg p \wedge q$ is the case. The semantic constraint $\mid\!\sim_2$ admits several different explanations amongst which is the sentence $(p \vee q) \wedge \neg(p \wedge q)$, which we may abbreviate by $p + q$. To see that $p + q \mid\!\sim_2 \neg p \wedge q$, note that although $10 \in Mod(p + q)$ and 10 is a counterexample to $p + q \models \neg p \wedge q$, 10 is not maximal in $Mod(\neg(\neg p \wedge q))$. Intuitively, the observation that the light is off while the fan is on has been explained by the conjecture that only one component can be on at a time.

6 Correlative Preferential Consequence

Consider $\mid\!\sim_3$. In this case the set of counterexamples $Mod(\varphi \wedge \neg\psi)$ is viewed as a subset of $X_3 = Mod(\varphi) \cup Mod(\neg\psi)$, and thus $\varphi \mid\!\sim_3 \psi$ if and only if $Mod(\varphi \wedge \neg\psi) \cap Max(Mod(\varphi) \cup Mod(\neg\psi)) = \varnothing$.

Recalling that $Mod(\varphi) \cup Mod(\neg\psi) = Mod(\psi \to \varphi)$, we get that

$$\varphi \mid\!\sim_3 \psi \text{ if and only if } Mod(\varphi \wedge \neg\psi) \subseteq S \setminus Max(\psi \to \varphi).$$

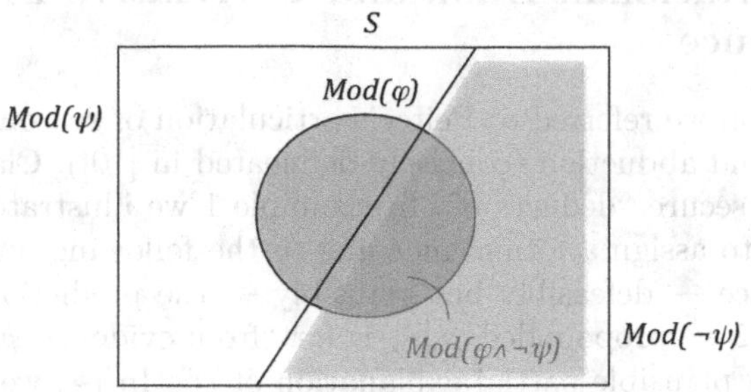

Fig. 2. The shaded region depicts $X_3 = Mod(\varphi) \cup Mod(\neg\psi)$

As far as the authors are aware, the relation $\mid\!\sim_3$ has not previously been studied, and its intuitive meaning, properties, and potential uses remain to be fully elucidated. Nevertheless, we are able to report some preliminary insights.

One striking aspect is that the counterexample states in $Mod(\varphi \wedge \neg\psi)$ are interdicted from the set of maximal models of $\psi \rightarrow \varphi$, giving the following surprising connection between $\mid\!\sim_3$ and the rational consequence relation $\mid\!\sim_1$:

$$\begin{aligned}
\varphi \mid\!\sim_3 \psi \quad & \text{if and only if} \quad Mod(\neg(\varphi \rightarrow \psi)) \subseteq S \setminus Max(\psi \rightarrow \varphi) \\
& \text{if and only if} \quad Max(\psi \rightarrow \varphi) \subseteq S \setminus Mod(\neg(\varphi \rightarrow \psi)) \\
& \text{if and only if} \quad Max(\psi \rightarrow \varphi) \subseteq Mod(\varphi \rightarrow \psi) \\
& \text{if and only if} \quad (\psi \rightarrow \varphi) \mid\!\sim_1 (\varphi \rightarrow \psi) \\
& \text{if and only if} \quad (\psi \rightarrow \varphi) \mid\!\sim_1 (\varphi \leftrightarrow \psi).
\end{aligned}$$

If $\varphi \mid\!\sim_3 \psi$ says that the truth of the conditional $\psi \rightarrow \varphi$ renders plausible the converse conditional $\varphi \rightarrow \psi$, one is reminded of psychological experiments showing that under some circumstances humans have a tendency to infer the converse from a conditional premiss in exactly this way [7, pages 51–54]. Perhaps $\mid\!\sim_3$ represents thought patterns deriving from mental models in the sense of Johnson-Laird — a topic for future research.

Further reflection reveals that, since $\preccurlyeq$ is a total preorder on S, the set $Max(\psi \rightarrow \varphi)$ from which counterexamples are interdicted must be one of the following three sets: $Max(\varphi)$, $Max(\neg\psi)$, or $Max(\varphi) \cup Max(\neg\psi)$. This allows us to prove that $\mid\!\sim_3$ is related to both $\mid\!\sim_1$ and $\mid\!\sim_2$ in an elegantly balanced way:

- If $Max(\psi \rightarrow \varphi) = Max(\varphi)$ then $Mod(\varphi \wedge \neg\psi)$ has no member in $Max(\varphi)$ and so $\varphi \mid\!\sim_1 \psi$
- If $Max(\psi \rightarrow \varphi) = Max(\neg\psi)$ then $Mod(\varphi \wedge \neg\psi)$ has no member in $Max(\neg\psi)$ and so $\varphi \mid\!\sim_2 \psi$
- If $Max(\psi \rightarrow \varphi) = Max(\varphi) \cup Max(\neg\psi)$ then both $\varphi \mid\!\sim_1 \psi$ and $\varphi \mid\!\sim_2 \psi$.

Summarising, if $\varphi \mid\!\sim_3 \psi$ then $\varphi \mid\!\sim_1 \psi$ or $\varphi \mid\!\sim_2 \psi$ or both.
Conversely, if both $\varphi \mid\!\sim_1 \psi$ and $\varphi \mid\!\sim_2 \psi$, then $\varphi \mid\!\sim_3 \psi$.
Overall: $(\mid\!\sim_1 \cap \mid\!\sim_2) \subseteq \mid\!\sim_3 \subseteq (\mid\!\sim_1 \cup \mid\!\sim_2)$.

7 More on Rational, Dual, and Correlative Preferential Consequence

In our introduction we referred to Peirce's articulation of reasoning into deduction, induction, and abduction (concisely delineated in [10]). Classical $\models$ is the canonical way to secure "deduction". In Example 1 we illustrated how $\varphi \mid\!\sim_1 \psi$ may be one way to assign a formal meaning to the following notion: "from evidence φ we induce — defeasibly but plausibly — the prediction that also ψ". And in Example 2 we propounded $\varphi \mid\!\sim_2 \psi$ for "from evidence ψ we abduce the hypothesis φ as a plausible partial explanation of ψ". In $\mid\!\sim_3$ we have a type of balanced, correlated combination of $\mid\!\sim_1$ and $\mid\!\sim_2$, roaming the interval between $\mid\!\sim_1 \cap \mid\!\sim_2$ and $\mid\!\sim_1 \cup \mid\!\sim_2$. But a sharper focus on φ as the given evidence versus ψ as the given evidence may enhance intuitive appreciation of the dual roles of $\mid\!\sim_1$ and $\mid\!\sim_2$.

When glossing $\varphi \mid\!\sim_1 \psi$ as "the information (evidence, observation) expressed by φ is given and affords, quite plausibly, that we have ψ too", φ is given, fixed, presumably reliable, and the actual state (if pertinent) is one in $Mod(\varphi)$. In contrast, ψ is now much more fluttery. Many different predictions may be underwritten by φ. When accepting defeasibility of prediction ψ by tolerating some of the counterexamples in $Mod(\varphi \wedge \neg\psi)$, the only dependable information we have against which to evaluate them sits embedded in $Mod(\varphi)$, which contains all of these counterexamples. So it is reasonable to tolerate only those counterexamples not maximally likely in $Mod(\varphi)$, i.e. those in $Mod(\varphi) \setminus Max(\varphi)$. The actual state (if pertinent) then would likely sit in $Max(\varphi)$ and not be a counterexample — a comforting thought.

When glossing $\varphi \mid\!\sim_2 \psi$ as "the information (evidence, observation) expressed by ψ is given and is afforded, quite plausibly, by the explanatory hypothesis φ", ψ is given, fixed, presumably reliable, and the actual state (if pertinent) is one in $Mod(\psi)$. Many hypotheses may constitute plausible if partial explanations of ψ, among them φ. (The monotonicity of $\mid\!\sim_2$ in this explanatory context seems agreeable: if φ plausibly and partially explains ψ, then so does $\varphi \wedge \alpha$. A next stage in one's abductive endeavour may then be the somewhat controversial search for the *best* explanation of ψ.) All counterexamples to $\varphi \models \psi$ falsify ψ, sit in $Mod(\neg\psi)$, but among them we tolerate only those states that do not add insult to injury by being very likely to occur. So to be tolerable a counterexample must not be "maximally bad", i.e. must not belong to $Max(\neg\psi)$.

In a correlative preferential entailment $\varphi \mid\!\sim_3 \psi$, we may be in a context where the agent's aim is neither to predict ψ from φ, nor to explain ψ by φ. No extra evidence supports either φ or ψ in an unbalanced way. When tolerating some counterexamples, the information in $\preccurlyeq$ now plays a balanced role with regard to proscribing states that support φ but violate ψ and is used to interdict those that are seriously embarrassing the entailment of ψ by φ by their maximally prominent presence in the crowd of kindred states which support φ or violate ψ.

The information embodied in $(S, \preccurlyeq, \ell)$ determines in a unique way all three consequence relations $\mid\!\sim_1$, $\mid\!\sim_2$, and $\mid\!\sim_3$ on the sentences of L_A. We now assume that $\ell : S \longrightarrow W_A$ is injective and then show that $\preccurlyeq$ can be recovered from each of $\mid\!\sim_1$ and $\mid\!\sim_2$.

A *literal* or *diagrammatic sentence* is any atomic sentence in A or the negation of such an atom. The *state description* or *diagram* of state $s \in S$ is the sentence $\delta(s)$ that is the conjunction of all the literals made true by $\ell(s)$. For instance, if $\ell(s) = 10$, then $\delta(s) = p \wedge \neg q$. Let s and t be any two (possibly equal) states. We demonstrate that any information about s and t in $\preccurlyeq$ can be retrieved from the corresponding relation $\mid\!\sim_1$ on sentences that are state descriptions or disjunctions of such. Remember that $Mod(\delta(s)) = \{s\}$. No undue cognitive torment is incurred when verifying the following:

$$s \prec t \text{ iff } \delta(s) \vee \delta(t) \mid\!\sim_1 \delta(t) \text{ and not } \delta(s) \vee \delta(t) \mid\!\sim_1 \delta(s)$$
$$s = t \text{ iff } \delta(s) \vee \delta(t) \mid\!\sim_1 \delta(s) \text{ and } \delta(s) \vee \delta(t) \mid\!\sim_1 \delta(t)$$
$$s \not\preccurlyeq t,\, t \not\preccurlyeq s, \text{ and } s \neq t \text{ iff not } \delta(s) \vee \delta(t) \mid\!\sim_1 \delta(s) \text{ and not } \delta(s) \vee \delta(t) \mid\!\sim_1 \delta(t).$$

So, when ℓ is one-to-one, then $\mid\!\sim_1$ harbours exactly the same information as $\preccurlyeq$. Since $\mid\!\sim_2$ is the dual or contrapositive relation of $\mid\!\sim_1$ ($\varphi \mid\!\sim_2 \psi$ iff $\neg\psi \mid\!\sim_1 \neg\varphi$), $\mid\!\sim_2$ on Boolean combinations of state descriptions also contains the same heuristic information as $\preccurlyeq$.

8 Contextual Rules

Consider $\mid\!\sim_4$. In this case the set of counterexamples $Mod(\varphi \wedge \neg\psi)$ is viewed as a subset of $X_4 = S$, and thus $\varphi \mid\!\sim_4 \psi$ if and only if $Mod(\varphi \wedge \neg\psi) \cap Max(S) = \varnothing$.

While this entailment relation is quite new, it has a simple intuitive basis. Recall that in the case of classical entailment we have, if $\top$ denotes your favourite tautology,

$$\varphi \models \psi \text{ if and only if } \top \models (\varphi \to \psi).$$

Now we observe that although, as previously noted, it is not in general the case that $\varphi \mid\!\sim_1 \psi$ if and only if $\top \mid\!\sim_1 (\varphi \to \psi)$, the latter condition is of independent interest, because:

$$\begin{aligned} \varphi \mid\!\sim_4 \psi \quad &\text{if and only if} \quad Mod(\varphi \wedge \neg\psi) \subseteq S \setminus Max(S) \\ &\text{if and only if} \quad Max(S) \subseteq Mod(\varphi \to \psi) \\ &\text{if and only if} \quad \top \mid\!\sim_1 (\varphi \to \psi). \end{aligned}$$

This relationship between $\mid\!\sim_4$ and $\mid\!\sim_1$ affords a new and surprising way to understand the former, namely as deduction from a knowledge base. Suppose an agent has background information expressed by a sentence κ. This background information can be encoded into a preference order on states — simply take the dichotomous total preorder on S that places all models of κ on the upper level and all nonmodels of κ on the lower level. Since $Max(S) = Max(\top) = Mod(\kappa)$ it follows that

$$\kappa \models \psi \quad \text{if and only if} \quad \top \mid\!\sim_1 \psi$$

where we have used the dichotomous preorder induced by κ for defining the rational consequence relation $\mid\!\sim_1$.

Hence in the context of background information κ we see that

$$\begin{aligned} \varphi \mid\!\sim_4 \psi \quad &\text{if and only if} \quad \top \mid\!\sim_1 (\varphi \to \psi) \\ &\text{if and only if} \quad \kappa \models (\varphi \to \psi). \end{aligned}$$

The real surprise lurks in the converse. Suppose we start (not with sentence κ, but) with some $\preccurlyeq$ and consider the corresponding $\mid\!\sim_4$. The simplification $\preccurlyeq^*$ of $\preccurlyeq$ into the dichotomous preference order on S with $Max(S)$ in the top level and $S \setminus Max(S)$ in the bottom level yields again the same $\mid\!\sim_4$. And $\preccurlyeq^*$ (or $\preccurlyeq$) yields the corresponding explicit object-language background informational sentence $\kappa = \bigvee\{\delta(s) \mid s \in Max(S)\}$. Now $\preccurlyeq^*$, $\mid\!\sim_4$, and κ harbour exactly the same information. And, against the background of this informational context, we may construe $\varphi \mid\!\sim_4 \psi$ as the conditional rule $\varphi \to \psi$.

9 Future Research

Impelled by the conviction that it would be a mistake to seek a single 'correct' consequence relation for the formalisation of common-sense reasoning and the belief that appropriate candidates would be supraclassical, we have described a coherent family of supraclassical consequence relations $\mid\!\sim_i$ within a single unifying framework.

In so doing, we tolerated all but the most habitual criminals in the set $Mod(\varphi \wedge \neg\psi)$ of counterexamples. Accordingly we may think of the family of $\mid\!\sim_i$ as the 'liberal' supraclassical relations, each of which includes inference-pairs (φ, ψ) as long as the counterexample states in $Mod(\varphi \wedge \neg\psi)$ are not maximal in the relevant superset of $Mod(\varphi \wedge \neg\psi)$.

There is an intriguing extension of this family. Tolerating only those criminals whose transgressions are very rare would deliver 'conservative' supraclassical relations, from each of which inference-pairs (φ, ψ) are excluded unless the counterexample states, if any, are all minimal in the relevant superset.

Recall the four supersets of $Mod(\varphi \wedge \neg\psi)$ within which counterexamples could be either 'not maximal' or 'minimal': $X_1 = Mod(\varphi)$, $X_2 = Mod(\neg\psi)$, $X_3 = Mod(\varphi) \cup Mod(\neg\psi)$, and $X_4 = S$.

As a generic symbol for a conservative supraclassical entailment relation we may use $\mid\!\approx$, with appropriate subscript. The conservative $\mid\!\approx_1$, $\mid\!\approx_2$, $\mid\!\approx_3$ and $\mid\!\approx_4$ would most naturally be defined by

$$\varphi \mid\!\approx_i \psi \text{ if and only if } Mod(\varphi \wedge \neg\psi) \subseteq Min(X_i).$$

However, there is a subtle problem with this. We would wish to allow the total preorder $\preccurlyeq$ on S to be $S \times S = \{(s, s') \mid s, s' \in S\}$, the relation of complete preferential equity between all states, in order to include the case of an agent with no discriminatory heuristic information at all. Unfortunately, $Min(X_i) = Max(X_i)$ in this case, which causes the constraint $Mod(\varphi \wedge \neg\psi) \subseteq Min(X_i)$ to violate the guiding intuition that counterexample states should certainly not be maximally typical or maximally likely in X_i.

We therefore introduce the notation $Min\ \overline{Max}\ (X_i) = Min(X_i) \setminus Max(X_i)$ and define the four conservative entailment relations by

$$\varphi \mathrel{|\!\!\approx}_i \psi \text{ if and only if } Mod(\varphi \wedge \neg\psi) \subseteq Min\ \overline{Max}\ (X_i).$$

In the limiting case of an agent with no heuristic information (i.e. when $\preccurlyeq$ is $S \times S$), we now have that both the liberal consequence relations $\mid\!\sim_i$ and the conservative consequence relations $\mathrel{|\!\!\approx}_i$ collapse to $\models$.

Precisely how the conservative consequence relations are related to the liberal consequence relations is as yet *terra* very much *incognita*.

References

1. Britz, K., Heidema, J., Labuschagne, W.: Semantics for dual preferential entailment. Journal of Philosophical Logic 38, 433–446 (2009)
2. Britz, K., Heidema, J., Varzinczak, I.: Constrained consequence. Logica Universalis 5, 327–350 (2011)
3. Freund, M.: On the notion of concept I. Artificial Intelligence 172, 570–590 (2008)
4. Freund, M.: On the notion of concept II. Artificial Intelligence 173, 167–179 (2009)
5. Hauser, N., Kloesel, C.: The Essential Peirce, vol. 1, pp. 1867–1893. Indiana University Press, Bloomington (1992)
6. Hauser, N., Kloesel, C.: The Essential Peirce, vol. 2, pp. 1893–1913. Indiana University Press, Bloomington (1998)
7. Johnson-Laird, P.: Mental Models. Harvard University Press, Cambridge (1983)
8. Kraus, S., Lehmann, D., Magidor, M.: Nonmonotonic reasoning, preferential models and cumulative logics. Artificial Intelligence 44, 167–207 (1990)
9. Lehmann, D., Magidor, M.: What does a conditional knowledge base entail? Artificial Intelligence 55, 1–60 (1992)
10. Rodrigues, C.T.: The method of scientific discovery in Peirce's philosophy: Deduction, induction, and abduction. Logica Universalis 5, 127–164 (2011)
11. Shoham, Y.: Reasoning about change: Time and causation from the standpoint of artificial intelligence. The MIT Press, Cambridge (1988)
12. van Benthem, J.: Logic and reasoning: Do the facts matter? Studia Logica 88, 67–84 (2008)
13. Zhang, D., et al.: Nonmonotonic reasoning and multiple belief revision. In: Proceedings IJCAI 1997, vol. 1, pp. 95–100. Morgan Kaufmann, Los Altos (1997)

Relative Expressiveness of Well-Founded Defeasible Logics

Michael J. Maher

School of Engineering and Information Technology
The University of New South Wales, Canberra
ACT 2600, Australia
m.maher@adfa.edu.au

Abstract. An approach to formulating relative expressiveness of defeasible logics was introduced in [14]. In this paper we address the relative expressiveness of the well-founded defeasible logics in the framework **WFDL** and their relationship to the defeasible logics in the framework **DL**. We show that, in terms of defeasible reasoning, the logics in **WFDL** have greater (or equal) expressiveness than those in **DL**, but it is not clear whether they have strictly greater expressiveness. We also show that different treatments of ambiguity lead to different expressiveness in **WFDL**, as it does in **DL**.

1 Introduction

Defeasible reasoning concerns reasoning where a chain of reasoning can be defeated (that is, not considered the basis of an inference) by another chain of reasoning (or, perhaps, several chains of reasoning). Defeasible logics are a class of non-monotonic logics designed to support defeasible reasoning. These logics have application in providing a computational representation of legal documents [13,22,1,9] and in legal reasoning [20,10]. They also have a role in the rule layer of the semantic web [4], and agent-based computing [6,11].

The defeasible logics we address are distinguished by their choices on three orthogonal issues. The first issue is the treatment of circular reasoning: when a proposition is not proved, but depends circularly on itself, should we consider the proposition unproved or undecided (neither proved nor unproved)? Treating the proposition as unproved implies a more powerful proof system, capable of detecting such cycles; we refer to such logics as *well-founded defeasible logics*, because they reflect the well-founded semantics of logic programming [8]. The most widely investigated defeasible logics regard the proposition as undecided, but a case is made in [18] that – at least in ontological reasoning – treating such propositions as unproved is more natural. The framework **DL** encompasses several logics that take the former viewpoint, while **WFDL** contains logics taking the latter view.

The second issue is one of *team defeat*: when there are competing claims (on inferring q or $\neg q$, say), should a single claim for q be required to overcome all competing claims in order to validate the inference, or is it sufficient that every claim for $\neg q$ is overcome by some claim for q, so that the claims for q, as a team, overcome all competing

S. Cranefield and A. Nayak (Eds.): AI 2013, LNAI 8272, pp. 338–349, 2013.

claims? The third issue addresses *ambiguity*, the situation where there is no resolution of the competing claims, so that neither q nor $\neg q$ can be derived. Should ambiguity block, so that inferences relying on q or $\neg q$ simply fail to apply, or should the fact that there are claims for q (say) that are not overcome by claims for $\neg q$ be allowed to influence later inferences, so that ambiguity propagates?

Relative expressiveness provides a way of evaluating features of a logic, to see whether the features can be imitated in another logic, or provide capabilities that cannot be imitated. Recently the relative expressiveness of defeasible logics in **DL** has been investigated [14,15]. That work showed that, within **DL**, different treatments of ambiguity lead to different expressiveness, while the use of teams or not does not affect expressiveness.

In this paper we study relative expressiveness in the well-founded defeasible logics of **WFDL**, and their relationship to defeasible logics in **DL**. We show that the **WFDL** logics are not immediately more expressive than the logics in **DL**: the different treatments of the monotonic component of the logics are a barrier. However, when we consider the defeasible reasoning component alone, we show that each of the logics of **WFDL** is more (or equal) expressive than the logics of **DL**. On the other hand, it is not clear that **WFDL** is strictly more expressive than **DL**, and we give an example suggesting that **DL** might have the capability to simulate **WFDL**. Within **WFDL**, we show that, as in **DL**, different treatments of ambiguity result in differences in expressiveness. We also show that the **WFDL** logics with individual defeat are equally as expressive as the corresponding logics that apply team defeat.

The paper is structured as follows. The next section provides an overview of defeasible logics. It is followed by a discussion and formulation of relative expressiveness. The next section investigates the simulation of **DL** in **WFDL**, while the following section addresses relative expressiveness within **WFDL**.

2 Defeasible Logic

In this section we can only present an outline of the defeasible logics we investigate. Further details can be obtained from [5] and the references therein. We address propositional defeasible logics, but the results should extend to a first-order language.

A defeasible theory is built from a language Σ of literals (which we assume is closed under negation) and a language Λ of labels. A *defeasible theory* $D = (F, R, >)$ consists of a set of facts F, a finite set of rules R, each rule with a distinct label from Λ, and an acyclic relation $>$ on Λ called the *superiority relation*. This syntax is uniform for all the logics considered here. Facts are individual literals expressing indisputable truths. Rules relate a set of literals (the body), via an arrow, to a literal (the head), and are one of three types: a strict rule, with arrow $\rightarrow$; a defeasible rule, with arrow $\Rightarrow$; or a defeater, with arrow $\leadsto$. Strict rules represent inferences that are unequivocally sound if based on definite knowledge; defeasible rules represent inferences that are generally sound. Inferences suggested by a defeasible rule may fail, due to the presence in the theory of other rules. Defeaters do not support inferences, but may impede inferences suggested by other rules. The superiority relation provides a local priority on rules. Strict or defeasible rules whose bodies are established defeasibly represent claims for the head of the rule to be concluded. The superiority relation contributes to the adjudication of these claims by an inference rule, leading (possibly) to a conclusion. Given a theory D, the corresponding languages are expressed by $\Sigma(D)$ and $\Lambda(D)$.

Defeasible logics derive conclusions that are outside the syntax of the theories. Conclusions may have the form $+dq$, which denotes that under the inference rule d the literal q can be concluded, or $-dq$, which denotes that the logic can establish that under the inference rule d the literal q cannot be concluded. The syntactic element d is called a tag. In general, neither conclusion may be derivable: q cannot be concluded under d, but the logic is unable to establish that. Tags $+\Delta$ and $-\Delta$ represent monotonic provability (and unprovability) where inference is based on facts, strict rules, and modus ponens. We assume these tags and their inference rules are present in every defeasible logic. What distinguishes a logic is the inference rule for defeasible reasoning. The four logics in **DL** correspond to four different pairs of inference rules, labelled ∂, δ, ∂^*, and δ^*; they produce conclusions of the form (respectively) $+\partial q$, $-\partial q$, $+\delta q$, $-\delta q$, etc. The inference rules δ and δ^* require auxiliary tags and inference rules, denoted by σ and σ^*, respectively. For each of the four principal defeasible tags d, the corresponding logic is denoted by $\mathbf{DL}(d)$.

There is not enough space to present all the inference rules for the logics in **DL**, so we focus on Δ, ∂ and ∂^* inference rules, presented in Figure 1. For the remaining inference rules and further properties of the logics in **DL**, see [5]. Some notation in the inference rules requires explanation. Given a literal q, its complement $\sim q$ is defined as follows: if q is a proposition then $\sim q$ is $\neg q$; if q has form $\neg p$ then $\sim q$ is p. We say q and $\sim q$ (and the rules with these literal in the head) *oppose* each other. R_s (R_{sd}) denotes the set of strict rules (strict or defeasible rules) in R. $R[q]$ ($R_s[q]$, etc) denotes the set of rules (respectively, strict rules) of R with head q. Given a rule r, $A(r)$ denotes the set of literals in the body of r.

The inference rules are presented in the form of the definition of a function $\mathcal{T}_D$ for a given theory D. Given a defeasible theory D, for any set of conclusions E, $\mathcal{T}_D(E)$ denotes the set of conclusions inferred from E using D and one application of an inference rule.

The inference rules for ∂ and ∂^* display the two sides of the team defeat issue: the rule r and rules t in the $+\partial$ inference rule form a team overcoming the opposing rules s whereas, in $+\partial^*$ the rule r alone must overcome the opposing rules s. For example, consider the following defeasible theory D on whether animals are mammals [2].

$$
\begin{array}{llll}
r_1 : monotreme \Rightarrow & mammal & \quad r_3 : laysEggs \Rightarrow \neg mammal \\
r_2 : \quad hasFur \Rightarrow & mammal & \quad r_4 : \quad hasBill \Rightarrow \neg mammal \\
\quad r_1 > r_3 & & \\
\quad r_2 > r_4 & &
\end{array}
$$

For a platypus, we have the facts: $monotreme$, $hasFur$, $laysEggs$, and $hasBill$. The rules r_3 and r_4 for $\neg mammal$ are over-ruled by, respectively, r_1 and r_2. Consequently, under inference with team defeat (∂ and δ), we conclude $+\partial mammal$ and $+\delta mammal$. Under inference without team defeat (∂^* and δ^*), there is no rule that overrules all the opposing rules. Consequently we cannot make any positive conclusion; we conclude $-\partial^* mammal$ and $-\partial^* \neg mammal$, and similarly for δ^*.

$+\Delta$) $+\Delta q \in \mathcal{T}_D(E)$ iff either
.1) $q \in F$; or
.2) $\exists r \in R_s[q]$ such that
.1) $\forall a \in A(r), +\Delta a \in E$

$-\Delta$) $-\Delta q \in \mathcal{T}_D(E)$ iff
.1) $q \notin F$, and
.2) $\forall r \in R_s[q]$
.1) $\exists a \in A(r), -\Delta a \in E$

$+\partial$) $+\partial q \in \mathcal{T}_D(E)$ iff either
.1) $+\Delta q \in E$; or
.2) The following three conditions all hold:
.1) $\exists r \in R_{sd}[q]$
$\forall a \in A(r), +\partial a \in E$, and
.2) $-\Delta{\sim}q \in E$, and
.3) $\forall s \in R[{\sim}q]$ either
.1) $\exists a \in A(s), -\partial a \in E$; or
.2) $\exists t \in R_{sd}[q]$ such that
.1) $\forall a \in A(t), +\partial a \in E$, and
.2) $t > s$.

$-\partial$) $-\partial q \in \mathcal{T}_D(E)$ iff
.1) $-\Delta q \in E$, and
.2) either
.1) $\forall r \in R_{sd}[q]$
$\exists a \in A(r), -\partial a \in E$; or
.2) $+\Delta{\sim}q \in E$; or
.3) $\exists s \in R[{\sim}q]$ such that
.1) $\forall a \in A(s), +\partial a \in E$, and
.2) $\forall t \in R_{sd}[q]$ either
.1) $\exists a \in A(t), -\partial a \in E$; or
.2) not($t > s$).

$+\partial^*$) $+\partial^* q \in \mathcal{T}_D(E)$ iff either
.1) $+\Delta q \in E$; or
.2) $\exists r \in R_{sd}[q]$ such that
.1) $\forall a \in A(r), +\partial^* a \in E$, and
.2) $-\Delta{\sim}q \in E$, and
.3) $\forall s \in R[{\sim}q]$ either
.1) $\exists a \in A(s), -\partial^* a \in E$; or
.2) $r > s$.

$-\partial^*$) $-\partial^* q \in \mathcal{T}_D(E)$ iff
.1) $-\Delta q \in E$, and
.2) $\forall r \in R_{sd}[q]$ either
.1) $\exists a \in A(r), -\partial^* a \in E$; or
.2) $+\Delta{\sim}q \in E$; or
.3) $\exists s \in R[{\sim}q]$ such that
.1) $\forall a \in A(s), +\partial^* a \in E$, and
.2) not($r > s$).

Fig. 1. Inference rules for Δ, ∂, and ∂^*

Both ∂ and ∂^* block ambiguity. Consider the following theory D.

$$\begin{array}{llll} r_1: & \Rightarrow p & r_3: & \Rightarrow q \\ r_2: & \Rightarrow \neg p & r_4: & \neg p \Rightarrow \neg q \end{array}$$

p and $\neg p$ are *ambiguous*: neither r_1 nor r_2 can overcome the other via the superiority relation. Thus $-\partial\neg p$ is inferred. Now, because the body of r_4 fails, there is no rule left to compete with r_3, and so $+\partial q$ is inferred. We also conclude $-\partial\neg q$; thus there is no ambiguity about q and $\neg q$, the ambiguity has been blocked. The same arguments apply for ∂^*.

On the other hand, δ and δ^* propagate ambiguity. $-\delta\neg p$ is inferred and consequently $-\delta\neg q$ is inferred. However, ambiguity propagating logics like $\mathbf{DL}(\delta)$ do not support a conclusion $+\delta q$. There is a possibility that $\neg p$ holds, given that r_2 was not overcome via the superiority relation but simply failed to overcome its competitor. Hence there is a possibility that $\neg q$ holds. And since r_3 cannot explicitly overcome r_4 via the superiority relation, the conclusion $+\delta q$ is not justified and, in fact, $-\delta q$ is concluded. This idea of "possibly holding" is called *support*; it is expressed by an auxiliary tag σ and defined by a corresponding inference rule in $\mathbf{DL}(\delta)$ (and, similarly, the auxiliary tag σ^* in $\mathbf{DL}(\delta^*)$). In the theory D above, among the conclusions are $+\sigma p$, $+\sigma\neg p$, $+\sigma\neg q$, and

$+\sigma q$. Since both q and $\neg q$ possibly hold, they are ambiguous and clearly the ambiguity has propagated.

2.1 Well-Founded Defeasible Logic

The inference rules for provability in **WFDL** are exactly the same as in **DL**, while the inference rules for unprovability extend those of **DL**. Hence the function $\mathcal{W}_D$ expressing one-step inference extends $\mathcal{T}_D$. Inference of unprovability in **WFDL** extends **DL** by the identification of "unfounded sets", representing sets of literals that are either directly unprovable in the **DL** sense or not provable and cyclically dependent on each other. In the latter case, **WFDL** detects unprovability that is not detectable by **DL**. Thus the central definitions in **WFDL** are those of a d-unfounded set, for each tag d. These definitions are inspired by the well-founded semantics of logic programs [8].

Consider a defeasible theory $D = (F, R, >)$. For any set of conclusions E and any tag d we define $+d_E = \{q \mid +dq \in E\}$ and $-d_E = \{q \mid -dq \in E\}$. We begin with Δ. A set S of literals is *Δ-unfounded* with respect to an extension E if: For every literal s in S, $s \notin F$ and for every strict rule $B \rightarrow s$ either

- $B \cap -\Delta_E \neq \emptyset$, or
- $B \cap S \neq \emptyset$

This definition extends the $-\Delta$ inference rule with an additional clause that captures the cyclic dependency. Indeed, all definitions of d-unfounded set are obtained in the same way: by adding the alternative that $B \cap S \neq \emptyset$. The corresponding definitions for defeasible inferences are more complex, since there are more factors that influence defeasible inference. Nevertheless, the basic idea is the same.

A set S of literals is *∂-unfounded* with respect to an extension E if: For every literal s in S, $s \in -\Delta$ and for every strict or defeasible rule $r_1 : A(r_1) \hookrightarrow s$ in D either

- $A(r_1) \cap -\partial_E \neq \emptyset$, or
- $A(r_1) \cap S \neq \emptyset$, or
- ${\sim}s \in +\Delta_E$, or
- there is a rule $r_2 : A(r_2) \hookrightarrow {\sim}s$ in D such that $A(r_2) \subseteq +\partial_E$ and for every rule strict or defeasible $r_3 : A(r_3) \hookrightarrow s$ in D either
  - $A(r_3) \cap -\partial_E \neq \emptyset$, or
  - $r_3 \not> r_2$.

Again notice the close relationship between the definition of ∂-unfounded set and the inference rule for ∂-unprovability in **DL**. The second disjunct has been added to that inference rule to capture cyclic dependency. The definition of other d-unfounded sets are obtained from the $-d$ inference rules in the same way.

Clearly the class of d-unfounded sets is closed under unions. Hence there is a greatest d-unfounded set wrt E, denoted by $U_D^d(E)$. We define $\mathcal{U}_D(E) = \{-dq \mid q \in U_D^d(E), d \text{ is a tag}\}$. The function $\mathcal{W}_D$ computes all the inferences that can be made from its argument in one inference step (either through a positive inference or identification of an unfounded set). We define $\mathcal{W}_D(E) = \mathcal{T}_D(E) \cup \mathcal{U}_D(E)$ and the least fixedpoint of $\mathcal{W}_D$ is the set of all conclusions from D, denoted by $WF(D)$. For any tag d and literal q, when $+dq \in WF(D)$ we also write $D \vdash_{WF} +dq$, and similarly for $-dq$.

3 Relative Expressiveness

[14] introduced a framework for addressing the relative expressiveness of defeasible logics. The framework identifies the greater (or equal) expressiveness of L_2 compared to L_1 with the ability to simulate any theory D in a logic L_1 by a theory $T(D)$ in the logic L_2, in the presence of an addition.

The *addition* of a theory A to a theory D is denoted by $D+A$. Addition is essentially the union of the theories, but we require $\Lambda(D) \cap \Lambda(A) = \emptyset$, so that the addition of theories preserves the property that distinct rules have distinct labels. This requirement also has the effect that a superiority statement in D cannot affect a rule in A, and vice versa. Let $D = (F, R, >)$ and $A = (F', R', >')$. Then $D + A = (F \cup F', R \cup R', > \cup >')$. $\Lambda(D+A) = \Lambda(D) \cup \Lambda(A)$ and $\Sigma(D+A) = \Sigma(D) \cup \Sigma(A)$.

A simulating theory $T(D)$ in general will involve additional literals, rules and labels beyond those of D. If additions A were permitted to affect these, the notion of simulation would become trivial, so we restrict additions to have only an indirect effect on $T(D)$, via $\Sigma(D)$. Given a theory D and a possible simulating theory $T(D)$, we say an addition A is *modular* if $\Sigma(A) \cap \Sigma(T(D)) \subseteq \Sigma(D)$, $\Lambda(D) \cap \Lambda(A) = \emptyset$, and $\Lambda(T(D)) \cap \Lambda(A) = \emptyset$. We will consider specific classes of additions but, for each class and any D and $T(D)$, only the modular additions in the class will be considered.

Since different logics involve different tags, conclusions from theories in different logics cannot be identical. For simulation it suffices that conclusions are equal modulo tags. Given logics L_1 and L_2, with principal tags d_1 and d_2, respectively, we say two conclusions α in L_1 and β in L_2 are *equal modulo tags* if α is $+d_1 q$ and β is $+d_2 q$ or α is $-d_1 q$ and β is $-d_2 q$.

Thus we have the following definition of simulation and relative expressiveness. For more discussion on the motivations for the definitions, see [14].

Definition 1. *Let C be a class of defeasible theories.*

We say D_1 in logic L_1 is simulated *by D_2 in L_2 with respect to a class C if, for every modular addition A in C, $D_1 + A$ and $D_2 + A$ have the same conclusions in $\Sigma(D_1 + A)$, modulo tags.*

We say a logic L_1 can be simulated by a logic L_2 with respect to a class C if every theory in L_1 can be simulated by some theory in L_2 with respect to additions from C.

We say L_2 is more (or equal) expressive *than L_1 if L_1 can be simulated by L_2 with respect to C.*

Different notions of relative expressiveness arise from different choices for C. There were two classes of additions investigated in [14]: the addition of *facts* (that is, A has the form $(F, \emptyset, \emptyset)$), and the addition of *rules* (that is, A has the form $(\emptyset, R, \emptyset)$). Simulation with respect to addition of rules is stronger than simulation with respect to addition of facts because any fact can equally be expressed as a strict rule with an empty body. We might also consider arbitrary additions, where A can be any defeasible theory, and no additions, where simulation is simply an implementation of the theory in a different logic.

The main results of [14,15] are that:

- all the **DL** logics have equal expressiveness, with respect to addition of *facts*
- the two ambiguity propagating logics in **DL** are equally expressive, with respect to addition of *rules*, as are the two ambiguity blocking logics in **DL**
- the ambiguity propagating logics and ambiguity blocking logics in **DL** are incomparably expressive, with respect to addition of *rules*
- when arbitrary additions are permitted, of the four defeasible logics in **DL**, none is more expressive than any other

4 Comparing the Expressiveness of WFDL and DL

In some ways, logics in **WFDL** are more powerful than those in **DL**. In terms of inference strength, it is shown in [16] that **WFDL**(∂) is strictly stronger than **DL**(∂), and this result extends to all other tags. Furthermore, the complexity of inference in **WFDL** is quadratic, as compared to linear complexity in **DL**. Thus it is of interest to see whether **WFDL** logics are more expressive than **DL** logics.

As we have seen, the difference between **DL** and **WFDL** is the ability of the latter to infer $-dq$ from loops such as (in the simplest case) $q \Rightarrow q$. As a result, as we might expect, **DL** logics cannot simulate **WFDL** logics with respect to addition of rules. It is less obvious that **WFDL** logics cannot simulate **DL** logics.

Theorem 1. *For every* $d, d' \in \{\partial, \partial^*, \delta, \delta^*, \sigma, \sigma^*\}$,

- **DL**(d') *cannot simulate* **WFDL**(d) *with respect to addition of rules.*
- **WFDL**(d') *cannot simulate* **DL**(d) *with respect to addition of rules.*

The proof uses the addition of a loop $q \Rightarrow q$, and shows that there is no way, within a theory $T(D)$, to counter the different treatment of loops in the two frameworks. Hence, according to the stronger formulation of relative expressiveness, the logics of **DL** and **WFDL** have incomparable expressiveness.

Even if we consider only addition of facts, the different treatment of loops in strict rules distinguishes **DL** logics from **WFDL**. For this reason, we consider a class of theories where this difference is not apparent, to see whether the incomparable expressiveness still holds. A defeasible theory D is *Δ-decisive* in **DL** if for every literal q in $\Sigma(D)$, $D \vdash +\Delta q$ or $D \vdash -\Delta q$. For such theories, the strict consequences are the same under **DL** and **WFDL**. Without this property, no simulation of a **DL** logic in a **WFDL** logic is possible.

Proposition 1. *Suppose* $D + A$ *is not Δ-decisive in* **DL**, *for some addition* A. *Then there is no simulation* $T(D)$ *in* **WFDL** *of* D *in* **DL** *with respect to addition of facts.*

Thus we require theories that remain Δ-decisive after the addition of arbitrary facts. That leads us to the following definitions. The *strict dependency graph* of a propositional defeasible theory D is a directed graph consisting of a vertex for each literal, and an edge from literal q to literal p iff there is a rule in D with head p in which q occurs in the body. A theory is *strict-acyclic* if the strict dependency graph is acyclic (that is, it forms a tree). Strict-acyclicity is sufficient to ensure that, even with additions of facts, the theory is Δ-decisive in **DL**.

Proposition 2. *Let D be any strict-acyclic defeasible theory and A be any set of facts. Then $D{+}A$ is Δ-decisive in* **DL**, *that is, for each $q \in \Sigma$, either $D{+}A \vdash +\Delta q$ or $D{+}A \vdash -\Delta q$. Furthermore, $D{+}A \vdash +\Delta q$ iff $D{+}A \vdash_{WF} +\Delta q$ and $D{+}A \vdash -\Delta q$ iff $D{+}A \vdash_{WF} -\Delta q$*

In the case where the addition is empty, this result covers more cases than Theorem 2.1(a) of [3] because of the use of a different dependency graph.

Strict-acyclicity seems close to being a syntactic characterization of the property that $D{+}A$ is Δ-decisive in **DL** for every addition of facts A. However, it is not a complete characterization, as the following theory D shows.

$$\begin{array}{l} a, c \rightarrow b \\ b, c \rightarrow a \\ c \quad\ \rightarrow b \end{array}$$

Here D is not strict-acyclic but is Δ-decisive in **DL** for every addition of facts A.

Now we establish the relative expressiveness of **DL** and **WFDL** logics when we focus on defeasible inference. We are able to identify a transformation of defeasible theories that proves to be a simulation of **DL** by **WFDL** wrt addition of facts, if we restrict ourselves to defeasible inference. The transformation employs new propositions, $a(q)$, $b(q)$, and $n(q)$, for each q in Σ, and new labels: r' for each $r \in R$, and $e(q), f(q), g(q), h(q)$ and $j(q)$, for each q in Σ.

Definition 2. *Let $D = (F, R, >)$ be a defeasible theory with language Σ. We define the transformation T of D to $T(D) = (F', R', >')$ as follows:*

1. *The facts of $T(D)$ are the facts of D. That is, $F' = F$.*
2. *Every strict rule of R is contained in R'. That is, $R_s \subseteq R'$.*
3. *For each strict or defeasible rule $r = B \hookrightarrow_r q$ in R, R' contains*

$$r' : B \Rightarrow a(q)$$

In addition, for every superiority statement $r > s$ in D, we have $r' >' s'$ in $T(D)$.

4. *For each defeater $r = B \leadsto q$ in R, R' contains*

$$r' : B \leadsto a(q)$$

5. *For each literal q in Σ, R' contains*

$$\begin{array}{rrl} e(q): & a(q) \Rightarrow & \neg n(q) \\ f(q): & \Rightarrow & n(q) \\ g(q): & n(q) \Rightarrow & \neg b(q) \\ h(q): & \Rightarrow & b(q) \\ j(q): & b(q) \Rightarrow & q \end{array}$$

and the superiority relation contains $e(q) > f(q)$ and $g(q) > h(q)$.

This transformation has linear complexity.

The transformation leaves the facts and strict rules unchanged. By Proposition 2, if we restrict our attention to strict-acyclic theories then the transformation provides a simulation of **DL** by **WFDL** for strict conclusions.

Part 5 of this transformation places several rules between the body and the head of rules in D that can be used to derive defeasible knowledge. Thus a rule r is broken into r' (in part 3) and $j(q)$ (in part 5) and the intervening rules $e(q)$, $f(q)$, $g(q)$, and $h(q)$. The effect of these intervening rules is prevent any d-unfounded sets except for those that are consequences of $-d$ inference in **DL**. At the same time, these rules are transparent to $+d$ and $-d$ inferences in **DL**.

These claims are formalized in the following lemma.

Lemma 1. *Let E be the set of conclusions drawn from $T(D)+A$ by* **DL**. *Then, for $d \in \{\partial, \partial^*, \delta, \delta^*, \sigma, \sigma^*\}$, the following statements hold:*

1. $a(q) \in +d_E$ *iff* $b(q) \in +d_E$
2. $a(q) \in -d_E$ *iff* $b(q) \in -d_E$
3. $n(q)$ *is in a d-unfounded set S wrt E iff* $a(q) \in +d_E$
4. $\neg n(q)$ *is in a d-unfounded set S wrt E iff* $a(q) \in S$ *or* $a(q) \in -d_E$
5. $\neg b(q)$ *is in a d-unfounded set S wrt E iff* $n(q) \in S$ *or* $n(q) \in -d_E$ *iff* $a(q) \in +d_E$
6. $b(q)$ *is in a d-unfounded set S wrt E iff* $n(q) \in +d_E$ *iff* $a(q) \in -d_E$ *iff* $b(q) \in -d_E$

These statements are easily verified using the inference rules in **DL** and the definition of d-unfounded sets.

As a consequence, we find that **WFDL**(d) can simulate **DL**(d) in defeasible reasoning (as well as in strict reasoning for our limited class of theories).

Theorem 2. *For strict-acyclic theories,* **WFDL**(d) *can simulate* **DL**(d) *with respect to addition of facts.*

Using results of [14,15] on simulation within **DL** we conclude that, under the restriction on theories, the logics of **WFDL** are more (or equal) expressive than the logics of **DL** with respect to addition of facts.

Theorem 3. *For strict-acyclic theories, and for every $d, d' \in \{\partial, \partial^*, \delta, \delta^*\}$:* **WFDL**$(d)$ *can simulate* **DL**(d') *with respect to addition of facts.*

We might expect that **WFDL** is strictly more expressive than **DL**, that is, that **DL** cannot simulate **WFDL**, but this is not as obvious as it first appears. For example, consider the theory D

$$a \Rightarrow b$$
$$b \Rightarrow a$$

This is a prime candidate for distinguishing **WFDL** from **DL** because it contains a loop, but the two rules in the loop appear necessary to support inferences when facts are added. In **WFDL**(∂) we have $-\partial a$ and $-\partial b$ as consequences of D, but in **DL** neither are consequences. When we add the fact a to D we can conclude $+\partial b$ and, symmetrically, when we add b we can conclude $+\partial a$.

At first glance, D looks difficult to simulate in **DL**, but consider the following theory D':

$$
\begin{array}{llll}
r_1 : a \rightarrow & a' & r_5 : \neg a' \leadsto \neg b \\
r_2 : b \rightarrow & b' & r_6 : \neg b' \leadsto \neg a \\
r_3 : & \Rightarrow \neg a' & r_7 : \quad a \Rightarrow \ b \\
r_4 : & \Rightarrow \neg b' & r_8 : \quad b \Rightarrow \ a
\end{array}
$$

with $r_3 > r_1$ and $r_4 > r_2$. In **DL**, from D' we have $+\partial\neg a'$ and $+\partial\neg b'$, using the superiority relation. Hence the defeaters r_5 and r_6 are applicable. Furthermore, $r_8 \not> r_6$ and $r_7 \not> r_5$, so we can infer $-\partial a$ and $-\partial b$.

If we add a to D' then we can infer $+\Delta a$ and $+\Delta a'$, and hence $-\partial\neg a'$. Thus the defeater r_5 is not applicable. Consequently, using r_7 we infer $+\partial b$. Similarly, adding the fact b allows us to infer $+\partial a$. Thus D' in $\mathbf{DL}(\partial)$ simulates D in $\mathbf{WFDL}(\partial)$ with respect to addition of facts.

It remains unclear whether the logics of **WFDL** are strictly more expressive than the logics of **DL** with respect to addition of facts.

5 The Relative Expressiveness of WFDL Logics

It was established in [14] that the two treatments of ambiguity have incomparable expressiveness wrt addition of rules in **DL**. This result carries across to **WFDL**. Consider the theory D, with rules

$$
\begin{array}{l}
r_1 : \ \Rightarrow \ p \\
r_2 : \ \Rightarrow \neg p
\end{array}
$$

and consider an addition A of rules

$$
\begin{array}{l}
r_3 : \qquad \Rightarrow \neg p \\
r_4 : \qquad \Rightarrow \ q \\
r_5 : \neg p \Rightarrow \neg q
\end{array}
$$

Inference from D and $D{+}A$ is the same whether team defeat or individual defeat is used, because the superiority relation is empty. Similarly, inference in **DL** and **WFDL** is the same, because there are no loops. D introduces ambiguity and the addition A forces identifiable behaviour of the treatment of ambiguity. In the ambiguity propagating logics we have $D{+}A \vdash_{WF} -\delta q$, while in the ambiguity blocking logics $D{+}A \vdash_{WF} +\partial q$. However, assuming there is a simulating theory D', A must be interpreted in the simulating logic, while producing a behaviour reflective of the simulated logic. This turns out not to be possible, for reasons similar to those in [14].

Theorem 4

- $\mathbf{WFDL}(\partial)$ *and* $\mathbf{WFDL}(\partial^*)$ *cannot simulate either* $\mathbf{WFDL}(\delta)$ *or* $\mathbf{WFDL}(\delta^*)$ *with respect to addition of rules.*
- $\mathbf{WFDL}(\delta)$ *and* $\mathbf{WFDL}(\delta^*)$ *cannot simulate either* $\mathbf{WFDL}(\partial)$ *or* $\mathbf{WFDL}(\partial^*)$ *with respect to addition of rules.*

Furthermore, none of the logics in **WFDL** is more expressive than the others, if we consider addition of arbitrary theories.

Proposition 3. *None of the logics* $\mathbf{WFDL}(\partial)$, $\mathbf{WFDL}(\partial^*)$, $\mathbf{WFDL}(\delta)$, *and* $\mathbf{WFDL}(\partial^*)$ *can simulate any of the others, with respect to addition of an arbitrary theory.*

Next, we show that well-founded logics with team defeat can simulate the corresponding logic with individual defeat. We use the same transformation as used in [15] for **DL** (Definition 9 of [15]). The part of the proof addressing $+d$ inferences is similar to that in [15], but the proof addressing $-d$ inferences is different since it must handle d-unfounded sets.

Theorem 5. *The logic* $\mathbf{WFDL}(\partial^*)$ *can be simulated by* $\mathbf{WFDL}(\partial)$, *and the logic* $\mathbf{WFDL}(\delta^*)$ *can be simulated by* $\mathbf{WFDL}(\delta)$, *with respect to addition of rules.*

Finally, for this section, we show that well-founded logics with individual defeat can simulate the corresponding logic with team defeat. Again, we use the transformation used in [15] (Definition 13) for the corresponding theorem for **DL**.

Theorem 6. *The logic* $\mathbf{WFDL}(\partial)$ *can be simulated by* $\mathbf{WFDL}(\partial^*)$, *and* $\mathbf{WFDL}(\delta)$ *can be simulated by* $\mathbf{WFDL}(\delta^*)$, *with respect to addition of rules.*

Combining Theorems 4, 5, and 6, we see that the logics of **WFDL** are divided into two classes of incomparable expressiveness (using the stronger formulation), and the two classes are characterized by their treatment of ambiguity.

6 Conclusion

We have established several relative expressiveness results for well-founded defeasible logics. In particular, we showed that the ambiguity propagating logics of **WFDL** have different expressiveness than the ambiguity blocking logics. We also showed the relationship between the logics of **WFDL** and the logics of **DL**, leaving open the question of whether **DL** can simulate **WFDL** with respect to addition of facts.

These results suggest that a logic from **WFDL** might be preferable to the corresponding logic from **DL**, since defeasible reasoning in a well-founded logic is at least as expressive as the corresponding logic from **DL**, unless it is important to be able to express undecidedness for monotonic provability; **WFDL** is unable to express such undecidedness. On the other hand, the use of team or individual defeat makes no difference to the expressiveness of the logic. Finally, the choice of treatment of ambiguity is significant to the expressiveness of the logic, but the choice should be based on other considerations since the two possibilities have incomparable expressiveness.

The results leave several open problems. Most directly, it is of interest whether all the relative expressiveness results for **DL** [15] carry over to **WFDL**. We have seen in this paper that many of them do. More broadly, the expressiveness relationship between **WFDL** logics and other defeasible logics, such as NDL [19], ADL [17] and Courteous Logic Programs [12] deserves investigation. Finally, and even more broadly, the expressiveness relationship between these defeasible logics and argumentation systems [7,21] is of interest.

References

1. Antoniou, G., Billington, D., Governatori, G., Maher, M.J.: On the modelling and analysis of regulations. In: Proc. Australasian Conf. on Information Systems, pp. 20–29 (1999)
2. Antoniou, G., Billington, D., Governatori, G., Maher, M.J.: Representation results for defeasible logic. ACM Trans. Comput. Log. 2(2), 255–287 (2001)
3. Antoniou, G., Billington, D., Governatori, G., Maher, M.J.: Embedding defeasible logic into logic programming. TPLP 6(6), 703–735 (2006)
4. Bassiliades, N., Antoniou, G., Vlahavas, I.P.: A defeasible logic reasoner for the semantic web. Int. J. Semantic Web Inf. Syst. 2(1), 1–41 (2006)
5. Billington, D., Antoniou, G., Governatori, G., Maher, M.J.: An inclusion theorem for defeasible logics. ACM Trans. Comput. 12(1), 6 (2010)
6. Dumas, M., Governatori, G., ter Hofstede, A.H.M., Oaks, P.: A formal approach to negotiating agents development. Electronic Commerce Research and Applications 1(2), 193–207 (2002)
7. Dung, P.M.: On the acceptability of arguments and its fundamental role in nonmonotonic reasoning, logic programming and n-person games. Artif. Intell. 77(2), 321–358 (1995)
8. Gelder, A.V., Ross, K.A., Schlipf, J.S.: The well-founded semantics for general logic programs. J. ACM 38(3), 620–650 (1991)
9. Governatori, G.: Representing business contracts in *ruleml*. Int. J. Cooperative Inf. Syst. 14(2-3), 181–216 (2005)
10. Governatori, G., Rotolo, A.: Changing legal systems: legal abrogations and annulments in defeasible logic. Logic Journal of the IGPL 18(1), 157–194 (2010)
11. Governatori, G., Rotolo, A., Padmanabhan, V.: The cost of social agents. In: Nakashima, H., Wellman, M.P., Weiss, G., Stone, P. (eds.) AAMAS, pp. 513–520. ACM (2006)
12. Grosof, B.N.: Prioritized conflict handling for logic programs. In: ILPS, pp. 197–211 (1997)
13. Grosof, B.N., Labrou, Y., Chan, H.Y.: A declarative approach to business rules in contracts: courteous logic programs in XML. In: ACM Conference on Electronic Commerce, pp. 68–77 (1999)
14. Maher, M.J.: Relative expressiveness of defeasible logics. TPLP 12(4-5), 793–810 (2012)
15. Maher, M.J.: Relative expressiveness of defeasible logics II. TPLP 13(4-5), 579–592 (2013)
16. Maher, M.J., Governatori, G.: A semantic decomposition of defeasible logics. In: AAAI/IAAI, pp. 299–305. AAAI Press (1999)
17. Maier, F., Nute, D.: Ambiguity propagating defeasible logic and the well-founded semantics. In: Fisher, M., van der Hoek, W., Konev, B., Lisitsa, A. (eds.) JELIA 2006. LNCS (LNAI), vol. 4160, pp. 306–318. Springer, Heidelberg (2006)
18. Maier, F., Nute, D.: Well-founded semantics for defeasible logic. Synthese 176(2), 243–274 (2010)
19. Nute, D.: Defeasible logic. In: Bartenstein, O., Geske, U., Hannebauer, M., Yoshie, O. (eds.) INAP 2001. LNCS (LNAI), vol. 2543, pp. 151–169. Springer, Heidelberg (2003)
20. Prakken, H.: Logical Tools for Modelling Legal Argument: A Study of Defeasible Reasoning in Law. Kluwer (1997)
21. Rahwan, I., Simari, G.: Argumentation in Artificial Intelligence. Springer (2009)
22. Reeves, D.M., Wellman, M.P., Grosof, B.N.: Automated negotiation from declarative contract descriptions. Computational Intelligence 18(4), 482–500 (2002)

Conjunctive Query Answering in $\mathcal{CFD}nc$: A PTIME Description Logic with Functional Constraints and Disjointness

David Toman and Grant Weddell

Cheriton School of Computer Science
University of Waterloo, Canada
{david,gweddell}@cs.uwaterloo.ca

Abstract. We consider conjunctive query answering and other basic reasoning services in $\mathcal{CFD}nc$, an alternative to the description logic $\mathcal{CFD}$ that retains the latter's ability to support PTIME reasoning in the presence of terminological cycles with universal restrictions over functional roles and also in the presence of functional constraints over functional role paths. In contrast, $\mathcal{CFD}nc$ replaces the ability to have conjunction on left-hand-sides of inclusion dependencies with the ability to have primitive negation on right-hand-sides. This makes it possible to say that primitive concepts must denote disjoint sets of individuals, a common requirement with many information sources.

1 Introduction

Scalability issues in reasoning over the semantic web have led the W3C to adopt two description logic (DL) fragments of OWL 2 that are designed to ensure PTIME complexity in the size of respective knowledge bases for a number of important reasoning problems. Called *profiles*, the DLs are $\mathcal{EL}$++ [2] and DL-Lite [1,4]. Medical ontologies were an important motivation for the former, whereas the latter was heavily influenced by a need to access information residing in data sources conforming to relational schema, particularly in cases where the schema has been derived via ER modelling.

Toman and Weddell proposed an alternative to DL-Lite called $\mathcal{CFD}$ that was designed to provide better support for data sources based on relational schema that include more extensive collections of dependencies such as primary and foreign keys [17]. The paper has shown that the problem of deciding *concept subsumption* in $\mathcal{CFD}$ had PTIME complexity, and therefore might qualify as a useful additional option for an OWL 2 profile. However, there are two issues with $\mathcal{CFD}$ that make it less attractive in this role: (1) unlike DL-Lite, it is not possible to say that two primitive concepts must denote disjoint sets of individuals or entities, a common requirement with many information sources, and (2) computing the certain answers to conjunctive queries is PSPACE-complete, even for simple queries of the form $\exists x.\mathrm{A}(x)$, where A is a primitive concept.

In this paper we introduce $\mathcal{CFD}nc$, an alternative to $\mathcal{CFD}$ that retains the latter's key abilities: supporting terminological cycles with universal restrictions over functional roles, and supporting a rich variety of functional constraints over functional role paths. In particular, $\mathcal{CFD}nc$ replaces the ability in $\mathcal{CFD}$ to have conjunction on left-hand-sides of inclusion dependencies with a new ability to have primitive negation on

S. Cranefield and A. Nayak (Eds.): AI 2013, LNAI 8272, pp. 350–361, 2013.

SYNTAX	SEMANTICS: "$(\cdot)^{\mathcal{I}}$"
$\mathsf{C} ::= \mathsf{A}$	$\mathsf{A}^{\mathcal{I}} \subseteq \triangle$
$\mid \neg \mathsf{A}$	$\triangle \setminus \mathsf{A}^{\mathcal{I}}$
$\mid \mathsf{C}_1 \sqcap \mathsf{C}_2$	$\mathsf{C}_1^{\mathcal{I}} \cap \mathsf{C}_2^{\mathcal{I}}$
$\mid \forall \mathsf{Pf}.\mathsf{C}$	$\{x : \mathsf{Pf}^{\mathcal{I}}(x) \in \mathsf{C}^{\mathcal{I}}\}$
$\mid \mathsf{A} : \mathsf{Pf}_1, \ldots, \mathsf{Pf}_k \to \mathsf{Pf}$	$\{x : \forall y \in \mathsf{A}^{\mathcal{I}}. \bigwedge_{i=1}^{k} \mathsf{Pf}_i^{\mathcal{I}}(x) = \mathsf{Pf}_i^{\mathcal{I}}(y) \Rightarrow \mathsf{Pf}^{\mathcal{I}}(x) = \mathsf{Pf}^{\mathcal{I}}(y)\}$

Fig. 1. $\mathcal{CFD}nc$ CONCEPTS

right-hand-sides (the same is also true for the original version of DL-Lite). This removes both problems with $\mathcal{CFD}$. In particular, we show that the following fundamental reasoning problems are in PTIME w.r.t. the size of the knowledge base: (1) *knowledge base consistency*, determining if at least one model exists for a given knowledge base; (2) *logical implication*, determining if a given inclusion dependency is logically entailed by the terminological component of a given knowledge base; (3) *instance checking*, determining if a given concept assertion is entailed by a given knowledge base; and (4) *conjunctive query answering*, computing certain answers for arbitrary conjunctive queries over a $\mathcal{CFD}nc$ knowledge bases. We also show that the combined complexity of CQ answering is complete for PSPACE.

Reasoning in DL-Lite, $\mathcal{EL}$, and their variants often relies on the existence of polynomially-sized *canonical models* (or *canonical structures* that closely resemble such models) to address the above reasoning tasks [10,12]. It is worth noting that $\mathcal{CFD}nc$ does not share this property: an equivalent of a canonical model for a $\mathcal{CFD}nc$ knowledge base is necessarily exponential in the size of the knowledge base.

We begin in the next section by introducing the syntax and semantics of $\mathcal{CFD}nc$ and talk about some of its key features and limitations. The problems above are the focus of Section 4 in which we appeal to an automata-based method for their resolution. This method is introduced in Section 3 where we consider the simpler problem of *concept satisfiability*. Computing certain answers for conjunctive queries is considered in Section 5, and a review of related work and summary comments then follow in Sections 6 and 7, respectively.

2 The Description Logic $\mathcal{CFD}nc$

A formal definition of $\mathcal{CFD}nc$ knowledge bases and the above reasoning problems now follows. Observe that the logic is based on *attributes* or *features* instead of the more common case of *roles* which can denote arbitrary binary relations. However, this is not really an issue, as $\mathcal{CFD}nc$ is ideal for expressing reification for predicates of arbitrary arity [13].

Definition 1 ($\mathcal{CFD}nc$ **Knowledge Bases).** *Let* F, PC *and* IN *be disjoint sets of (names of) attributes, primitive concepts and individuals, respectively. A* path function Pf *is a word in* F^* *with the usual convention that the empty word is denoted by id and concatenation by ".".* Concept descriptions *are defined by the grammar on the left-hand-side of*

Figure 1 in which occurrences of "A" denote primitive concepts. A concept produced by the "A : $\mathsf{Pf}_1, \ldots, \mathsf{Pf}_k \to \mathsf{Pf}$" production of this grammar is called a path functional dependency *(PFD). In addition, to avoid undecidability [16], any occurrence of a PFD must adhere to one of the following two forms:*

$$\begin{array}{l} 1.\ \mathrm{A} : \mathsf{Pf}_1, \ldots, \mathsf{Pf}\,.\,\mathsf{Pf}_i, \ldots, \mathsf{Pf}_k \to \mathsf{Pf}\ \ or \\ 2.\ \mathrm{A} : \mathsf{Pf}_1, \ldots, \mathsf{Pf}\,.\,\mathsf{Pf}_i, \ldots, \mathsf{Pf}_k \to \mathsf{Pf}\,.f \end{array} \tag{1}$$

Metadata and data in a $\mathcal{CFD}nc$ knowledge base $\mathcal{K}$ are respectively defined by a TBox *$\mathcal{T}$ consisting of a finite set of* inclusion dependencies *of the form* $\mathrm{A} \sqsubseteq \mathrm{C}$*, and by an* ABox *$\mathcal{A}$ consisting of a finite set of* concept assertions *of the form* $\mathrm{A}(a)$ *and* path function assertions *of the form* $\mathsf{Pf}_1(a) = \mathsf{Pf}_2(b)$*, where* A *is a primitive concept,* C *an arbitrary concept,* $\{\mathsf{Pf}_1, \mathsf{Pf}_2\} \subseteq \mathsf{F}^*$ *and where* $\{a, b\} \subseteq \mathsf{IN}$.

Semantics is defined in the standard way with respect to a structure $(\triangle, (\cdot)^{\mathcal{I}})$, where $\triangle$ is a domain of "objects" and $(\cdot)^{\mathcal{I}}$ an interpretation function that fixes the interpretation of primitive concepts A *to be subsets of $\triangle$, attributes f to be total functions on $\triangle$, and individuals a to be elements of $\triangle$. The interpretation is extended to path expressions by interpreting id, the empty word, as the identity function $\lambda x.x$, concatenation as function composition, and to derived concept descriptions* C *as defined in Figure 1.*

An interpretation satisfies an inclusion dependency $\mathrm{A} \sqsubseteq \mathrm{C}$ *if* $\mathrm{A}^{\mathcal{I}} \subseteq \mathrm{C}^{\mathcal{I}}$*, a concept assertion* $\mathrm{A}(a)$ *if* $a^{\mathcal{I}} \in \mathrm{A}^{\mathcal{I}}$ *and a path function assertion* $\mathsf{Pf}_1(a) = \mathsf{Pf}_2(b)$ *if* $\mathsf{Pf}_1^{\mathcal{I}}(a^{\mathcal{I}}) = \mathsf{Pf}_2^{\mathcal{I}}(b^{\mathcal{I}})$*. An interpretation satisfies a knowledge base $\mathcal{K}$ if it satisfies each inclusion dependency and assertion in $\mathcal{K}$.* □

The conditions imposed on PFDs in (1) distinguish, for example, PFDs of the form $\mathrm{C} : f \to id$ and $\mathrm{C} : f \to g$ from PFDs of the form $\mathrm{C} : f \to g.h$. This is necessary to retain PTIME complexity for the reasoning problems [9,16] and does not impact the modelling utility of $\mathcal{CFD}nc$ for formatted legacy data sources. It remains possible, for example, to capture arbitrary keys or functional dependencies in a relational schema.

3 TBox and Concept Satisfiability

It is easy to see that every $\mathcal{CFD}nc$ TBox $\mathcal{T}$ is consistent (by setting all primitive concepts to be interpreted as the empty set). For other reasoning tasks, such as concept satisfiability and knowledge base consistency, it is convenient to assume by default, and without loss of generality, that $\mathcal{CFD}nc$ knowledge bases are given in a normal form.

Lemma 2 (TBox and ABox Normal Forms). *For every $\mathcal{CFD}nc$ TBox $\mathcal{T}$, there exists an equivalent TBox $\mathcal{T}'$ that adheres to the following (more limited) grammar for $\mathcal{CFD}nc$ concept descriptions.*

$$\mathrm{C} ::= \mathrm{A} \mid \neg\mathrm{A} \mid \forall f.\mathrm{A} \mid \mathrm{A} : \mathsf{Pf}_1, \ldots, \mathsf{Pf}_k \to \mathsf{Pf}$$

Also, for every ABox $\mathcal{A}$, there exists an equivalent ABox $\mathcal{A}'$ containing only assertions of the form $f(a) = b$ and $a = b$. □

Obtaining $\mathcal{T}'$ and $\mathcal{A}'$ from an arbitrary knowledge base $\mathcal{K}$ is achieved by a straightforward introduction of auxiliary names for intermediate concept descriptions and individuals (e.g., see defn. of *simple concepts* in [15,16]); the normalized TBox and ABox are linear in the size of the inputs.

Definition 3 (A Transition Relation for $\mathcal{T}$). *Let $\mathcal{T}$ be a $\mathcal{CFD}nc$ TBox in normal form. We define a transition relation $\delta(\mathcal{T})$ over the set of states $S = \mathsf{PC} \cup \{\neg \mathrm{A} \mid \mathrm{A} \in \mathsf{PC}\}$ and the alphabet F as follows:*

$$\mathrm{A}_1 \xrightarrow{\epsilon} \mathrm{A}_2 \in \delta(\mathcal{T}) \text{ if } \mathrm{A}_1 \sqsubseteq \mathrm{A}_2 \in \mathcal{T}$$
$$\mathrm{A}_1 \xrightarrow{\epsilon} \neg\mathrm{A}_2 \in \delta(\mathcal{T}) \text{ if } \mathrm{A}_1 \sqsubseteq \neg\mathrm{A}_2 \in \mathcal{T}$$
$$\mathrm{A}_1 \xrightarrow{f} \mathrm{A}_2 \in \delta(\mathcal{T}) \text{ if } \mathrm{A}_1 \sqsubseteq \forall f.\mathrm{A}_2 \in \mathcal{T}$$

where ϵ is the empty letter transition and $f \in \mathsf{F}$. □

The transition relation allows us to construct *non-deterministic finite automata* (NFA) that can be used for various reasoning problems that relate to a $\mathcal{CFD}nc$ TBox $\mathcal{T}$. Note that we follow common practice in automata theory and use ϵ for the empty letter in transition relations.[1]

Lemma 4. *Let $M = (S, \{\mathrm{A}\}, \{\mathrm{B}\}, \delta(\mathcal{T}))$ be an NFA with the set of states S (as above), start state* A, *final state* B, *and transition relation $\delta(\mathcal{T})$. Then $\mathcal{T} \models \mathrm{A} \sqsubseteq \forall \mathsf{Pf}.\mathrm{B}$ whenever $\mathsf{Pf} \in \mathcal{L}(M)$.*

Proof (sketch) For $\mathsf{Pf} \in \mathcal{L}(M)$ there must be a run

$$\mathrm{A} = \mathrm{A}_0 \xrightarrow{l_1} \mathrm{A}_1 \xrightarrow{l_2} \mathrm{A}_2 \cdots \mathrm{A}_{k-1} \xrightarrow{l_k} \mathrm{A}_k = \mathrm{B}$$

in M where $l_i \in \mathsf{F} \cup \{\epsilon\}$ and such that $\mathsf{Pf} = l_1.l_2.\cdots.l_k$. It follows from the definition of $\delta(\mathcal{T})$ that $\mathrm{A}_{i-1} \xrightarrow{l_i} \mathrm{A}_i$ exists if $\mathrm{A}_{i-1} \sqsubseteq \mathrm{A}_i$, for $l_i = \epsilon$, or $\mathrm{A}_{i-1} \sqsubseteq \forall l_i.\mathrm{A}_i$, for $l_i \in \mathsf{F}$ (and hence these dependencies are trivially implied by $\mathcal{T}$). The claim then follows by simple transitive reasoning, all necessary cases derive from the fact that

$$\{\mathrm{B}_1 \sqsubseteq \forall \mathsf{Pf}.\mathrm{B}_2, \mathrm{B}_2 \sqsubseteq \forall \mathsf{Pf}'.\mathrm{B}_3\} \models \mathrm{B}_1 \sqsubseteq \forall \mathsf{Pf}.\mathsf{Pf}'.\mathrm{B}_3,$$

and the lemma then follows by induction on the length of the run. □

Note that the converse implication in this lemma may not hold, e.g., when A is inconsistent with respect to $\mathcal{T}$.

The problem of *concept satisfiability* asks, for a given concept C and TBox $\mathcal{T}$, if there exists an interpretation $\mathcal{I}$ for $\mathcal{T}$ in which $\mathrm{C}^{\mathcal{I}}$ is non-empty. Such problems can be reduced to the case where C is a primitive concept A by simply augmenting $\mathcal{T}$ with $\{\mathrm{A} \sqsubseteq \mathrm{C}\}$, where A is a fresh primitive concept.

Given a primitive concept A and TBox $\mathcal{T}$, one can test for primitive concept satisfiability by using the following NFA, denoted $\mathsf{nfa}^a_{\mathrm{B}}(\mathcal{T}, \{\mathrm{A}(a)\})$:

$$(S \cup \{a\}, \{a\}, \{\mathrm{B}\}, \delta(\mathcal{T}) \cup \{a \xrightarrow{\epsilon} A\}),$$

with states given by primitive concepts, their negations, and a distinguished node a, with start state a, with final state $\mathrm{B} \in S$, and with transition relation $\delta(\mathcal{T}) \cup \{a \xrightarrow{\epsilon} A\}$.

Theorem 5 (Concept Satisfiability). A *is satisfiable with respect to the TBox $\mathcal{T}$ if and only if $\mathcal{L}(\mathsf{nfa}^a_{\mathrm{B}}(\mathcal{T}, \{\mathrm{A}(a)\})) \cap \mathcal{L}(\mathsf{nfa}^a_{\neg\mathrm{B}}(\mathcal{T}, \{\mathrm{A}(a)\})) = \emptyset$ for every* $\mathrm{B} \in \mathsf{PC}$.

Proof (sketch) For a primitive concept $\mathrm{B} \in \mathsf{PC}$, a word Pf in the intersection language of the two automata above is a witness of the fact that $\mathsf{Pf}^{\mathcal{I}}(a^{\mathcal{I}}) \in \mathrm{B}^{\mathcal{I}}$ and $\mathsf{Pf}^{\mathcal{I}}(a^{\mathcal{I}}) \in$

[1] Another option would have been to use id for this purpose, but we thought, on balance, that this would hinder readability.

$\neg B^{\mathcal{I}}$ must hold in every model of $\mathcal{T}$, for reasons analogous to the proof of Lemma 4, which leads to a contradiction since Pf is a (total) function.

Conversely, if no such word exists then one can construct a *deterministic* finite automaton from $\mathsf{nfa}^a_{\mathrm{B}}(\mathcal{T}, \{\mathrm{A}(a)\})$, using the standard subset construction, in which no state containing both B and ¬B is reachable from the start state $\{a\}$. Unfolding the transition relation of this automaton, starting from the state $\{a\}$, labelling nodes by the concepts associated with the automaton's states, and adding missing features to complete trees in which no primitive concept is true for any node, yields a tree interpretation that satisfies $\mathcal{T}$ (in particular in which all PFD constraints are satisfied vacuously) and whose root a provides a witness for satisfiability of A. □

Since the $|\mathsf{nfa}^a_{\mathrm{B}}(\mathcal{T}, \{\mathrm{A}(a)\})| \in \mathcal{O}(|\mathcal{T}|)$ and all the needed automata operations can be implemented in PTIME [7], the following result is immediate.

Corollary 6. *Concept satisfiability with respect to $\mathcal{CFD}nc$ TBoxes is in PTIME.*

Note that it is impossible to *precompute* all inconsistent concepts since this would require consideration of all possible *types* over PC (i.e., finite subsets of primitive concepts), a process essentially equivalent to constructing an equivalent deterministic automaton which can require exponential time [7].

4 ABox Reasoning and $\mathcal{K}$ Satisfiability

The automata-based approach to *concept satisfiability* can be extended to the more general problem of knowledge base consistency. Intuitively, each ABox individual a must be linked to the TBox automaton in a fashion similar to how the "prototypical object" a was linked in Section 3. This idea leads to the following definition:

Definition 7 (A Transition Relation for $\mathcal{A}$). *Let $\mathcal{A}$ be a $\mathcal{CFD}nc$ ABox in normal form. We create a transition relation $\delta(\mathcal{A})$ for an* nfa *over the set of states $S = \mathsf{PC} \cup \{a \mid a \text{ in } \mathcal{A}\}$ and the alphabet* F *as follows:*

$$\begin{aligned} a \xrightarrow{\epsilon} a &\in \delta(\mathcal{A}) \text{ if } a \text{ appears in } \mathcal{A},\\ a \xrightarrow{\epsilon} \mathrm{A} &\in \delta(\mathcal{A}) \text{ if } \mathrm{A}(a) \in \mathcal{A},\\ a \xrightarrow{f} b &\in \delta(\mathcal{A}) \text{ if } f(a) = b \in \mathcal{A} \text{ and}\\ a \xrightarrow{\epsilon} b, b \xrightarrow{\epsilon} a &\in \delta(\mathcal{A}) \text{ if } a = b \in \mathcal{A}. \end{aligned}$$

where ϵ is the empty letter transition. □

Observe that we have used ϵ transitions to simulate equality assertions in $\mathcal{A}$. This is justified, e.g., by considering the ABox individuals to be nominals.

(*aside on notation*) Hereon, we write "$n \overset{\mathsf{Pf}}{\rightsquigarrow} m$ in δ" if $\mathsf{Pf} \in \mathcal{L}(\mathsf{nfa}(S, \{n\}, \{m\}, \delta))$, where S is a set of states (that will be clear from the context), m and n are states in S, and δ is a NFA transition relation over S (also be clear from context). (*end of aside*)

Unfortunately, taking $\delta(\mathcal{T}) \cup \delta(\mathcal{A})$ alone as the transition relation of an NFA and then testing for consistency of every ABox individual (as in Theorem 5) is not sufficient as the following cases illustrate. The problems raised by each case will be addressed by defining rules that impose additional *closure* conditions on the transition relation. To begin, we need to ensure that ABox assertions $f(a) = b$ are functional:

Example 8 (Path Function Assertions). *Consider the ABox $\mathcal{A} = \{f(a) = b, f(a) = c\}$. Clearly $b^{\mathcal{I}}$ must equal $c^{\mathcal{I}}$ in any model $\mathcal{I}$ of a knowledge base that includes $\mathcal{A}$.* □

To remedy this, we define a *functionality rule* for $\delta(\mathcal{T}, \mathcal{A})$ as follows:

if $a \overset{f}{\rightsquigarrow} b$ and $a \overset{f}{\rightsquigarrow} c$ in $\delta(\mathcal{T}, \mathcal{A})$ **then** $\{b \overset{\epsilon}{\rightarrow} c, c \overset{\epsilon}{\rightarrow} b\} \subseteq \delta(\mathcal{T}, \mathcal{A})$.

Next, we need to ensure that ABox assertions of the form $f(a) = b$ are coherent with TBox assertions $\mathrm{A} \sqsubseteq \forall f.\mathrm{B}$ with respect to concept memberships of a and b:

Example 9 (ABox and Value Restrictions). *Consider the TBox $\mathcal{T} = \{\mathrm{A} \sqsubseteq \forall f.\mathrm{B}\}$ and an ABox $\mathcal{A} = \{f(a) = b, \mathrm{A}(a)\}$. Clearly, in any model $\mathcal{I}$ of the knowledge base $(\mathcal{T}, \mathcal{A})$, $b^{\mathcal{I}}$ must be an element of $\mathrm{B}^{\mathcal{I}}$. However,* B cannot *be reached from b in $\delta(\mathcal{T}) \cup \delta(\mathcal{A})$, and therefore an automaton based on such a transition relation alone cannot reflect the correct concept membership of b.* □

We define a *coherence rule* for the transition relation $\delta(\mathcal{T}, \mathcal{A})$ to remedy this as follows:

if $a \overset{f}{\rightsquigarrow} b$, $a \overset{\epsilon}{\rightsquigarrow} A$, and $A \overset{f}{\rightsquigarrow} B$ in $\delta(\mathcal{T}, \mathcal{A})$ **then** $b \overset{\epsilon}{\rightarrow} B \in \delta(\mathcal{T}, \mathcal{A})$.

Finally, PFDs are no longer trivially satisfied as the ABox $\mathcal{A}$ may not be tree shaped.

Example 10 (ABox and PFDs). *Consider $\mathcal{A} = \{A(a), B(b), f(a) = c, f(b) = c\}$.*

- *A TBox $\mathcal{T} = \{\mathrm{A} \sqsubseteq \mathrm{B} : f \to id\}$ implies that the individuals a and b must denote the same domain element.*
- *A TBox $\mathcal{T} = \{\mathrm{A} \sqsubseteq \mathrm{B} : f \to g\}$ implies that there must be an additional (anonymous) individual d such that $g(a) = d$ and $g(b) = d$.*

Note that the PFD $\mathrm{A} \sqsubseteq \mathrm{B} : f.g \to id$ is also *violated by the pair of individuals a and b, this despite the fact that neither of these two individuals is the origin of an* explicit *$f.g$ path in $\mathcal{A}$: since features are interpreted as total functions, the individual c must have an "outgoing" g feature, and therefore a and b must agree on $f.g$.* □

A remedy for these cases is obtained by defining a *PFD closure rule* for the transition relation $\delta(\mathcal{T}, \mathcal{A})$ for each PFD $\mathrm{A} \sqsubseteq \mathrm{B} : \mathsf{Pf}_1, \ldots, \mathsf{Pf}_k \to \mathsf{Pf} \in \mathcal{T}$. The rule will refer to the following auxiliary functions.

$\mathsf{match}(a, b, \mathsf{Pf}, \delta(\mathcal{T}, \mathcal{A}))$: Returns *true* if there is a (possibly empty) prefix Pf' of Pf such that $a \overset{\mathsf{Pf}'}{\rightsquigarrow} c$ and $b \overset{\mathsf{Pf}'}{\rightsquigarrow} c$ in $\delta(\mathcal{T}, \mathcal{A})$ for some individual c; it returns *false* otherwise.

$\mathsf{expf}(a, \mathsf{Pf}, \delta(\mathcal{T}, \mathcal{A}))$: Returns the minimal set of transitions (by creating new individuals) such that $a \overset{\mathsf{Pf}}{\rightsquigarrow} c$ in $\delta(\mathcal{T}, \mathcal{A})$ holds for some c.

$\mathsf{mkeq}(a, b, \mathsf{Pf}, \delta(\mathcal{T}, \mathcal{A}))$: Returns $\{c \overset{\epsilon}{\rightarrow} d, d \overset{\epsilon}{\rightarrow} c\}$ where, for some individuals c and d, we have $a \overset{\mathsf{Pf}}{\rightsquigarrow} c$ and $b \overset{\mathsf{Pf}}{\rightsquigarrow} d$ in $\delta(\mathcal{T}, \mathcal{A})$.

The PFD closure rule is then defined as follows:

if $\{a \overset{\epsilon}{\rightsquigarrow} A, b \overset{\epsilon}{\rightsquigarrow} B\} \subseteq \delta(\mathcal{T}, \mathcal{A})$ **and**
  $\mathsf{match}(a, b, \mathsf{Pf}_i, \delta(\mathcal{T}, \mathcal{A}))$, for $0 < i \leq k$, **and** not $\mathsf{match}(a, b, \mathsf{Pf}, \delta(\mathcal{T}, \mathcal{A}))$
then $\mathsf{expf}(a, \mathsf{Pf}, \delta(\mathcal{T}, \mathcal{A})) \subseteq \delta(\mathcal{T}, \mathcal{A})$, $\mathsf{expf}(b, \mathsf{Pf}, \delta(\mathcal{T}, \mathcal{A})) \subseteq \delta(\mathcal{T}, \mathcal{A})$, and
  $\mathsf{mkeq}(a, b, \mathsf{Pf}, \delta(\mathcal{T}, \mathcal{A})) \subseteq \delta(\mathcal{T}, \mathcal{A})$

The rules enable one to define a transition relation for an NFA that captures reasoning in the knowledge base $(\mathcal{T}, \mathcal{A})$ as follows.

Definition 11 (Transition Relation $\delta(\mathcal{T}, \mathcal{A})$). *Let $\delta(\mathcal{T}, \mathcal{A})$ be the smallest transition relation containing $\delta(\mathcal{T})$ and $\delta(\mathcal{A})$ that is closed under the functionality, coherence, and PFD closure rules.* □

Note that $\delta(\mathcal{T}, \mathcal{A})$ is constructed by applying the closure rules to $\delta(\mathcal{T}) \cup \delta(\mathcal{A})$ until a fixpoint is reached (in a polynomial number of steps). We use $\delta(\mathcal{T}, \mathcal{A})$ as the transition function for the NFA $\mathsf{nfa}^a_{\mathrm{B}}(\mathcal{T}, \mathcal{A})$ with the start state $\{a\}$ and final state B (similarly to Section 3).

Theorem 12 (Knowledge Base Consistency). *A knowledge base $(\mathcal{T}, \mathcal{A})$ is consistent if and only if*

$$\mathcal{L}(\mathsf{nfa}^a_{\mathrm{B}}(\mathcal{T}, \mathcal{A})) \cap \mathcal{L}(\mathsf{nfa}^a_{\neg\mathrm{B}}(\mathcal{T}, \mathcal{A}))$$

is empty for all primitive concepts $B \in \mathsf{PC}$ and all ABox individuals a in $\mathcal{A}$.

Proof (sketch) Assume $\mathsf{Pf} \in \mathcal{L}(\mathsf{nfa}^a_{\mathrm{B}}(\mathcal{T}, \mathcal{A})) \cap \mathcal{L}(\mathsf{nfa}^a_{\neg\mathrm{B}}(\mathcal{T}, \mathcal{A}))$ for some path function Pf, individual a and primitive concept B, and that $\mathcal{I} \models (\mathcal{T}, \mathcal{A})$. Composing all the assertions corresponding to the transitions in $\delta(\mathcal{T}, \mathcal{A})$ along the runs corresponding to Pf in the two automata, however, implies that $\mathsf{Pf}^{\mathcal{I}}(a^{\mathcal{I}}) \in \mathrm{B}^{\mathcal{I}}$ and $\mathsf{Pf}^{\mathcal{I}}(a^{\mathcal{I}}) \in \neg\mathrm{B}^{\mathcal{I}}$ (similarly to Lemma 4); a contradiction as interpretations of path functions are functional.

For the other direction we define an interpretation $\mathcal{I}$ as follows: let $\lceil a \rceil$ be an representative of the equivalence class $\{a \mid a \overset{\epsilon}{\rightsquigarrow} b, b \overset{\epsilon}{\rightsquigarrow} a \text{ in } \delta(\mathcal{T}, \mathcal{A})\}$ and let $\mathrm{PF}(a)$ denote

$$\{f.\mathsf{Pf} \mid a \overset{f}{\rightsquigarrow} b \text{ not in } \delta(\mathcal{T}, \mathcal{A})\} \text{ for any individual } b\}.$$

Then set

- $\Delta^{\mathcal{I}} = \bigcup_{a \text{ in } \mathcal{A}} \{\lceil a \rceil . id\} \cup \{\lceil a \rceil . \mathsf{Pf} \mid \mathsf{Pf} \in \mathrm{PF}(a)\}$;
- $a^{\mathcal{I}} = \lceil a \rceil . id$;
- $\mathrm{A}^{\mathcal{I}} = \{\lceil a \rceil . \mathsf{Pf} \mid a \overset{\mathsf{Pf}}{\rightsquigarrow} \mathrm{A} \text{ in } \delta(\mathcal{T}, \mathcal{A})\}$; and
- $f^{\mathcal{I}} = \{(\lceil a \rceil . id, \lceil b \rceil . id) \mid a \overset{f}{\rightsquigarrow} b \text{ in } \delta(\mathcal{T}, \mathcal{A})\} \cup \{(\lceil a \rceil . \mathsf{Pf}, \lceil a \rceil . \mathsf{Pf}.f) \mid \lceil a \rceil . \mathsf{Pf}, \lceil a \rceil . \mathsf{Pf}.f \in \Delta^{\mathcal{I}}\}$.

It is immediate that $\mathcal{I} \models \mathcal{A}$ since $\delta(\mathcal{A}) \subseteq \delta(\mathcal{T}, \mathcal{A})$ and we corrected for all violations of PFDs. By inspecting $\mathcal{T}$ it is also easy to see that $\mathcal{I} \models \mathcal{T}$. □

Note that the core of this construction is again the subset construction for NFA determinization (cf. Theorem 5) where the TBox-ABox interactions are facilitated by the closure rules. What remains is to show that knowledge base consistency can be checked in PTIME.

Lemma 13. *$|\delta(\mathcal{T}, \mathcal{A})|$ is polynomial in $|\mathcal{T}| + |\mathcal{A}|$.*

Proof (sketch) The number of individuals in $\delta(\mathcal{T}, \mathcal{A})$ is bounded by $|\mathcal{A}| + 2|\mathcal{T}||\mathcal{A}|^2$ since the PFD closure rule can add at most two new individuals per pair of individuals in $\mathcal{A}$ and PFD in $\mathcal{T}$. Thus, since the number of states is polynomial in $|\mathcal{T}| + |\mathcal{A}|$, the number of transitions in $\delta(\mathcal{T}, \mathcal{A})$ is also at most polynomial in $|\mathcal{T}| + |\mathcal{A}|$. □

This result, together with the argument we made for concept satisfiability with respect to a TBox, yields a PTIME algorithm for KB consistency. Since we do not assume the unique name assumption, the problem is also PTIME-hard (we have Horn-SAT embedded in reasoning with the PFDs alone).

Corollary 14. *Knowledge base consistency for $\mathcal{CFD}nc$ is PTIME-complete.* □

4.1 Logical Implication

Now we consider the logical implication questions of the form $(\mathcal{T}, \mathcal{A}) \models \mathrm{C}(a)$, $(\mathcal{T}, \mathcal{A}) \models \mathsf{Pf}_1(a) = \mathsf{Pf}_2(b)$, and ultimately $\mathcal{T} \models \mathrm{A} \sqsubseteq \mathrm{C}$. Since C can be a complex concept and $\mathcal{CFD}nc$ is not closed under negation, logical implication must be resolved by asking several separate questions by exhaustively applying the following simplification rules:

$$\mathsf{Simp}(\forall\,\mathsf{Pf}\,.\mathrm{C}_1 \sqcap \mathrm{C}_2) \to \mathsf{Simp}(\forall\,\mathsf{Pf}\,.\mathrm{C}_1) \cup \mathsf{Simp}(\forall\,\mathsf{Pf}\,.\mathrm{C}_2)$$
$$\mathsf{Simp}(\forall\,\mathsf{Pf}\,.\forall\,\mathsf{Pf}'\,.\mathrm{C}_1) \to \mathsf{Simp}(\forall\,\mathsf{Pf}\,.\,\mathsf{Pf}'\,.\mathrm{C}_1)$$

obtaining a set of irreducible concepts of the forms $\forall\,\mathsf{Pf}\,.\mathrm{A}$, $\forall\,\mathsf{Pf}\,.\neg\mathrm{A}$, and $\forall\,\mathsf{Pf}\,.\mathrm{A} : \mathsf{Pf}_1, \ldots, \mathsf{Pf}_k \to \mathsf{Pf}'$. We call the irreducible concepts obtained by $\mathsf{Simp}(\mathrm{C})$ *simplifications* of the concept C. Note that $|\,\mathsf{Simp}(\mathrm{C})| \in \mathcal{O}(|\mathrm{C}|^2)$.

Lemma 15. *$(\mathcal{T}, \mathcal{A}) \models \mathrm{C}(a)$ ($\mathcal{T} \models \mathrm{A} \sqsubseteq \mathrm{C}$) if and only if $(\mathcal{T}, \mathcal{A}) \models \mathrm{D}(a)$ ($\mathcal{T} \models \mathrm{A} \sqsubseteq \mathrm{D}$, respectively) for all $\mathrm{D} \in \mathsf{Simp}(\mathrm{C})$.*

Proof (sketch) By observing that the each step of simplifications preserves logical implication. □

The simplified logical implication questions can now be reduced in a natural way to $\mathcal{CFD}nc$ knowledge base satisfiability as follows:

Theorem 16 (Instance Checking)

1. *$(\mathcal{T}, \mathcal{A}) \models \forall\,\mathsf{Pf}\,.\mathrm{A}(a)$ iff $(\mathcal{T}, \mathcal{A} \cup \{\forall\,\mathsf{Pf}\,.\neg\mathrm{A}(a)\})$ is not satisfiable.*
2. *$(\mathcal{T}, \mathcal{A}) \models \forall\,\mathsf{Pf}\,.\neg\mathrm{A}(a)$ iff $(\mathcal{T}, \mathcal{A} \cup \{\forall\,\mathsf{Pf}\,.\mathrm{A}(a)\})$ is not satisfiable.*
3. *$(\mathcal{T}, \mathcal{A}) \models (\forall\,\mathsf{Pf}\,.\mathrm{A} : \mathsf{Pf}_1, \ldots, \mathsf{Pf}_k \to \mathsf{Pf}')(a)$ iff*
$$(\mathcal{T}, \mathcal{A} \cup \{\mathsf{Pf}(a) = b, \mathrm{A}(c), \mathrm{D}(\mathsf{Pf}'(b)), \neg\mathrm{D}(\mathsf{Pf}'(c)\} \cup \bigcup_{0<i\leq k} \{\mathsf{Pf}_i(b) = \mathsf{Pf}_i(c)\})$$
is not satisfiable, for b and c fresh individuals and D a fresh primitive concept.
4. *$(\mathcal{T}, \mathcal{A}) \models (\mathsf{Pf}_1(a) = \mathsf{Pf}_2(b))$ iff $(\mathcal{T}, \mathcal{A} \cup \{\mathrm{D}(\mathsf{Pf}_1(a)), \neg\mathrm{D}(\mathsf{Pf}_2(b))\}$ is not satisfiable, where D a fresh primitive concept.* □

For *logical implication* questions of the form $\mathcal{T} \models \mathrm{A} \sqsubseteq \mathrm{C}$, where C is irreducible, simply replace the ABox $\mathcal{A}$ in the above by $\{\mathrm{A}(a)\}$. The results then follow by virtue of the first three cases in the preceding theorem. Overall, we have the following:

Corollary 17. *Both instance checking and logical implication for $\mathcal{CFD}nc$ are in PTIME.*

5 Conjunctive Queries

A *conjunctive query* (CQ) is an expression of the form $\{\bar{x} \mid \exists\bar{y}.\text{BODY}\}$ where BODY is a conjunction of atomic formulas of the form $\mathrm{C}(x)$ and $f(x) = y$ for C a $\mathcal{CFD}nc$ concept description not containing PFDs, and x and y are variables among $\bar{x} \cup \bar{y}$. We often conflate the BODY of the query with the set of its atomic conjuncts. We call the variables $\bar{x}$ the *answer variables*. An *answer to a CQ* φ w.r.t. a KB $\mathcal{K}$ is a vector of individuals $\bar{a} \subseteq \mathsf{IN}$ such that $\mathcal{K} \models \varphi(\bar{a})$ where $\varphi(\bar{a})$ is a formula obtained from φ by substituting $\bar{x}$ by $\bar{a}$. We assume that CQs are connected; otherwise we simply process each component separately.

To compute answers for a CQ φ we use the notion of *CQ folding*; we assume (w.l.o.g.) that exactly one concept is associated with each variable in BODY of φ:

Definition 18. *Let φ be a CQ. We define a set $\mathsf{Fold}(\varphi)$ to be the least set of CQ that contains φ and is closed under the following two rules.*

1. *If $\{\bar{x} \mid \exists\bar{y}.\text{BODY}\} \in \mathsf{Fold}(\varphi)$, $\{f(x) = y, \mathrm{C}(y)\} \subseteq$ BODY and y does not appear elsewhere in BODY nor in $\bar{x}$, then $\{\bar{x} \mid \exists\bar{y}.\text{BODY} - \{f(x) = y, \mathrm{C}(y)\} \cup \{\forall f.(\mathrm{C})(x)\}\} \in \mathsf{Fold}(\varphi)$.*
2. *If $\{\bar{x} \mid \exists\bar{y}.\text{BODY}\} \in \mathsf{Fold}(\varphi)$ and $\{f(x) = y, f(x') = y\} \subseteq$ BODY, then $\{\bar{x} \mid \exists\bar{y}.\text{BODY}\}[x/x'] \in \mathsf{Fold}(\varphi)$;*

The intuition behind this definition is that to find query answers it is now sufficient to *match* the queries in $\mathsf{Fold}(\varphi)$ explicitly against the (extended) ABox (nodes that denote individuals in $\delta(\mathcal{T}, \mathcal{A})$) and verify correct concept membership for these nodes as prescribed by the query as possible matches outside of this ABox are reduced to instance checks against concepts obtained by the *folding* process.

Lemma 19. *Let φ be a CQ with at least one answer variable. Then $\bar{a}$ is an answer to φ over $\mathcal{K} = (\mathcal{T}, \mathcal{A})$ if and only if there is a mapping $\mu : \bar{x} \cup \bar{y} \to \mathsf{SA}$ where SA is the set of states of the $\mathsf{nfa}(\mathcal{T}, \mathcal{A})$ that correspond to objects, such that*

1. *$\mu(x)$ is a state corresponding to an individual in $\mathcal{A}$ for $x \in \bar{x}$ an answer variable;*
2. *$\mu(x) \overset{f}{\rightsquigarrow} \mu(y) \in \delta(\mathcal{T}, \mathcal{A})$ for all $f(x) = y \in$ BODY; and*
3. *$\mu(x)$ satisfies concept C for all $\mathrm{C}(x) \in$ BODY (as in Theorem 16),*

for at least one $\{\bar{x} \mid \exists\bar{y}.\text{BODY}\} \in \mathsf{Fold}(\varphi)$.

Proof (sketch) Observing that $\delta(\mathcal{T}, \mathcal{A})$ restricted to SA is essentially a part of the minimal model of $\mathcal{K}$ (as $\mathcal{K}$ is essentially Horn) and that every element of $\mathsf{Fold}(\varphi)$ implies φ, it is easy to see that whenever (1-3) are satisfied, there will be a match of φ in the minimal model and thus $\bar{a}$ will be an answer. Conversely, if a match of φ in a minimal model exists yielding $\bar{a}$ answer, then part of the match will be realized in $\delta(\mathcal{T}, \mathcal{A})$ restricted to o SA (as at least one variable must be bound to an ABox individual) and the reminder of the match must be forest-like. Hence one of the queries in $\mathsf{Fold}(\varphi)$ will match in $\delta(\mathcal{T}, \mathcal{A})$ and the remaining conjuncts will be verified using the *folded* concepts in that query in condition (3). □

For CQ without answer variables, we need an additional step that checks whether the query (when equivalent to a concept) matches in the tree part of every interpretation of $\mathcal{K}$. We use the following construction: Let C be a $\mathcal{CFD}nc$ concept not containing a PFD and $\mathsf{Simp}(\mathrm{C}) = \{\forall\,\mathsf{Pf}_1.\mathrm{L}_1, \ldots, \forall\,\mathsf{Pf}_k.\mathrm{L}_k\}$. We define an automaton

$$M(\mathrm{C}) = \mathsf{nfa}^s_{F_1}(\mathcal{T}, \mathcal{A}') \times \ldots \times \mathsf{nfa}^s_{F_k}(\mathcal{T}, \mathcal{A}').$$

where $\mathcal{A}' = \mathcal{A} \cup \{f_i(s) = a_i \mid a_i \text{ an individual in } \mathcal{A}\}$ such that s and f_i do not appear in $\mathcal{T} \cup \mathcal{A}$ and, for $0 < j \leq k$, F_j is a set of final states: those states m in $\mathsf{nfa}(\mathcal{T}, \mathcal{A}')$ for which $m \overset{\mathsf{Pf}_j}{\rightsquigarrow} \mathrm{L}_j$

Lemma 20. *Let φ be a CQ without answer variables. Then $\mathcal{K} \models \varphi$ if and only if conditions (2) and (3) in Lemma 19 are met, or if there is a concept* C $\in \mathsf{Fold}(\varphi)$ *such that $M(\mathrm{C})$ is nonempty.*

Proof (sketch) The first condition is similar to Lemma 19, the second allows for queries that can be folded into a concept to be realized completely outside of the (extended) ABox. Nonemptiness of $M(\mathrm{C})$ indeed corresponds to finding an object that makes the query true in the minimal model. □

The constructions used in Lemmas 19 and 20 also yield the upper bounds; the lower bounds follow from reductions from graph reachability, Horn-SAT, and the DFA intersection problem [11], respectively. For the final lower bound, one uses a knowledge base in which the query is not directly satisfied by any ABox individual. Note that PSPACE-hardness holds even for CQs of the form $\exists x.(A_1(x) \wedge \ldots \wedge A_k(x))$.

Corollary 21. *Data complexity for CQ query answering over $\mathcal{CFD}nc$ KB is in PTIME. CQ query answering over $\mathcal{CFD}nc$ KB is NLOGSPACE-hard for data complexity (in $|\mathcal{A}|$), PTIME-hard in $|\mathcal{T} + \mathcal{A}|$, and PSPACE-hard for combined complexity.*

6 Related Work

PFDs in $\mathcal{CFD}nc$ were first introduced and studied in the context of graph-oriented data models such as RDF and its refinements [8,18]. Subsequently, an FD concept constructor was proposed and incorporated in Classic [3], an early DL with PTIME reasoning capabilities, without changing the complexity of its implication problem. We mentioned earlier that removing the conditions imposed on PFDs by (1) makes logical implication EXPTIME-complete [9] and general reasoning undecidable [16].

We also mentioned earlier that relaxing the syntactic restrictions for left-hand sides of inclusion dependencies often causes the loss of PTIME complexity for some of the reasoning problems of $\mathcal{CFD}nc$. Here are several cases worth noting.

- Allowing conjunction "$\sqcap$" yields the logic $\mathcal{CFD}^{\perp}$ and therefore makes logical implication PSPACE-complete [17].
- Allowing conjunction and value restriction "$\forall$" makes logical implication EXPTIME-complete [9].

In [6], the authors consider a DL with functional dependencies and a general form of keys added as additional varieties of dependencies, called a *key box*. They show that their dialect is undecidable for DLs with inverse roles, but becomes decidable when unary functional dependencies are disallowed. This line of investigation is continued in the context of PFDs and inverse features, with analogous results [14]. Subsequently, Calvanese et al. have shown how DL-Lite can be extended with a path-based variety of identification constraints analogous to PFDs without affecting the complexity of reasoning problems [5].

For CQ answering, it is worth observing that the transition function $\delta(\mathcal{T}, \mathcal{A})$ contains a *completion* of the ABox $\mathcal{A}$ with respect to concept subsumptions. Hence despite of data complexity laying outside AC_0, a variant of the *combined approach* to query answering [10,12] seems to be feasible.

7 Summary

We have presented the DL logic $\mathcal{CFD}nc$, a variation on the logic $\mathcal{CFD}$ with the following notable properties.

- $\mathcal{CFD}nc$ retains what we believe are the most important features of $\mathcal{CFD}$: its ability to capture terminological cycles with universal restrictions over functional roles and its ability to capture a rich variety of functional constraints over functional role paths.
- In contrast to $\mathcal{CFD}$, the logic adds an ability to express disjointness of atomic concepts.
- Also in contrast to $\mathcal{CFD}$, the logic supports important reasoning services in PTIME: determining knowledge base consistency, deciding logical implication and instance checking.

There are a number of open issues and directions for continued research. The consequences of allowing $\mathcal{CFD}nc$ concept constructors other that conjunction on the left-hand-side of inclusion dependencies is, to the best of our knowledge, open. In particular, this includes values restrictions "$\forall$", negated primitive concepts "$\neg$A" and PFDs.

One enhancement to $\mathcal{CFD}nc$ that we believe is straightforward, and that would considerably enhance its utility for modelling RDF data sources, would be to allow roles and role inclusion axioms of either the form "$f \sqsubseteq R$" or the form "$R_1 \sqsubseteq R_2$" to be included in $\mathcal{CFD}nc$ TBoxes, and then to allow roles to be mentioned in conjunctive queries. We conjecture that allowing $\mathcal{EL}$ role constructors on right-hand-sides of inclusion dependencies in $\mathcal{CFD}nc$ would also be possible without damage to its PTIME capabilities.

References

1. Artale, A., Calvanese, D., Kontchakov, R., Zakharyaschev, M.: The DL-Lite family and relations. J. Artif. Intell. Res (JAIR) 36, 1–69 (2009)
2. Baader, F., Brandt, S., Lutz, C.: Pushing the $\mathcal{EL}$ Envelope. In: Proc. Int. Joint Conf. on Artificial Intelligence (IJCAI), pp. 364–369 (2005)
3. Borgida, A., Weddell, G.: Adding Uniqueness Constraints to Description Logics (Preliminary Report). In: Bry, F., Ramakrishnan, R., Ramamohanarao, K. (eds.) DOOD 1997. LNCS, vol. 1341, pp. 85–102. Springer, Heidelberg (1997)

4. Calvanese, D., de Giacomo, G., Lembo, D., Lenzerini, M., Rosati, R.: Tractable Reasoning and Efficient Query Answering in Description Logics: The DL-Lite Family. Journal of Automated Reasoning 39(3), 385–429 (2007)
5. Calvanese, D., De Giacomo, G., Lembo, D., Lenzerini, M., Rosati, R.: Path-Based Identification Constraints in Description Logics. In: Proc. of the 11th Int. Joint Conf. on Principles of Knowledge Representation and Reasoning (KR), pp. 231–241 (2008)
6. Calvanese, D., De Giacomo, G., Lenzerini, M.: Identification Constraints and Functional Dependencies in Description Logics. In: Proc. Int. Joint Conf. on Artificial Intelligence (IJCAI), pp. 155–160 (2001)
7. Hopcroft, J.E., Ullman, J.D.: Introduction to Automata Theory, Languages and Computation. Addison-Wesley (1979)
8. Ito, M., Weddell, G.: Implication Problems for Functional Constraints on Databases Supporting Complex Objects. Journal of Computer and System Sciences 49(3), 726–768 (1994)
9. Khizder, V.L., Toman, D., Weddell, G.: On Decidability and Complexity of Description Logics with Uniqueness Constraints. In: Van den Bussche, J., Vianu, V. (eds.) ICDT 2001. LNCS, vol. 1973, pp. 54–67. Springer, Heidelberg (2000)
10. Kontchakov, R., Lutz, C., Toman, D., Wolter, F., Zakharyaschev, M.: The combined approach to query answering in DL-Lite. In: KR (2010)
11. Kozen, D.: Lower bounds for natural proof systems. In: Proceedings of the 18th Annual Symposium on Foundations of Computer Science, pp. 254–266. IEEE Computer Society (1977)
12. Lutz, C., Toman, D., Wolter, F.: Conjunctive query answering in the description logic EL using a relational database system. In: Proc. Int. Joint Conf. on Artificial Intelligence (IJCAI), pp. 2070–2075 (2009)
13. Toman, D., Weddell, G.: On Reasoning about Structural Equality in XML: A Description Logic Approach. Theoretical Computer Science 336(1), 181–203 (2005)
14. Toman, D., Weddell, G.: On the Interaction between Inverse Features and Path-functional Dependencies in Description Logics. In: Proc. Int. Joint Conf. on Artificial Intelligence (IJCAI), pp. 603–608 (2005)
15. Toman, D., Weddell, G.: On Keys and Functional Dependencies as First-Class Citizens in Description Logics. In: Furbach, U., Shankar, N. (eds.) IJCAR 2006. LNCS (LNAI), vol. 4130, pp. 647–661. Springer, Heidelberg (2006)
16. Toman, D., Weddell, G.E.: On keys and functional dependencies as first-class citizens in description logics. J. Autom. Reasoning 40(2-3), 117–132 (2008)
17. Toman, D., Weddell, G.E.: Applications and extensions of ptime description logics with functional constraints. In: Proc. Int. Joint Conf. on Artificial Intelligence (IJCAI), pp. 948–954 (2009)
18. Weddell, G.: A Theory of Functional Dependencies for Object Oriented Data Models. In: International Conference on Deductive and Object-Oriented Databases, pp. 165–184 (1989)

Propositionalisation of Multi-instance Data Using Random Forests

Eibe Frank and Bernhard Pfahringer

Department of Computer Science, University of Waikato
{eibe,bernhard}@cs.waikato.ac.nz

Abstract. Multi-instance learning is a generalisation of attribute-value learning where examples for learning consist of labeled bags (i.e. multi-sets) of instances. This learning setting is more computationally challenging than attribute-value learning and a natural fit for important application areas of machine learning such as classification of molecules and image classification. One approach to solve multi-instance learning problems is to apply propositionalisation, where bags of data are converted into vectors of attribute-value pairs so that a standard propositional (i.e. attribute-value) learning algorithm can be applied. This approach is attractive because of the large number of propositional learning algorithms that have been developed and can thus be applied to the propositionalised data. In this paper, we empirically investigate a variant of an existing propositionalisation method called TLC. TLC uses a single decision tree to obtain propositionalised data. Our variant applies a random forest instead and is motivated by the potential increase in robustness that this may yield. We present results on synthetic and real-world data from the above two application domains showing that it indeed yields increased classification accuracy when applying boosting and support vector machines to classify the propositionalised data.

1 Introduction

Multi-instance learning is a generalisation of standard propositional learning–also called attribute-value learning–first introduced in [6]. Whereas propositional learning represents each example as one fixed-size vector of attribute-value pairs, multi-instance learning uses a bag of such vectors to represent examples and class labels are only associated with entire bags. The original learning assumption for multi-instance learning presented in [6] is applicable to two-class classification problems only and states that there is at least one vector in a positive bag that causes that bag to be a positive one. Negative bags are assumed to not contain any such "positive" vectors. Later work [18] has generalised this so-called multi-instance assumption to allow for arbitrary minimum and maximum counts of "positive" vectors as the necessary and sufficient condition for a positive bag.

Algorithms for multi-instance learning can be grouped into three classes. First there are dedicated new algorithms, with the most prominent one being Diversity Density [11]. Secondly, standard propositional learners like decision trees or

S. Cranefield and A. Nayak (Eds.): AI 2013, LNAI 8272, pp. 362–373, 2013.

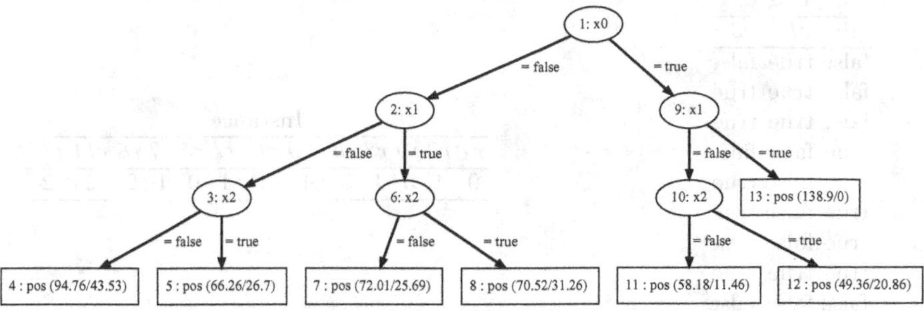

Fig. 1. Unpruned decision tree used for propositionalisation. In (x/y), x gives the total weight of all instances at the leaf node, and y gives the weight of all misclassified instances. Each leaf node is labeled *pos*, which means that for each leaf the sum of weights for positive instances is greater than the sum of weights for negative instances.

support vector machines, can be adapted–sometimes this is called upgraded–to deal with multi-instance data. Typical examples are MITI [3] and MISVM [1]. Thirdly, instead of adapting the algorithm, the data can be adapted, or propositionalised, to turn it into a standard propositional representation. As multi-instance learning is a special case of relational learning, any propositionalisation method from relational learning [9] could be applied, but generic relational learning methods often do not scale well. Therefore specialised multi-instance propositionalisation methods, inspired by general relational algorithms, have attracted some interest. One such method is PROPER [15], which is a specialisation of RelAggs [10]. On the other hand, the TLC algorithm [18] introduced a genuinely new way of propositionalisation, based on hierarchically partitioning the full instance space into sub-regions. The algorithm presented and analysed in this paper is a simple, yet effective extension of TLC.

The next section will describe this extension. Some implementation aspects are discussed in Section 3. Section 4 provides insights into the algorithm's behaviour by applying it to a synthetic multi-instance problem and Section 5 evaluates the method using a number of standard multi-instance benchmark datasets. Section 6 provides some pointers for future work and conclusions.

2 Propositionalisation Using Random Forests

The multi-instance learning method presented in this paper is a simple extension of the Two-Level Classification (TLC) method as presented in [18]. TLC uses a heuristic approach to partition the instance space into regions. Once this has been done, a bag of instances (i.e. an example for learning in a multi-instance dataset) is propositionalised by counting how many instances of the bag fall into each region. These counts are attribute values in the propositionalised problem (i.e. there is one attribute for each region in the partition). Once each bag of data has been propositionalised in this form, and each bag's classification has

Bag		
x0	*x1*	*x2*
false	true	false
false	true	true
false	true	true
true	false	false
true	false	true
true	false	true
true	false	false
true	true	true
false	false	false

⇒

Instance												
r1	*r2*	*r9*	*r3*	*r6*	*r10*	*r13*	*r4*	*r5*	*r7*	*r8*	*r11*	*r12*
9	4	5	1	3	4	1	1	0	1	2	2	2

Fig. 2. Bag of instances and propositionalised form

been attached to its propositionalised form, a standard single-instance learning algorithm for classification problems can be applied to the data.

The motivation for using this approach is that by dividing the instance space into regions and measuring occupancy, it becomes possible to describe the distribution of a bag's instances in instance space. This provides an alternative to simpler propositionalisation approaches that compute summary statistics such as the mean and standard deviation of the attribute values in a bag. In this manner, it is possible to preserve more information when propositionalising.

The question is how to define the regions in the instance space. TLC uses a standard single-instance decision tree to obtain a partition. To learn this tree, all instances from all bags are joined into a single dataset, discarding bag membership information, and labeled by their bag's class label. To make sure that large bags receive as much weight as small bags, each instance in this dataset is weighted by $\frac{1}{|X|} \times \frac{N}{b}$, where X is is the bag the instance comes from, N is the number of instances in the joined data, and b is the number of bags in the original dataset. In this manner, the sum of weights for the instances in the new dataset is N.

Let us consider an illustrative example, where we generated synthetic multi-instance data with three Boolean attributes $x0$, $x1$, and $x2$, for each bag. We generated 100 bags of instances, where each bag had between one and 10 instances, with equal probability for each bag size. When sampling instances, the joint probability distribution over the attributes was the uniform distribution, making all combinations of attribute values equally likely. The classification of each bag was determined as follows. A bag received the class label "positive" if it contained at least one instance for which both $x0$ and $x1$ had the value true. If it did not contain any instance with this property, the bag was labeled "negative". This relationship is an example of the classic (or "standard") assumption for multi-instance learning given in [6].

We then applied the above process to generate a partitioning, using an unpruned decision tree grown using information gain (based on the `REPTree` classifier with option `-P` in WEKA [8]). The resulting tree is shown in Figure 1. The leaf nodes in the tree show the majority class, which is "positive" in all cases, determined by examining the sum of weights of the instances in each class. This tree defines 13 regions, one for each node in the tree. Note that *all* nodes in the tree are used to define regions, not just the leaf nodes. In this example problem, (leaf) node 13 is the key region, because a bag is positive if and only if it has an instance in this region, but in general any node (or even combination of nodes) in the tree may need to be considered to determine class membership of a bag. Thus, occupancy counts for all regions are used as attribute values in the propositionalised data, so that the single-instance learner applied to the propositionalised data can identify the salient relationship.

Figure 2 shows an example bag and its propositionalised version, propositionalised using the tree in Figure 1. The class label has been omitted in this case, but the bag (and the resulting instance) are both positive because of membership in region 13. Note that the tree is explored in a breadth-first manner to generate the attribute values for the instance.

The hypothesis we investigate in this paper is that for large and messy real-world data a single tree may not be sufficient to obtain a robust learning method for multi-instance data. When considering standard classification problems, it is well-known that ensembles of trees such as random forests [5] outperform a single tree in terms of predictive performance. Here, we want to use an ensemble of trees for propositionalisation. The basic process is the same: the multi-instance dataset is converted into a single-instance dataset by attaching each instance's bag label to the instance and reweighting the instance, just as in TLC described above. Then, rather than learning a single tree, we learn an ensemble of trees using the random forest method, for example, using the `RandomForest` class in WEKA. If tree i in the ensemble of trees has l_i nodes (internal nodes + leaf nodes), then the ensemble defines $\sum_i l_i$ regions. To propositionalise a bag of instances, we then simply calculate the occupancy counts for all of these regions and create a feature vector of size $\sum_i l_i$. This vector is then labeled with the bag's label and can be processed using a single-instance learner.

It is clear that the resulting propositionalised instances have many more attributes than in the single-tree-based TLC method. The size of the feature vector is determined by the size of the ensemble. Our hypothesis is that larger ensembles, and thus feature vectors, will generally lead to improved classification accuracy when applying a learning algorithm to the propositionalised data.

3 Implementation in the WEKA Workbench

The WEKA machine learning workbench has support for multi-instance data in recent versions of the software, facilitated by the availability of relation-valued attributes: each bag of instances is stored as the value of a relation-valued attribute. For the experiments reported in this paper, a new `PartitionGenerator` interface has been added to WEKA that is implemented by several tree learners. It has a method that returns an array with counts that indicates, for a given instance, in which regions of the partition this instance is present. A tree learner may fill in this array by traversing a tree in a breadth-first fashion. In the case of a tree ensemble, the vectors for the individual trees are simply concatenated. WEKA now also has a `PartitionMembershipFilter` that can apply any `PartitionGenerator` to a given dataset to obtain these vectors for all instances. In conjunction with a new `MultiInstanceWrapper` filter in the `MultiInstanceFilters` package for WEKA 3.7, this filter can be applied to multi-instance data. When this is done, the vectors for all instances in a bag are simply added together by this filter to yield a vector of membership counts for a bag.

For added convenience, the `MultiInstanceLearning` package contains a new classifier implementation called `TLC` (for Two-Level Classification [18]) that applies the filtering process, using the above two filters, based on a particular `PartitionGenerator` specified as a parameter, and then runs a standard single-instance classifier on the propositionalised data. This single-instance classifier can also be specified as a parameter. It is thus straightforward to run systematic experiments with different partition generators and single-instance classifiers. For the experiments in this paper, WEKA's `RandomForest` class and `REPTree` decision tree learner were modified to implement the `PartitionGenerator` interface and the modified code is now part of the official WEKA code repository.

4 A Synthetic Problem

Our hypothesis is that propositionalisation using a random forest yields a more robust classifier than propositionalisation based on a single decision tree. In this section, we test this hypothesis empirically by introducing different levels of noise and data redundancy in a very simple synthetic learning problem. In this learning problem, there is a single numeric attribute that completely determines the classification of a bag: if this attribute has at least one positive value in the bag concerned, the bag is classified "positive"; otherwise, its class label is "negative". As we want to test the robustness of learning algorithms, we modify this deterministic relationship by first duplicating the attribute and its values for a particular bag to yield n copies and then taking a certain percentage of these copies for the particular bag concerned and replacing all their positive attribute values by their additive inverses (i.e. attribute value x becomes $-x$ if $x > 0$), *without* changing the class label of the bag. Hence, the modified data has n attributes per bag instead of one, where some of the attributes of positive bags may be corrupted and do not correctly indicate that the bag is positive. A learning algorithm must thus exploit the redundancy in these attributes to achieve maximum accuracy and cannot rely on a single attribute alone.

The exact set-up of the experiment is as follows. Based on a particular seed for the pseudo random number generator, we generate 100 bags containing between one and five instances each, where each bag size is given equal probability. For each bag, we first generate uncorrupted attribute values by sampling from the uniform distribution over the range $[-0.5, 0.5)$. If one of these attribute values is positive, the bag's class label is set to "positive", otherwise it is set to "negative". Once the class label has been determined, n copies of the uncorrupted attribute values for a bag are generated to yield n attributes. Then, a biased coin is flipped

n times, where this coin has probability p of coming up heads. If heads is the result of the coin flip, all positive attribute values in the corresponding copy of the attribute for the bag concerned (if any) are replaced by their additive inverse to introduce non-determinism in the relationship between this attribute's values and the bag's classification. In this way, all negative bags in the data have n identical attributes, but the n attributes of a positive may bag differ, depending on the value of the chosen probability p.

To measure accuracy of a learning algorithm on this data, we apply stratified 10-fold cross-validation to estimate the value of the kappa statistic, which can be viewed as a normalised version of classification accuracy that is particularly useful when the classes are unbalanced. Kappa is computed as follows:

$$\kappa = \frac{a - a_r}{1 - a_r},$$

where a is the estimated classification accuracy of the learning algorithm we want to evaluate, and a_r is the expected accuracy of a random classifier that assigns instances randomly to classes in such a manner that it assigns the same number of instances to each class as the learning algorithm we are evaluating. If kappa is greater than zero, the learning algorithm exhibits accuracy greater than what would be expected by assigning classifications randomly to the bags occuring in the test folds of the cross-validation. A value of one is the maximum that can be achieved.

We compare propositionalisation using unpruned decision trees grown using the information gain (based on `REPTree` with option `-P` in WEKA) to propositionalisation using a random forest of size 10 (based on `RandomForest` with option `-K 1` in WEKA). LogitBoost with decision stumps and 100 boosting iterations was used as the learning algorithm for the propositionalised data (`LogitBoost` with option `-I 100` in WEKA). Figures 3 and 4 show the results obtained for different numbers of attributes (i.e. values of n) and two different noise levels (i.e. values of p). Figure 3 shows results for the case where $p = 0.1$ and Figure 4 shows results for the case where $p = 0.3$. Each point in the plot corresponds to an average over 10 different runs of the experiments, where data was generated from scratch for each run using a different seed for the pseudo random number generator, followed by 10-fold cross-validation on this fresh data. The error bars correspond to 95%-level confidence intervals.

The graphs show that both propositionalisation methods benefit from redundancy in the data: accuracy increases as the number of attributes increases. However, at the higher noise level we consider ($p = 0.3$), accuracy levels out earlier when using a single, deterministically-grown decision tree. In contrast, propositionalisation using random forests benefits more from adding redundancy in the input by including more attributes: we can see that for $p = 0.3$, and more than four attributes ($n > 4$), the random-forest-based method achieves a level of accuracy that is statistically significantly higher than that obtained using a single tree—the 95%-level confidence intervals for the two methods do not overlap.

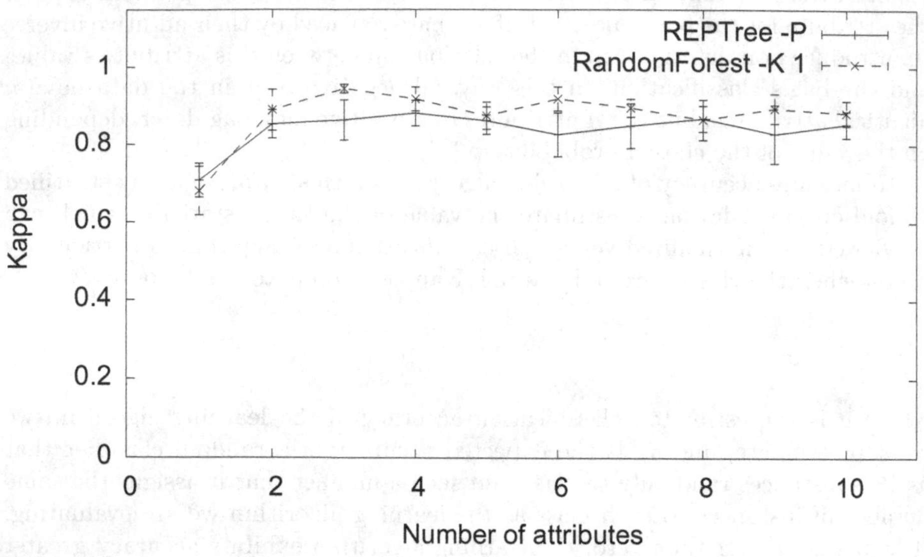

Fig. 3. Average kappa for $p = 0.1$

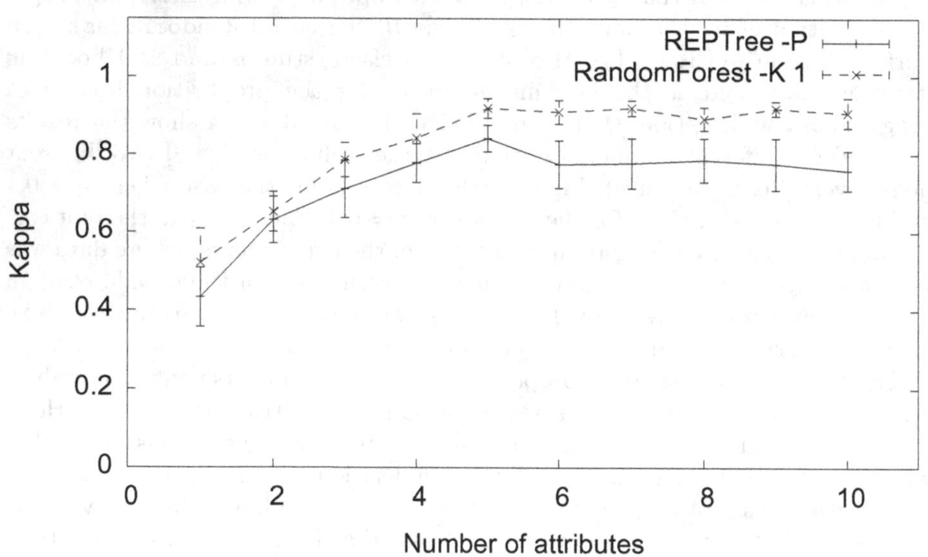

Fig. 4. Average kappa for $p = 0.3$

Table 1. 10-times 10-fold cross-validated classification accuracy obtained using linear support vector machines

Dataset	Decision Tree	Forest (10 trees)	Forest (50 trees)	Forest (100 trees)
musk1	84.4±13.2	87.7±10.7	89.1±10.6	89.4± 10.3
musk2	76.0±12.9	80.7±11.6	80.4±11.1	80.5± 10.9
mutagenesis3-atoms	86.5± 8.5	86.6± 8.1	86.7± 8.0	86.7± 8.1
mutagenesis3-bonds	86.1± 7.3	86.7± 7.5	86.7± 7.6	87.1± 7.5
mutagenesis3-chains	83.4± 8.9	84.5± 9.0	84.8± 9.6	85.3± 9.3
thioredoxin	88.4± 4.2	88.7± 4.6	89.3± 4.2	89.6± 4.2
suramin	65.0±45.2	61.0±46.9	57.0±47.7	55.0± 47.9
elephant	80.9± 9.8	86.0± 8.0	88.8± 7.4 ∘	88.9± 7.0 ∘
fox	63.0± 9.2	62.7± 9.1	65.0±10.0	66.2± 8.1
tiger	80.1± 9.1	81.1± 9.6	83.6± 8.2	84.3± 7.9
bikes	79.6± 4.6	81.6± 4.2	83.8± 4.3 ∘	84.3± 3.8 ∘
cars	67.8± 4.7	72.4± 4.4 ∘	75.4± 4.8 ∘	76.1± 4.4 ∘
people	78.4± 4.6	80.7± 4.5	82.5± 3.9 ∘	83.4± 3.9 ∘

∘ statistically significant improvement

5 Experiments on Real-world Data

To evaluate the performance of the random-forest-based propositionalisation approach, we performed experiments on benchmark multi-instance datasets that have previously been used in the literature. For propositionalisation, we used unpruned decision trees grown deterministically using information gain (`REPTree` with option `-P` in WEKA), and random forests with 10, 50, and 100 trees (`RandomForest` with options `-I 10`, `-I 50`, and `-I 100` in WEKA). Classification accuracy was estimated using 10-times 10-fold stratified cross-validation. Propositionalisation was performed separately based on each of the 100 training sets of the repeated cross-validation, so that the test data was never used in the propositionalisation process. We evaluated two learning algorithms in conjunction with the propositionalisation methods: linear support vector machines with $C = 1$ (`SMO` with option `-no-checks -N 2` in WEKA), using WEKA's `NonSparseToSparse` filter to create input data in sparse format, and boosted decision stumps, using 100 boosting iterations (`LogitBoost` in WEKA with option `-I 100`). The corrected resampled paired t-test [13] was used to establish statistical significance when considering observed differences in estimated accuracy, with a significance level of 0.05. All experiments were performed using the WEKA Experimenter interface [8].

The experimental results are shown in Tables 1 and 2. They include results for the two well-known *musk* datasets [6], where the task is to determine whether a molecule is active based on its geometric properties. Another task included is mutagenicity prediction [16], which was considered for multi-instance tree and rule learning in [19], based on three different representations of molecules as bags of instances *muta-atoms, muta-bonds* and *muta-chains.* We also include the *thioredoxin* protein identification task [17] and the *suramin* data, which is another drug activity prediction problem: identifying suramin [4] analogues that can act as anti-cancer agents. Image classification is another important application area

Table 2. 10-times 10-fold cross-validated classification accuracy obtained using LogitBoost with 100 boosting iterations

Dataset	Decision Tree	Forest (10 trees)	Forest (50 trees)	Forest (100 trees)
musk1	84.6±12.9	86.2±12.0	88.4±10.4	88.4± 11.4
musk2	76.5±12.6	80.9±10.8	82.5±11.7	82.7± 12.2
mutagenesis3-atoms	85.5± 9.3	86.4± 8.7	86.0± 8.7	85.5± 8.3
mutagenesis3-bonds	86.7± 7.3	87.7± 7.1	87.6± 7.5	87.2± 7.0
mutagenesis3-chains	87.7± 8.1	86.5± 8.6	87.4± 7.6	87.8± 8.3
thioredoxin	90.2± 5.0	90.6± 4.7	91.2± 4.8	91.9± 4.2
suramin	49.5±47.9	56.5±48.0	51.5±48.4	56.0± 48.3
elephant	79.8± 9.4	83.7± 8.5	86.3± 7.5	86.4± 7.0
fox	63.0±10.0	62.4±11.2	65.9±10.9	64.4± 10.3
tiger	78.2± 8.8	81.4± 9.0	83.2± 8.5	83.5± 9.2
bikes	79.7± 4.5	81.2± 4.0	82.3± 4.7	83.2± 4.6 ∘
cars	70.1± 5.1	72.9± 3.9	74.8± 5.0 ∘	75.1± 4.5 ∘
people	77.3± 4.6	79.6± 4.1	80.4± 4.1	81.0± 3.8 ∘

∘ statistically significant improvement

of multi-instance learning methods. We include two sets of content-based image classification datasets. The first set consists of the *elephant, fox* and *tiger* [1] datasets and the second one contains the *bikes, cars* and *people* datasets. The latter set is based on Ohta-based features as in [12], and derived from the GRAZ02 dataset [14].

Table 1 shows that, when applied in conjunction with a linear support vector machine, using random forests for propositionalisation is preferable to using a single deterministic decision tree. This is particularly apparent in the case of the image classification datasets. Noteworthy improvements in predictive accuracy are obtained for all six image classification problems when using 100 trees in the random forests, and in four cases the improvement is statistically significant. The results also show that accuracy generally improves as more trees are included in the random forests. Again, bigger improvements are obtained in the image classification datasets. This is consistent with our hypothesis that random-forest-based classification can better exploit redundancy in the input data because the features in the image classification datasets are likely to be highly redundant.

Table 2 shows a similar picture when using 100 boosted decision stumps instead of linear support vector machines. Using random forests with 100 trees instead of a single unpruned decision tree yields higher estimated accuracy for 12 out of the 13 datasets. In the case of the three image classification datasets *bikes, cars*, and *people*, the improvement in accuracy is statistically significant.

Comparing the performance of boosted stumps and support vector machines when using propositionalisation based on random forests with 100 trees, we can see that support vector machines produce better accuracy on the image classification datasets, whereas there is no clear difference on the other datasets. Given that 100 boosted decision stumps can only test a maximum of 100 regions in the partitioned instance space, this indicates that high accuracy on the image classification datasets requires consultation of more than 100 regions to yield accurate classifications.

Table 3. Average training time in seconds in 10-times 10-fold cross-validated classification, obtained using linear support vector machines

Dataset	Decision Tree	Forest (10 trees)	Forest (50 trees)	Forest (100 trees)
musk1	0.1±0.1	0.1± 0.1	0.4± 0.0 ∘	1.0± 0.1 ∘
musk2	0.9±0.1	2.2± 0.3 ∘	13.7± 2.0 ∘	36.3± 6.1 ∘
mutagenesis3-atoms	0.1±0.0	0.8± 0.1 ∘	9.5± 0.7 ∘	62.9± 18.4 ∘
mutagenesis3-bonds	0.1±0.0	2.2± 0.1 ∘	23.3± 2.2 ∘	127.7± 27.7 ∘
mutagenesis3-chains	0.3±0.0	4.2± 0.2 ∘	59.0±13.1 ∘	263.3± 47.7 ∘
thioredoxin	3.1±0.1	55.9± 2.9 ∘	665.2±93.5 ∘	1919.3±145.9 ∘
suramin	0.0±0.0	0.2± 0.0 ∘	0.8± 0.1 ∘	1.8± 0.1 ∘
elephant	0.3±0.0	0.8± 0.0 ∘	7.5± 0.4 ∘	41.4± 18.3 ∘
fox	0.3±0.0	0.9± 0.1 ∘	10.4± 0.6 ∘	73.2± 27.3 ∘
tiger	0.2±0.0	0.6± 0.0 ∘	5.8± 0.3 ∘	21.3± 2.8 ∘
bikes	1.0±0.1	5.9± 0.2 ∘	105.8±17.5 ∘	400.0± 75.4 ∘
cars	1.4±0.1	9.5± 0.4 ∘	213.8±34.6 ∘	729.8±133.4 ∘
people	0.9±0.1	5.0± 0.2 ∘	76.6±13.6 ∘	317.9± 61.3 ∘

∘ statistically significant degradation

It is instructive to compare the accuracy obtained in the experiments presented here to that obtained in [2] (Table 4, semi-random ensemble), which evaluated random forests of size 100 grown using a modified version of the MITI algorithm [3], a tree inducer designed for multi-instance learning. Exactly the same experimental protocol, based on the same 10-times 10-fold cross-validation runs, was applied in [2]. Propositionalisation using 100 trees, applied in conjunction with linear support vector machines (Table 1), produces higher estimated accuracy for eight of the twelve datasets considered both here and in [2], and lower accuracy for four datasets. Overall, accuracy obtained using propositionalisation appears very competitive.

Note that, computationally, support vector machines are well suited for the propositionalised data: the propositionalised bags yield very sparse feature vectors because most regions defined by a decision tree will not contain any instances of any particular bag. Hence, most attribute values in the propositionalised data will be zero, yielding sparse attribute vectors. Sparse vectors can be dealt with very efficiently in support vector machines because dot products of sparse vectors can be computed by iterating over the non-zero elements in the vectors only. WEKA supports data in sparse format, where only non-zero values in the instances are explicitly represented, and the `NonSparseToSparse` filter can be used to create this data.

Tables 3 and 4 show training times, including the propositionalisation process, averaged over the 10 runs of 10-fold cross-validation. It can be seen that using random forests to propositionalise the data significantly increases training time in all cases. One reason is that a tree ensemble needs to be grown, rather than a single tree. (Note that this process can be parallelised.) Another reason is that the instances in the propositionalised data have many more attributes when using random forests than when using a single deterministic tree because an ensemble of trees is used instead of a single tree and a single tree in a random forest is generally larger than a single deterministically grown tree, where attribute selection using information gain aims to minimise tree size. The results also show

Table 4. Average training time in seconds in 10-times 10-fold cross-validated classification, obtained using LogitBoost with 100 boosting iterations

Dataset	Decision Tree	Forest (10 trees)	Forest (50 trees)	Forest (100 trees)
musk1	0.2±0.1	0.7±0.1 ∘	3.5± 0.2 ∘	7.6± 0.5 ∘
musk2	1.1±0.2	5.5±0.6 ∘	30.9± 3.5 ∘	70.6± 8.7 ∘
mutagenesis3-atoms	0.6±0.0	9.7±0.4 ∘	53.7± 1.9 ∘	129.3± 11.1 ∘
mutagenesis3-bonds	0.9±0.1	14.4±0.5 ∘	82.9± 3.2 ∘	215.2± 23.6 ∘
mutagenesis3-chains	1.2±0.1	21.6±0.8 ∘	134.7± 8.5 ∘	427.2± 37.2 ∘
thioredoxin	7.0±0.3	106.2±3.6 ∘	946.0±59.2 ∘	2641.3±126.3 ∘
suramin	0.0±0.0	0.2±0.0 ∘	1.1± 0.1 ∘	2.5± 0.2 ∘
elephant	0.6±0.1	9.2±0.3 ∘	49.6± 1.3 ∘	116.6± 9.4 ∘
fox	0.7±0.1	11.0±0.4 ∘	60.5± 1.3 ∘	147.2± 11.3 ∘
tiger	0.5±0.1	8.0±0.4 ∘	42.6± 1.3 ∘	95.7± 7.0 ∘
bikes	7.1±0.2	90.5±1.4 ∘	598.0±12.0 ∘	1545.8± 38.2 ∘
cars	10.7±0.3	136.8±1.7 ∘	952.5±23.0 ∘	2381.3±131.4 ∘
people	6.0±0.2	76.2±1.1 ∘	496.3± 7.7 ∘	1208.0± 49.9 ∘

∘ statistically significant degradation

that applying a linear support vector machine is faster than applying boosting due to the fact that sparse data can be processed efficiently.

6 Conclusions

Multi-instance learning is an interesting and useful generalisation of propositional learning. This paper has presented a simple, yet effective extension of the TLC propositionalisation method that grows random forests for propositionalising multi-instance data. The new method's increased robustness with regard to noise in the input was demonstrated with a synthetic example, and a comprehensive evaluation on benchmark datasets representing image and molecule classification problems also shows improved accuracy for the new method, albeit at the cost of a considerable increase in runtime.

The standard random forest method applied in this paper chooses split points on numeric attributes deterministically when considering these attributes for splitting. However, as the instances' class labels used in this process are simply taken to be their bags' labels, they may be incorrect. Hence, it would be interesting to apply a method that chooses split points randomly. This is what the Extra-Trees algorithm [7] for growing a tree ensemble does. Applying it to propositionalisation of multi-instance data is a promising avenue for future research.

References

1. Andrews, S., Tsochantaridis, I., Hofmann, T.: Support vector machines for multiple-instance learning. In: Proc. Conf. on Neural Information Processing Systems, pp. 561–568. MIT Press (2003)
2. Bjerring, L., Frank, E.: Beyond trees: Adopting MITI to learn rules and ensemble classifiers for multi-instance data. In: Wang, D., Reynolds, M. (eds.) AI 2011. LNCS, vol. 7106, pp. 41–50. Springer, Heidelberg (2011)

3. Blockeel, H., Page, D., Srinivasan, A.: Multi-instance tree learning. In: Proc. 22nd Int. Conf. on Machine Learning, pp. 57–64. ACM (2005)
4. Braddock, P.S., Hu, D.E., Fan, T.P., Stratford, I., Harris, A.L., Bicknell, R.: A structure-activity analysis of antagonism of the growth factor and angiogenic activity of basic fibroblast growth factor by suramin and related polyanions. Br. J. Cancer 69(5), 890–898 (1994)
5. Breiman, L.: Random forests. Machine Learning 45(1), 5–32 (2001)
6. Dietterich, T.G., Lathrop, R.H., Lozano-Perez, T.: Solving the multiple instance problem with axis-parallel rectangles. Artificial Intelligence 89(1-2), 31–71 (1997)
7. Geurts, P., Ernst, D., Wehenkel, L.: Extremely randomized trees. Mach. Learn. 63(1), 3–42 (2006)
8. Hall, M., Frank, E., Holmes, G., Pfahringer, B., Reutemann, P., Witten, I.H.: The WEKA data mining software: an update. SIGKDD Explor. 11(1), 10–18 (2009)
9. Kramer, S., Lavrač, N., Flach, P.: Propositionalization approaches to relational data mining. In: Relational Data Mining, pp. 262–286. Springer (2000)
10. Krogel, M.-A., Rawles, S., Železný, F., Flach, P.A., Lavrač, N., Wrobel, S.: Comparative evaluation of approaches to propositionalization. In: Horváth, T., Yamamoto, A. (eds.) ILP 2003. LNCS (LNAI), vol. 2835, pp. 197–214. Springer, Heidelberg (2003)
11. Maron, O., Lozano-Pérez, T.: A framework for multiple-instance learning. In: Proc. Conf. on Neural Information Processing Systems, pp. 570–576. MIT Press (1998)
12. Mayo, M.: Effective classifiers for detecting objects. In: Proc. 4th Int. Conf. on Computational Intelligence, Robotics, and Autonomous Systems (2007)
13. Nadeau, C., Bengio, Y.: Inference for the Generalization Error. Machine Learning 52(3), 239–281 (2003)
14. Opelt, A., Pinz, A., Fussenegger, M., Auer, P.: Generic object recognition with boosting. IEEE Transaction on Pattern Analysis and Machine Intelligence 28(3), 416–431 (2006)
15. Reutemann, P., Pfahringer, B., Frank, E.: A toolbox for learning from relational data with propositional and multi-instance learners. In: Webb, G.I., Yu, X. (eds.) AI 2004. LNCS (LNAI), vol. 3339, pp. 1017–1023. Springer, Heidelberg (2004)
16. Srinivasan, A., Muggleton, S., King, R., Sternberg, M.: Mutagenesis: ILP experiments in a non-determinate biological domain. In: Proc. 4th Int Workshop on Inductive Logic Programming, pp. 217–232. GMD (1994)
17. Wang, C., Scott, S., Zhang, J., Tao, Q., Fomenko, D., Gladyshev, V.: A study in modeling low-conservation protein superfamilies. Tech. rep., Department of Comp. Sci., University of Nebraska-Lincoln (2004)
18. Weidmann, N., Frank, E., Pfahringer, B.: A two-level learning method for generalized multi-instance problems. In: Lavrač, N., Gamberger, D., Todorovski, L., Blockeel, H. (eds.) ECML 2003. LNCS (LNAI), vol. 2837, pp. 468–479. Springer, Heidelberg (2003)
19. Chevaleyre, Y., Zucker, J.-D.: Solving multiple-instance and multiple-part learning problems with decision trees and rule sets. Application to the mutagenesis problem. In: Stroulia, E., Matwin, S. (eds.) AI 2001. LNCS (LNAI), vol. 2056, pp. 204–214. Springer, Heidelberg (2001)

An Effective Method for Imbalanced Time Series Classification: Hybrid Sampling

Guohua Liang

The Centre for Quantum Computation and Intelligent Systems,
Faculty of Engineering and IT, University of Technology, Sydney NSW 2007 Australia
Guohua.Liang@student.uts.edu.au
http://www.qcis.uts.edu.au

Abstract. Most traditional supervised classification learning algorithms are ineffective for highly imbalanced time series classification, which has received considerably less attention than imbalanced data problems in data mining and machine learning research. Bagging is one of the most effective ensemble learning methods, yet it has drawbacks on highly imbalanced data. Sampling methods are considered to be effective to tackle highly imbalanced data problem, but both over-sampling and under-sampling have disadvantages; thus it is unclear which sampling schema will improve the performance of bagging predictor for solving highly imbalanced time series classification problems. This paper has addressed the limitations of existing techniques of the over-sampling and under-sampling, and proposes a new approach, hybrid sampling technique to enhance bagging, for solving these challenging problems. Comparing this new approach with previous approaches, over-sampling, SPO and under-sampling with various learning algorithms on benchmark data-sets, the experimental results demonstrate that this proposed new approach is able to dramatically improve on the performance of previous approaches. Statistical tests, Friedman test and Post-hoc Nemenyi test are used to draw valid conclusions.

Keywords: Hybrid sampling, over-sampling, under-sampling, imbalanced data, time series data, ensemble learning, and classification.

1 Introduction

In data mining research, mining time series data is one of the most challenging problems [1], and the imbalanced data problem is a fundamental classification problem [2]. Most traditional supervised classification learning algorithms are ineffective for highly imbalanced time series classification (HITSC) [3]. Due to its challenging issues of high dimensionality, large scale, and uneven class distribution among different classes, and considering the sequence of the numerical attributes carrying special information as whole instead of individual attributes [4, 5], it has received considerably less attention than imbalanced data problems in data mining and machine learning research. HITSC refers to a situation in which the proportions of the training examples of time series

S. Cranefield and A. Nayak (Eds.): AI 2013, LNAI 8272, pp. 374–385, 2013.

data are varied significantly among different classes. This study mainly focuses on imbalanced binary time series classification (TSC), e.g., the proportion of positive examples that are far fewer than the proportion of negative examples in the training data of the TSC.

Bagging [6] was introduced by Breiman in 1996. Previous research shows that bagging can improve the performance of individual classifiers if base learners are unstable [6–9], but it has a limitation for solving highly imbalanced data problems. Sampling techniques are considered to be one of the most effective ways to tackle highly imbalanced problems, but since both over-sampling and under-sampling techniques have their limitations, it is unclear which sampling schema is able to enhance the performance of bagging. These challenging issues have motivated me to propose a new approach, hybrid-sampling (H-Sampling) techniques, to enhance bagging, for solving HITSC problems.

The proposed new H-sampling approach randomly over-samples the positives and under-samples the negatives to half of the original training size, $\frac{|P|+|N|}{2}$, respectively, to generate a set of balanced bootstrap samples from the original training set. This set of balanced bootstrap samples is used to train a set of classifiers; then each test example is predicted by a set of trained classifiers; lastly, the final prediction of each test example is made by the majority votes of these predictions of the set of trained classifiers. Comparing the performance of this new approach with previous approaches [10, 3, 5],the over-sampling method SPO and under-sampling method with various algorithms on the benchmark data-sets, the experimental results demonstrate that the proposed new approach, H-sampling to enhance bagging, is superior to previous approaches [10, 3, 5], and dramatically improves the performance of previous approaches. Statistical tests, Friedman and post-hoc Nemenyi tests for comparing the performance of multiple learning methods over multiple benchmark data-sets are applied to draw valid conclusions.

The key contributions of this paper are as follows. (1) This paper addresses the limitations of the existing over-sampling and under-sampling techniques, and proposes a new approach, H-sampling technique to enhance bagging, for improving the performance of prediction models to solve the HITSC problems. (2) Empirically comparing the performance of this new approach with previous approaches on the benchmark data-sets, the experimental results demonstrate that the new approach, H-Sampling integrating the unstable base learner, decision trees $J48$ with bagging, is effective for solving the HITSC problems and is dramatically superior to previous approaches: the over-sampling method, SPO and the under-sampling method with KNN.

The paper is organized as follows. Section 2 presents an outline of the proposed new approach. Section 3 shows related work. Section 4 presents the evaluation measures. Sections 5 and 6 provide the experimental setting and experimental analysis. Section 7 concludes this work.

Algorithm 1. H-Sampling Bagging

Input:

D, original training set, containing $|P|$ positive and $|N|$ negative instances;
a learning scheme (algorithm, e.g., $J48$);

Output: A composite model, C^*.

Method:

for $i = 1$ **to** k **do**

Create balanced bootstrap samples of size $|D_i|$ sub-sets, $|D_i| = |P_i| + |N_i|$ where P_i and N_i are randomly drawn with replacement from original training set, P and N, respectively:
$|P_i| = |N_i| = \frac{(|P|+|N|)}{2}$ and;

end

return *a set of bootstrap samples D_i (containing k bootstrap samples);*

Train each base classifier model C_i from D_i;

To use the composite model, C^* for a test set T on an instance x where its true class label is y:

$C^*(x) = \arg\max_y \sum_i \delta\,(C_i(x) = y)$

Delta function $\delta(.) = 1$ if argument is true, else 0.

2 Hybrid Sampling Approach

Algorithm 1 outlines the proposed new approach, H-sampling integrating unstable learner decision trees $J48$ with bagging. This new approach is different from previous approaches [10, 3, 5] because H-sampling reduces the disadvantage of under-sampling, loosing to much important information for training, and the disadvantages of over-sampling, over-fitting, high computational cost and longer training time. This new approach, H-sampling, randomly selects the positives and the negatives to the balanced point at half of the original training size, $\frac{|P|+|N|}{2}$. For example, the positives are randomly selected with replacement from the entire positive class to the size of the balanced point; the negatives are randomly selected with replacement from the negative class of original training set to the size of the balanced point.

For the proposed prediction model, suppose the size of an ensemble is k, a set of classifiers C_i (for i=1 to k) is built from a set of balanced bootstrap samples D_i; each new test example is classified by a set of classifiers C_i, and the final prediction is made by majority votes to aggregate the predictions of the set of classifiers C_i by using a delta function $\delta(.) = 1$ if the prediction of C_i is a true class label, else the delta function $\delta(.) = 0$.

Majority votes, aggregating the set of predicted class labels, use the delta function to vote for a class and the class label obtaining the highest number of votes is considered as the output of the final prediction.

3 Related Work

This paper proposes a new approach, H-sampling integrating unstable learner decision trees $J48$ with bagging for solving HITSC problems. This new approach is different to previous approaches [10, 3, 5] because both over-sampling and under-sampling techniques have disadvantages. This new approach not only reduces the limitations of over-sampling and under-sampling techniques, but also enhances bagging to effectively improve the performance of the previous approaches for solving HITSC problems.

The main disadvantages of over-sampling are that over-sampling dramatically increases the computational cost of training and training time, and may cause over-fitting, even though it maintains the important information for training, because additional large number of new positive examples with high dimensional features are generated to balance the training set for HITSC [3]. The main disadvantages of under-sampling may lose important and useful information for training and may degrade the performance of the prediction models, even though it significantly reduces the computational cost of training, because only a proportion of the majority class examples are selected to train prediction models.

In earlier research, a structure-preserving over-sampling (SPO) [10] method with support vector machines (SVM) was proposed for solving HITSC problems; it achieves better results than other over-sampling methods and state-of-the-art methods in TSC, based on a comparison of the average values of two evaluation measures, F_{value} and Geometric mean (G_{mean}), without statistical analysis to support this conclusion. The study compared SPO with over-sampling methods, which include repeating (REP), SMOTE [11] (SMO), Borderline_SMOTE [12] (BoS), ADASYN [13] (ADA), and DataBoost [14] (DB); and with state-of-the-art methods in TSC, which include Easy Ensemble [15] (Easy), BalanceCascade [15] (Bal), One nearest neighbor classifier using Euclidean distance [16] (1NN), and One nearest neighbor classifier using dynamic time warping distance [17] (1NN_DW).

Our other work [3] proposed an under-sampling technique integrated with SVM, which is more efficient than other more complicated approaches, such as SPO with SVM for HITSC. However, it is unclear whether the under-sampling method with various supervised learning algorithms is more effective than the over-sampling method, SPO, and the under-sampling technique integrated with SVM for HITSC.

Our previous work [5] conducted an empirical evaluation of the performance of over-sampling methods (e.g., the complex SPO [10]) and under-sampling with various supervised learning algorithms selected from Weka [18], such as Sequential Minimal Optimization (SMO) of SVM, decision trees ($J48$), Random Tree (RTree), K Nearest Neighbor (KNN) with default parameter setting K=1, and Multi-layer Proceptron (MLP). The experimental results indicate that the under-sampling technique with KNN achieves better results than the existing complicated SPO method for ITSC.

3.1 Statistical Tests

Friedman and post-hoc Nemenyi tests are applied to compare the performance of the multiple learning methods on multiple data-sets, where it is inappropriate to compare their average value, because the average values are susceptible to outliers [19, 5]. Therefore, average rank is preferred for evaluating the performance of multiple learning methods. This work therefore performs statistical tests to evaluate the performance of the multiple learning methods on multiple data-sets. The Friedman test is utilized to obtain the average rank of the performance of the multiple learning methods on multiple data-sets; the post-hoc Nemenyi test is utilized to check whether there is a statistically significant difference between the learning methods at a 95% confidence interval.

4 Evaluation Metrics

The estimated overall accuracy is an ineffective evaluation measure for the imbalanced classification task [5, 20–22], so two evaluation measures are used for this study: F_{value} and G_{mean}.

Table 1 presents a confusion matrix for a binary classification problem; the columns represent the predicted class, and the rows represent the actual class. The evaluation measures are derived from the confusion matrix as follows:

Table 1. Confusion matrix for a binary classification problem

	Predicted Positives	Predicted Negatives
Actual Positives (P)	True Positive (TP)	False Negative (FN)
Actual Negatives (N)	False Positive(FP)	True Negative (TN)

$$TPR = \frac{TP}{TP + FN} \tag{1}$$

$$TNR = \frac{TN}{TN + FP} \tag{2}$$

$$recall = \frac{TP}{TP + FN} \tag{3}$$

$$precision = \frac{TP}{TP + FP} \tag{4}$$

$$F_{value} = \frac{2recall * precision}{recall + precision} \tag{5}$$

$$G_{mean} = \sqrt{TPR * TNR} \tag{6}$$

5 Experimental Setup

Java platform is used to implement the new approach, H-sampling technique integrated unstable learner, decision trees J48 [18] with bagging, and to investigate the performance of the new approach and previous approaches. 31 bootstrap samples are used in the ensemble. A 10-trial 10-fold cross-validation evaluation is performed for this study. The Friedman test is used for the calculation of average rank.

Table 2. Time series data-sets [5]

Data-sets		Data Information					Class Information	
		TS	Instances			Ratio	Previous	Altered
Index	Name	Length	$(P^+\&N^-)$	P^+	N^-	P^+/N^-	Class	class
1	Adiac	176	781	23	758	0.0303	37	2
2	S-Leaf	128	1125	75	1050	0.0714	15	2
3	Wafer	152	7164	762	6402	0.0119	2	2
4	FaceAll	131	2250	112	2138	0.0524	14	2
5	Yoga	426	3300	1530	1770	0.8644	2	2

5.1 Data-Sets

Table 2 shows a summary of the characteristics of the five time series data-sets from the public UCR time series repository [23], which were used as the benchmark data-sets of previous work [10, 3, 5].

6 Experimental Results Analysis

This section contains two sub-sections: 6.1 comparison of the performance of over-sampling, under-sampling with various algorithms and H-sampling methods on HITSC; and 6.2 comparison of the performance of other learning methods, SPO, under-sampling with various algorithms, and H-sampling methods for HITSC.

6.1 Comparison of the Performance of Over-sampling, Under-sampling, and Hybrid-sampling Methods

Table 3 presents a comparison of the performance of this new approach, H-sampling to enhance bagging, with previous approaches, over-sampling methods and the under-sampling with various algorithms based on the F_{value} and G_{mean} measures. The experimental results indicate that this new approach, H-sampling to enhance bagging, achieves the best performance with F_{value} across all over-sampling methods and the under-sampling with various algorithms on average value and average rank of F_{value}. This new approach achieves the highest average value 0.962 with smallest standard deviation (STD) 0.031 and the best average rank 1.4, respectively, which are the best results across all methods; while KNN

Table 3. Comparison of the performance of over-sampling, under-sampling methods with different learning algorithms, and H-sampling to enhance bagging based on the evaluation metrics F_{value} and G_{mean}

Metrics	Data-set	Results from Previous Research [10]						Results from previous Work [5]					This Work
		Over-sampling Methods						Under-sampling					H-sampling
	Name	REP	SMO	BoS	ADA	DB	SPO	SVM	J48	RTree	KNN	MLP	H-Bagging
F_{value}	Adiac	0.375	0.783	0.783	0.783	0.136	0.963	0.967	0.883	0.903	0.918	0.947	**0.975**
	S-Leaf	0.761	0.764	0.764	0.759	0.796	0.796	0.841	0.820	0.849	0.836	0.786	**0.932**
	Wafer	0.962	0.968	0.968	0.967	0.977	0.982	0.891	0.929	0.956	**0.999**	0.933	0.980
	FaceAll	0.935	0.935	0.935	0.935	0.890	0.936	0.957	0.876	0.863	0.909	0.919	**0.995**
	Yoga	0.710	0.729	0.721	0.727	0.689	0.702	0.744	0.771	0.811	0.807	0.780	**0.926**
	AverageValue	0.740	0.836	0.834	0.834	0.698	0.876	0.880	0.856	0.876	0.894	0.873	**0.962**
	STD	0.236	0.108	0.110	0.109	0.332	0.122	0.110	0.061	0.055	0.075	0.083	**0.031**
	AverageRank	8.90	6.90	7.30	7.70	8.70	4.50	7.40	7.80	6.40	4.40	6.60	**1.40**
	CD	7.45											
G_{mean}	Adiac	0.480	0.831	0.831	0.831	0.748	**0.999**	0.957	0.910	0.920	0.958	0.975	0.989
	S-Leaf	0.800	0.861	0.861	0.849	0.898	0.898	0.902	0.809	0.812	0.887	0.856	**0.976**
	Wafer	0.965	0.969	0.970	0.970	0.980	0.984	0.903	0.907	0.956	**0.998**	0.937	0.988
	FaceAll	0.950	0.950	0.950	0.950	0.948	0.957	0.966	0.870	0.860	0.929	0.925	**0.997**
	Yoga	0.741	0.756	0.750	0.755	0.724	0.735	0.630	0.807	0.803	0.808	0.774	**0.976**
	AverageValue	0.787	0.783	0.872	0.871	0.860	0.915	0.872	0.861	0.870	0.916	0.893	**0.985**
	STD	0.197	0.088	0.090	0.089	0.117	0.108	0.138	0.051	0.067	0.073	0.079	**0.009**
	AverageRank	9.30	6.80	6.90	7.20	7.50	4.10	6.60	8.60	8.20	4.20	7.20	**1.40**
	CD	7.45											

with the under-sampling method achieves the average value 0.894 with STD 0.075 and average rank 4.40, respectively, which is the second best across all methods on F_{value}.

On average value and average rank of the G_{mean} measure, this new approach, H-sampling to enhance bagging achieves the highest average value 0.985 with smallest STD 0.009 and lowest average rank 1.40, respectively, which is the best across all the compared methods; while, the SPO over-sampling method achieves average value 0.915 with STD 0.108 and average rank 4.1, respectively, which is the second best across all the compared methods on average rank of the G_{mean} measure, whereas KNN with the under-sampling method achieves average value 0.916 with STD 0.073 and average rank 3.4, respectively, which is the second best across all the compared methods on average of the G_{mean} measure. The results highlighted in red indicate the correction of the previous work [10, 5].

Figs 1 and 2 present a comparison of this new approach, H-sampling to enhance bagging, with previous approaches, over-sampling and under-sampling with various algorithms, with the Nemenyi test, where the x-axis indicates the ranking order of the sampling methods; the y-axis indicates the average rank of the F_{value} and G_{mean} performance, respectively, and the vertical bars indicate the "Critical Difference". Groups of sampling methods that are no significantly different at a 95% confidence interval are indicated when the vertical bars overlap. Comparing the performance of this new approach with previous approaches, over-sampling [10] and under-sampling with various algorithms [24], based on F_{value} and G_{mean}, H-sampling with bagging has the best average rank on both measures. KNN with the under-sampling method has the second best average rank of F_{value}; while the SPO over-sampling method has the second best average rank of G_{mean}. Statistical tests indicate that there is no statistically significant difference at a 95% confidence interval between over-sampling SPO, under-sampling KNN, and H-sampling with Bagging on the average rank of F_{vlaue} and G_{mean}; however, there is statistically significant difference at a 95% confidence interval between H-sampling and two over-sampling methods, DB and REP on F_{vlaue} measure, and between H-sampling and two over-sampling methods,J48 and REP on G_{mean} measure.

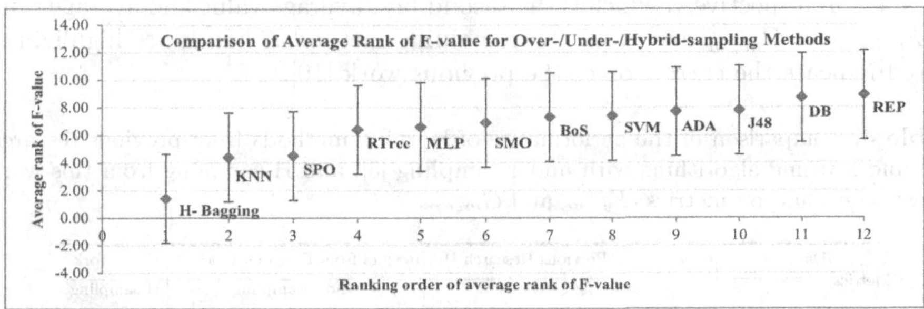

Fig. 1. Comparison of average rank of the F_{value} with the Nemenyi test for the over-sampling, SPO, under-sampling, and H-sampling methods, where the x-axis indicates the ranking order of all the sampling methods with learning algorithms, the y-axis indicates the average rank of the F_{value}, and the vertical bars indicate the "Critical Difference"

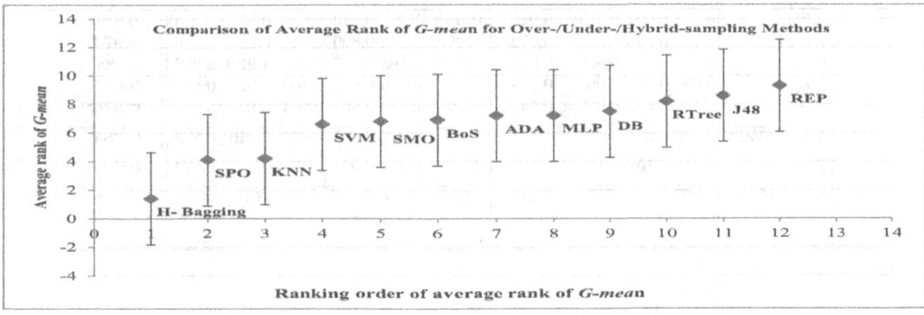

Fig. 2. Comparison of average rank of the G_{mean} with the Nemenyi test for all the over-sampling, under-sampling, and H-Bagging methods, where the x-axis indicates the ranking order of all the sampling methods with learning algorithms, the y-axis indicates the average rank of the G_{mean}, and the vertical bars indicate the "Critical Difference"

6.2 Comparison of the Performance Learning Methods, Over-sampling SPO, Under-sampling, and H-sampling Methods

Table 4 presents a comparison of the performance of previous work (learning methods [10] and the under-sampling with various algorithms [5]), and this work H-sampling with bagging based on F_{value} and G_{mean} evaluation measures. The experimental results indicate that H-sampling with bagging achieves the best performance on F_{value} and G_{mean} across all previous approaches, and H-sampling methods on average value of 0.962 and 0.985, and average rank of 1.40 and 1.40, respectively, which is the best average value and average rank of F_{value} and G_{mean} across all previous learning methods [10] and under-sampling method [5]; KNN achieves an average value of 0.894 and 0.916, and an average rank of 3.0 and 2.4, respectively, which is the second best average value and average rank of F_{value} and G_{mean} across all the remaining methods.The results highlighted in red indicate the correction of the previous work [10].

Table 4. Comparison of the performance of learning methods from previous research [10] and learning algorithms with under-sampling [5], and H-sampling from this work based on evaluation metrics: F_{value} and G_{mean}

Metrics	Data-set	Results from Previous Research [10]					Results from Previous Work [5]					This Work
		Learning Methods					Under-sampling					H-sampling
	Name	Easy	Bal.	1NN	1NN_DW	SPO	SVM	J48	RTree	KNN	MLP	H-Bagging
F_{value}	Adiac	0.534	0.348	0.800	0.917	0.963	0.967	0.883	0.903	0.918	0.947	**0.975**
	S-Leaf	0.521	0.578	0.716	0.429	0.796	0.841	0.820	0.849	0.836	0.786	**0.932**
	Wafer	0.795	0.954	0.949	0.857	0.982	0.891	0.929	0.956	**0.999**	0.933	0.980
	FaceAll	0.741	0.625	0.802	0.959	0.936	0.957	0.876	0.863	0.909	0.919	**0.995**
	Yoga	0.356	0.689	0.652	0.710	0.702	0.744	0.771	0.811	0.807	0.780	**0.926**
	AverageValue	0.589	0.639	0.784	0.774	0.876	0.880	0.856	0.876	0.894	0.873	**0.962**
	STD	0.179	0.218	0.112	0.215	0.122	0.092	0.061	0.055	0.075	0.083	**0.031**
	AverageRank	10.4	9	8.4	7.2	4.6	4.6	6.6	4.6	3.8	5.4	**1.4**
	CD	7.00										
G_{mean}	Adiac	0.782	0.897	0.875	0.920	**0.999**	0.957	0.910	0.920	0.958	0.975	0.989
	S-Leaf	0.721	0.898	0.798	0.572	0.898	0.902	0.809	0.812	0.887	0.856	**0.976**
	Wafer	0.817	0.970	0.953	0.870	0.984	0.903	0.907	0.956	**0.998**	0.937	0.988
	FaceAll	0.792	0.918	0.983	0.985	0.957	0.966	0.870	0.860	0.929	0.925	**0.997**
	Yoga	0.464	0.688	0.695	0.741	0.735	0.630	0.807	0.803	0.808	0.774	**0.976**
	AverageValue	0.713	0.874	0.861	0.818	0.915	0.872	0.861	0.870	0.916	0.893	**0.985**
	STD	0.145	0.108	0.117	0.164	0.108	0.113	0.051	0.067	0.073	0.079	**0.009**
	AverageRank	10.80	6.50	7.20	7.10	3.50	7.10	7.20	6.50	3.20	5.50	**1.40**
	CD	7.00										

Figs 3 and 4 present a comparison of the performance of previous work (learning methods and the under-sampling method with various algorithms) and this new approach, H-sampling to enhance bagging, using the Nemenyi test, where the x-axis indicates the ranking order of the learning methods and learning algorithms; the y-axis indicates the average rank of F_{value} and G_{mean} performance, respectively, and the vertical bars indicate the "Critical Difference".

Groups of learning methods and learning algorithms that are no statistically significant difference at a 95% confidence interval are indicated when the vertical bars overlap. Comparing the previous approaches [10, 5] and this approach, H-sampling to enhance bagging, based on F_{value} and G_{mean}, H-sampling with bagging has the best average rank on both measures. KNN with under-sampling method has the second best average rank of F_{value} and G_{mean}. The statistical test results demonstrate that H-sampling with bagging method is statistically significantly better than 1NN, Bal. and Easy on F_{value}, and better than Easy on G_{mean} at a 95% confidence interval; however, there is no statistically significant difference between this new approach, H-sampling with bagging and previous approaches, over-sampling SPO and under-sampling KNN at a 95% confidence interval on both the F_{value} and G_{mean} measures.

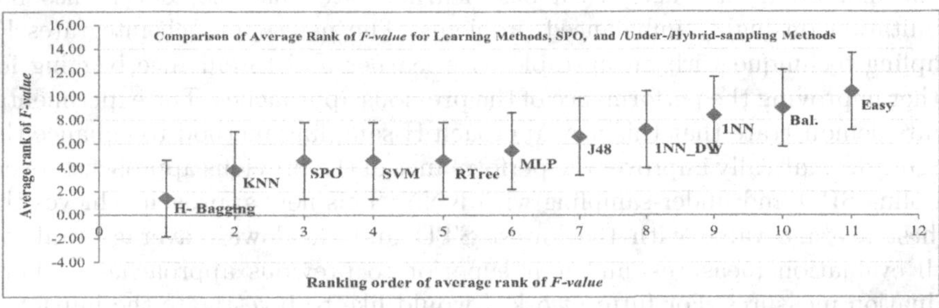

Fig. 3. Comparison of average rank of the F_{value} metric with the Nemenyi test for the learning methods, SPO, under-sampling, and H-sampling methods, where the x-axis indicates the ranking order of all the learning methods and sampling methods with learning algorithms, the y-axis indicates the average rank of F_{value}, and the vertical bars indicate the "Critical Difference"

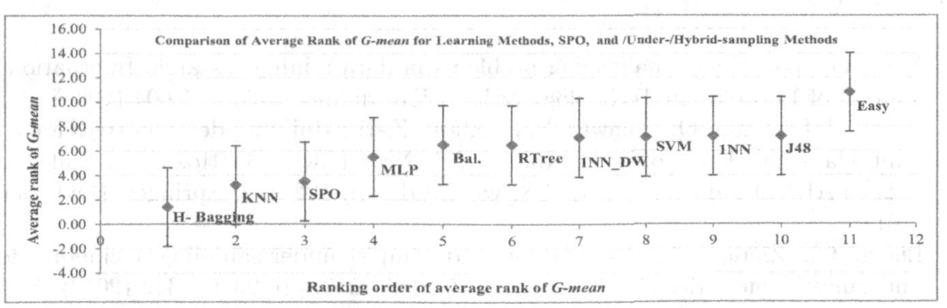

Fig. 4. Comparison of average rank of the G_{mean} metric with the Nemenyi test for the learning methods, SPO, under-sampling, and H-sampling methods, where the x-axis indicates the ranking order of all the learning methods, sampling methods with learning algorithms, the y-axis indicates the average rank of G_{mean}, and the vertical bars indicate the "Critical Difference"

7 Conclusion

This paper has addressed the limitations of existing techniques of over-sampling and under-sampling methods, and proposed a new approach, H-sampling schema to enhance bagging for improving the performance of previous approaches. It has empirically compared this new approach with the previous approaches of over-sampling, SPO and under-sampling with various algorithms based on two evaluation measures, F_{value} and G_{mean} on benchmark data-sets. Statistical tests are used to draw valid conclusions.

This new approach, H-sampling, reduces the computational cost and training time of over-sampling by using fewer positives in training, and increases the capability of under-sampling by using more negatives for training. Bagging is one of most effective ensemble learning methods for improving the performance of the individual classifiers when base learners are unstable, but it also has a limitation on highly imbalanced problems. The new approach integrates H-sampling technique with an unstable base learner $J48$ to enhance bagging for further improving the performance of the previous approaches. The experimental results demonstrate that this new approach H-sampling method to enhance the bagging dramatically improves the performance of the previous approaches, over-sampling SPO and under-sampling with KNN. This new approach achieves the highest average value with the lowest STD and the lowest average rank on both evaluation measures, and it is superior to previous approaches on both evaluation measures. For future work, I would like to investigate the impact of the performance of H-sampling integrating bagging with other base learners: unstable learners and stable learners.

Acknowledgments. Special thanks to Dr. Tony Bagnall for generously sharing his ARFF format time series data-sets.

References

1. Yang, Q., Wu, X.: 10 challenging problems in data mining research. International Journal of Information Technology & Decision Making 5(4), 597–604 (2006)
2. Hoens, T.R., Qian, Q., Chawla, N.V., Zhou, Z.-H.: Building decision trees for the multi-class imbalance problem. In: Tan, P.-N., Chawla, S., Ho, C.K., Bailey, J. (eds.) PAKDD 2012, Part I. LNCS, vol. 7301, pp. 122–134. Springer, Heidelberg (2012)
3. Liang, G., Zhang, C.: An efficient and simple under-sampling technique for imbalanced time series classification. In: CIKM 2012, pp. 2339–2342 (2012)
4. Hidasi, B., Gáspár-Papanek, C.: ShiftTree: An interpretable model-based approach for time series classification. In: Gunopulos, D., Hofmann, T., Malerba, D., Vazirgiannis, M. (eds.) ECML PKDD 2011, Part II. LNCS, vol. 6912, pp. 48–64. Springer, Heidelberg (2011)
5. Liang, G., Zhang, C.: A comparative study of sampling methods and algorithms for imbalanced time series classification. In: Thielscher, M., Zhang, D. (eds.) AI 2012. LNCS, vol. 7691, pp. 637–648. Springer, Heidelberg (2012)
6. Breiman, L.: Bagging predictors. Machine Learning 24(2), 123–140 (1996)

7. Quinlan, J.: Bagging, boosting, and c4.5. In: Proceedings of the 13th National Conference on Artificial Intelligence, pp. 725–730 (1996)
8. Bauer, E., Kohavi, R.: An empirical comparison of voting classification algorithms: Bagging, boosting, and variants. Machine Learning 36(1), 105–139 (1999)
9. Dietterich, T.: An experimental comparison of three methods for constructing ensembles of decision trees: Bagging, boosting, and randomization. Machine Learning 40(2), 139–157 (2000)
10. Cao, H., Li, X., Woon, Y., Ng, S.: SPO: Structure preserving oversampling for imbalanced time series classification. In: Proceedings of the IEEE 11th International Conference on Data Mining, ICDM 2011, pp. 1008–1013 (2011)
11. Chawla, N., Bowyer, K., Hall, L., Kegelmeyer, W.: SMOTE: Synthetic minority over-sampling technique. Journal of Artificial Intelligence Research 16(1), 321–357 (2002)
12. Han, H., Wang, W.-Y., Mao, B.-H.: Borderline-SMOTE: A new over-sampling method in imbalanced data sets learning. In: Huang, D.-S., Zhang, X.-P., Huang, G.-B. (eds.) ICIC 2005. LNCS, vol. 3644, pp. 878–887. Springer, Heidelberg (2005)
13. He, H., Bai, Y., Garcia, E., Li, S.: ADASYN: Adaptive synthetic sampling approach for imbalanced learning. In: IEEE International Joint Conference on Neural Networks, IJCNN 2008, pp. 1322–1328. IEEE (2008)
14. Guo, H., Viktor, H.L.: Learning from imbalanced data sets with boosting and data generation: The DataBoost-IM approach. ACM SIGKDD Explorations Newsletter 6(1), 30–39 (2004)
15. Liu, X.Y., Wu, J., Zhou, Z.H.: Exploratory undersampling for class-imbalance learning. IEEE Transactions on Systems, Man, and Cybernetics, Part B: Cybernetics 39(2), 539–550 (2009)
16. Wei, L., Keogh, E.: Semi-supervised time series classification. In: Proceedings of the 12th ACM SIGKDD International Conference on Knowledge Discovery and Data Mining, pp. 748–753. ACM (2006)
17. Xi, X., Keogh, E., Shelton, C., Wei, L., Ratanamahatana, C.A.: Fast time series classification using numerosity reduction. In: Proceedings of the 23rd International Conference on Machine Learning, ICML 2006, pp. 1033–1040 (2006)
18. Witten, I., Frank, E.: Data Mining: Practical Machine Learning Tool and Techniques. Morgan Kaufmann (2005)
19. Demšar, J.: Statistical comparisons of classifiers over multiple data sets. Journal of Machine Learning Research 7, 1–30 (2006)
20. Liang, G.: An investigation of sensitivity on bagging predictors: An empirical approach. In: 26th AAAI Conference on Artificial Intelligence, pp. 2439–2440 (2012)
21. Liang, G., Zhu, X., Zhang, C.: The effect of varying levels of class distribution on bagging with different algorithms: An empirical study. International Journal of Machine Learning and Cybernetics (2012), `http://link.springer.com/article/10.1007%2Fs13042--012--0125--5`
22. Liang, G., Zhang, C.: Empirical study of bagging predictors on medical data. In: 9th Australian Data Mining Conference, AusDM 2011, pp. 31–40 (2011)
23. Keogh, E., Zhu, Q., Hu, B., Hao, Y., Xi, X., Wei, L., Ratanamahatana, C.A.: The UCR Time Series Classification/Clustering homepage (2011), `http://www.cs.ucr.edu/~eamonn/time_series_data/`
24. Liang, G., Zhu, X., Zhang, C.: An empirical study of bagging predictors for imbalanced data with different levels of class distribution. In: Wang, D., Reynolds, M. (eds.) AI 2011. LNCS, vol. 7106, pp. 213–222. Springer, Heidelberg (2011)

Computer Aided Diagnosis of ADHD Using Brain Magnetic Resonance Images

B.S. Mahanand[1], R. Savitha[2], and S. Suresh[2]

[1] Department of Information Science and Engineering
Sri Jayachamarajendra College of Engineering, Mysore, India
bsmahanand@sjce.ac.in

[2] School of Computer Engineering
Nanyang Technological University, Singapore
{savi0001,ssundaram}@ntu.edu.sg

Abstract. This paper presents a pilot study on the development of an automated diagnostic tool for Attention Deficiency Hyperactivity Disorder (ADHD) based on regional anatomy of the child brain. For the pilot study, amygdala and cerebellar vermis are chosen from magnetic resonance images obtained from ADHD-200 consortium data set. These regions play a vital role in the control of emotional response and behavior/locomotion, respectively. The images are preprocessed, registered by transforming each image to the space of the population average. The gray matter tissue probability values of amygdala and cerebellar vermis are obtained by applying a region-of-interest mask. These values are then used to train a Projection Based Learning algorithm for a Meta-cognitive Radial Basis Function Network (PBL-McRBFN) for the diagnosis of ADHD and prediction of its subtype. Performance results show that the PBL-McRBFN diagnoses ADHD and predicts its subtypes based on these regions with an accuracy of approx. 65% and 62%, respectively.

Keywords: Attention Deficient Hyperactivity Disorder, Meta-cognitive Radial Basis Function Network, Projection Based Learning, Region-of-Interest, Magnetic Resonance Imaging.

1 Introduction

Attention Deficiency Hyperactivity Disorder (ADHD) is a childhood neuropsychiatric disorder that is comorbid with other neurological disorders. It is estimated that about 5% of school-going children world wide are affected by ADHD [1]. The Diagnostic and Statistical Manual of Mental disorders, IV edition, Text Revision defines ADHD as a persistent and age-inappropriate pattern of inattention, hyperactivity or both [2]. Although both genetic and environmental factors are known to influence the onset of ADHD [3], the cause of the disorder is not completely understood. The studies in literature that aim at understanding the neurobiology of ADHD in childhood is extensively surveyed in [4].

Neuroimaging based regional anatomical study of the human brain can provide us insights to the neurobiology of ADHD in childhood. Magnetic Resonance

S. Cranefield and A. Nayak (Eds.): AI 2013, LNAI 8272, pp. 386–395, 2013.

Imaging (MRI) is one of the most important neuroimaging procedure that provides accurate information about the anatomy of the brain. Further, MRI based studies on the pathophysiology of ADHD in the brains of children show significant decrease in total cerebral volume [5], abnormalities in the frontostriatal areas, temporoparietal lobes, basal ganglia, corpus callosum [6], amygdala, and thalamus [7], abnormal/delayed cortical development [8] etc. Therefore, there is a need to develop automated diagnostic tools based on regional anatomical study to understand the onset and development of ADHD. This paper reports a pilot study on the development of an automated diagnostic tool that performs regional anatomical study of the brain to understand the influence of the various regions in the onset and development of ADHD. This pilot study reports the diagnostic results from the analysis of amygdala and cerebellar vermis that are known to play a key role in control of emotional response and behavior/locomotion, respectively.

As we are interested in conducting a regional anatomical study, we use the Region-of-Interest (ROI) methods to extract voxels of the regions [9]. In this paper, we conduct an automated diagnostic study of ADHD based on regional anatomy of the most frequently reported regions for ADHD, namely, amygdala and cerebellar vermis, obtained from the ADHD-200 data set [10]. First, the ROI of these regions are defined using the Wake Forest University Pick-atlas [11]. The voxels of the ROI thus obtained are then used as input features to train a Projection Based Learning algorithm of a Meta-cognitive Radial Basis Function Network (PBL-McRBFN) classifier [12] to perform ADHD diagnosis. Based on the diagnostic study, the subjects are classified either as Typically Developing Controls (TDC) or ADHD. Finally, the ADHD subjects are classified as belonging to either the inattentive and hyperactive type (combined type) or the inattentive type using the PBL-McRBFN classifier.

A meta-cognitive neural network is a network that has been developed based on the the best human learning strategy, namely, self-regulated learning [13]. A meta-cognitive neural network has a cognitive component that represents knowledge and a meta-cognitive component that monitors and controls the knowledge represented by the cognitive component. Recently, a number of meta-cognitive learning algorithms have been developed and are available in the literature [14–17]. In [18], the various models of human meta-cognition have been reviewed in detail. Of these various models, the model proposed by Nelson and Narens [19] is the most comprehensive and imitable in machine learning. Several machine learning algorithms have been developed based on this Nelson and Narens model of human meta-cognition for real-valued neural networks [20, 12], complex-valued neural networks [21–23] and neuro-fuzzy inference systems [24, 25]. It can be seen from these studies that a meta-cognitive network with self-regulated learning outperform the networks without meta-cognition. As the ADHD 200 competition data set is a real-valued data set, we use the real-valued meta-cognitive neural network in our work. Of the real-valued meta-cognitive neural networks, the projection based learning algorithm of the meta-cognitive radial basis function network [12] performs classification better than the meta-cognitive neural network [20] due to its ability to store past knowledge and to use the knowledge at a later stage of learning. Hence, in this work, we employ the projection based learning algorithm of the meta-cognitive radial basis function network [12] for the diagnosis of ADHD and identification of its subtype.

McRBFN is a meta-cognitive neural network with a cognitive component and a meta-cognitive component that monitors and controls the cognitive component. For each sample in the data set, the meta-cognitive component chooses suitable learning strategies, depending on the relative knowledge of the sample with respect to the cognitive component. Thus, it realizes the best human learning strategy of 'self-regulated learning' by deciding *what-to-learn*, *when-to-learn* and *how-to-learn* for the cognitive component. McRBFN realizes *what-to-learn* by deleting samples that are similar to the knowledge already learnt by the network, *when-to-learn* by reserving samples for future use and *how-to-learn* by using the sample to add a neuron or to update the network parameters. When a neuron is used in the learning process, a projection based learning algorithm is used to estimate the output weights for a fixed hidden neuron parameters. The hidden neurons are Gaussian in nature and its parameters are fixed based on sample overlapping conditions. Thus, the PBL algorithm begins with zero hidden neurons, and adds neurons, updates the network parameters until an optimum network structure is obtained.

The paper is organized as follows: Section 2 presents the ROI extraction method used and the description of the PBL-McRBFN classifier. In Section 3, the experimental study on the ability of PBL-McRBFN classifier to diagnose ADHD and predict its subtypes is presented. Finally, Section 4 summarizes the main conclusions from the study.

2 Materials and Methods

This section describes the data and the machine learning tools used in the ADHD diagnostic study. The diagnostic tool takes the whole brain MR images of the subjects, extracts the features from the ROI of the chosen regions, which are then used for automated ADHD diagnosis using a PBL-McRBFN classifier.

2.1 Data Set

In our study, the ADHD-200 consortium data set has been used [10]. This data set is a collection of brain MR images of 941 subjects from 8 participating members of the consortium. The subjects include 581 TDC and 360 ADHD subjects, of which 210 are both inattentive and attention deficient, 137 are inattentive and 13 are hyperactive. All MR images are processed with the Diffeomorphic Anatomical Registration Through Exponentiated Lie Algebra (DARTEL) [26] using the Statistical Parametric Mapping (SPM) software package based on the Burner pipeline of the ADHD-200 consortium data set.

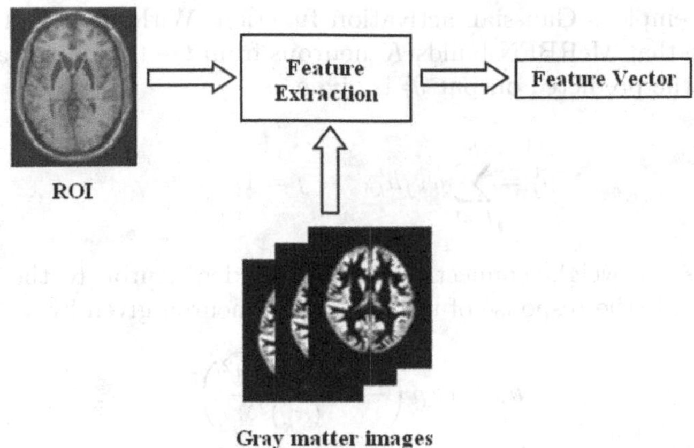

Fig. 1. ROI based Feature Extraction

2.2 ROI Based Feature Extraction

A feature extraction approach based on the Region-of-Interest (ROI) method is employed in this work. In essence, ADHD patients lack control over their emotional response and/or their behaviour and locomotion. Hence, amygdala that plays a key role in processing emotions and cerebellar vermis that controls behavior/locomotion are chosen for this study. It is also reported in the literature that there are pronounced structural changes in amygdala of ADHD patients [27] and that smaller volume of cerebellar vermis is attributed to symptoms of hyperactivity and inattention [28]. The ROI based feature extraction process is explained in Fig. 1. The ROI masks of amygdala and cerebellar vermis are defined using the Wake Forest University Pickatlas [11]. The gray matter tissue probability values are extracted as features from the modulated and normalized grey matter images, using the generated ROI masks. The extracted features are then used as an input to the PBL-McRBFN classifier.

2.3 PBL Algorithm for McRBFN Classifier

Given a stream of training data samples, $\{(\mathbf{x}^1, c^1), \cdots, (\mathbf{x}^t, c^t), \cdots\}$, where $\mathbf{x}^t \in \Re^m$ is an m-dimensional input of the tth sample, and $c^t \in [1, n]$ is its class label, and n is the total number of classes. The coded class labels $\mathbf{y}^t \in \Re^n$ are given by:

$$y_j^t = \begin{cases} 1 \text{ if } & c^t = j \\ -1 \text{ otherwise} \end{cases} \quad j = 1, \cdots, n \tag{1}$$

The objective of McRBFN classifier is to approximate the function $\mathbf{x}^t \in \Re^m \rightarrow \mathbf{y}^t \in \Re^n$. McRBFN has two components, namely, the cognitive component and the meta-cognitive component.

Cognitive Component of McRBFN is a radial basis function network. The hidden layer employs Gaussian activation function. Without loss of generality, let us assume that McRBFN builds K neurons from $t-1$ training samples. For an input $\mathbf{x}^t$, the predicted output $\widehat{y}_j^t$ is given as

$$\widehat{y}_j^t = \sum_{k=1}^{K} w_{kj} h_k^t, \qquad j = 1, \cdots, n \tag{2}$$

Where w_{kj} is the weight connecting the kth hidden neuron to the jth output neuron and h_k^t is the response of the kth hidden neuron given by

$$h_k^t = exp\left(-\frac{\|\mathbf{x}^t - \boldsymbol{\mu}_k^l\|^2}{(\sigma_k^l)^2}\right) \tag{3}$$

Where $\boldsymbol{\mu}_k^l \in \Re^m$ and $\sigma_k^l \in \Re^+$ are the center and width of the kth hidden neuron, and the superscript l is the class of the kth hidden neuron.

Next, we present the projection based learning algorithm of McRBFN.

Projection Based Learning Algorithm works on the principle of minimization of error function given by the hinge loss and finds the optimal network output parameters for which the error function is minimum. For t consecutive samples, the error function is

$$J(\mathbf{W}) = \frac{1}{2}\sum_{i=1}^{t}\sum_{j=1}^{n}\begin{cases} 0 & \text{if } y_j^i \widehat{y}_j^i > 1 \\ \left(y_j^i - \widehat{y}_j^i\right)^2 & \text{otherwise} \end{cases} \tag{4}$$

The optimal output weights ($\mathbf{W}^* \in \Re^{K \times n}$) are estimated such that $J(\mathbf{W}^*)$ is minimum, i.e., $\mathbf{W}^* := arg \min_{\mathbf{W} \in \Re^{K \times n}} J(\mathbf{W})$ Thus, $\mathbf{W}^*$ is obtained by equating the first order partial derivative of $J(\mathbf{W})$ with respect to the W to 0, and are hence determined using $\mathbf{W}^* = \mathbf{A}^{-1}\mathbf{B}$ [12], where the elements of $\mathbf{A}$ and $\mathbf{B}$ are defined as $a_{kp} = \sum_{i=1}^{t} h_k^i h_p^i$ and $b_{pj} = \sum_{i=1}^{t} h_p^i y_j^i$, respectively. Here, $k, p = 1, \cdots, K; j = 1, \cdots, n$.

Meta-Cognitive Component of McRBFN monitors the cognitive component through the knowledge measures and controls the learning of the cognitive component through the self-regulated thresholds. The meta-cognitive component measures the knowledge of the cognitive component with respect the current sample t using the predicted class label ($\widehat{c}^t$), maximum hinge loss error (E^t), confidence of the classifier ($\hat{p}(c^t|\mathbf{x}^t)$) and the class-wise significance (ψ_c) that are defined in [12]. Based on these knowledge measures, the meta-cognitive component chooses the best of the following learning strategies for the current training sample. These strategies address the basic principles of self-regulated human learning (i.e., *what-to-learn*, *when-to-learn* and *how-to-learn*).

Sample Delete Strategy: A sample is deleted from the training data set if $\widehat{c}^t == c^t$ AND $\hat{p}(c^t|\mathbf{x}^t) \geq \beta_d$, where β_d is the delete threshold that controls the number of samples participating in the learning process, and is usually selected in the range [0.9, 0.95]. The sample delete strategy avoids over-training and reduces computational effort.

Neuron Growth Strategy: A sample is used to add a neuron to the network if ($\hat{c}^t \neq c^t$ OR $E^t \geq \beta_a$) AND $\psi_c(\mathbf{x}^t) \leq \beta_c$, where β_c is the knowledge measurement threshold and β_a is the self-adaptive addition threshold. The β_c and β_a allows samples with significant knowledge for learning first and rest for fine tuning. The thresholds β_c and β_a are usually selected in the range [0.3, 0.7] and [1.3, 1.7], respectively. β_a is self-adapted according to $\beta_a := \delta\beta_a + (1-\delta)E^t$, where δ is the slope that controls rate of self-adaptation and is set close to 1.

When a new neuron is added, the parameters of the neuron are initialized based on the neuron's overlap with inter/intra-class. The overlap is measured using the Euclidean distance of the sample to the nearest hidden neuron in the intra and inter classes. If the neuron does not overlap with other neurons in any class, then the parameters are initialized as $\boldsymbol{\mu}^c_{K+1} = \mathbf{x}^t$; $\sigma^c_{K+1} = \kappa\sqrt{{\mathbf{x}^t}^T\mathbf{x}^t}$, where κ is an overlap factor of the hidden units, which lies in the range $0.5 \leq \kappa \leq 1$. However, if the neuron overlaps with neurons in intra/inter class, the center is shifted to reduce overlapping and hence, misclassification, as shown in [12]. In the PBL-McRBFN, the knowledge of the past samples is stored in the network as neurons center/width and their associated class label are used to initialize the output weight of the new neuron. When a neuron is added to McRBFN, the output weights are initialized according to:

$$\begin{bmatrix} \mathbf{W}^t_K \\ \mathbf{w}^t_{K+1} \end{bmatrix} = \left[\begin{array}{c|c} \mathbf{A}^{t-1}_{K\times K} & \mathbf{a}^T_{K+1} \\ \hline \mathbf{a}_{K+1} & a_{K+1,K+1} \end{array}\right]^{-1} \begin{bmatrix} \mathbf{B}^{t-1}_{K\times n} \\ \mathbf{b}_{K+1} \end{bmatrix} \tag{5}$$

Where $\mathbf{W}^t_K$ is the output weight matrix for K hidden neurons, and $\mathbf{w}^t_{K+1}$ is the vector of output weights for new hidden neuron after learning from t^{th} sample. It must be noted that the size of matrix $\mathbf{A}$ is increased from $K \times K$ to $(K+1) \times (K+1)$ according to:

$$\mathbf{A}^t = \left[\begin{array}{c|c} \mathbf{A}^{t-1}_{K\times K} + (\mathbf{h}^t)^T\, \mathbf{h}^t & \mathbf{a}^T_{K+1} \\ \hline \mathbf{a}_{K+1} & a_{K+1,K+1} \end{array}\right] \tag{6}$$

Where $\mathbf{h}^t = [h^t_1, h^t_2, \cdots, h^t_K]$ is a vector of the existing K hidden neurons response for new (t^{th}) training sample. The elements of $\mathbf{a}^T_{K+1}$ and $a_{K+1,K+1}$ is computed according to $a_{K+1,p} = \sum_{i=1}^{K+1} h^i_{K+1}h^i_p$, $p = 1, \cdots, K$; and $a_{K+1,K+1} = \sum_{i=1}^{K+1} h^i_{K+1}h^i_{K+1}$, respectively. Also, the size of $\mathbf{B}$ is increased from $K \times n$ to $(K+1) \times n$

$$\mathbf{B}^t_{(K+1)\times n} = \begin{bmatrix} \mathbf{B}^{t-1}_{K\times n} + (\mathbf{h}^t)^T (\mathbf{y}^t)^T \\ \mathbf{b}_{K+1} \end{bmatrix} \tag{7}$$

and $\mathbf{b}_{K+1} \in \Re^{1\times n}$ is assigned as $b_{K+1,j} = \sum_{i=1}^{K+1} h^i_{K+1}\tilde{y}^i_j$, $j = 1, \cdots, n$, Where $\tilde{y}^i$ is the pseudo-output for the ith hidden neuron ($\boldsymbol{\mu}^l_i$).

Parameter Update Strategy: The current (t^{th}) training sample is used to update the output weights of the cognitive component ($\mathbf{W}_K = [\mathbf{w}_1, \mathbf{w}_2, \cdots, \mathbf{w}_K]^T$)

if $c^t == \widehat{c}^t$ AND $E^t \geq \beta_u$, where β_u is the self-adaptive parameter update threshold that is selected in the interval of [0.4, 0.7]. The β_u is adapted based on the hinge loss error as $\beta_u := \delta\beta_u + (1-\delta)E^t$. When a sample is used to update the network parameters, the $\mathbf{A} \in \Re^{K \times K}$ matrix is updated as $\mathbf{A}^t = \mathbf{A}^{t-1} + (\mathbf{h}^t)^T \mathbf{h}^t$, and hence the output weights are updated as $\mathbf{W}_K^t = \mathbf{W}_K^{t-1} + (\mathbf{A}^t)^{-1} (\mathbf{h}^t)^T (\mathbf{e}^t)^T$, where $\mathbf{e}^t$ is hinge loss for t^{th} sample.

Sample Reserve Strategy: If the sample does not satisfy any of the above criteria, then the current sample is pushed to the rear end of data stream for future use.

This learning is continued for all the samples in the training data set.

3 Results

In this section, we study the performance of PBL-McRBFN in diagnosing ADHD and identifying its subtypes using the ADHD-200 consortium data set. First, the ability of PBL-McRBFN in diagnosing ADHD is studied with the features chosen from amygdala and cerebellar vermis. Next, its ability to distinguish combined type Vs inattentive type of ADHD based on the individual regions is studied. Sensitivity and specificity are used as performance measures.

First, we present the diagnostic results of PBL-McRBFN in classifying the subjects as TDC or ADHD patients. The number of features in the selected regions obtained by ROI, the number of neurons (K), and the training and testing sensitivities (Sens) and specificities (Spec) are tabulated in Table 1. It was observed that these regions can differentiate between TDC and ADHD on the training and testing samples by an accuracy of about 95% and 65%, respectively. The training specificity and sensitivity of the PBL-MCRBFN classifier for these regions are as high as 0.993 and 0.996, respectively. Therefore, it can be inferred that the classifier is capable of differentiating TDC and ADHD subjects accurately. In the testing data set, while the classification results with amygdala has high specificity of 0.766 (and hence, low type I error), the results with cerebellar vermis has high sensitivity of 0.714 (and hence, low type II error). The accuracy of the PBL-McRBFN classifier is greater than the accuracy reported in the literature on the complete data set with a SVM classifier [29]. The sensitivity and specificity of the SVM classifier has been reported as 0.33 and 0.79, respectively. Thus, it can be inferred based on these results that the PBL-McRBFN classifier is capable of diagnosing ADHD with higher accuracy using the features in amygdala and cerebellar vermis.

Further, we study the ability of PBL-McRBFN to predict the subtype of ADHD. A subject with ADHD can be either inattentive or hyperactive or both. As the ADHD-200 data set has only 13 samples for the hyperactive group, we solve this problem of diagnosing the subtype as either inattentive or inattentive and hyperactive (combined type). The training and testing performances of PBL-McRBFN in predicting the subtype of ADHD is also presented in Table 1. It is observed that the performance of PBL-McRBFN in classifying the subtype of ADHD is approximately 62%, which is greater than the results reported in [29].

Table 1. PBL-McRBFN Performance: TDC Vs ADHD

Study	Region	No. of features	K	Training		Testing	
				Sens	Spec	Sens	Spec
ADHD Vs TDC	Amygdala	1050	300	0.886	0.993	0.506	0.766
	Cerebellar Vermis	6358	355	0.996	0.905	0.714	0.553
Subtype	Amygdala	1050	154	0.991	0.950	0.730	0.510
	Cerebellar Vermis	6358	155	1	0.931	0.346	0.755

Further, it can be observed that while the classification results with amygdala has high sensitivity (and hence, low type I error), the results with cerebellar vermis has high specificity (and hence, low type II error). In this context, sensitivity refers to the ability to classify combined type accurately and specificity refers to the ability to classify inattentive type accurately. Thus, it can be inferred that the PBL-McRBFN classifier can predict the subtype of ADHD with higher accuracy based on the features obtained from amygdala and cerebellar vermis.

4 Conclusion

Results of a pilot study on the development of an automated diagnostic tool for Attention Deficiency Hyperactivity Disorder (ADHD) based on regional anatomy obtained from the MR images of child brain have been presented. MR images obtained from ADHD-200 data set are preprocessed and the region-of-interest of amygdala and cerebellar vermis are obtained. The gray matter tissue probability values thus obtained are used as features to train a PBL-McRBFN classifier for the diagnosis of ADHD and the prediction of its subtype. It is observable from the performance results that amygdala and cerebellar vermis are sufficient for the diagnosis. Extending the study to other regions of the human brain to understand the pathophysiology of ADHD is a scope for future work.

Acknowledgment. We thank the ADHD-200 consortium for making MRI data available.

References

1. Polanczyk, G., de Lima, M.S., Horta, B.L., Biederman, J., Rohde, L.A.: The worldwide prevalence of ADHD: a systematic review and metaregression analysis. American Journal of Psychiatry 164(6), 942–948 (2007)
2. Diagnostic and statistical manual of mental disorders, IV edition, Text Revision. American Psychiatric Association, Washington D. C (2000)
3. Banaschewski, T., Becker, K., Scherag, S., Franke, B., Coghill, D.: Molecular genetics of attention-deficit/hyperactivity disorder: An overview. European Child and Adolescent Psychiatry 19(3), 237–257 (2010)
4. Cortese, S.: The neurobiology and genetics of Attention-Deficit/Hyperactivity Disorder (ADHD): What every clinician should know. European Journal of Paediatric Neurology 16(5), 422–433 (2012)

5. Cherkasova, M.V., Hechtman, L.: Neuroimaging in attention deficit hyperactivity disorder: beyond the frontostriatal circuitry. Canadian Journal of Psychiatry 54(10), 651–664 (2009)
6. Giedd, J.N., Rapoport, J.L.: Structural MRI of pediatric brain development: what have we learned and where are we going? Neuron. 67(5), 728–734 (2010)
7. Ivanov, I., Bansal, R., Hao, X., Zhu, H., Kellendonk, C., Miller, L., Sanchez-Pena, J., Miller, A.M., Chakravarty, M.M., Klahr, K., Durkin, K., Greenhill, L.L., Peterson, B.S.: Morphological abnormalities of the thalamus in youths with attention deficit hyperactivity disorder. Americal Journal of Psychiatry 167(4), 397–408 (2010)
8. Shaw, P., Lerch, J., Greenstein, D., Sharp, W., Clasen, L., Evans, A., Giedd, J., Castellanos, F.X., Rapoport, J.: Longitudinal mapping of cortical thickness and clinical outcome in children and adolescents with attention-deficit/hyperactivity disorder. Archives of General Psychiatry 63(5), 540–549 (2006)
9. Jack, C.R., Petersen, R.C., O'Brien, P.C., Tangalos, E.G.: MR-based hippocampal volumetry in the diagnosis of Alzheimer's disease. Neurology 42(1), 183–188 (1992)
10. Milham, P.M., Damien, F., Maarten, M., Stewart, H.M.: The ADHD-200 consortium: A model to advance the translational potential of neuroimaging in clinical neuroscience. Frontiers in Systems Neuroscience 6, 1–5 (2012)
11. Maldjian, J.A., Laurienti, P.J., Kraft, R.A., Burdette, J.H.: An automated method for neuroanatomic and cytoarchitectonic atlas-based interrogation of fMRI data sets. NeuroImage 19(3), 1233–1239 (2003)
12. Babu, G.S., Suresh, S.: Meta-cognitive rbf network and its projection based learning algorithm for classification problems. Applied Soft. Computing Journal 13(1), 654–666 (2013)
13. Joysula, D.P., Vadali, H., Donahue, J., Hughes, F.C.: Modeling meta-cognition for learning in artificial systems. In: World Congress on Nature and Biologically Inspired Computing, pp. 1419–1424 (2009)
14. Suresh, S., Dong, K., Kim, H.: A sequential learning algorithm for self-adaptive resource allocation network classifier. Neurocomputing 73(16-18), 3012–3019 (2010)
15. Mahanand, B.S., Suresh, S., Sundararajan, N., Kumar, M.A.: Identification of brain regions responsible for Alzheimer's disease using a self-adaptive resource allocation network. Neural Networks 32, 313–322 (2012)
16. Babu, G.S., Suresh, S.: Sequential projection-based metacognitive learning in a radial basis function network for classification problems. IEEE Transactions on Neural Networks and Learning Systems 24(2), 194–206 (2013)
17. Babu, G.S., Suresh, S., Mahanand, B.S.: A novel PBL-McRBFN-RFE approach for identification of critical brain regions responsible for Parkinson's disease. Expert Systems with Applications (2013), doi:10.1016/j.eswa.2013.07.073
18. Subramanian, K., Suresh, S., Sundararajan, N.: A meta-cognitive neuro-fuzzy inference system (McFIS) for sequential classification problems. IEEE Transactions Fuzzy Systems (2013), doi:10.1109/TFUZZ.2013.2242894
19. Nelson, T.O., Narens, L.: Metamemory: A theoretical framework and new findings. Psychology of Learning and Motivation 26, 125–173 (1990)
20. Babu, G.S., Suresh, S.: Metacognitive neural network for classification problems in a sequential learning framework. Neurocomputing 81(1), 86–96 (2011)
21. Suresh, S., Savitha, R., Sundararajan, N.: A sequential learning algorithm for complex valued self regulating resource allocation network-CSRAN. IEEE Transactions Neural Networks 22(7), 1061–1072 (2011)

22. Savitha, R., Suresh, S., Sundararajan, N.: Metacognitive learning in a fully complex-valued radial basis function neural network. Neural Computation 24(5), 1297–1328 (2012)
23. Savitha, R., Suresh, S., Sundararajan, N.: A meta-cognitive learning algorithm for a fully complex-valued relaxation network. Neural Networks 32, 209–218 (2012)
24. Subramanian, K., Suresh, S.: A meta-cognitive sequential learning algorithm for neuro-fuzzy inference system. Applied Soft. Computing 12(11), 3603–3614 (2012)
25. Suresh, S., Subramanian, K.: A sequential learning algorithm for meta-cognitive neuro-fuzzy inference system for classification problems. In: Proceedings of International Joint Conference on Neural Networks, pp. 2507–2512 (2011)
26. Ashburner, J.: A fast diffeomorphic image registration algorithm. NeuroImage 38(1), 95–113 (2007)
27. Frodl, T., Skokauskas, N.: Meta-analysis of structural MRI studies in children and adults with attention deficit hyperactivity disorder indicates treatment effects. Acta Psychiatrica Scandinavica 125(2), 114–126 (2012)
28. Bledsoe, J.C., Semrud-Clikeman, M., Pliszka, S.R.: Neuroanatomical and neuropsychological correlates of the cerebellum in children with attention-deficit/ hyperactivity disorder-combined type. Journal of the American Academy of Child and Adolescent Psychiatry 50(6), 593–601 (2011)
29. Colby, J.B., Rudie, J.D., Brown, J.A., Douglas, P.K., Cohen, M.S., Shehzad, Z.: Insights into multimodal imaging classification of ADHD. Frontiers in Systems Neuroscience 6(59), 1–18 (2012)

Evaluating Sparse Codes on Handwritten Digits

Linda Main, Benjamin Cowley, Adam Kneller, and John Thornton

Institute for Integrated and Intelligent Systems, Griffith University, QLD, Australia
{l.main,j.thornton,a.kneller}@griffith.edu.au,
benjamin.cowley@griffithuni.edu.au

Abstract. Sparse coding of visual information has been of interest to the neuroscientific community for many decades and it is widely recognised that sparse codes should exhibit a high degree of statistical independence, typically measured by the kurtosis of the response distributions. In this paper we extend work on the hierarchical temporal memory model by studying the suitability of the augmented spatial pooling (ASP) sparse coding algorithm in comparison with independent component analysis (ICA) when applied to the recognition of handwritten digits. We present an extension to the ASP algorithm that forms synaptic receptive fields located closer to their respective columns and show that this produces lower Naïve Bayes classification errors than both ICA and the original ASP algorithm. In evaluating kurtosis as a predictor of classification performance, we also show that additional measures of dispersion and mutual information are needed to reliably distinguish between competing approaches.

1 Introduction

Investigating the statistical properties of sparse image representations has been a focus of computational neuroscience research for several decades. The motivating hypothesis is that the mammalian neocortex forms models of the environment based on the statistics of visual sensory experience. Sparse coding approaches have been used to demonstrate the neural selectivity of V1 simple cells [16], the neural responses at later stages of visual processing [1], and can be plausibly implemented within the hierarchically structured layers of visual cortex [7].

In previous work [19] it was shown that an augmented version of the spatial pooler algorithm proposed in [6] exhibits desirable properties of sparsity and is a useful feature detector when applied to temporal sequences of images. In the current paper we further investigate this *augmented spatial pooler* (ASP) by comparing the statistical independence of ASP sparse codes with codes generated using *independent component analysis* (ICA). We chose ICA as a comparison algorithm as it is explicitly designed to extract statistically independent signals [9] and is also used in several state-of-art unsupervised feature detection algorithms (e.g. [11]). In order to characterise the nature of the statistical independence, we compare ASP and ICA sparse codes using several statistical measures of sparseness, e.g. kurtosis, dispersion and mutual information. Our focus on statistical

S. Cranefield and A. Nayak (Eds.): AI 2013, LNAI 8272, pp. 396–407, 2013.

independence is motivated by the assumption that codes with greater statistical independence will better represent the independent causes in the world that generate the raw input of the coding process. We test this assumption by comparing the performance of the two approaches using a *Naïve Bayes classifier*. Our choice of Naïve Bayes is motivated by the expectation that sparse codes with greater statistical independence will produce a better Naïve Bayes classification, i.e. because Naïve Bayes classification itself assumes independence.

The ASP algorithm is one component of the *Hierarchical Temporal Memory* (HTM) model of the neocortex proposed and developed over the last eight years by Jeff Hawkins and various collaborators [6, 7]. The HTM model is a synthesis of general concepts of neocortical function and morphology, and is based on the view that the uniform appearance of the neocortex signifies a similarly uniform functional process [7]. The model is structured as a hierarchical network of identical processing units comprising two functions: a spatial pooler (SP) and a temporal pooler (TP), which cooperate to encode sensory input into temporal sequences [6]. The poolers are modeled as collections of columns, where each column comprises a set of neurons and their associated dendrites and synapses. This columnar structure is modeled on the cortical mini-column, which Mountcastle has described as a basic functional unit within the neocortex [15]. In the spatial pooling function, the columns become active or inactive through inter-column competition and inhibition, which allows HTMs to self organise in direct response to the input while producing sparse distributed representations. The TP then makes Bayesian-like inferences about the temporal ordering of the codes produced by the SP which are fed back into the hierarchy in the form of predictions. This predictive function of the TP distinguishes HTMs from other hierarchical Bayesian models, e.g. [13]. Furthermore, HTM's flexibility and efficiency in processing temporal sequences makes it suitable for online data streams, unlike other sparse coding approaches such as ICA [10].

The current research extends existing work on evaluating the HTM model by studying the relative suitability of the ASP algorithm in comparison with ICA in the domain of handwritten digit recognition. Previous work [20] compared kurtosis measures for ASP and ICA on natural images on the assumption that higher kurtosis would equate to greater statistical independence and hence would produce better feature encoders. Here we question the suitability of kurtosis as the sole indicator of the performance of a feature detector and investigate alternate measures of dispersion and mutual information, while using Naïve Bayes as a measure of relative performance. We also introduce an extension to the ASP algorithm that clusters synapses more closely to their respective columns and show how this can produce a lower Naïve Bayes classification error.

In the remainder of the paper we provide details of the ASP algorithm and our own extensions, followed by a brief description of the FastICA algorithm we used for comparison purposes. We then provide details of the statistical measures used to evaluate the ASP and ICA sparse codes and present the results of our experiments on the well known MNIST handwritten digits dataset. Finally the significance of these results are analysed and discussed.

2 Augmented Spatial Pooling

The basic functional unit of the HTM poolers is a *column* which consists of a collection of cells and dendrites that form synaptic connections with other columns and with an input layer. The HTM temporal pooler connects individual cells within a column to cells within other columns and the spatial pooler connects an entire column to the elements of an input layer. As the paper is concerned with SP sparse codes, we shall not consider the temporal pooler further. Instead, we shall treat a column as a single *coding unit* (CU) with dendrites that only synapse with the individual elements of an input layer.

Unlike more traditional neural networks, the SP connections are not multiplicatively weighted to determine the strength of the input signal. Instead, the synapses are *potential* and assigned a *permanence value* which indicates whether the synapse is *connected* (i.e. active), or potential and inactive. It is only when the permanence value of a synapse passes a certain threshold that the synapse becomes connected and the value of the input to which it is connected is then passed on by the dendrite. The activity level for a column is then calculated as the sum of the input from all its connected synapses.

Each column is positioned above a single pixel of the input image and no pixel has more than one column above it. A set of dendrites from each column is mapped to a subset of the pixels of the image, where each dendrite forms a single synaptic connection to a single pixel. This topographic layout allows for the calculation of the Euclidean distances from a column to each of its potential synapses. The column's list of potential synapses is then ordered by these distances, and this ordering is used when determining the initial synapse permanence values. The area that bounds a column's connected synapses is termed the column's *receptive field*, and the mean receptive field size of all columns is the size of the *inhibition area*. The sparse encoding of an input pattern is produced via a process of competition whereby the more active columns within a given inhibition area inhibit (or switch off) their less active neighbours.

Learning in the spatial pooler is based on how well the column synapses match (or overlap) the input to which they are connected. This is achieved by modifying the synapse permanence values of the columns which win the inhibition competition – synapses connected to active input have their permanence values increased, while synapses connected to inactive input have their permanence values decreased. In addition, columns that fail to reach a minimum average activation threshold are able to competitively form new synapses with the input (for full details see [19]).

Initialising the Spatial Pooler Synapses: The process of initialising the columns and their synapses is presented in Algorithm 1. The first step is to randomly select a percentage of the image pixels to which potential synapses will be mapped according to the value of $P(potential)$ (in the current implementation, this is set to 0.1). The next step is to probabilistically set the perm(s) permanence values for each of the potential synapses for all columns according to the original ASP algorithm [20] or according to two modified approaches (ASP+M and ASP+G) detailed below.

The last step of initialisation is to calculate each column's receptive field area (line 13) and the inhibition area size (line 15) to be used by all columns during the inhibition competition. After each training instance is presented to the system the receptive fields of all columns and the inhibition area are recalculated.

Algorithm 1. initialiseColumns(*columns*, *method*, *m*)

```
1: for each column c in columns do
2:    for each synapse s in column c do
3:       if random(0,1) ≤ P(potential) then
4:          if method is ASP then
5:             perm(s) = threshold + (random(0.0, 0.1) − distance(s))
6:          else if method is ASP+M then
7:             perm(s) = threshold + (random(0.0, 0.1) − (distance(s) × m))
8:          else if method is ASP+G then
9:             perm(s) = gaussianPDF(distance(s) × m, 0, σ) + random(-0.05, 0.05)
10:         end if
11:      end if
12:   end for
13:   totalReceptiveFieldAreas += calculateReceptiveFieldArea(c)
14: end for
15: inhibitionArea = totalReceptiveFieldAreas / numCols
```

ASP+M: In the HTM specifications [6] initial synapse permanence values are randomly set to within a small range of the permanence *threshold* and are linearly biased such that synapses which are closer to their column are more likely to have an initial permanence value at or above the threshold. In the original ASP algorithm [19] the synapse permanence values have a potential range of 0 to 1 with the threshold set at 0.2 and initial permanences bound to be within 0.1 of the threshold. We use these values in our current study (see lines 4–5).

By way of extension, we also experimented with a multiplier m that increases the range of possible initial permanence values, such that they are no longer within the 0.1 bound. This multiplier increases the probability of synapses close to their column having permanence values above the threshold, and decreases the probability for those further from their column. We name this method ASP+M (see lines 6–7).

For both ASP and ASP+M, we standardise the distance(s) measure for each synapse to range between 0.0 and 0.1. In ASP this distance is subtracted from a random number that also ranges from 0.0 to 0.1 and the result is added to the preset *threshold* value of 0.2. In ASP+M, we simply multiply the normalised distance value by a value m and proceed as for ASP (see line 7).

ASP+G: The second extension to ASP replaces the linear bias with a Gaussian distribution centred on each column, where synapses furthest from the column are least likely to have their permanence values set above the connection threshold. The initial permanence values are then randomly selected from within a boundary of 0.05 of the Gaussian probability density function. We name this method ASP+G (see lines 8–9).

ASP+G converts a synapse's distance(s) from its column into a number of standard deviations from the mean of a Gaussian, such that a column's most distant synapse is set to m standard deviations (in the experimental study we set $m = 5$). We then calculate the probability of this distance using a Gaussian probability density function (PDF) with mean 0 and standard deviation σ. The synapse's *permanence* value equals this probability plus a random noise value between ± 0.05 (see line 9).

3 Independent Component Analysis

Our experimental study compares ASP sparse codes with codes generated using independent component analysis (ICA). ICA belongs to the class of *blind source separation* (BSS) methods aimed at separating data into maximally independent features. It is based on the assumption that the sources are non-Gaussian and statistically independent [17]. However, particularly in the case of natural images, the independence assumption has been criticised and has led to the development of *independent subspace analysis* (ISA) which, in addition to performing ICA, also attempts to find independent *groups* of features [9]. In order to equitably compare ASP codes with the higher order codes learned by ISA, a hierarchical implementation of spatial pooling would also be required. However, such a hierarchical implementation runs the risk of obfuscating the underlying performance of the two algorithms. In contrast, the lower order functionality of ICA is more directly comparable to a single layer ASP and allows for unmodified statistical measurements. For these reasons we have not included ISA in the study.

ICA may be considered a multivariate, parallel version of projection pursuit which seeks to maximise the kurtosis, or alternatively minimise the mutual information, of separated signals. Maximising kurtosis is based on the assumption that a mixed signal, which is often Gaussian in nature, is composed of independent non-Gaussian signals. Further, ICA assumes that any Gaussian signal may contain noise, which is first separated from the mixed signal by applying *principle component analysis* (PCA) [17]. This may result in reducing the dimensions of the mixed signal, and consequently, ICA will extract fewer signals than suggested by the dimensions of the raw input.

The Matlab-based FastICA implementation of ICA, which we use in this study, employs the Kullback-Leibler divergence as a measure of the distance between an extracted signal and a Gaussian signal, and seeks to maximise that distance in the gradient descent search [10]. As FastICA explicitly maximises non-Gaussianity (i.e. kurtosis) during the learning process, we expect the sparse codes it produces will display high kurtosis.

4 Statistical Properties of Sparse Codes

In this study we are interested in the statistical properties of ASP and ICA sparse codes and apply a series of measurements to enable a quantitative comparison between them. For the purpose of these comparisons, we define a *coding unit* (CU) as a column in ASP and a basis function in ICA. For both algorithms, the CUs are the constituents of the sparse responses to the image data.

Willmore, Mazer and Gallant [21] describe a range of statistical measures used to characterise sparse codes and note that different studies used similar terminology but in different contexts, making the comparison of the models and their sparseness measurements problematic. To resolve some of the confusion, they clearly define the different interpretations of sparseness and present a taxonomy of these concepts. The first classification they describe distinguishes *overall activity* from the *shapes of response distributions.* Overall activity measures the mean firing rates of neurons and is intuitively suitable for characterising the behaviour of biological neurons rather than computational models of neurons. As the sparse CUs of ASP and ICA are theoretical *collections* of neurons, we do not attempt similar measurements in this study. Instead we investigate the shapes of response distributions which can be measured in two distinct dimensions: *lifetime* measures that characterise the response distribution of single CUs to many images and *population* measures that characterise the response distribution of an entire collection of CUs to a single image instance. The following sections detail several such measures that we will use in the experimental comparison:

4.1 Lifetime Measures

Lifetime Sparseness: A response distribution where CUs are primarily inactive but occasionally respond strongly is defined as having high lifetime sparseness. A response of this kind may be characterised by measuring the kurtosis of the response distribution. To compare the lifetime sparseness of ASP and ICA sparse codes, we calculate the average kurtosis of each CU's responses to an entire set of input images using the method in [22].

Maximisation of Information: Willmore et al. [21] state that if the visual cortex maximises information, then the response distribution should be exponential in shape. In this respect we have fitted an exponential model, of the form ae^{bx}, to the response distribution of each CU of ASP and ICA. We use the goodness of fit (measured by the root mean square error, RMSE) of the model to characterise how well the CU's response distribution matches the exponential distribution. A close fit will result in a low RMSE and indicate that the sparse codes of the distribution have a high degree of information.

4.2 Population Measures

Population Sparseness: A sparse code where the population response distribution to a single image has high kurtosis is referred to as having high population sparseness, i.e. only a small proportion of the CUs is active at any given time. We calculate the population kurtosis of a single image using the method in [22].

Mutual Information Minimisation: Given two random variables, X and Y, the *mutual information*, $I(X;Y)$, between these variables is:

$$I(X;Y) = \sum_{x,y} P_{XY}(x,y) \log \frac{P_{XY}(x,y)}{P_X(x)P_Y(y)} \quad (1)$$

where $P_X(x)$ and $P_Y(y)$ are the probability mass functions of X and Y respectively, and $P_{XY}(x,y)$ is their joint probability mass function. If we consider a single CU to be a random variable whose distribution of responses to a set of M images constitutes its probability mass function, then we may calculate the average mutual information between all pairs of CUs as follows:

$$averageMutualInformation = \frac{2}{(N-1)N} \sum_{i=1}^{N-1} \sum_{j=i+1}^{N} I(i;j) \quad (2)$$

where $I(i;j)$ is the mutual information between CUs i and j computed as above, and $\frac{2}{(N-1)N}$ is the number of distinct pairs possible from N CUs. This gives us a measure of the average amount of information that is shared between CUs. High mutual information values indicate that the sparse codes have a high level of redundancy, whereas an efficient system will have low redundancy producing a low mutual information value [3].

Redundancy Minimisation: As a second measure of redundancy, we consider the degree of load sharing by counting the number of sparse codes in which the CUs participate. Following [22], for each image we set a threshold value equal to the standard deviation of the response distribution for that image. CUs whose response magnitude for that image is greater than the threshold value are considered 'on' and CUs with response magnitudes less than the threshold value are considered 'off'. The binarised response of CU c to image i is then calculated as:

$$binarisedActivity_i^c = \begin{cases} 1 & \text{if } a_i^c > \sigma_i \\ 0 & \text{otherwise} \end{cases} \quad (3)$$

where a_i^c is the response magnitude of CU c to image i, and σ_i is the standard deviation of the CU population's response distribution to image i. The participation level of a CU c is then the total count of the sparse codes in which it is considered 'on'.

Dispersion: The response of a CU to an image will vary from strongly positive, where an image matches the CU's receptive field, through zero where the image is orthogonal to the receptive field, to strongly negative where the image is the inverse of the receptive field. If an image set has a high variance along the axis of the CU's receptive field then the responses of the unit over the set will show a high variance between the positive and negative extremes. This indicates that the CU's receptive field is well matched to some of the variations in the image set and the CU is therefore appropriate for encoding differences between members of this set [23]. Conversely, a low variance indicates that the variations in the image set are largely orthogonal to the CU's receptive field and that the CU is not appropriate for encoding variations in the image set.

By comparing the response variances of all CUs in a set, we can obtain a measure of how evenly the coding is *dispersed*. Willmore, Watters and Tolhurst [23] suggest visualising and quantifying these differences using scree plots. After binarising the ASP and ICA CU activities we calculate and normalise the variance of all units such that the maximum is 1. We plot the variances in rank order, highest to lowest, and use the area under the curve as a measure of dispersion. A low area indicates that there are a few high variance CUs that are encoding the majority of the variations in the image set, i.e. the coding is concentrated, not dispersed. Conversely, a large area shows there are many high variance CUs sharing the encoding and the resulting codes have greater dispersion.

5 Experimental Evaluation and Discussion

As earlier spatial pooler versions have been tested on character recognition problems (e.g. [5]) we chose to test our modified spatial poolers, ASP+M and ASP+G, and ICA on the well known MNIST handwritten digits dataset [12]. The set comprises 70,000 handwritten images (28 × 28 pixels) of the digits 0–9, split into 60,000 training images and 10,000 test images. ASP and FastICA use random seeds, so we executed all experiments five times using different seeds, and report the average of these executions. All tests were conducted on a Dell Optiplex 990 3.10 GHz Intel Core i5 processor with 16 Gb of 1333 MHz DDR3 RAM running Windows 7 Enterprise v. 6.1 and Matlab v. 7.12 (R2011a). For classification we used the Matlab Statistics Toolbox Naïve Bayes classifier.

Varying Initial Receptive Field Size: We tested our modified ASP algorithms on a range of settings in order to investigate the effect of the different methods for initialising ASP's synapse permanence values. ASP+M was tested using multiplier values 0–6 with 4 giving the highest classification. ASP+G was tested by altering the standard deviation of the Gaussian distribution from 0.1 to 1.8 in steps of 0.1, with 0.9 producing the best classification result. Table 1 summarises the mean Naïve Bayes classification accuracies, and population and lifetime kurtosis of the response distributions for ASP, ASP+M and ASP+G. Setting the ASP+M multiplier to 0 is equivalent to removing the linear bias and setting synapse permanence values randomly without respect to their distance from their column. Results for this setting are included in Table 1 as ASP− and show that randomly selecting active synapses without any topographic reference to their column produce the worst classification accuracy (82.65%), population kurtosis (38.30) and lifetime kurtosis (49.87).

The ASP+M multiplier causes active synapses to be more clustered around their column than for ASP, and could explain ASP+M's higher accuracy of 88.62

Table 1. Summary of results for ASP, ASP+M, ASP+G and ICA

Algorithm	ASP−	ASP	ASP+M	ASP+G	ICA
Accuracy (%)	82.65	84.39	88.62	89.45	86.12
Population Kurtosis	38.30	47.80	73.79	65.48	48.13
Lifetime Kurtosis	49.87	60.85	98.43	83.86	4864.19
Mutual Information	—	—	—	0.0082	0.0063
Goodness of Fit (RMSE)	—	—	—	7.26	1190.10
Dispersion Area	—	—	—	121.06	177.20

(compared to ASP's accuracy of 84.39). The more clustered receptive fields of ASP+M (with a mean $1:4.975$ per column ratio of 8 active synapses to 39.80 receptive field pixels) compared to ASP (with a mean $1:11.39$ per column ratio of 13 active synapses to 148.10 receptive field pixels) allows ASP+M to encode more specific features than ASP. Both ASP's population kurtosis (47.80) and lifetime kurtosis (60.85) are lower than ASP+M's (73.79 and 98.43 respectively) indicating that smaller and more densely clustered receptive fields produce codes exhibiting greater sparseness and higher classification accuracies.

In contrast, ASP+G has approximately the same ratio of active synapses to receptive field size as ASP (6 synapses to 61.30 pixels or $1:10.21$), but achieves this ratio using approximately half as many synapses covering a much smaller mean receptive field. While the smaller more clustered receptive fields of ASP+M are sensing more specific features than the larger receptive fields of ASP+G, the denser synaptic connections of the ASP+M columns suggest insufficient information is being sensed to uniquely encode the classes. However, ASP+G's lower population kurtosis of 65.48 and lower lifetime kurtosis of 83.86 compared to ASP+M (the highest of the ASP methods) would lead us to expect ASP+G to have lower accuracy than ASP+M, whereas the reverse is the case. This suggests that kurtosis is not a perfect predictor of accuracy.

Comparison with ICA: Comparing the lifetime kurtosis of ASP+G and ICA (83.86 and 4,864.19 respectively) to their population kurtosis (65.48 for ASP+G and 48.13 for ICA) further supports the argument that using only kurtosis as a performance indicator is inappropriate. The very high lifetime kurtosis of ICA would lead us to expect a correspondingly high accuracy, whereas ASP+G actually outperforms ICA. This anomalous result led us to manually inspect the binarised ICA response distributions, revealing that an average of 10 ICA CUs had only responded to a single image and were therefore acting as *instance detectors* rather than as members of a feature detector. In addition, ICA's response distributions had an average of 36 overly sparse codes (i.e. responses with 5 or less CUs) and were typically encoded by the same CUs within the distribution, which explains the disproportionately high lifetime kurtosis. Response distributions with CU behaviours of this nature are indicative of a model having overfitted to the data.

When we consider that ICA iteratively seeks to extract the *most independent* signal, it is to be expected that it would extract instance detectors when applied to the relatively simple images of handwritten digits. As ICA was designed to perform sparse encoding of natural images, and is now overfitting this simpler dataset, we consider ICA to be a less general sparse encoder than ASP, which competitively encodes a wider range of image types.

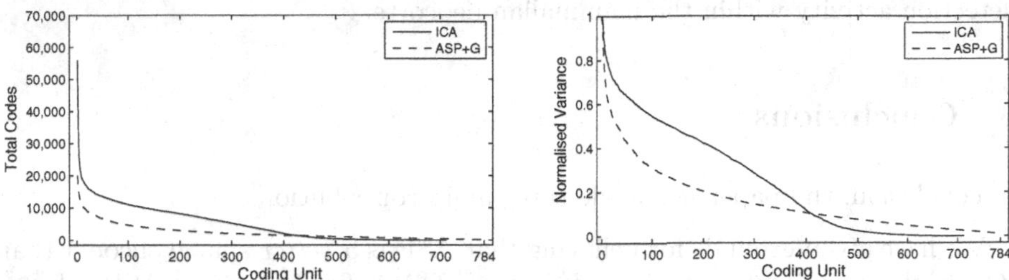

Fig. 1. Left: Coding unit (CU) participation counts for ASP+G and ICA. **Right:** Dispersion area for ASP+G and ICA.

Next, we consider the degree of redundancy by measuring the mutual information shared between CUs. It is reasonable to expect that CUs of an overfitted model will have very low mutual information because they are active in isolation or with very few other CUs. This goes against the theoretical expectation that CUs should share minimal information so as to reduce redundancy. Clearly, some mutual information is needed to ensure the CUs are acting as members of feature detectors and not as instance detectors. We find that the mutual information for ASP+G and ICA (at 0.0082 and 0.0063 respectively) supports this interpretation of the response distributions, when considered *together* with the high lifetime kurtosis and lower accuracy of ICA.

From our alternative dispersion measure using participation counts (see Figure 1), we find further support for overfitting by ICA. The high peak participation for ICA is caused by a mean of 8 CUs which are active in more than 20,000 sparse codes. At the other end of the graph we notice a mean of 13 CUs which *fail to participate in any sparse code*, which indicates that ICA has learned redundant basis functions. The mean activity for ICA is 1,358.53, considerably larger than that of ASP+G (365.16), showing that ASP+G produces more dispersed codes, as encoding is more evenly distributed. This is a result of the underlying ASP algorithm's competitive nature which is explicitly designed to disperse the encoding. In contrast, the dispersion measure of [23] (graphed in Figure 1) indicates the ICA response distributions (177.20) are more dispersed than ASP+G (121.06).

When we measured the degree of information retained in the codes (by fitting an exponential curve to the response distribution and calculating the RMSE of the fit), the RMSE for ASP+G was 7.26 compared to 1,190.10 for ICA. This indicates ASP+G is encoding more salient class features, whereas ICA's high RMSE demonstrates its loss of class specific information as it encodes instances.

Finally, a t-test, with p-value of 0.05, performed on ASP+G and ICA classification accuracies rejects the null hypothesis, indicating that the difference in mean accuracies between ASP+G and ICA is statistically significant.

As ASP runs 8.36 times faster than ICA (with a mean convergence time of 299 seconds compared with 2,582 seconds for ICA), and is simpler to implement than the complex gradient descent approach of ICA (for which the neural mechanisms have yet to be elucidated), we consider ASP a more plausible model of feature detection activity within the mammalian neocortex.

6 Conclusions

In conclusion, the paper has made three main contributions:

1. We have given grounds for believing that ASP is a *better* sparse encoder than ICA in the domain of character recognition. This is firstly because ASP is faster (i.e. more efficient) than ICA at generating codes, which also suggests the ASP strategy is more biologically plausible. Secondly, ASP produces better Naïve Bayes classification accuracy, suggesting that, in practice, ASP is a more useful feature detector than ICA for domains such as character recognition, where the ICA basis functions can degenerate into simple instance detectors.
2. We have shown that kurtosis on its own is not necessarily the best way to measure the statistical independence of sparse codes and that additional measures of dispersion and mutual information are needed to give a reliable picture of the potential performance of an encoder.
3. We have shown there is an important relationship between the synapse initialisation strategy used for ASP and the subsequent classification accuracy of the generated codes. Specifically we have shown that a Gaussian distribution that strongly clusters synapses within a short distance of their associated column is better than the alternative linear and linear multiplicative strategies tested.

In future work we plan to investigate both class-conditional ICA and ISA approaches to character recognition by developing analogous class-conditional and hierarchical ASP algorithms. We also plan to extend our work to the domain natural images and, upon integration of a TP, to temporal data such as sound.

References

1. Carlson, E.T., Rasquinha, R.J., Zhang, K., Connor, C.E.: A sparse object coding scheme in area v4. Current Biology 21(4), 288–293 (2011)
2. Felleman, D.J., Van Essen, D.C.: Distributed hierarchical processing in the primate cerebral cortex. Cerebral Cortex 1(1), 1–47 (1991)
3. Friston, K.: Learning and inference in the brain. Neural Networks 16(9), 1325–1352 (2003)
4. Fu, M., Yu, X., Lu, J., Zuo, Y.: Repetitive motor learning induces coordinated formation of clustered dendritic spines in vivo. Nature 483(7387), 92–95 (2012)

5. George, D., Hawkins, J.: A hierarchical Bayesian model of invariant pattern recognition in the visual cortex. In: Proceedings of the International Joint Conference on Neural Networksm, IJCNN 2005, pp. 1812–1817 (2005)
6. Hawkins, J., Ahmad, S., Dubinsky, D.: Hierarchical temporal memory including HTM cortical learning algorithms. Tech. rep., Numenta, Inc, Palto Alto (2011), `https://www.groksolutions.com/technology.html#cla-whitepaper`
7. Hawkins, J., Blakeslee, S.: On intelligence. Henry Holt, New York (2004)
8. Hawkins, J., George, D.: Hierarchical temporal memory: Concepts, theory and terminology. Tech. rep., Numenta, Inc, Palto Alto (2006), `http://www.numenta.com/htm-overview/education/Numenta_HTM_Concepts.pdf`
9. Hyvärinen, A., Hurri, J., Hoyer, P.: Natural Image Statistics: A probabilistic approach to early computational vision. Springer-Verlag, New York Inc. (2009)
10. Hyvärinen, A., Karhunen, J., Oja, E.: Independent Components Analysis. John Wiley and Sons, Inc., New York (2001)
11. Le, Q.V., Zou, W.Y., Yeung, S.Y., Ng, A.Y.: Learning hierarchical invariant spatio-temporal features for action recognition with independent subspace analysis. In: 2011 IEEE Conference on Computer Vision and Pattern Recognition (CVPR), pp. 3361–3368 (2011)
12. LeCun, Y., Cortes, C.: MNIST handwritten digit database. AT&T Labs (1998), `http://yann.lecun.com/exdb/mnist`
13. Lee, T.S., Mumford, D.: Hierarchical Bayesian inference in visual cortex. Journal of the Optical Society of America A 20(7), 1434–1448 (2003)
14. Malone, B.J., Kumar, V.R., Ringach, D.L.: Dynamics of receptive field size in primary visual cortex. Journal of neurophysiology 97(1), 407–414 (2007)
15. Mountcastle, V.B.: Introduction to the special issue on computation in cortical columns. Cerebral Cortex 13(1), 2–4 (2003)
16. Olshausen, B.A., et al.: Emergence of simple-cell receptive field properties by learning a sparse code for natural images. Nature 381(6583), 607–609 (1996)
17. Stone, J.V.: Independent component analysis. Wiley Online Library (2004)
18. Thornton, J., Main, L., Srbic, A.: Fixed frame temporal pooling. In: Thielscher, M., Zhang, D. (eds.) AI 2012. LNCS, vol. 7691, pp. 707–718. Springer, Heidelberg (2012)
19. Thornton, J., Srbic, A.: Spatial pooling for greyscale images. International Journal of Machine Learning and Cybernetics 4, 207–216 (2013)
20. Thornton, J., Srbic, A., Main, L., Chitsaz, M.: Augmented spatial pooling. In: Wang, D., Reynolds, M. (eds.) AI 2011. LNCS, vol. 7106, pp. 261–270. Springer, Heidelberg (2011)
21. Willmore, B.D., Mazer, J.A., Gallant, J.L.: Sparse coding in striate and extrastriate visual cortex. Journal of Neurophysiology 105(6), 2907–2919 (2011)
22. Willmore, B., Tolhurst, D.J.: Characterizing the sparseness of neural codes. Network: Computation in Neural Systems 12(3), 255–270 (2001)
23. Willmore, B., Watters, P.A., Tolhurst, D.J.: A comparison of natural-image-based models of simple-cell coding. Perception-London 29(9), 1017–1040 (2000)

Minimum Message Length Ridge Regression for Generalized Linear Models

Daniel F. Schmidt and Enes Makalic

Centre for MEGA Epidemiology, The University of Melbourne
Carlton, VIC 3053, Australia
{dschmidt,emakalic}@unimelb.edu.au

Abstract. This paper introduces an information theoretic model selection and ridge parameter estimation criterion for generalized linear models based on the minimum message length principle. The criterion is highly general in nature, and handles a range of target distributions, including the normal, binomial, Poisson, geometric and gamma distributions. Estimation of the regression parameters, the ridge hyperparameter and the set of covariates associated with the targets is all performed within the same framework by minimisation of the message length. Experiments on simulated and real data suggest that the criterion is competetive with, and often superior to, the corrected Akaike information criterion in terms of both parameter estimation and model selection tasks.

1 Introduction

In conventional Gaussian-linear regression modelling we make the assumption that the targets $\mathbf{y} = (y_1, \ldots, y_n)' \in \mathbb{R}^n$ are normally distributed, with variance τ, and mean μ_i given by

$$\mu_i = \bar{\mathbf{x}}_i \boldsymbol{\beta} + \alpha, \tag{1}$$

where $\bar{\mathbf{x}}_i' \in \mathbb{R}^k$ is a vector of features, $\boldsymbol{\beta} \in \mathbb{R}^k$ is a vector of regression coefficients, and $\alpha \in \mathbb{R}$ is the intercept parameter. It is typically the case that we do not believe the targets to be normally distributed; for example, the targets may be non-negative integers or binary variables. The generalized linear model (GLM) [1] framework was developed to easily extend linear models to alternative target distributions. In this paper we restict attention to distributions which satisfy

$$\mathrm{E}\left[y|\mu\right] = \mu, \tag{2}$$

$$\mathrm{var}[y|\mu, \phi] = \phi\, v(\mu), \tag{3}$$

where $\phi > 0$ is a dispersion parameter which in many cases will simply be equal to one, and $v(\cdot)$ is a variance function that depends only on the mean μ. Defining $\boldsymbol{\psi} = (\alpha, \boldsymbol{\beta}')'$ as the vector of regression coefficients and $\eta_i(\boldsymbol{\psi}) \equiv \eta_i = \bar{\mathbf{x}}_i \boldsymbol{\beta} + \alpha$ as the linear predictor, the GLM approach specifies $f(\mu_i) = \eta_i$, where $f(\cdot)$ is called a *link function*; that is, a GLM specifies the conditional mean as a suitable

S. Cranefield and A. Nayak (Eds.): AI 2013, LNAI 8272, pp. 408–420, 2013.

(monotic, isomorphic) function of the linear predictor. The function $f^{-1}(\eta_i) = \mu_i$ is usually known as the inverse-link function.

In general, the regression coefficients α and $\boldsymbol{\beta}$ are unknown, and we only have access to the data $\mathbf{y}$ and the covariates $\mathbf{X} = (\bar{\mathbf{x}}_1', \ldots, \bar{\mathbf{x}}_n')'$. The task is then to estimate the regression coefficients on the basis of the data alone. There exists a large range of estimation strategies available for GLMs, and a particularly popular approach is *ridge regression* [2]. This is a regularisation procedure that is known to improve estimation accuracy in the presence of colinearity in the covariates. The (generalized) ridge regression procedure estimates $\alpha, \boldsymbol{\beta}$ by solving

$$\left\{\hat{\alpha}, \hat{\boldsymbol{\beta}}\right\} = \underset{\alpha \in \mathbb{R}, \boldsymbol{\beta} \in S(c)}{\arg\min} \left\{ -\sum_{i=1}^{n} \log p(y_i | \mu_i; \phi) \right\} \tag{4}$$

where $p(\cdot)$ is the chosen target distribution and $S(c)$ is the set of permissible regression coefficients, defined by

$$S(c) = \left\{ \boldsymbol{\beta} \in \mathbb{R}^k : \boldsymbol{\beta}' \boldsymbol{\Sigma} \boldsymbol{\beta} \leq c \right\}, \tag{5}$$

with $\boldsymbol{\Sigma} \in \mathbb{R}^{k \times k}$ a positive-definite matrix. The hyperparameter c determines the amount of "freedom" the estimator has to fit the data; for a sufficiently large choice of c the ridge estimator reduces to the regular maximum likelihood estimator, while smaller values of c result in estimates that are "shrunk" towards the origin. It is usual to estimate c by minimisation of an information criteria such as Akaike's information criterion (AIC), or by a resampling procedure such as cross-validation. It is also possible to interpret the ridge estimator in a Bayesian manner, in which the regularisation term arises due to the choice of a multivariate normal prior distribution over the regression coefficients $\boldsymbol{\beta}$.

In this paper we exploit the Bayesian interpretation to use the minimum message length (MML) principle to estimate the regularisation hyper-parameter; furthermore, because of the nature of the MML principle, we can also use the same criterion to perform feature selection, i.e., to choose which columns of $\mathbf{X}$ are associated with the targets $\mathbf{y}$. The result is a *single, highly general criterion for the statistical inference of generalized linear models that is applicable to a wide range of target distributions*, and has excellent performance in terms of both parameter estimation and model selection.

2 Inference by Minimum Message Length

Minimum message length (MML) [3,4,5] is an information theoretic principle of inductive inference based on the connections between statistical inference and data compression. The key idea underlying the MML principle is that if a statistical model compresses data, then the model has (with high probability) captured regularities and structure in the data. The MML principle advocates selecting the model that most compresses the data (i.e., the one with the shortest "message length") as the most plausible explanation of the data. As any compressed representation of data must also be decompressable, the details of the statistical model used to encode the data must also be part of the compressed data string. Thus, more complex models inflate the message length by a greater amount, and this acts to naturally balance model complexity against the goodness of fit of the model, and automatically guards against the problem of overfiting the data.

In general, the calculation of the exact (strict) message length is an NP-hard problem [6]. There exists a range of approximations to the exact message length that are less computationally intensive [5]; the most widely used of these is the Wallace–Freeman approximation (MML87) [4]. Let $\boldsymbol{\theta} \in \Theta$ denote the continuous parameters of a statistical model, $p(\mathbf{y}|\boldsymbol{\theta})$ denote the likelihood of the data $\mathbf{y}$ conditional on the parameters $\boldsymbol{\theta}$, and let $\pi(\boldsymbol{\theta})$ denote a Bayesian prior distribution over Θ that will be used to model the continuous parameters. The MML87 message length for data $\mathbf{y}$ and model $\boldsymbol{\theta}$ is given by

$$I(\mathbf{y}, \boldsymbol{\theta}) = -\log p(\mathbf{y}|\boldsymbol{\theta}) + \frac{1}{2}\log|\mathbf{J}(\boldsymbol{\theta})| - \log \pi(\boldsymbol{\theta}) + c(k) \tag{6}$$

where $\mathbf{J}(\boldsymbol{\theta})$ is the Fisher information matrix, k is the number of continuous model parameters and

$$c(k) = -\frac{k}{2}\log(2\pi) + \frac{1}{2}\log(k\pi) - 0.5772.$$

To estimate a model using MML87, we search for the $\boldsymbol{\theta}$ that minimises (6). Under certain regularity conditions of the likelihood $p(\mathbf{y}|\boldsymbol{\theta})$ and prior distribution $\pi(\boldsymbol{\theta})$ the MML87 message length is very close to the exact strict message length [5]. The aim of this paper is to apply the MML87 approximation to the problem of ridge estimation and model selection in the context of generalized linear models.

Ridge estimation in the MML framework is equivalent to allowing the prior distribution to depend on a *hyperparameter*, and extending the estimation procedure to include this new hyperparameter. Previous work [7] has shown that inference of hyperparameters may be done within the MML87 framework, and this technique has been applied to linear regression with a normal target distribution and a special choice of ridge prior in [8]. MML has been previously applied to linear models with normal targets [5] and binomial targets [9], and both these cases essentially depend on special types of ridge priors. To some extent, the MML criterion presented in this paper generalises this previous work, as it allows for general ridge estimation and a large number of target distributions.

3 MML GLM Ridge Regression

To compute message lengths using the MML87 approximation (6) we require: (i) the negative log-likelihood function; (ii) prior distributions over all parameters; and (iii) an appropriate Fisher information matrix. Define the full vector of parameters for a GLM as $\boldsymbol{\theta} = (\alpha, \boldsymbol{\beta}', \phi)'$, where ϕ may be constrained to $\phi = 1$ for some target distributions, and define $\boldsymbol{\psi} = (\alpha, \boldsymbol{\beta}')'$ as the vector of regression parameters. It is usual to assume that the targets are independent random variables, conditional on the features, so that the likelihood function can be factorised into the product

$$p(\mathbf{y}|\boldsymbol{\theta};\mathbf{X}) = \prod_{i=1}^{n} p(y_i|\boldsymbol{\theta};\bar{\mathbf{x}}_i). \tag{7}$$

To implement ridge regression within a Bayesian context the required prior distribution for the $\boldsymbol{\beta}$ coefficients is a multivariate normal with mean $\mathbf{0}_k$ and variance-covariance matrix $(\phi/\lambda)\boldsymbol{\Sigma}^{-1}$. Scaling the covariance matrix by the dispersion parameter ϕ greatly simplifies the resulting estimates of α and $\boldsymbol{\beta}$ as they become independent of the estimate of ϕ. As the origin holds no special meaning for the intercept we choose a uniform distribution for α. The priors for α and $\boldsymbol{\beta}$, conditional on ϕ and λ are:

$$\pi(\boldsymbol{\psi}|\phi,\lambda) = \pi_{\boldsymbol{\beta}}(\boldsymbol{\beta}|\phi,\lambda)\cdot\pi_{\alpha}(\alpha|\phi), \tag{8}$$

$$\pi_{\boldsymbol{\beta}}(\boldsymbol{\beta}|\phi,\lambda) = \left(\frac{\lambda}{2\pi\phi}\right)^{\frac{k}{2}}\cdot|\boldsymbol{\Sigma}|^{\frac{1}{2}}\cdot\exp\left(-\frac{\lambda\boldsymbol{\beta}'\boldsymbol{\Sigma}\boldsymbol{\beta}}{2\phi}\right), \tag{9}$$

$$\pi_{\alpha}(\alpha|\phi) \propto \frac{1}{\sqrt{\phi}}. \tag{10}$$

Due to the fact that we condition on ϕ in (9) and (10), we may first estimate α, $\boldsymbol{\beta}$, and then subsequently estimate ϕ (if required). The prior for α is improper, and must technically be restricted to some subset of $\mathbb{R}$; the particular choice of subset is not important as the α parameter is common to all GLMs and the normalisation term will simply increase all message lengths by a constant amount. Suitable priors for ϕ are discussed in Section 3.1.

In the ridge regression framework, the regularisation hyper-parameter λ is not considered to be a *a priori* known; rather, it is estimated from the data along with the other model parameters. This can introduce some problems into the standard MML87 message length, as the assumption of a "flat" prior distribution is violated when λ becomes very large, and the resulting normal distribution becomes tightly concentrated around the origin. To address this problem we use the "corrected" form of the Fisher information matrix that takes into account the curvature of the prior. To correct the Fisher information matrix, Wallace proposed a clever procedure in the case of conjugate likelihood and prior distributions, in which the model parameters are treated as "fake" data, and the Fisher information is calculated using both the real and "fake" data (see [5], pp. 236–237 for further details).

The likelihood (7) is not, in general, conjugate with the prior distribution (9). However, it is well known that the likelihood of many common GLMs can be approximated around some point, $\boldsymbol{\psi}_0 = (\alpha_0, \boldsymbol{\beta}_0')'$, by a multivariate normal distribution with appropriate mean and covariance matrix; such approximations form the basis of the efficient iteratively reweighted least squares procedure for maximum likelihood estimation of GLM regression coefficients. Define $\boldsymbol{\mu}_0 = f^{-1}(\mathbf{X}\boldsymbol{\beta}_0 + \alpha_0\mathbf{1}_n)$; the approximate negative log-posterior for $\boldsymbol{\psi}$, up to constants independent of $\boldsymbol{\psi}$, is then given by

$$-\log p(\boldsymbol{\psi}|\mathbf{y},\phi,\lambda) \approx \left(\frac{1}{2\phi}\right)(\mathbf{z}(\boldsymbol{\mu}_0) - \mathbf{X}\boldsymbol{\beta} - \alpha\mathbf{1}_n)'\mathbf{W}(\boldsymbol{\mu}_0)(\mathbf{z}(\boldsymbol{\mu}_0) - \mathbf{X}\boldsymbol{\beta} - \alpha\mathbf{1}_n) + \left(\frac{\lambda}{2\phi}\right)\boldsymbol{\beta}'\boldsymbol{\Sigma}\boldsymbol{\beta}, \quad (11)$$

where $\mathbf{z}(\cdot)$ is a vector-valued function with entries

$$z_i(\boldsymbol{\mu}) = f(\mu_i) + (y_i - \mu_i)\left(\frac{\partial f(\mu_i)}{\partial \mu_i}\right), \quad (12)$$

and $\mathbf{W}(\boldsymbol{\mu}_0) = \text{diag}(\mathbf{w}(\boldsymbol{\mu}_0))$ is an $(n \times n)$ diagonal matrix, where $\mathbf{w}(\cdot)$ is a vector valued function with entries

$$w_{i,i}(\boldsymbol{\mu}) = \left(\frac{1}{v(\mu_i)}\right)\left(\frac{\partial f(\mu_i)}{\partial \mu_i}\right)^{-2}. \quad (13)$$

The functions $f(\mu_i)$, $f^{-1}(\eta_i)$ and $\partial f(\mu_i)/\partial \mu_i$ for several common choices of link function are given in Table 1, and the variance function $v(\mu_i)$ for a range of distributions is given in Table 2.

The likelihood term in the approximation (11) is conjugate with the normal prior density for the coefficients, and we may now view the prior $\pi_{\boldsymbol{\beta}}(\boldsymbol{\beta}|\phi,\lambda)$ as the posterior of some uninformative prior $\pi_0(\boldsymbol{\beta})$ and a likelihood of k "prior samples", all equal to zero, with design matrix $\mathbf{X}_0$ satisfying $\mathbf{X}_0'\mathbf{X}_0 = (\lambda/\phi)\boldsymbol{\Sigma}$. This yields a "corrected" Fisher information matrix for the regression parameters $\boldsymbol{\psi}$ of the form

$$\mathbf{J}(\boldsymbol{\psi}|\phi,\lambda) = \left(\frac{1}{\phi}\right)\left((\mathbf{1}_n,\mathbf{X})'\,\mathbf{W}(\boldsymbol{\mu})\,(\mathbf{1}_n,\mathbf{X}) + \lambda\mathbf{S}\right), \quad (14)$$

where

$$\mathbf{S} = \begin{pmatrix} 0 & \mathbf{0}_k' \\ \mathbf{0}_k & \boldsymbol{\Sigma} \end{pmatrix}, \quad (15)$$

and $\boldsymbol{\mu} = f^{-1}(\mathbf{X}\boldsymbol{\beta} + \alpha\mathbf{1}_n)$. The "correction" has the effect of increasing the determinant of (14) for increasing λ; that is, the tighter the prior becomes around the origin, the larger the determinant of the corrected Fisher. In contrast, in the limit as $\lambda \to 0$ (and the normal prior (9) converges to a uniform distribution over $\boldsymbol{\beta}$) the corrected Fisher information reduces to the standard, "uncorrected" Fisher information.

3.1 Coding ϕ

Some target distributions require the coding of an extra dispersion parameter ϕ. This can be largely treated in a unified manner irrespective of the specific details of the target distribution by choosing the prior distribution $\pi_\phi(\cdot)$ to be the co-ordinate wise reference prior, i.e.,

$$\pi_\phi(\phi) \propto \sqrt{J(\phi)/n}, \quad (16)$$

Table 1. Commonly used link functions and their derivatives and inverses; $\eta_j = \bar{\mathbf{x}}_j\boldsymbol{\beta}+\alpha$ is the linear predictor

	Link Function, $f(\mu_j)$	$\partial f(\mu_j)/\partial \mu_j$	Inverse Link, $f^{-1}(\eta_j)$
Identity	$\eta_j = \mu_j$	$\frac{\partial \eta_j}{\partial \mu_j} = 1$	$\mu_j = \eta_j$
Logit	$\eta_j = \log\left(\frac{\mu_j}{1-\mu_j}\right)$	$\frac{\partial \eta_j}{\partial \mu_j} = \frac{1}{\mu_j(1-\mu_j)}$	$\mu_j = \frac{1}{1+\exp(-\eta_j)}$
Log	$\eta_j = \log(\mu_j)$	$\frac{\partial \eta_j}{\partial \mu_j} = \frac{1}{\mu_j}$	$\mu_j = \exp(\eta_j)$

where $J(\phi)$ is the Fisher information for ϕ. Due to the fact that the distributions considered in Table 2 are parameterised in terms of orthogonal mean and dispersion parameters, the Fisher information for (ψ_i, ϕ) is zero for all $i = 1, \ldots, k+1$. The determinant of the full Fisher information matrix can then be written as the product

$$|\mathbf{J}(\boldsymbol{\theta}; \lambda)| = |\mathbf{J}(\boldsymbol{\psi}|\phi; \lambda)| \cdot J(\phi). \tag{17}$$

This decomposition, coupled with the choice of reference prior (16) dramatically simplifies the MML87 codelength for ϕ by cancelling the $J(\phi)$ terms present in the determinant of the Fisher information (17) and the prior (16).

3.2 Complete Message Length for a GLM

Two message length formulae are required: one for the case in which $k > 0$ (i.e., there is at least one covariate included in the model), and one for the special case in which $k = 0$. We now cover these two cases seperately. It is important to note that these formulae allow for the comparison of regression models with different numbers of covariates, as long as all the models under consideration have the same target distribution; to compare between models with different target distributions, additional constants are required.

Message Length when $k > 0$. In this case, the model parameters are α, $\boldsymbol{\beta}$ and ϕ (if required). The prior (9) for $\boldsymbol{\beta}$ depends on λ, which is treated as an unknown hyperparameter that must be estimated from the data. Therefore, we also need to transmit λ to the receiver. There exists a procedure to determine the optimum codelength for hyperparameters in the MML framework [7], but it is difficult to apply to generalized linear models; instead, the codelength for λ is approximated by the usual asymptotic formula, i.e., $I(\lambda) = (1/2)\log n$. As there is only a single hyperparameter the suboptimal coding of λ is not expected to have any large effect on the resulting MML inferences.

The total number of free parameters is equal to $m = k + 2$ if ϕ is a free parameter, and $m = k + 1$ if ϕ is constrained to a constant for the target distribution under consideration (see Table 2 for details). Using (9), (10), (14), (16) and (17) in (6) yields

Table 2. Commonly used distributions and their variance functions; μ_j is the appropriate mean function; we define $y_j \in \{0,1\}$ and $0^0 = 1$ for the binomial likelihood; κ is the inverse of the shape parameter in the case of the Gamma distribution and ξ is the inverse of the shape parameter in the case of the inverse-Gaussian distributions

	Link	PDF, $p(y_j\|\boldsymbol{\theta}; \bar{\mathbf{x}}_j)$	$v(\mu_j)$	ϕ
Normal	Identity	$\left(\frac{1}{2\pi\tau}\right)^{\frac{1}{2}} \exp\left(-\frac{(y_j-\mu_j)^2}{2\tau}\right)$	1	τ
Binomial	Logit	$\mu_j^{y_j}(1-\mu_j)^{(1-y_j)}$	$\mu_j(1-\mu_j)$	1
Poisson	Log	$\frac{\mu_j^{y_j}\exp(-\mu_j)}{\Gamma(y_j+1)}$	μ_j	1
Geometric	Log	$\frac{\mu_j^{y_j}}{(\mu_j+1)^{y_j+1}}$	$\mu_j^2+\mu$	1
Gamma	Log	$\frac{y_j^{\frac{1}{\kappa}-1}\exp\left(-\frac{y_j}{\kappa\mu_j}\right)}{(\kappa\mu_j)^{\frac{1}{\kappa}}\,\Gamma\left(\frac{1}{\kappa}\right)}$	μ_j^2	κ
Inverse-Gaussian	Log	$\left(\frac{1}{2\pi\xi y_j^3}\right)^{\frac{1}{2}} \exp\left(-\frac{(y_j-\mu_j)^2}{2\xi\mu_j^2 y_j}\right)$	μ_j^3	ξ

$$I(\mathbf{y}, \boldsymbol{\theta}, \lambda; \mathbf{X}) = -\log p(\mathbf{y}|\boldsymbol{\theta}; \mathbf{X}) + \frac{1}{2}\log|\left((\mathbf{1}_n, \mathbf{X})'\,\mathbf{W}(\boldsymbol{\mu})\,(\mathbf{1}_n, \mathbf{X}) + \lambda\mathbf{S}\right)|$$
$$-\frac{1}{2}\log|\boldsymbol{\Sigma}| + \frac{k}{2}\log\left(\frac{2\pi}{\lambda}\right) + \left(\frac{\lambda}{2\phi}\right)\boldsymbol{\beta}'\boldsymbol{\Sigma}\boldsymbol{\beta} + (1/2)\log n + c(m) + \text{const} \tag{18}$$

where $\boldsymbol{\mu} = f^{-1}(\mathbf{X}\boldsymbol{\beta} + \alpha\mathbf{1}_n)$, $\mathbf{S}$ is given by (15), $\mathbf{W}(\boldsymbol{\mu})$ is given by (13) and const denotes constant terms independent of $\boldsymbol{\theta}$, λ and $\mathbf{y}$.

Message Length when $k = 0$. In this case, no covariates are being used to model the data $\mathbf{y}$ and the model parameters are simply the intercept α and the dispersion parameter ϕ (if required). The total number of parameters is then $m = 2$ if ϕ is a free parameter, or $m = 1$ otherwise. As $\boldsymbol{\beta}$ is not being transmitted, there is no requirement to transmit the hyperparameter λ, and the message length simplifies to

$$I(\mathbf{y}, \alpha) = -\log p(\mathbf{y}|\boldsymbol{\theta}; \mathbf{X}) + \frac{1}{2}\log \mathbf{1}_n'\mathbf{W}(\boldsymbol{\mu})\mathbf{1}_n + c(m) + \text{const}, \tag{19}$$

where $\boldsymbol{\mu} = f^{-1}(\alpha\mathbf{1}_n)$.

4 Estimating ψ, ϕ and λ

Finding the exact estimates for $\hat{\psi} = (\hat{\alpha}, \hat{\beta}')'$ that minimise (18) is computationally expensive due to the presence of the log-determinant of the Fisher information. To avoid this problem, the posterior mode, or maximum a posteriori (MAP) estimates will be used as a surrogate for the exact MML estimates; for moderate to large sample sizes, the difference between the MML and MAP estimates is expected to be small. Furthermore, for case of normal and gamma target distributions, the MAP and MML estimators exactly coincide. This is easy to verify by noting that the corrected Fisher information matrix (14) in both of these cases is independent of ψ.

The posterior mode estimates may be obtained by using the well-known iteratively reweighted least-squares (IRLS) algorithm [10]. Although this algorithm is usually used to obtain the maximum likelihood estimates, it is easily adapted to find ridge estimates through the use of data augmentation. This is done by defining a new, augmented, design matrix

$$\mathbf{X}_A = \begin{pmatrix} \mathbf{1}_n & \mathbf{X} \\ \mathbf{0}_k & \mathrm{diag}(\sqrt{v_1}, \ldots, \sqrt{v_k})\mathbf{E}' \end{pmatrix}, \tag{20}$$

where $\mathbf{v}$ are the eigenvalues of $\boldsymbol{\Sigma}$, and $\mathbf{E}$ is a matrix whose columns are the eigenvectors of $\boldsymbol{\Sigma}$. In the common case of $\boldsymbol{\Sigma} = \mathbf{I}_k$, we have $\mathbf{v} = \mathbf{1}_k$ and $\mathbf{E} = \mathbf{I}_k$.

The algorithm begins by initialising the estimate of the conditional mean vector with suitable starting values:

$$\hat{\boldsymbol{\mu}}_\lambda \leftarrow \begin{cases} \mathbf{y}/2 + 1/4 & \text{(Binomial)} \\ \mathbf{y} + 1/4 & \text{(Poisson)} \\ \mathbf{y} & \text{(Otherwise)} \end{cases} \tag{21}$$

The IRLS ridge algorithm then procedes as follows:

1. Form the augmented weight matrix and "data" vector using (12) and (13) :

$$\mathbf{W}_A(\hat{\boldsymbol{\mu}}_\lambda) \leftarrow \begin{pmatrix} \mathbf{W}(\hat{\boldsymbol{\mu}}_\lambda) & \mathbf{0}_{n\times k} \\ \mathbf{0}_{k\times n} & \lambda \mathbf{I}_k \end{pmatrix}, \quad \mathbf{z}_A(\hat{\boldsymbol{\mu}}_\lambda) \leftarrow \begin{pmatrix} \mathbf{z}(\hat{\boldsymbol{\mu}}_\lambda) \\ \mathbf{0}_k. \end{pmatrix} \tag{22}$$

2. Update the estimates of the regression coefficients:

$$\hat{\psi}_\lambda \leftarrow \left(\mathbf{X}_A'\mathbf{W}_A(\hat{\boldsymbol{\mu}}_\lambda)\mathbf{X}_A\right)^{-1}\mathbf{X}_A'\mathbf{W}_A(\hat{\boldsymbol{\mu}}_\lambda)\mathbf{z}_A(\hat{\boldsymbol{\mu}}_\lambda) \tag{23}$$

3. Update the estimate of the conditional mean vector:

$$\hat{\boldsymbol{\mu}}_\lambda \leftarrow f^{-1}(\mathbf{X}\hat{\boldsymbol{\beta}}_\lambda + \hat{\alpha}_\lambda \mathbf{1}_n)$$

4. If the change in estimates is sufficiently small, terminate. Otherwise, go to Step 1.

An advantage of conditioning the prior (9) for $\boldsymbol{\beta}$ on ϕ is that the estimating equation for ψ, given by (23), is independent of ϕ. As the choice of ϕ has no effect on the MAP estimate of the coefficients ψ, we may first estimate ψ using the above procedure, and once a suitable estimate has been obtained, we may subsequently use it to estimate ϕ.

4.1 Estimating ϕ

Once we have obtained the MAP estimate for the model coefficients, $\hat{\psi}_\lambda$, we may estimate the dispersion parameter ϕ, if necessary. An initial estimate for ϕ can then be obtained by minimising the approximate negative log posterior (11) for ϕ, yielding

$$\hat{\phi}_\lambda \approx \left(\frac{1}{n}\right)\left[(\mathbf{z}(\hat{\boldsymbol{\mu}}_\lambda) - \hat{\boldsymbol{\beta}}_\lambda \mathbf{X} - \hat{\alpha}_\lambda \mathbf{1}_n)'\mathbf{W}(\hat{\boldsymbol{\mu}}_\lambda)(\mathbf{z}(\hat{\boldsymbol{\mu}}_\lambda) - \hat{\boldsymbol{\beta}}_\lambda \mathbf{X} - \hat{\alpha}_\lambda \mathbf{1}_n) + \lambda \hat{\boldsymbol{\beta}}'_\lambda \boldsymbol{\Sigma} \hat{\boldsymbol{\beta}}_\lambda\right]. \tag{24}$$

In the case of the normal distribution ($\phi \equiv \tau$), this estimate is the exact MML estimate for the noise variance. In the case of the gamma and inverse Gaussian distributions, this estimate may be close for large sample sizes, but will not in general be equal to the exact MML estimate. We now detail how to find the MML estimate in these two cases.

Gamma Regression. In this case, $\phi \equiv \kappa$, which plays the role of the inverse of the shape parameter found in the usual parameterisation of the gamma distribution. Let $(\hat{\mu}_\lambda)_i$ denote the i-th co-ordinate of the conditional mean vector estimate $\hat{\boldsymbol{\mu}}_\lambda$. The MML estimate of κ may be obtained by minimising

$$\left(\frac{1}{\kappa}\right)\sum_{i=1}^n \left(\frac{y_i}{(\hat{\mu}_\lambda)_i} + \log(\hat{\mu}_\lambda)_i\right) + \left(\frac{\kappa-1}{\kappa}\right)\sum_{i=1}^n \log y_i + \left(\frac{n}{\kappa}\right)\log\kappa + n\log\Gamma\left(\frac{1}{\kappa}\right). \tag{25}$$

Closed form solutions for the MML estimate do not exist, and they must be found numerically. The approximate estimate (24) is a suitable starting point for a numerical minisation procedure.

Inverse Gaussian Regression. In this case, $\phi \equiv \xi$, which plays the role of the inverse of the shape parameter found in the usual parameterisation of the inverse Gaussian distribution. An exact estimate may be obtained by minimising the message length; this is given by

$$\hat{\xi}_\lambda = \left(\frac{1}{n}\right)\sum_{i=1}^n \frac{(y_i - (\hat{\mu}_\lambda)_i)^2}{y_i\,(\hat{\mu}_\lambda)_i^2}, \tag{26}$$

where $(\hat{\mu}_\lambda)_i$ denotes the i-th component of the conditional mean estimate $\hat{\boldsymbol{\mu}}_\lambda$.

4.2 Estimating λ

The regularisation parameter λ may also be estimated from the data by minimisation of the message length. Due to the use of the "corrected" Fisher information matrix, the MML87 message length is valid even for very large λ, and the MML estimate may be obtained by solving

$$\hat{\lambda} = \underset{\lambda \in \mathbb{R}_+}{\arg\min} \left\{ I(\mathbf{y}, \hat{\boldsymbol{\theta}}_\lambda, \lambda; \mathbf{X}) \right\},$$

where $\hat{\boldsymbol{\theta}}_\lambda = (\hat{\boldsymbol{\psi}}'_\lambda, \hat{\phi}_\lambda)'$, and $\hat{\boldsymbol{\psi}}_\lambda$ and $\hat{\phi}_\lambda$ are the estimates for $\boldsymbol{\psi}$ and ϕ, conditional on λ, obtained using the procedures described Sections 4 and 4.1.

5 Selecting Covariates

One of the strengths of MML is that minimisation of the message length can be used to estimate both continuous model parameters, as well as perform model selection. In the setting of GLMs, the most common model selection problem is identifying which covariates from a design matrix are associated with the target. The MML ridge scheme developed in this paper can easily be adapted to perform model selection. Let $\mathbf{X} = (\mathbf{x}_1, \ldots, \mathbf{x}_q)$ denote the complete, $(n \times q)$ design matrix, where $\mathbf{x}_i \in \mathbb{R}^n$, let $\gamma \subset \{1, \ldots, q\}$ index a particular subset of covariates, and let $k_\gamma = |\gamma|$ denote the number of covariates in the subset. We can then define a sub-design matrix by

$$\mathbf{X}_\gamma = \left(\mathbf{x}_{\gamma_1}, \ldots, \mathbf{x}_{\gamma_{k_\gamma}}\right).$$

For the message to be decodable, the particular subset γ being used must encoded; a prior over Γ is therefore required. If nothing is known *a priori* about the likelihood of any covariate being included in the final model, a prior that treats all subset sizes equally likely is appropriate [8]. This yields a codelength of

$$I(\gamma) = \log \binom{q}{k_\gamma} + \log(q+1).$$

The MML estimate of γ is then found by solving

$$\hat{\gamma} = \underset{\gamma \in \Gamma}{\arg\min} \left\{ I(\mathbf{y}, \hat{\boldsymbol{\theta}}_\lambda, \hat{\lambda}; \mathbf{X}_\gamma) + I(\gamma) \right\},$$

where $I(\mathbf{y}, \hat{\boldsymbol{\theta}}_\lambda, \hat{\lambda}; \mathbf{X}_\gamma)$ is given by either (18) or (19), depending on k_γ.

6 Experiments

The MML GLM ridge criterion was compared to the corrected Akaike Information Criterion (AIC$_c$) in both parameter estimation and model selection experiments. The AIC$_c$ has previously been shown to perform well when applied to regression models, even in the case of small sample sizes [11]. Given a particular λ, the AIC$_c$ score for the model is

$$\text{AIC}_c(\mathbf{y}; \lambda, \mathbf{X}) = -\log p(\mathbf{y}|\hat{\boldsymbol{\theta}}_\lambda; \mathbf{X}) + \hat{k}_\lambda \left(\frac{n}{n - \hat{k}_\lambda - 1} \right), \qquad (27)$$

where

$$\hat{k}_\lambda = \text{Tr}\left(\mathbf{X} \left(\mathbf{X}'_A \mathbf{W}_A(\hat{\boldsymbol{\mu}}_\lambda) \mathbf{X}_A \right)^{-1} \mathbf{X}' \mathbf{W}(\hat{\boldsymbol{\mu}}_\lambda) \right)$$

is the degrees-of-freedom of the fitted regression model, $\mathbf{X}_A$ is the augmented design matrix given by (20), $\mathbf{W}(\hat{\boldsymbol{\mu}}_\lambda)$ is the weight matrix given by (13), $\mathbf{W}_A(\hat{\boldsymbol{\mu}}_\lambda)$ is given by (22), $\hat{\boldsymbol{\theta}}_\lambda$ are the MAP estimates of $\boldsymbol{\theta}$ and $\hat{\boldsymbol{\mu}}_\lambda = f^{-1}(\mathbf{X}\hat{\boldsymbol{\beta}}_\lambda + \hat{\alpha}_\lambda \mathbf{1}_n)$. The AIC$_c$ estimate of λ is found by minimising (27).

Table 3. Median ratios of the Kullback–Leibler (KL) divergences obtained by the MML and AIC$_c$ estimates over the KL divergence obtained by the maximum likelihood estimates

		$\rho = 0.1$		$\rho = 0.5$		$\rho = 0.9$	
	n	AIC$_c$	MML	AIC$_c$	MML	AIC$_c$	MML
Normal	25	0.66	0.47	0.57	0.45	0.31	0.35
	50	0.91	0.70	0.86	0.71	0.64	0.62
	100	0.96	0.85	0.95	0.84	0.83	0.79
	250	0.98	0.94	0.98	0.94	0.93	0.90
Binomial	25	0.03	0.05	0.02	0.06	0.02	0.03
	50	0.56	0.22	0.51	0.19	0.27	0.19
	100	0.82	0.69	0.77	0.64	0.63	0.56
	250	0.96	0.91	0.94	0.89	0.89	0.88
Poisson	25	0.77	0.58	0.83	0.62	0.46	0.39
	50	0.94	0.96	0.96	0.95	0.92	0.89
	100	1.00	1.00	0.99	0.98	0.97	0.97
	250	1.00	1.00	1.00	1.00	1.00	1.00

6.1 Parameter Estimation Simulations

The performance of both the MML ridge estimates and the AIC$_c$ ridge estimates were compared to the maximum likelihood estimates on simulated data. At each of the 1,000 iterations of the simulation, a vector of $k = 10$ "true" regression coefficients was sampled from a normal distribution, $\beta_i^* \sim N(0,1)$, and a design matrix of $n = \{25, 50, 100, 250\}$ samples was generated from a multivariate normal distribution with a mean of zero, and covariance structure $\mathrm{E}\left[x_{i,j} x_{i,k}\right] = \rho^{|j-k|}$, where $\rho = \{0.1, 0.5, 0.9\}$. Targets of the chosen distribution (normal, binomial, Poisson) were then generated using the regression coefficients $\boldsymbol{\beta}^*$ and generated design matrix. Maximum likelihood, MML and AIC$_c$ were used to estimate the regression coefficients, with $\boldsymbol{\Sigma} = \mathbf{I}_k$, and Kullback–Leibler (KL) divergences [12] from the true model were calculated for all three estimates.

The median ratios of the KL divergence obtained by the MML and AIC$_c$ estimates over the KL divergence obtained by the maximum likelihood estimates are presented in Table 3. In all cases the ratio is less than or equal to one, and in many cases is substantially smaller than one, indicating that ridge regression offers an excellent alternative to maximum likelihood estimation. The improvements are generally larger for higher levels of correlation, which is expected given the nature of ridge regularisation. The MML estimates are competetive with, or superior to, the AIC$_c$ estimates in all cases, and in the case of normal regression models MML is superior in all but one case.

6.2 Model Selection Experiments on Real Data

The MML and AIC$_c$ ridge procedures were also tested in terms of model selection on several real datasets. Three datasets were chosen (two from the UCI machine learning repository [13], and one previously analysed in [14]): (i) the Pima indians dataset (binary targets, $q = 8$ covariates, $n = 768$ samples); (ii)

Table 4. Kullback–Leibler divergences for three real datasets estimated by cross-validation

	Pima Indians		Diabetes		Boston Housing	
n	AIC$_c$	MML	AIC$_c$	MML	AIC$_c$	MML
25	3.775	1.063	1.659	1.404	5.991	3.895
50	1.552	0.641	1.281	1.233	3.692	3.370
100	0.542	0.528	1.151	1.144	3.200	3.181
250	0.501	0.501	1.101	1.103	3.057	3.099

the diabetes data (normal targets, $q = 10$, $n = 442$); and (iii) the Boston housing data (normal targets, $q = 14$, $n = 506$). Each dataset was randomly split into training and testing samples, and to make the task more difficult, four extra noise covariates generated from a standard normal distribution were appended to each training sample. MML and AIC$_c$ were used to select a subset of the candidate regressors based on the training sample, with the potential subsets being determined from the path generated by the Lasso procedure [15]. The testing sample was subsequently used to assess the predictive performance of the criteria, measured in terms of mean KL divergence. Each test was repeated 100 times. The results are presented in Table 4, and show that MML is competetive with, or superior to, AIC$_c$ for all three datasets, and for all sample sizes. The performance difference is especially noticable for the Pima indians dataset.

References

1. Nelder, J.A., Wedderburn, R.W.M.: Generalized linear models. Journal of the Royal Statistical Society. Series A (General) 135(3), 370–384 (1972)
2. Hoerl, A., Kennard, R.: Ridge regression. In: Encyclopedia of Statistical Sciences, vol. 8, pp. 129–136. Wiley, New York (1988)
3. Wallace, C.S., Boulton, D.M.: An information measure for classification. Computer Journal 11(2), 185–194 (1968)
4. Wallace, C.S., Freeman, P.R.: Estimation and inference by compact coding. Journal of the Royal Statistical Society (Series B) 49(3), 240–252 (1987)
5. Wallace, C.S.: Statistical and Inductive Inference by Minimum Message Length. Information Science and Statistics. Springer (2005)
6. Farr, G.E., Wallace, C.S.: The complexity of strict minimum message length inference. Computer Journal 45(3), 285–292 (2002)
7. Makalic, E., Schmidt, D.F.: Minimum message length shrinkage estimation. Statistics & Probability Letters 79(9), 1155–1161 (2009)
8. Schmidt, D.F., Makalic, E.: MML invariant linear regression. In: Nicholson, A., Li, X. (eds.) AI 2009. LNCS, vol. 5866, pp. 312–321. Springer, Heidelberg (2009)
9. Makalic, E., Schmidt, D.F.: MML logistic regression with translation and rotation invariant priors. In: Thielscher, M., Zhang, D. (eds.) AI 2012. LNCS, vol. 7691, pp. 878–889. Springer, Heidelberg (2012)
10. McCullagh, P., Nelder, J.A.: Generalized Linear Models, 2nd edn. Chapman & Hall/CRC (1989)

11. McQuarrie, A.D.R., Tsai, C.L.: Regression and Time Series Model Selection. World Scientific (1998)
12. Kullback, S., Leibler, R.A.: On information and sufficiency. The Annals of Mathematical Statistics 22(1), 79–86 (1951)
13. Asuncion, A., Newman, D.: UCI machine learning repository (2007)
14. Efron, B., Hastie, T., Johnstone, I., Tibshirani, R.: Least angle regression. The Annals of Statistics 32(2), 407–451 (2004)
15. Tibshirani, R.: Regression shrinkage and selection via the Lasso. Journal of the Royal Statistical Society (Series B) 58(1), 267–288 (1996)

Ultimate Order Statistics-Based Prototype Reduction Schemes

A. Thomas and B. John Oommen*

School of Computer Science, Carleton University, Ottawa, Canada, K1S 5B6
smithasam2007@yahoo.com, oommen@scs.carleton.ca

Abstract. The objective of Prototype Reduction Schemes (PRSs) and Border Identification (BI) algorithms is to reduce the number of training vectors, while simultaneously attempting to guarantee that the classifier built on the reduced design set performs as well, or nearly as well, as the classifier built on the original design set. In this paper, we shall push the limit on the field of PRSs to see if we can obtain a classification accuracy comparable to the optimal, by condensing the information in the data set into a *single training* point. We, indeed, demonstrate that such PRSs exist and are attainable, and show that the design and implementation of such schemes work with the recently-introduced paradigm of Order Statistics (OS)-based classifiers. These classifiers, referred to as Classification by Moments of Order Statistics (CMOS) is essentially anti-Bayesian in its *modus operandus*. In this paper, we demonstrate the power and potential of CMOS to yield single-element PRSs which are either "selective" or "creative", where in each case we resort to a non-parametric or a parametric paradigm respectively. We also report a single-feature single-element creative PRS. All of these solutions have been used to achieve classification for real-life data sets from the UCI Machine Learning Repository, where we have followed an approach that is similar to the Naïve-Bayes' (NB) strategy although it is essentially of an anti-Naïve-Bayes' paradigm. The amazing facet of this approach is that the training set can be reduced to *a single* pattern from each of the classes which is, in turn, determined by the CMOS features. It is even more fascinating to see that the scheme can be rendered operational by using the information in a *single feature* of such a *single data point*. In each of these cases, the accuracy of the proposed PRS-based approach is very close to the optimal Bayes' bound and is almost comparable to that of the SVM.

Keywords: Prototype Reduction Schemes, Classification using Order Statistics (OS), Moments of OS.

1 Introduction

In traditional non-parametric classification, the training patterns play a significant role in the classification process. This is because a decision boundary is obtained by considering *all* the samples in the training set. However, modern

* *Chancellor's Professor*; *Fellow: IEEE* and *Fellow: IAPR*. This author is also an *Adjunct Professor* with the University of Agder in Grimstad, Norway.

S. Cranefield and A. Nayak (Eds.): AI 2013, LNAI 8272, pp. 421–433, 2013.

rapid advancements in this field have led to the development of efficient classification methods in which the schemes achieve the classification based on a *subset* of the training patterns. A Prototype Reduction Schemes (PRS) is a generic method for reducing the number of training vectors, without affecting the performance of the classifier built on the reduced design set [1–4]. Instead of considering all the training patterns for the classification, a subset of the whole set is selected based on certain criteria. The training is then performed on this reduced set, which is also called the "Reference" set. More recent advances have involved the use of Border Identification (BI) algorithms [5–8] to choose these prototypes from the so-called "border" points of the various classes.

Traditionally, a good PRS can reduce the size of the training set to a small percentage (for example, 10%) of the original set. But how small can one make this reduced set? Is it possible to, at least conceptually, reduce the set of prototypes to contain *only a single element* from each class. The aim of this paper is to investigate this issue both conceptually and from a practical perspective. Indeed, we shall demonstrate that we can push and attain the limit on the field of PRSs to obtain a classification accuracy comparable to the optimal, by condensing the information in the data set into a *single training* point. Apart from showing that such a PRS exists and is attainable, we shall also show that the design and implementation of such a mechanism relies on the recently-introduced paradigm of Order Statistics (OS)-based classifiers.

One should, of course, mention that the new point obtained by invoking the PRS is not necessarily a member of the original data set. Rather, it can be an artificially created point, representative of the training set, as perceived from the perspective of the data sets OSs.

We now consider another facet of a typical PRS-based PR solution. Whenever a practitioner designs a PRS, he works with the premise that *all* features are crucial for the classification. The problem that is "dual" to the PRS problem is the following: Apart from reducing the size of the "Reference" set, is it possible to also reduce the number of features utilized within the latter. This paper addresses both of these issues simultaneously. To be specific, we state that the OS-based PRS scheme that we propose has the fascinating property that it can be rendered operational by using the information in a *single feature* of the *single data point* obtained using an OS-based computation. Indeed, in each of these cases, the accuracy of this approach is very close to the optimal Bayes' bound and is almost comparable to that of the SVM. In a nutshell, this is the fundamental contribution of this paper, and we are not aware of any reported comparable results.

To put this paper in the right context, a word about these OS-based classifiers is not out of place [9–11]. Almost all the well-known classifiers involved in pattern classification are based on a Bayesian principle which aims to maximize the *a posteriori* probability, where they have been characterized by their respective indicators such as their means, variances etc.. In the field of PR, however, there are some families of indicators that have noticeably been uninvestigated, specifically those related to its Order Statistics (OS). The interesting point about these indicators is that some of them are quite unrelated to the traditional moments themselves, and in spite of this, have not been used in achieving PR. The main question that has earlier excited our interest is whether these indicators/indices possess any potential in PR.

The salient differences between the traditional Bayesian paradigm and the newly-proposed OS-based anti-Bayesian paradigm can be highlighted as below. Consider Figure 1, where for simplicity, we have used unit-lengthed intervals to display the span of the two class-conditional distributions. Whenever a testing sample comes from these distributions, the CMOS will compare the testing sample with the *higher*-order 2-OS, $E[\mathbf{x}_{2,2}]$ of the first distribution, i.e., $\frac{2}{3}$, and with with the *lower*-order 2-OS $E[\mathbf{x}_{1,2}]$ of the second distribution, i.e., $h + \frac{1}{3}$, and the sample will be labeled with respect to the class which minimizes the corresponding quantity, as shown in Figure 1. We emphasize that the comparison is not made with the *means* of the two distributions, but with certain non-central outlier-like points, rendering it "Anti"-Bayesian. Observe that for the above rule to work, we must enforce the ordering of the OS of the two distributions, and this requires that $\frac{2}{3} < h + \frac{1}{3} \implies h > \frac{1}{3}$. The case when this condition is not satisfied, and the details of CMOS have been explained in [9–11].

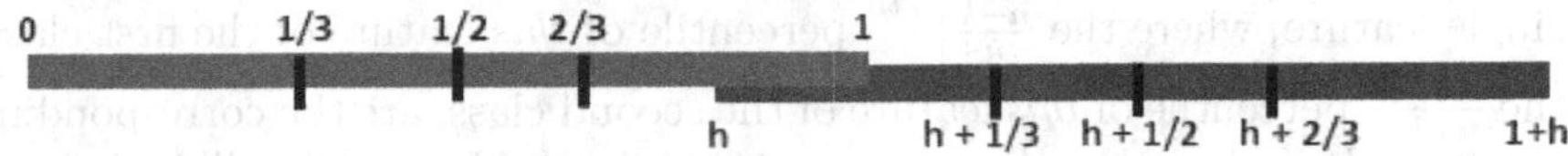

Fig. 1. A schematic of OS-based Anti-Bayesian Classification

This paper takes this concept to the next level, i.e., to that concerning PRSs.

From an overall perspective, we now discuss how we are to achieve our goal to reduce the cardinality of the OS-based PRS to be unity for each class. First of all, we know that PRSs can be broadly classified as being "selective" or "creative" [12]. A "selective" PRS yields as its output a set of prototypes which are *chosen* from the original training points. As opposed to this, a "creative" PRS *creates* a set of artificial points which may not be found in the original training set, and these points are thereafter used in the classification.

We first study the task of designing "selective" OS-based PRSs in Section 4. Since, at this juncture, we are not willing to assume a distributional form for the corresponding features, we are forced to work with the non-parametric representation that the training data captures. By working with the multi-dimensional non-parametric form of the data, and by thereafter invoking an OS-based paradigm, we are able to obtain a *single* prototype with which we can accomplish efficient classification. This *single* prototype is, as a vector, a "created" point, although, in every single dimension, the value is "selected" from the actual training sample that is closest to the value specified by the OS value.

Two versions of this strategy have been proposed, namely, the first which considers the entire vectorial form of the resultant prototype (in Section 4.1), and the second which invokes a majority vote by considering the OS-based classification of the individual features. The latter, which is a *Scalar*-Based Selective PRS, has been described in Section 4.2. It is worth mentioning that the classification results obtained by both these methods – both of which involve only a *single* prototype – are quite satisfactory, and are comparable, though understandably, marginally inferior, to those obtained from a NB or SVM strategy.

After investigating selective PRSs, we subsequently consider the task of designing "creative" OS-based PRSs in Section 5. In this case, we assume a distributional form for the corresponding features, and so we proceed to work with the parametric representation that the training data captures. By working with a multi-dimensional parametric form of the data, and by thereafter invoking an OS-based paradigm, we succeed in obtaining a *single* prototype in the "Reference" set, which can be used for classification. This process has been explained in Section 5.1. As in the non-parametric case, we have also developed a *Scalar*-Based Creative PRS in Section 5.2. Again, it is worth mentioning that the classification results obtained from both these parametric strategies (i.e., the vector, and the majority-voted individual-feature based) are quite satisfactory, and comparable, though marginally inferior, to those obtained from a NB or SVM strategy.

The final concluding contribution is actually far more ambitious. It consists of using only a *single* feature of a *single* prototype. In this case, in Section 6, we have designed a "creative" PRS scheme which merely includes the OS-based points of a single feature, where the ${\frac{n-k+1}{n+1}}^{th}$ percentile of *this* feature of the first class, and the ${\frac{k}{n+1}}^{th}$ percentile of *this* feature of the second class, are the corresponding "prototypes". It is clear that the accuracy of this *scalar*-based OS will be inferior to that of the corresponding *vector*-based OS. However, astonishingly enough, the accuracy does not degrade significantly – the resultant classifier still yields an accuracy that is acceptable considering the fact that one requires only a single *scalar* comparison to achieve the classification.

The reader must observe that the intent of this paper is not to compare the resultant classification accuracies with those obtained from an entire ensemble of classification methodologies. Rather, our aim is to show that we can obtain very efficient classification by merely using a single (vector or scalar) prototype which is either selected or created. Thus, we have compared our proposed scheme with only *three* standard algorithms which have been universally considered as benchmarks. We believe that the results presented here conclusively demonstrate the power of our contribution.

1.1 Contributions of This Paper

The novel contributions of this paper are:

- We propose a "selective" PRS which can be metaphorically perceived to be the "Ultimate" selective PRS because, by using a non-parametric paradigm, it reduces the size of the "Reference" set to be a *single* pattern from each class, which is thereafter utilized in the classification;
- We also propose a "creative" PRS which can be considered to be the "Ultimate" creative PRS because, by invoking a parametric paradigm, it also reduces the size of the "Reference" set to be a *single* pattern from each class;
- In both of the above cases, we have also shown that it is possible to derive a majority-based PRS which fuses the classification results of the various

features of the *single* d-dimensional prototype. The classification accuracies of these fused scalar schemes are marginally worse than those of the corresponding vector-based algorithms;
- We have also shown that it is possible to derive a single scalar prototype, i.e., one which involves only a a *single* feature of a *single* d-dimensional vector. The classification accuracy of this single-scalar PRS is marginally worse than that of the vector-based methods;
- In every case, we demonstrate, by testing the algorithms on real-life data sets from the UCI repository, that the new PRS-based classification schemes yield accuracies comparable to the traditional NB classifiers, and even the SVM, even though the computations needed are, really, of an atomic magnitude.

In the interest of space, the formal algorithms for all these strategies cannot be included here. But in the interest of completeness, as a representative example, we have included the formal algorithm for for one of these strategies, namely for the Ultimate *Vector*-based Creative PRS in Section 5.1.

We conclude this section by remarking that, to the best of our knowledge, analogous results have been unreported in the literature.

2 CMOS-Based Classification: The Generic Classifier

The multi-dimensional OS-based classifier is based on its uni-dimensional counterpart developed earlier. Since its understanding is crucial to this paper, it is briefly explained here.

Consider a 2-class problem with classes ω_1 and ω_2, where their class-conditional densities are $f_1(x)$ and $f_2(x)$ respectively (i.e, their corresponding distributions are $F_1(x)$ and $F_2(x)$ respectively). If we perform a classification based on ν_1 and ν_2, the *medians* of the distributions, this is equivalent to the strategy in which the task is performed based on a *single* OS. For all symmetric distributions, this classification accuracy attains the Bayes' accuracy – which is not too astonishing because the median is identical to the mean. But the intriguing aspect emerges when we use higher order OS that are not located centrally (close to the means), but rather *distant* from the means. Indeed, for uni-dimensional OS-based PR, our methodology is based on considering the n-order OSs, and comparing the testing sample with the $n-k$ OS of the first distribution and the kth OS of the second. By considering the entire spectrum of the possible values of k, the results in and showed that the specific value of k is usually not so crucial. Further, if these symmetric pairs of the OS are used in PR, the classification based on *these* attains the optimal Bayes' bound for a large number of symmetric distributions of the exponential family. The PR is near-optimal when the distributions are asymmetric.

Theses results were generalized for multi-dimensional distributions by invoking a Naïve-Bayes' approach, which essentially implies that that the first moments of the OS in each of the dimensions are uncorrelated.

With this as the background, we shall now consider how we can derive single-element OS-based PRSs which can be used to design classifiers for real-life data. Since our solutions have been tested on both artificial and real-life data-sets, we shall, in the interest of continuity, briefly describe the sets that we have used.

3 Experimental Data Sets

3.1 Artificial Data Sets

For a *prima facie* testing of artificial data, we generated two classes that obeyed Gaussian distributions. To do this, we made use of a Uniform $(0, 1)$ random variable generator to generate data values that follow a Gaussian distribution. The expression $\mathbf{z} = \sqrt{-2ln(u_1)}\ cos(2\pi u_2)$ is known to yield data values that follow $N(0, 1)$ [13]. Thereafter, by using the technique described in [14], one can generate Gaussian random vectors which possess any arbitrary mean and covariance matrix. The means of the classes were $[2\ 2\ 2\ 2\ 2]^T$ and $[-2\ -2\ -2\ -2\ -2]^T$ respectively, and the covariances of the two classes were identical and had the form[1]:

$$\Sigma = \begin{bmatrix} a^2 & b & 0 & a & \alpha ab \\ b & 2a+3b & 0 & b & a \\ 0 & 0 & 1 & 0 & 0 \\ a & b & 0 & 2a+3b & b \\ \alpha ab & a & 0 & b & b^2 \end{bmatrix}$$

This rendered the classes to have an optimal linear classifier. With regard to the cardinality of the data set, each of the classes had 200 instances in the corresponding 5-dimensional space.

3.2 Real-Life Setup

The data sets [15] used in this study have two classes, and the number of attributes varies from 4 up to 32. The data sets are given in Table 1.

4 OS-Based "Selective" PRSs Using a Non-parametric Perspective

In this section, we discuss the problem of designing a "Selective" OS-based PRS. Since we are ultimately going to select a training sample, at this juncture, we take the position that we are not willing to assume a *distributional form* for the corresponding features. Consequently, we are forced to work with the non-parametric representation that the training data captures. This implies that one has to resort to a non-parametric avenue in which we are able to compute the corresponding prototypes by approximating the distribution using a multi-dimensional kernel. Although a generalized kernel could be used for this phase, in the interest of simplicity, for a *prima facie* case, we have opted to use a simplistic bin-based approach. Once the histogram of the features has been obtained in each dimension, the training sample that lies closest to the point representing the $\frac{n-k+1}{n+1}^{th}$ percentile of the first distribution and the $\frac{k}{n+1}^{th}$ percentile of the

[1] In our experiments, we set $a = 5$, $b = 4$, and $\alpha = 0.4$.

Table 1. The Real-life data sets used in our experiments, where C, I and R represent Categorical, Integer and Real Respectively

Data set	No. Instances	No. Attributes	No. Classes	Attribute Type
WOBC	699	9	2	I
WDBC	569	32	2	Real
WDBC	569	32	2	R
Diabetes	768	8	2	I, R
Hepatitis	155	19	2	C, I, R
Iris	150	4	3	Real
Mushroom	8124	22	2	C
Statlog (Heart)	270	13	2	C, R
Statlog (Australian Credit)	690	14	2	C, I, R
Vote	435	16	2	C, I

second distribution of the given data sets is *selected* to be the prototype of interest. Indeed, by using these *selected* patterns as vector prototypes – *a single one from each class* – one can now achieve classification. One should observe that this *single* prototype is, as a vector, a "created" point, although, in every single dimension, the value is "selected" from the actual training sample that is closest to the value specified by the OS value.

Although the specific value of k is not so crucial [9–11], in this paper, as mentioned earlier, we have set $k = 1$, implying that we have, in each dimension, worked with the pattern that falls at the $\frac{2}{3}$ percentile of the first distribution and the pattern that falls at the $\frac{1}{3}$ percentile of the second.

To obtain the final PRS, we can envision two methodologies, namely where the computations are vector-based or scalar-based, which are described below.

4.1 The *Vector*-Based Selective OS-Based PRS

The *Vector*-based selective OS-based PRS is obtained by comparing the testing sample with the prototype procured by the above process. Such a comparison can be achieved using any metric, but for the sake of simplicity, we have utilized the well-known Euclidean norm.

The proposed method has been rigorously tested on the various artificial and real-life data sets obtained from the UCI repository [15] described above. It has also been compared with other well-known schemes including the NB, SVM, and the kNN. In order to obtain the results, the algorithms were executed 50 times with the 10-fold cross validation scheme. The results are tabulated in Table 2. To ensure standardization , the performance of the benchmark classifiers are taken from [16–18]. By examining the table of results (see Column 6), we can see that the proposed algorithm can achieve a comparable classification when compared to the other traditional classifiers, which is particularly impressive because once the *single* prototype has been computed after the training phase, the testing is done by exactly two vector-based computations (one for each class), comparing the testing sample with the resultant prototypes. For example, for the Breast Cancer data set, we can see that the new approach yielded a accuracy of 95.06% which should be compared to the accuracies of the SVM (96.99%), NB (96.40%) and the kNN (96.60%). The reader will observe that the classification accuracies for all the data sets is commendable except for the "Diabetes" set. This is because, for this data set, the approximation of the distributions using simplistic histograms in the d-dimensional space is rather crude. Superior results are obtained in this case when we resort to obtaining the OS-based points using the criteria explained in Section 5.1.

4.2 The *Scalar*-Based Selective OS-Based PRS

In the *Scalar*-based selective OS-based PRS, the patterns are treated as a group of scalars and a classification is performed for each dimension. Thereafter, the final determination of the identity of the testing sample is achieved based on a majority vote. The scalar-based selective CMOS has been tested on the various artificial and real-life data sets and the results are tabulated in Table 2. If we examine the table (see Column 8), one can see that the approach yields a near optimal accuracy for the all the data sets except the Diabetes data set, which, as before has a poor accuracy for all the classifiers, and for which the histogram leads to a very crude approximation. For example, if we consider the Hepatitis data set, the proposed approach yields an accuracy of 81% while the traditional classifiers yields 84.54% (SVM), 82.58% (NN) and 83.19% (NB), which is still quite astonishing considering that all the information in the entire training set has been crystallized into a single prototype *distant from the mean.*

We now move on to present the vector and scalar-based "Creative" PRSs in which the Reference set has only a single element.

5 A CMOS-Based "Creative" PRS Using a Parametric Perspective

We now consider the task of designing a "creative" OS-based PRS, where we again aim to attain the goal that the cardinality of the Reference set is unity. Since we are now willing to permit the option of assuming a distributional form for the corresponding features, we have chosen to resolve this fundamental issue by invoking a strategy analogous to a Naïve-Bayes' approach, although it, really, is of an *anti*-Naïve-Bayes' paradigm. As a Naïve-Bayes' strategy requires the uncorrelation of the features, if we consider a k-OS CMOS, we need to determine, for every feature, the $\frac{n-k+1}{n+1}^{th}$ percentile of the first distribution and the $\frac{k}{n+1}^{th}$ percentile of the second distribution. From an anti-Naïve-Bayes'perspective, we can obtain the corresponding values of all of the features by assuming a Gaussian[2] distribution for all the features. The OS-based PRS that we thus propose

[2] Any other member of the exponential family described in [9] could have just as well been used. We have chosen to use the Gaussian distribution because it is more general than the others, and involves the means and the variances of the features.

here again consists of the *single created* point in the d-dimensional space characterized by the location of the $\frac{n-k+1}{n+1}^{th}$ percentile of the first distribution and the $\frac{k}{n+1}^{th}$ percentile of the second distribution. As shown in [9], for the value of $k = 1$, the 2-OS CMOS positions for the classes that follow a Gaussian distributions can be expressed as $u_1 = \mu_1 - \frac{\sigma}{\sqrt{2\pi}}$ and $u_2 = \mu_2 + \frac{\sigma}{\sqrt{2\pi}}$. We thus opt to use these expressions to obtain the corresponding CMOS positions, whence the vector and scalar-based PRS schemes are derived.

5.1 The *Vector*-Based "Creative" OS-Based PRS

For this approach also, we consider the possibility of perceiving the training set as vectors or as scalars. The *Vector*-based "Creative" OS-based PRS considers the final prototype as a vector, which has been artificially created as a new pattern by resorting to the expressions for u_1 and u_2. The testing sample is then compared with the *single* OS-based prototype, and the identity is determined with regard to how distant it is from the latter. Since the individual variances are known, this distance is computed using the Mahalanobis distance.

The formal algorithm for this approach is given in Algorithm 1.

Algorithm 1. `Vector_based_Creative_PRS`(T, TP)

Input:

T: The training set, comprising of elements T_1 and T_2 from classes ω_1 and ω_2 respectively.
TP: the testing set

Output:

Classification for TP

Method:
Training
for i = 1 to d **do**
  Estimate mean of T_1 as μ_{1i} and mean of T_2 as μ_{2i}
  Estimate the standard deviations of T_1 and T_2 as σ_{1i} and σ_{2i}
end for
for i = 1 to d **do**
  Determine the ith component of $\mathbf{u_1}$, $u_{1i} = \mu_{1i} - \frac{\sigma_{1i}}{\sqrt{2\pi}}$
  Determine the ith component of $\mathbf{u_2}$, $u_{2i} = \mu_{2i} + \frac{\sigma_{2i}}{\sqrt{2\pi}}$
end for
End_Training

Testing
for all $\mathbf{x} \in TP$ **do**
  if $M\_Dist(\mathbf{u_1}, \mathbf{x}) < M\_Dist(\mathbf{u_2}, \mathbf{x})$ **then**
    Assign $\mathbf{x}$ to class ω_1
  else
    Assign $\mathbf{x}$ to class ω_2
  end if
end for

End_Testing
End Algorithm

Table 2. Classification of Real-life data sets by CMOS

Data set	Traditional Classifiers			CMOS Classifier			
	NB	NN	SVM	Vector		Scalar	
				Creative	Selective	Creative	Selective
WOBC	96.40	96.60	96.99	96.94	95.06	94.35	92.06
WDBC	92.97	96.66	97.71	93.43	90.07	89.25	86.82
Diabetes	73.11	71.90	73.84	73.76	65.74	76.74	43.41
Hepatitis	83.19	82.58	84.54	76.67	75.13	81.87	81.00
Iris	95.13	96.00	96.67	94.4	92.50	93.80	77.80
AU Credit	87.40	85.90	85.51	94.76	84.21	83.03	48.19
Heart	83.00	84.40	85.60	84.59	83.93	77.11	60.67
Vote	94.29	90.23	94.33	93.43	91.0	89.10	85.36

The vector-based *Creative* CMOS has been tested for the same data sets as before, and the results are tabulated in Table 2. From the table (see Column 5), we can conclude that the new approach is comparable with the other well-used and well-established classifiers. This approach achieves "almost" optimal classification when compared to the traditional classifiers. For example, if we consider the classification of the Breast Cancer data set, we see that Algorithm achieves 96.94% accuracy as opposed to the 96.99% of SVM, 96.40% of NB and 96.6% of NN. One can see that the difference in the accuracies is almost negligible. For the other data sets too, this approach attains a near-optimal classification when compared to the traditional classifiers, even though there is only a single element in the Reference set, and the testing involves only two vector comparisons.

5.2 The *Scalar*-Based "Creative" OS-Based PRS

In this approach, each pattern was considered as a vector, and the distance calculations were based on the Mahalanobis metric. As in the case of the selective scheme described in Section 4.2, a similar classification can be achieved by considering the various feature values as scalars and by accomplishing the task by computing the majority vote.

The scalar-based creative CMOS has also been tested on the various artificial and real-life data sets and the results are tabulated in Table 2 (see Column 7). Again, an examination of the table shows that the classification results are near-optimal. For example, if we consider the Vote data set, the proposed approach yields an accuracy of 93.43% while the traditional classifiers yields 94.33% (SVM), 90.24% (NN) and 94.29% (NB). Observe that the prototype-based NN performs even better than the traditional NN which involves the entire training set, which is quite astonishing considering that all the information in the entire training set has been crystallized into a single newly-created prototype.

6 Classification Based on One Selected Feature

In this section we have embarked on an even far more ambitious goal which consists of seeing if we could do the classification by using only a *single* feature of a *single* prototype. To achieve this goal, we have operated with the philosophy proposed in Section 5 and designed a "creative" vector PRS. But rather than use all the components of the vector in the classification, we have merely chosen the OS-based points of a *single feature*, where the $\frac{n-k+1}{n+1}^{th}$ percentile of *this* feature of the first class, and the $\frac{k}{n+1}^{th}$ percentile of *this* feature of the second class, are the corresponding "prototypes" (where we have, as usual, used $k = 1$).

The proposed approach of has been tested on the artificial and real-life data sets described earlier, and the results are tabulated in Table 3. If we closely investigate the table, one can see that the method attains a comparable classification when compared to the traditional classifiers. Specifically, for the Diabetes data set, if the classification is performed based on the OS positions of the feature *Plasma Glucose Concentration*, an accuracy of 73.63% is attained as opposed to the accuracy of 73.84% attained by SVM . The reader should not be surprised that the accuracies are not always so outstanding. However, astonishingly enough, the accuracy does not degrade significantly – the resultant classifier still yields an accuracy that is acceptable considering the fact that one requires only two *scalar* comparisons to achieve the classification.

Table 3. Classification of Artificial and Real-life data sets using the Scalar-based *Creative* CMOS involving only a single dimension

Data set	SVM	Dimension	Feature	CMOS
Artificial Set	98.75	3	A3	98.475
WOBC	96.99	2	Uniformity of Cell Size	93.04
WDBC	97.71	27	Worst Compactness	91.29
Diabetes	73.84	2	Plasma Glucose Concentration	73.63
Hepatitis	84.54	12	Ascites	83.93
Iris	96.67	4	Petal Width	95.5
AU Credit (Statlog)	92.1	7	A9	84.84
Heart (Statlog)	85.60	2	Chest Pain Type	78.52
Vote	94.33	4	Physician-fee-freeze	95.40

7 Conclusions

Almost all the well-known classifiers involved in pattern classification are based on a Bayesian principle which aims to maximize the *a posteriori* probability. Quite recently, a new paradigm, known as CMOS, the classification by moments of Order Statistics, has been introduced to attain the same task, but with a counter-intuitive philosophy as compared to the Bayesian principle. In [10], the foundational theory of the CMOS was introduced, and a generic classifier that can be used for any distribution was provided. The applications of CMOS on various uni-dimensional distributions of the exponential family were included in [9]. The results of [9] were extended for multi-dimensional distributions in [11].

In this paper, we have demonstrated the power and potential of CMOS to yield single-element PRSs which are either "selective" or "creative", where in each case we resort to a non-parametric or a parametric paradigm respectively. We have derived a single-feature single-element creative PRS. All of these solutions have been used to achieve classification for artificial and real-life data sets from the UCI Machine Learning Repository. All of the reported algorithms yield an acceptable accuracy when compared to many of the established benchmark methods. It is even more fascinating to see that our paradigm performs favorably by using the information in a *single feature* of such a *single data point.*

References

1. Garcia, S., Derrac, J., Cano, J.R., Herrera, F.: Prototype Selection for Nearest Neighbor Classification: Taxonomy and Empirical Study. IEEE Transactions on Pattern Analysis and Machine Intelligence 34(3), 417–435 (2012)
2. `http://sci2s.ugr.es/pr/` (April 18, 2013)
3. Kim, S., Oommen, B.J.: On Using Prototype Reduction Schemes and Classifier Fusion Strategies to Optimize Kernel-Based Nonlinear Subspace Methods. IEEE Transactions on Pattern Analysis and Machine Intelligence 27, 455–460 (2005)
4. Triguero, I., Derrac, J., Garcia, S., Herrera, F.: A Taxonomy and Experimental Study on Prototype Generation for Nearest Neighbor Classification. IEEE Transactions on Systems, Man and Cybernetics - Part C: Applications and Reviews 42, 86–100 (2012)
5. Duch, W.: Similarity Based Methods: A General Framework for Classification, Approximation and Association. Control and Cybernetics 29(4), 937–968 (2000)
6. Foody, G.M.: Issues in Training Set Selection and Refinement for Classification by a Feedforward Neural Network. In: Proceedings of IEEE International Geoscience and Remote Sensing Symposium, pp. 409–411 (1998)
7. Foody, G.M.: The Significance of Border Training Patterns in Classification by a Feedforward Neural Network using Back Propogation Learning. International Journal of Remote Sensing 20(18), 3549–3562 (1999)
8. Li, G., Japkowicz, N., Stocki, T.J., Ungar, R.K.: Full Border Identification for Reduction of Training Sets. In: Bergler, S. (ed.) Canadian AI. LNCS (LNAI), vol. 5032, pp. 203–215. Springer, Heidelberg (2008)
9. Oommen, B.J., Thomas, A.: Optimal Order Statistics-based "Anti-Bayesian" Parametric Pattern Classification for the Exponential Family. Pattern Recognition (2013) (accepted for Publication)
10. Thomas, A., Oommen, B.J.: The Fundamental Theory of Optimal "Anti-Bayesian" Parametric Pattern Classification Using Order Statistics Criteria. Pattern Recognition 46, 376–388 (2013)
11. Thomas, A., Oommen, B.J.: Order Statistics-based Parametric Classification for Multi-dimensional Distributions (submitted for publication 2013)
12. Kim, S., Oommen, B.J.: A brief Taxonomy and Ranking of Creative Prototype Reduction Schemes. Pattern Analysis and Applications 6, 232–244 (2003)
13. Devroye, L.: Non-Uniform Random Variate Generation. Springer, New York (1986)
14. Fukunaga, K.: Introduction to Statistical Pattern Recognition, 2nd edn. Academic Press, San Diego (1990)
15. Frank, A., Asuncion, A.: UCI Machine Learning Repository (2010), `http://archive.ics.uci.edu/ml` (April 18, 2013)

16. http://www.is.umk.pl/projects/datasets.html (April 18, 2013)
17. Karegowda, A.G., Jayaram, M.A., Manjunath, A.S.: Cascading K-means Clustering and k-Nearest Neighbor Classifier for Categorization of Diabetic Patients. International Journal of Engineering and Advanced Technonlogy 01, 147–151 (2012)
18. Salama, G.I., Abdelhalim, M.B., Elghany Zeid, M.A.: Breast Cancer Diagnosis on Three Different Datasets using Multi-classifiers. International Journal of Computer and Information Technology 01, 36–43 (2012)

Group Recommender Systems: A Virtual User Approach Based on Precedence Mining

Venkateswara Rao Kagita, Arun K. Pujari, and Vineet Padmanabhan

School of Computer and Information Sciences
University of Hyderabad, Hyderabad - 500046, India
venkateswar.rao.kagita@gmail.com, {akp,vineet}cs@uohyd.ernet.in

Abstract. The recommendation framework based on precedence mining as outlined in [3] is limited to personal recommendation and cannot be trivially extended for group recommendation scenario. In this paper, we extend the precedence mining model for group recommendation by proposing a novel way of defining a virtual user by taking `transitive precedence relation` into account. We obtained experimental results for different combinations of parameter settings and for different group-sizes on `MovieLens` data-set based on our virtual-user model. We show that our framework has better performance in terms of `precision and recall` when compared with other methods.

1 Introduction

Broadly speaking, recommender systems are based on one of two strategies (1) *Content-based* filtering (2) *Collaborative* filtering. The content filtering approach [4,6] creates a profile for each user or product to characterize its nature. On the other hand, collaborative filtering [4], relies only on past user behavior without requiring the creation of explicit profiles and therefore certain patterns of consumption of items exhibited by the whole set of users, U, is not captured. The recent precedence mining model proposed by Parameswaran et.al. [3] gets over this shortcomings and attempts to capture pairwise precedence relation occurring frequently among all users. Recommender systems based on precedence relations is concerned with mining precedence relations among items consumed by users and thereafter recommends new items having high precedence probability score. Though precedence mining, as demonstrated by Parameswaran et. al., can be an important approach for recommender systems, current research is mostly limited to personal recommendation. In this work we extend the concept of precedence probability to a group by introducing a virtual user that can more effectively represent a group. Traditionally a virtual user represents all users of the group and the profile of a virtual user contains the set of common items consumed by members of the group wherein the items are considered in any order. We argue that *most often there can be very few items utilized by all members of the group for non-cohesive group and therefore common item strategy may not work well in such situations.*

S. Cranefield and A. Nayak (Eds.): AI 2013, LNAI 8272, pp. 434–440, 2013.

2 Precedence Mining

Let $O = \{o_1, o_2, \ldots, o_n\}$ be the set of items and $U = \{u_1, u_2, \ldots, u_m\}$ be the set of users. $profile(u_j)$ is sequence of items known to have been consumed previously by user u_j. O_j is the set of items consumed by u_j. For given O, U, k and a target user $u \in U$, the Personal Recommender System (PRS) is concerned with recommending k items for consumption by user u. The recommended items are absent in $profile(u)$, i.e., are not used previously by u. In the case of a Group Recommender System(GRS) the idea is to recommend k items to a group $G \subseteq U$ of users based on past consumptions of members of the group for given O, U and k. Any recommender system (personal or group) aims at selecting items for recommendation such that these items are expectedly preferred to other items by the user for whom it is recommended. Define $support_i$ as the number of users consumed item o_i. We define p_{ij} as the number of users having consumed item o_i preceding o_j in their profiles. The precedence probability for items o_i and o_j, denoted as $PP(o_i|o_j)$ represents the probability of o_i preceding o_j. We define $PP(o_i|o_j) = \frac{p_{ij}}{support_i}$ and $Score(o_i, u_j) = \frac{support_i}{n} \times \prod_{o_l \in O_j} PP(o_l|o_i)$ We introduce the concept of transitive precedence relation. Intuitively, transitive precedence relation attempts to capture the relationship between objects o_i and o_j based on the o_i preceding o_l and o_l preceding o_j. In other words, o_i preceding o_j transitively through o_l.

3 Virtual User: Our Approach

There are several ways of extending a personal recommender system to a group recommender system [1,2,5]. Common approaches of group recommendation are either based on (1) *Merging strategy* (2) *Virtual user strategy.* Merging strategy can be implemented in three ways- `Merged profiles`, `merging recommendation` and `merging score`. For instance, a trivial way of computing the merged score for a group $G \subseteq U$ is to take the aggregated score of individual members of the group. Thus $Score_M(o_i, G) = F_{u_j \in G}(Score(o_i, u_j))$, where F is an aggregate function. We take F to be AVG in this study. In Virtual user strategy [6] the idea is to create a virtual user which in some way represents the group interests mediated in an integrated profile. The virtual user $v(G)$ represents the whole group G. $profile(v(G))$ is generated from $profile(u_j), u_j \in G$. One popular approach to integrate individual user profile is to consider all the common items. In this model, the profile of the virtual user, $profile(v(G))$, is computed from the set $\bigcap_{u_j \in G} O_j$ arranging its elements in any order. *We compute the score for every object using* $profile(v(G))$. We discuss below our proposal of creating a virtual user that can more effectively represent a group. In a group recommender scenario an item could fall into one of the three categories (1) it is an item that is used previously by each member of the group (2) it is an item that is used previously by some (but not all) members of the group and (3) it is an item that is not used by any member of the group. Items in first category are the common behavior exhibited by all members of the group. Diversity in the group is captured in usage of items in the second category. We define two ways of calculating weights for virtual user. (1) **Virtual User by thresholding**: Define weight for an item consumed by an individual user as

$$weight(o_i, u_j) = \begin{cases} 1 & \text{if } o_i \in O_j \\ score(o_i, u_j) & otherwise \end{cases}$$

Define weight of an item for group G as $weight(o_i, G) = \sum_{u_j \in G} weight(o_i, u_j)/|G|$. We construct profile for the virtual user $v(G)$ of group G as follows. An item $o_i \in profile(v(G))$, if $weight(o_i, G) \geq \nu$ where ν is a predefined threshold.

Our proposal of construction of $profile(v(G))$ can also be viewed as an *enhanced common item* approach. The set of consumed items for group G is $\bigcup_{u_j \in G} O_j$ and the set of common items consumed by all members of the group is $\bigcap_{u_j \in G} O_j$. The items of $\bigcup_{u_j \in G} O_j \setminus \bigcap_{u_j \in G} O_j$ with high likelihood of consumption by members of the group are appended to $\bigcap_{u_j \in G} O_j$ constitutes the profile of the virtual user. The effectiveness of this strategy is dependent on user-specified parameter ν and proper tuning of value of ν for application in hand is crucial. For a very high value of ν, our strategy is equivalent to common item strategy. And for very small values of ν, the method may yield erroneous result. (2) **Virtual User by Weight:** In order to avoid parameter tuning, we propose another model of virtual user strategy below. We define *weighted-profile* of a user as a sequence of items and associated weights. The conventional definition of *profile* can be mapped to *weighted-profile* in a straight forward manner by assigning weight 1, if the item is previously consumed and 0 if it is not consumed. Given the *weighted-profile(u)* for all users $u \in U$, we compute the *weighted-profile(v(G))* of the virtual user of group G by taking *weight(o, v(G))* for $o \in \bigcup_{u_j \in G} O_j$. Define $Score(o_i, v(G)) = \frac{support_i}{n} \times \prod_{o_l} weight(o_l, v(G)) \times PP(o_l|o_i)$. Define **transitive precedence probability** between objects o_a, o_b and o_c as $PP(o_a|o_b) \times PP(o_b|o_c)$.

Proposition 1. *Virtual user with weight strategy considers transitive precedence probability.*

Consider the $Score(o_k, v(G))$. It can be expanded as given below. In the derivation we use c to denote common items and nc to denote non-common items.

$$\begin{aligned} Score(o_k, v(G)) &= \frac{support_k}{n} \times \prod_{o_l \in O_j} PP(o_l|o_k) \times weight(o_l, v(G)) \\ &= \frac{support_k}{n} \times \underbrace{\prod_{c} PP(o_l|o_k)}_{A} \times \underbrace{\prod_{nc} PP(o_l|o_k) \times weight(o_l, v(G))}_{B} \end{aligned}$$

Term A is precedence probability of an unused item and common item. Term B is defined as follows.

$$= \prod_{l \in nc} PP(o_l|o_k) \times weight(o_l, v(G))$$

$$= \prod_{l \in nc} X + \frac{1}{|G|} \times \frac{support_l}{n} \sum_{u} \prod_{o_s \in u} PP(o_s|o_l) \times PP(o_l|o_k)$$

From the derivation, a different interpretation of *Score* for the proposed virtual strategy emerges. For common items, the direct precedence probabilities are taken in the product term. But for items in the second category, precedence probabilities of two terms are paired together and these pair-terms are of the form $PP(o_a|o_b)PP(o_b|o_c)$, where o_a, o_b and o_c are items of first, second and third categories, respectively. This situation arises for those users who did not consume o_b. It is mentioned earlier that the items in the second category capture the diversity in the group. In our proposed model of virtual user strategy, the items in first category are treated exactly as these are for any user in the group. It is natural to treat common items of the group in this manner. The precedence probabilities between items in first and third categories are computed according to the original definition. The items in the second category are used to compute transitive precedence probabilities between items in first and third categories. Similarly, if an item in second category is used then we take its direct pairwise precedence probability with items in third category. But if the item is not used then transitive precedence probability of items in first category with items in third category are taken through such an item.

Considering all precedence probabilities in the product term for score computation may yield misleading results as one instance of precedence probability with small value will pull down the overall score. There can be two strategies to compute score by neglecting small precedence probabilities.

(1) **Fixed-I Model:** The first strategy is to select top I of the precedence probability values. Hence, the definition is modified as follows. $Score(o_i, u_j) = \frac{support_i}{n} \times \prod^{(\mathrm{I})}_{o_l \in O_j} PP(o_l|o_i)$ where $\prod^{(I)}$ of p numbers $a_1, a_2, \ldots, a_p$ means multiplication of top I highest values of the set $\{a_1,\ a_2, \ldots, a_p\}$.

(2) **Variable-I Model:** The second strategy is to select a threshold value τ such that the precedence probability values which exceed the threshold are only selected for multiplication. Hence, the definition is modified as follows. $Score(o_i, u_j) = \frac{support_i}{n} \times \prod^{(\tau)}_{o_l \in O_j} PP(o_l|o_i)$. Similarly, $\prod^{(\tau)}$ of p numbers $a_1, a_2, \ldots, a_p$ means selecting values of the set $\{a_1,\ a_2,\ \ldots,\ a_p\}$ which exceed threshold τ for multiplication.

4 Experimental Analysis

In this section we report our experimental analysis to study the performance of different strategies introduced in the earlier sections. In our experiments we

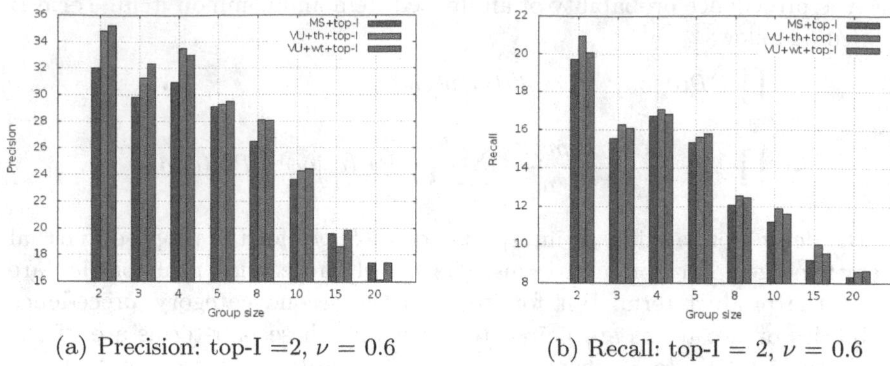

(a) Precision: top-I =2, $\nu = 0.6$ (b) Recall: top-I = 2, $\nu = 0.6$

Fig. 1. Precision and Recall comparison by varying group sizes for *Fixed -I model*

used the Movie-Lens dataset for the evaluation, which consists of 100,000 ratings given by 943 users for 1682 movies. The range of ratings is in between 1(bad)and 5(excellent). For each experiment we vary the group size from 2 to 20 generated randomly. For each group size 50 instances are randomly generated and for each instance all the three methods are run and the performance is compared based on the measures described above, namely *precision* and *recall*. These quantities are averaged over 50 instances for fixed group sizes. For methods which require threshold we experimented with different values of threshold. For the sake of ease of comprehension we only give the graphs for some selected values of the threshold. However our observation is that these values are representative sets for entire domain of the threshold. In order to calculate the accuracy we used 50% of the data as user history and compute the recommendation for the remaining 50% of the movies. We compare recommendation of proposed algorithms with actual usage for the last 50% of the movies. In our experiments we have taken k value as 10. In previous sections we introduced two different ways of computing scores for resulting profile of virtual user. These scores are based on *Fixed -I* and *Variable -I* model. Simple computation of score by taking into account precedence probabilities of all pairs will not yield correct recommendation and hence are not considered in our experiment. In the foregoing discussion, it is seen that none of these four strategies of score computation can be directly preferred over others. There can be situations where one way computing score may turn out to be preferable over any other. In the first round of our experiments we study the performance of merging score with virtual user strategies for each of the two methods of score computation. In other words for a problem instance we run merging score algorithm with *Fixed -I model* and *Variable -I model.* We do the same for *Virtual user with threshold* and *Virtual user with weights.* The output of all these three methods with two models of score computation are compared by taking 50 instances generated randomly for different values of group size. Figure 1 compares the precision and recall values with *Fixed -I model* (top-I). It can be seen that the precision and recall for virtual user strategy is

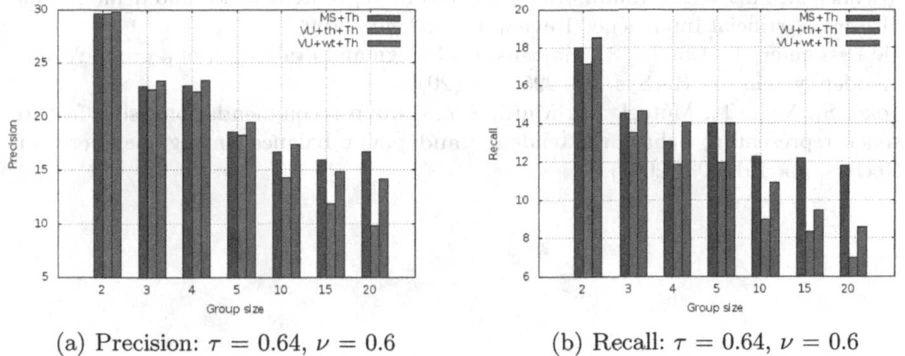

(a) Precision: $\tau = 0.64$, $\nu = 0.6$

(b) Recall: $\tau = 0.64$, $\nu = 0.6$

Fig. 2. Precision and Recall comparison by varying group sizes for *Variable-I model*

always better than merging scores strategy. Figure 2 depicts the precision (2(a)) and recall (2(b)) comparison of three strategies for *Variable-I model.* It can be seen from the graphs that precision and recall for a *Virtual user with weight* is better than *Merging score* strategy at smaller group sizes. But as the group size increases performance of *Merging score* increases over the other two approaches.

5 Conclusions

The main result of this paper is a new way of creating a virtual-user that can more effectively represent a group. To achieve this end we proposed two methods based on virtual-user strategy called *virtual user by weight* and *virtual user by thresholding.* Virtual user by weight strategy takes care of the transitive precedence relationship among a group of items. On the other hand virtual-user by thresholding takes care of items that are usually ignored by the traditional virtual-user-model. In order to efficiently calculate the *score* of items so as to recommend the item with the highest score for a target group we outline two models called *Fixed-I* and *Variable-I.* We experimented our models extensively for different combinations of parameter settings and for different group sizes on Movie-Lens data. We show that our framework has better performance in terms of precision and recall when compared with other methods.

References

1. Baskin, J.P., Krishnamurthi, S.: Preference aggregation in group recommender systems for committee decision-making. In: RecSys, pp. 337–340 (2009)
2. Berkovsky, S., Freyne, J.: Group-based recipe recommendations: analysis of data aggregation strategies. In: RecSys, pp. 111–118 (2010)
3. Parameswaran, A.G., Koutrika, G., Bercovitz, B., Molina, H.G.: Recsplorer: recommendation algorithms based on precedence mining. In: SIGMOD, pp. 87–98 (2010)

4. Michael, J.: Pazzani. A framework for collaborative, content-based and demographic filtering. Artificial Intelligence Review 13, 393–408 (1999)
5. De Pessemier, T., Dooms, S., Martens, L.: Design and evaluation of a group recommender system. In: RecSys, pp. 225–228 (2012)
6. Seko, S., Yagi, T., Motegi, M., Muto, S.Y.: Group recommendation using feature space representing behavioral tendency and power balance among members. In: RecSys, pp. 101–108 (2011)

A New Paradigm for Pattern Classification: Nearest *Border* Techniques

Yifeng Li[1], B. John Oommen[2], Alioune Ngom[1], and Luis Rueda[1]

[1] School of Computer Science, University of Windsor, Canada
{li11112c,angom,lrueda}@uwindsor.ca

[2] School of Computer Science, Carleton University, Canada
Also *Adjunct Professor*, University of Agder, Grimstad, Norway
oommen@scs.carleton.ca

Abstract. There are many paradigms for pattern classification. As opposed to these, this paper introduces a paradigm that has not been reported in the literature earlier, which we shall refer to as the Nearest *Border* (NB) paradigm. The philosophy for developing such a NB strategy is as follows: Given the training data set for each class, we shall first attempt to create borders for each individual class. After that, we advocate that testing is accomplished by assigning the test sample to the class *whose border it lies closest to*. This claim is actually counter-intuitive, because unlike the centroid or the median, these border samples are often "outliers" and are, really, the points that represent the class the least. However, we have formally proven this claim, and the theoretical results have been verified by rigorous experimental testing.

Keywords: Pattern Classification, Border Identification, SVM.

1 Introduction

The problem of classification in machine learning can be quite simply described as follows: If we are given a limited number of training samples, and if the class-conditional distributions are unknown, the task at hand is to predict the class label of a new sample with minimum risk. Within the generative model, one resorts to modeling the class-conditional distributions $p(\boldsymbol{x}|w_i)$ and priors $p(w_i)$ and $p(\boldsymbol{x})$, and then computing the *a posteriori* distribution $p(w_i|\boldsymbol{x}) = \frac{p(\boldsymbol{x}|w_i)p(w_i)}{p(\boldsymbol{x})}$ after the testing sample arrives. The strength of this strategy is that one obtains an optimal performance if the assumed distributions are the same as the actual one. The limitation, of course, is that it is often difficult, if not impossible, to compute. The alternative is to directly approximate the posterior distribution itself. This paper advocates such a philosophy.

The goal of this paper is to present a new paradigm in pattern recognition, which we shall refer to as the Nearest *Border* (NB) paradigm. This archetype possesses similarities to many of the well-established methodologies in pattern recognition, and can also be seen to include many of *their* salient facets/traits.

S. Cranefield and A. Nayak (Eds.): AI 2013, LNAI 8272, pp. 441–446, 2013.

There are four family algorithms that are most closely related to our NB paradigm. They include i) Prototype Reduction (PR) schemes [6], ii) Border Identification (BI) algorithms [6], iii) "Anti-Bayesian" Order-Statistics (OS) based algorithms [6], and iv) Support Vector Machines (SVMs) [8].

The novel contributions of this paper are the following:

1. We propose a new pattern recognition paradigm, the Nearest *Border* paradigm, in which we create borders for each individual class, and where testing is accomplished by assigning the test sample to the class whose border it lies closest to.
2. Our paradigm falls within the family of PR schemes, because it yields a reference set which is a small subset of original training patterns. The testing is achieved by *only* utilizing the latter.
3. Our paradigm falls within the family of BI methods.
4. The Nearest *Border* paradigm is essentially "anti-Bayesian" in its salient characteristics. This is because the testing is not done based on central concepts such as the centroid or the median, but by comparisons using these border samples, which are often "outliers" and which, in one sense, represent the class the least.
5. The Nearest *Border* paradigm is closely related to the family of SVMs, because the computations and optimization used are similar to those involved in deriving SVMs.

2 Method

2.1 The Theory of the Nearest Border (NB) Paradigm

We assume that we are dealing with a classification problem involving g classes: $\{\omega_1, \cdots, \omega_g\}$. For any specific class ω_i, we define a region $\mathcal{R}_i$ that is described by the function $f_i(\boldsymbol{x}) = 0$ (which we shall refer to as its "border"), where $\mathcal{R}_i = \{\boldsymbol{x} | f_i(\boldsymbol{x}) > 0\}$. We describe $\mathcal{R}_i$ in this manner so that it is able to capture the main mass of the probability distribution $p_i(\boldsymbol{x}) = p(\boldsymbol{x}|\omega_i)$. All points that lie outside of $\mathcal{R}_i$, are said to fall in its "outer" region, $\bar{\mathcal{R}}_i$, where $\bar{\mathcal{R}}_i = \{\boldsymbol{x}|f_i(\boldsymbol{x}) < 0\}$. These points are treated as outliers as far as class ω_i is concerned. The function $f_i(\boldsymbol{x})$ is crucial to our technique because it explicitly defines the region $\mathcal{R}_i$. Formally, the function $f_i(\boldsymbol{x})$ must be defined in such a way that:

1. $f_i(\boldsymbol{x})$ is the *signed distance* from the point $\boldsymbol{x}$ to the border such that $f_i(\boldsymbol{x}) > 0$ if $\boldsymbol{x} \in \mathcal{R}_i$, and $f_i(\boldsymbol{x}) < 0$ if $\boldsymbol{x} \in \bar{\mathcal{R}}_i$;
2. If $f_i(\boldsymbol{x}_1) > f_i(\boldsymbol{x}_2)$, then $p_i(\boldsymbol{x}_1) > p_i(\boldsymbol{x}_2)$;
3. If $f_i(\boldsymbol{x}) > f_j(\boldsymbol{x})$, then $p(w_i|\boldsymbol{x}) > p(w_j|\boldsymbol{x})$.

In order to predict the class label of a new sample $\boldsymbol{x}$, we calculate its signed distance from each class, and thereafter assign it to the class with the minimum distance. In other words, we invoke the softmax rule: $j = \arg\max_{i=1}^{g} f_i(\boldsymbol{x})$.

The main challenge that we face in formulating, designing and implementing such a NB theory lies in the complexity of conveniently and accurately procuring such borders. The reader will easily see that this is equivalent to the problem of identifying functions $\{f_i(\boldsymbol{x})\}$ that satisfy the above constraints. Although a host of methods to do this are possible, in this paper, we propose one that identifies the boundaries using the one-class SVM.

2.2 NB Classifiers: The Implementations of the NB Paradigm

The basic Nearest Centroid (NC) approach only uses the means of the class-conditional distribution, and this is the reason why it is not effective for the scenario when the variances of the various classes are very different. The NC scheme can be extended to allow different class variance by using, for example, Gaussian Mixture Model. The difficulty of extending any linear model, e.g. SVM, from its two-class formulation to its corresponding multi-class formulation, lies in the fact that a hyperplane always partitions the feature space into two "open" subspaces, implying that this can lead to ambiguous regions that may be generated by some extensions of the two-class regions for the multi-class case. The most popular schemes to resolve this are the one-against-all (using a softmax function) and one-against-one solutions.

As an one-class model, the work based on Tax and Duin's Support Vector Domain Description (SVDD or one-class SVM) [5] aims to find a closed hypersphere in the feature space that captures the main part of the distribution. By examining the corresponding SVM, we see that the hypersphere obtained by the SVDD is the estimate of the features' *Highest Density Region* (HDR). In particular, for the univariate distribution, the estimation of the *Highest Density Interval* (HDI) involves searching for the threshold p^* that satisfies: $\int_{x:p(x|D)>p^*} p(x|D)dx = 1-\alpha$. The $(1-\alpha)\%$ HDI is defined as $C_\alpha(p^*) = \{x : p(x|D) \geq p^*\}$. If we now define the *Central Interval* (CI) by the interval:

$$C_\alpha(l,u) = \{x \in (l,u)|P(l \leq x \leq u|D) = 1-\alpha, P(x \leq l) = \frac{\alpha}{2}, P(x \geq u) = \frac{\alpha}{2}\},$$

one will see that, for symmetric unimodal univariate distribution, HDI coincides with the CI. However, for nonsymmetric univariate distributions, the HDI is smaller than the CI. For known distributions, the CI can be estimated by the corresponding quantile. However, for unknown distributions, the CI can be estimated by a Monte Carlo approximation (or by the histogram, or the *Order Statistics*). However, in the context of this paper, we remark that by virtue of Vapnik's principle, it is not necessary to estimate the density by invoking a non-parametric method. For multivariate distributions, we can estimate the $(1-\alpha)\%$ HDR $C_\alpha(f)$ by using the equation:

$$\min_f \int_{f(x)\geq 0} 1dx, \text{ s.t. } \int_{x:f(x)\geq 0} p(x|D)dx = 1-\alpha. \quad (1)$$

We shall refer to this optimal contour $f^*(x) = 0$ as the $(1-\alpha)$-border/contour.

Our idea for classification is the following: We can learn a hypersphere using SVDD for each class in the feature space in order to describe the border of this class. We then calculate the distance from a unknown sample to the border of each class and assign it to the class with the minimum distance. The training phase of our approach is to learn the hypersphere $f_i(\boldsymbol{x}) = 0$ for each class. The prediction phase then involves assigning the unknown sample $\boldsymbol{x}$ using the rule: $j = \arg\max_{i=1}^{g} f_i(\boldsymbol{x})$. In particular, we note that:

1. $f_i(\boldsymbol{x}) \in \mathbb{R}$ is the signed distance of $\boldsymbol{x}$ from the corresponding boundary;
2. For points inside the ith hypersphere, $f_i(\boldsymbol{x}) > 0$;
3. For points outside the hypersphere, $f_i(\boldsymbol{x}) < 0$. Further, the larger $f_i(\boldsymbol{x})$ is, the closer it is to class ω_i, and the higher the value of $p(w_i|\boldsymbol{x})$ is. From the parameters of $f_i(\boldsymbol{x})$, we can see that $f_i(\boldsymbol{x})$ considers both mean and variance of the distribution. It can be further enhanced by the *normalized distance* through the operation of dividing it by R_i (the radius of the hypersphere), that is $\frac{f_i(\boldsymbol{x})}{R_i}$.

We refer to this approach as the *Nearest Border approach based on Hyper-Sphere* (NB-HS). Hereafter, the hypersphere based NB using the un-normalized and normalized decision rules will be denoted by *ν-NB*, and *ν-NBN*, respectively, where ν is the upper bound of the fraction of outliers and the lower bound of the fraction of the support vectors in SVDD. As the number of training samples increases to infinity, these two bounds converge to ν. However, in practice, we usually have a very limited number of training samples. In order to obtain ν which corresponds to the α fraction of outliers, firstly, we need to let $\nu = \alpha$, and then reduce ν gradually until the α fraction of outliers are obtained. This variant of NB will be named the *α-NB* in the subsequent sections.

3 Experimental Results

The NB schemes were rigorously tested. Our computational experiments can be divided into two segments. First, we verify the capability of our method on three artificial data sets. Then, we statistically compared our approach with benchmark classifiers on 17 well-known real-life data sets.

Accuracy on Synthetic Data: We verified our methods on three synthetic data sets described as follows and shown in Fig. 1. Each data set has four classes and 100 two-dimensional points in each class. In the *SameVar* data, all classes have the same variance, while in *DiffVar*, the classes have different variances. *NonLinear* is a nonlinear data set.

For the artificial data, we compared our method with the Naive Bayes [2], 1-NN [2], NC [7], and SVM [4] classifiers. Linear kernel was used for the NBs, NC, and SVM on the first two data sets, and the *Radial Basis Function* (RBF) kernel was used on the last one. We ran a 3-fold cross-validation on each data set 20 times. The mean accuracies and standard deviations are shown in Fig. 2a.

On the *SameVar* data, first, we can see that there is no significant difference between the ν-NB and ν-NBN, and α-NB. All of them yielded an almost-equivalent accuracy as the Naive Bayes. Second, it can be seen from Fig. 1a that the NB was able to identify the centers of each class accurately. The borders

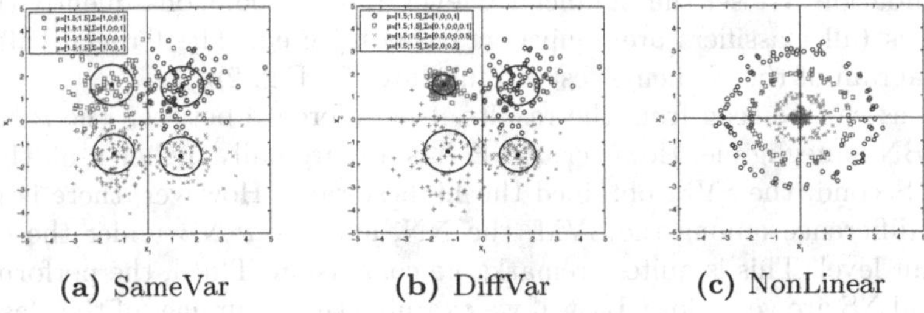

(a) SameVar (b) DiffVar (c) NonLinear

Fig. 1. Plots of the synthetic data sets

have the same volume, which demonstrates that the NB can identify the borders consistent with the variances. The NB approaches yielded an accuracy similar to the NC, which is reasonable because the identical variance of all classes is of no consequence to the NB. Third, the NN and SVM do not obtain comparable results, because the distance measure of the NN is affected by noise, and the SVM is not able to "disentangle" each class well using an one-versus-all scheme.

On the *DiffVar* data, first, we see that the results again confirm that the NB can identify the borders consistent with the variances (see Fig. 1b). The mean accuracies of all the NB approaches were very close to the Naive Bayes classifier. However, the NC yielded worse results than the NB. This is because the variance information helped the NB, while the NC scheme did not consider it.

Finally, for the *NonLinear* data, first, we affirm that all our NB methods and the SVM yielded comparably good results. Second, the Naive Bayes did not work well this time, because the data is not Gaussian. Further, the kernel NC was not competent either, because the data in the high-dimensional feature space have different variances for all the classes.

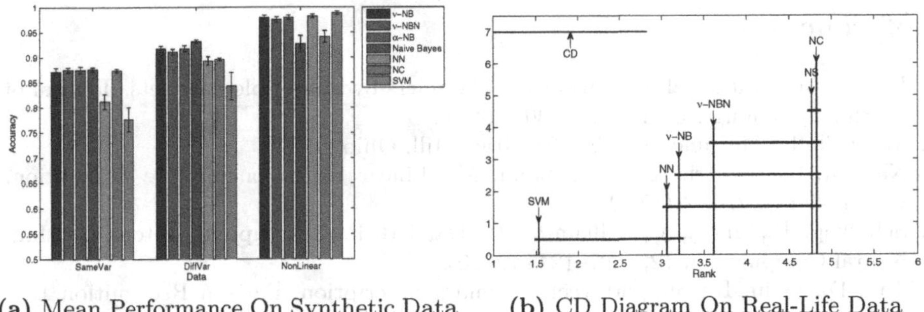

(a) Mean Performance On Synthetic Data (b) CD Diagram On Real-Life Data

Fig. 2. Performance on synthetic and real-life data

Statistical Test on Real-Life Data: In order to test the performance of our NB methods, we compared them with benchmark classifiers on 17 real-life data sets. The benchmark methods included were the 1-NN, NC, Nearest Subspace (NS) [3], and the SVM. We used the RBF kernel in our classifiers. We applied the Friedman test with Nemenyi test as a post-hoc test [1] on the accuracies of 3-fold cross-validation. We set the significance level to $\alpha = 0.05$. Consequently the null hypothesis (all classifiers are equivalent) was rejected. The Crucial-Difference (CD) diagram of the Nemenyi test is illustrated in Fig. 2b.

First, as can be seen from the results, the difference between the ν-NB and the ν-NBN is negligible. However, ν-NB has a marginally higher rank than the ν-NBN. Second, the SVM obtained the highest rank. However, there is no significant difference among the SVM, the NN, and the ν-NB under the current significant level. This is quite a remarkable conclusion. Third, the performances of NC and NS are very close. Last, if we examine the accuracies of the classifiers, we can clearly identify two distinct groups: {SVM, NN, ν-NB, ν-NBN}, and {NC, NS}, demonstrating that our newly-introduced NB schemes are competitive to the best reported algorithms in the literature.

4 Conclusions and Future Work

In this paper, we introduced a new paradigm for classification which has not been reported in the literature. We refer to it as the Nearest *Border* paradigm. We emphasize that our methodology is actually counter-intuitive, because unlike the centroid or the median, these border samples are often "outliers" and are, indeed, the points that represent the class the least. The theoretical results have been verified by rigorous experimental testing. We preliminarily assume that the class-conditional distribution is unimodal and homoscedastic in feature space. We will focus on a method which is able to learn the border of complex distributions, for example using hyperellipse, local learning, or mixture models.

Acknowledgments. We acknowledge the valuable suggestions from the reviewers. This research is support by Canadian NSERC Grants #RGPIN228117-2011.

References

1. Demsar, J.: Statistical comparisons of classifiers over multiple data sets. Journal of Machine Learning Research 7, 1–30 (2006)
2. Mitchell, T.: Machine Learning. McGraw Hill, Ohio (1997)
3. Naseem, I., Togneri, R., Bennamoun, M.: Linear regression for face recognition. PAMI 32(11), 2106–2112 (2010)
4. Scholkopf, B., Smola, A., Williamson, B., Bartlett, P.: New support vector algorithm. Neural Computation 12, 1207–1245 (2000)
5. Tax, D., Duin, R.: Support vector domain description. Pattern Recognition Letters 20, 1191–1199 (1999)
6. Thomas, A., Oommen, B.J.: The fundamental theory of optimal "anti-Bayesian" parametric pattern classification using order statistics criteria. Pattern Recognition 46, 376–388 (2013)
7. Tibshirani, R., Hastie, T., Narasimhan, B., Chu, G.: Class prediction by nearest shrunken centroids, with applications to DNA microarrays. Statistical Science 18(1), 104–117 (2003)
8. Vapnik, V.: Statistical Learning Theory. Wiley-IEEE Press, New York (1998)

Learning Polytrees with Constant Number of Roots from Data

Javad Safaei[1], Ján Maňuch[1,2], and Ladislav Stacho[2]

[1] Department of Computer Science, University of British Columbia, Vancouver, Canada
{jsafaei,jmanuch}@cs.ubc.ca

[2] Department of Mathematics, Simon Fraser University, Burnaby, Canada
{jmanuch,lstacho}@sfu.ca

Abstract. Chow and Liu [2] has shown that learning trees that maximize likelihood score given data can be done in polynomial time. A generalization of directed trees are polytrees. However, Dasgupta [3] has proved that learning maximum likelihood polytrees from data (and even approximation of the optimal result with a constant ratio) is NP-Hard. Therefore, researchers have focused on learning maximum likelihood polytrees with a constant number of roots. Gaspers et al. [5] have presented such an algorithm with complexity $O(mn^{3k+4})$ using matroid theory. We present a direct combinatorial algorithm with complexity $O(mn^{3k+1})$.

1 Introduction

The problem of learning Bayesian networks from data has been studied extensively. Chickering [1] has shown that learning a Directed Acyclic Graph (DAG) that maximizes the Bayesian score given data is NP-complete, even when the number of parents for each node is at most two. Research has concentrated on finding types of DAGs for which the learning problem is polynomially solvable. In their seminal work, Chow and Liu [2] showed that learning maximum likelihood (*ML*) trees from data is polynomial by proposing an algorithm with complexity $O(mn^2 + n^2 \log(n))$, where m is the number of data set vectors and n the number of variables or vertices of the DAG. Edmonds [4] independently proposed a similar algorithm referring to this problem as the optimal branching problem, and the resulting directed tree as a branching. Tarjan [6] improved the complexity of this learning algorithm to $O(mn^2)$.

Dasgupta [3] studied another type of DAGs, *polytrees*, i.e., directed graphs with no undirected loops, and showed that learning maximum likelihood polytrees is also an NP-complete problem, even if each node has at most two parents. In addition, he showed that finding approximation with a constant ratio of the optimal solution is also NP-complete. A natural question is whether this problem becomes tractable if we restrict number of polytree roots. Gaspers et al. [5] define an optimization problem in that from a decomposable scoring function a

S. Cranefield and A. Nayak (Eds.): AI 2013, LNAI 8272, pp. 447–452, 2013.

k-branching polytree should be learned, where a k-branching polytree is a polytree that with at most k arc removals is transferred to a directed forest. It is easy to observe that k-branching problem is equivalent with learning polytrees with up to $k+1$ roots. Gaspers et al. [5] have reduced this problem to the Matroid intersection problem and show it can be solved in $O(n^{3k+4})$. In the setting of maximum likelihood optimization based on data, this complexity becomes $O(mn^{3k+4})$, where m is the size of data. As our main result, we give a different algorithm for this problem, which works in time $O(mn^{3k+1})$.

2 Background

In this section, we will introduce all necessary concepts and results required to formally define the problem and state our results. We start with a description of the input: the data set and the optimal model, and finish with tree models.

2.1 Data Set Form

The data set D is a set of vectors $D = \{D_1, D_2, \ldots, D_m\}$, where each vector has exactly n dimensions. The data set can be represented as an $m \times n$ matrix D, where $d_{j,i}$ is the value of the j-th vector at dimension i. For each $1 \leq i \leq n$, the i-th component of the vectors represents a random variable X_i that takes a distinct value x with probability $\mathbf{P}(X_i = x)$. A joint probability distribution over data can be written as $\mathbf{P}(\mathbf{X} = \mathbf{x})$, where $\mathbf{X} = \{X_1, X_2, \ldots, X_n\}$ is the vector of random variables, and $\mathbf{x} = \{x_1, x_2, \ldots, x_n\}$ is a vector of values. We call $\mathbf{P}$ the *empirical* probability distribution if we estimate it from data as follows:

$$\mathbf{P}(X_i = x) = \frac{\sum_{j=1}^{m} \langle d_{j,i} = x \rangle}{m}, \quad \mathbf{P}(\mathbf{X} = \mathbf{x}) = \frac{\sum_{j=1}^{m} \langle D_j = \mathbf{x} \rangle}{m}, \tag{1}$$

where $\langle\rangle$ is the indicator function returning 1 if condition holds, and 0 otherwise.

2.2 Maximum Likelihood Model Scoring for DAGs

In this paper, we use maximum likelihood scoring for Bayesian networks, as follows:

$$\vec{\mathcal{G}}^* = \arg\max_{\vec{G}} P(D|\vec{G}), \tag{2}$$

where $\vec{G}$ is a DAG and $\vec{\mathcal{G}}^*$ is the set of all maximum likelihood DAGs. If we consider the input vectors in the data set D independently, then the likelihood probability $P(D|\vec{G})$ can be written as:

$$P(D|\vec{G}) = \prod_{j=1}^{m} P(D_j|\vec{G}), \tag{3}$$

The probability of a data given a DAG $\vec{G}$ on n nodes is defined as:

$$P(D_j|\vec{G}) \stackrel{def}{=} \mathbf{P}_{\vec{G}}(D_j) \stackrel{def}{=} \prod_{i=1}^{n} \mathbf{P}(X_i = d_{j,i}|\mathbf{\Pi}_i = \pi_{j,i}) \quad (4)$$

where $\mathbf{\Pi}_i$ denotes the set of all parents of the node X_i in $\vec{G}$, and $\pi_{j,i}$ is a set of all their values in the vector D_j. $\mathbf{P}_{\vec{G}}(D_j)$ is also called *factorized* form of distribution $\mathbf{P}$ *with respect to* $\vec{G}$.

2.3 Tree Models and Directionality

Chow and Liu [2] in their seminal work showed that if we limit DAGs to trees, then the maximum likelihood tree is equal to the maximum weighted spanning tree, *MST*, of the complete graph with vertex set $\mathbf{X}$ and edges weighted by mutual information of their end points, which can be found in time $O(n^2(\log n + m))$. To learn maximum likelihood polytree, we need to distinguish merging edges from other edges. Formally, *merging nodes* in a DAG are nodes with more than one parent, and *merging edges* are all incoming edges to a merging node. Let $\mathrm{ME}(\vec{G})$ denote the set of all merging edges of $\vec{G}$. Skeleton of a directed graph $\vec{G}$ is the undirected graph that contains all edges of $\vec{G}$ without directionality, and we denote it by $\bar{G}$.

Proposition 1. *(Root Selection in DAGs, [7]) If two different DAGs $\vec{G}$, $\vec{G'}$ have the same skeleton (i.e., $\bar{G} = \bar{G'}$) and the same set of merging edges (i.e., $\mathrm{ME}(\vec{G}) = \mathrm{ME}(\vec{G'})$), then $\mathbf{P}_{\vec{G}}(\mathbf{X}) = \mathbf{P}_{\vec{G'}}(\mathbf{X})$.*

We will use Proposition 1, later in the next section to show that after picking merging edges, the orientation of the remaining edges (as long as they do not introduce any new merging edges) does not affect the likelihood of any dataset given the polytree.

3 Learning Polytrees with a Constant Number of Roots

If a polytree has k roots, the number of merging nodes can vary from 1 to $k-1$. Without loss of generality, we may assume that $X_1, \ldots, X_L$ are all and only merging nodes, where $L < k$. Let $\mathbf{X}_M = \{X_1, X_2, \ldots, X_L\}$, and $\mathbf{\Pi}_M = \bigcup_{\ell=1}^{L} \mathbf{\Pi}_\ell$, where each $\mathbf{\Pi}_\ell$ is the parent set of node X_ℓ. We have the following proposition for these sets.

Proposition 2. *In a polytree $\vec{F}$ with $k > 1$ roots the following properties hold:*

$$2 \le |\mathbf{\Pi}_\ell| \le k, \quad \sum_{\ell=1}^{L} |\mathbf{\Pi}_\ell| = L + k - 1 \quad (5)$$

Our algorithm enumerates all possible combinations of the merging nodes and their parents, i.e., all possible sets of merging edges. These edges will form a forest, which is a subset of all edges in the polytree, and we will also refer to it as a *sub-polytree forests* (of nodes $\mathbf{X}_M$ and $\mathbf{\Pi}_M$). For each such selection of merging edges, we will run the MST algorithm on the remaining edges, similar to the one used by [2], with one exception: we do not allow components in the subgraph without merging edges that contain merging nodes to merge together. This will guarantee that we can orient these newly added edges so that no new merging edges are created. There might be multiple ways how to orient edges produces by MST algorithm, but by Proposition 1, they all yield the same likelihood score.

To determine the complexity of our algorithm, we need to upper bound the number of sub-polytree forests. First, we will bound the number of valid size vectors $(|\mathbf{\Pi}_1|, |\mathbf{\Pi}_2|, \ldots,)$, i.e., vectors satisfying Equations (5), and then we bound the number of merging edges selections yielding the desired sizes of parents sets. The number of valid vectors is exactly the number $p_L(k-L-1)$ of unordered partitions $a_1 + \cdots + a_L = k - L - 1$, where $a_\ell = |\mathbf{\Pi}_\ell| - 2$. Since there is no exact formula for $p_L(k-L-1)$, we will upper bound it by the number of ordered partitions $q_L(k-L-1) = \binom{k-2}{L-1}$, cf. [8].

For each such partition of parents set sizes, we have several choices how to pick the merging nodes and their parents from all n nodes. Let us denote the total number of sub-polytree forests for polytrees with n nodes and k roots by $T(n,k)$. Since there are $\binom{n}{L}$ ways to choose the merging nodes and $\binom{n-1}{|\mathbf{\Pi}_\ell|} = \binom{n-1}{a_\ell+2}$ ways to choose the parents of node X_ℓ, we have:

$$\begin{aligned} T(n,k) &\leq \sum_{L=1}^{k-1} \sum_{a_1+\cdots+a_L=k-L-1} \binom{n}{L}\binom{n-1}{a_1+2}\binom{n-1}{a_2+2}\cdots\binom{n-1}{a_L+2} \\ &\leq \sum_{L=1}^{k-1} q_L(k-L-1,L)n^{2L+k-1} \quad \text{(by Proposition 2)} \\ &\leq n^{k+1}\sum_{L=1}^{k-1}(n^2)^{L-1}\binom{k-2}{L-1} = n^{k+1}(1+n^2)^{k-2} \\ &\in O\left(n^{3k-3}\right) \quad \text{(for a constant } k\text{)} \end{aligned} \tag{6}$$

Equation (6) overcounts the number of valid sub-polytree forests. In particular, it is possible that selected merging edges $E = \cup_{\ell=1}^{L} \mathbf{\Pi}_\ell \times \{X_\ell\}$ create a loop, i.e., the selection is not valid, but it is still included in the count $T(n,k)$. Proposition 2 can be used to prune some of the invalid cases.

Theorem 1. *(Maximum Likelihood Polytree with $k+1$ Roots) A polytree with $k+1$ roots that maximizes log likelihood scoring function, Equation* (3), *for data set D can be found in time $O(mn^{3k+1})$.*

The main result in Gaspers et al. [5] can be summarized as follows:

Theorem 2. *(Learning k-branching, [5]) Learning a k-branching that maximizes any decomposable function that can be computed in time $O(n)$ for each polytree, can be solved in time $O(n^{3k+4})$.*

The assumption of Theorem 2 is that decomposable objective function can be computed in time $O(n)$ for each polytree, while in our case the log likelihood function is decomposable but requires $O(nm)$ time to be computed for each polytree. Therefore, if we apply Theorem 2 to our log likelihood scoring function (cf. Equation (3)), we get an algorithm running in time $O(mn^{3k+4})$.

It is easy to see that learning k-branching is equivalent to learning all polytrees from one, two, and up to $k+1$ roots. More precisely, each polytree with one root is a branching, and a polytree with $k+1$ roots has $L \leq k$ merging nodes with $L+k$ merging edges (by Proposition 2), and by deleting k of them ($|\mathbf{\Pi}_\ell| - 1$ of them for each merging node X_ℓ), all merging nodes become ordinary nodes (with one parent) and we obtain a directed tree, or a branching. It follows that we can solve k-branching problem with the maximum likelihood function in time $O(mn^{3k+1})$ which improves the results in [5] by a factor $O(n^3)$.

4 Experiments

We run our proposed algorithm with $k = 2, 3$ roots on a data set of peptides to analyze dependencies of different amino acids in these peptides. This data set is curated with the help of Kinexus Bioinformatics Corporation[1]. In our case peptides are subsequences of 20 different amino acids of length 9. We run our experiment on two different set of peptides:

1. 803 peptides that are phosphorylated by protein kinase PKC (uniprot key P17252).
2. 1000 randomly selected peptides that are phosphorylated by some protein kinase.

For each of these two data sets different polytrees are generated. Table 1 shows their maximum natural log likelihood scores normalized by the number of peptides. Table shows results for 5 different methods: "MWST" algorithm by Chow and Liu [2] which computes a directed tree, "MWST skeleton" heuristic methods for 2 and 3 roots and "optimal" method is our algorithm. In "MWST skeleton" heuristic method, we first fixed the skeleton of the tree by MWST and then tested all 2 (3) possible roots polytrees with this skeleton. The results show that increasing number of roots from two to three does not increase the log likelihood score significantly, while training time and number of trees that needs to be checked increases exponentially. This experiment, was run on a 64 bit PC, with a processor of four i5 2.67 GHz CPUs and 6 GB of RAM.

Note that normalized log likelihood score for PKC protein kinase is higher than the one for random peptides, which is expected since the peptides phosphorylated by PKC are more similar to each other and they are better fitted in

---

[1] `www.kinexus.ca`

Table 1. Natural log likelihood scores of two data sets with respect the polytree computed by different methods. The first column for each data set, contains the log likelihood score normalized by number of peptides used in the data set. The second column shows the computation time in seconds. The third column shows the number of examined polytrees.

Algorithm	803 peptides for PKC kinase			1000 randomly selected peptides		
	Score	Time	# polytrees	Score	Time	# polytrees
MWST 1 root	-19.15	0.14	1	-21.46	0.04	1
MWST skeleton 2 roots	-18.14	1.06	9	-20.40	1.12	8
MWST skeleton 3 roots	-17.26	2.86	23	-19.34	2.76	18
Optimal 2 roots	-18.02	27.47	252	-20.37	35.97	252
Optimal 3 roots	-16.96	2551.49	23184	-19.29	3235.37	23184

any tree dependency model (e.g. undirected or directed with 2 or 3 roots) than the randomly selected peptides set which are more divergent.

Acknowledgment. This work was supported in part by Natural Sciences and Engineering Research Council of Canada and the MITACS Accelerate Internship Program.

References

[1] Chickering, D.M.: Learning Bayesian networks is NP-complete. In: Learning from data, pp. 121–130. Springer (1996)
[2] Chow, C., Liu, C.: Approximating discrete probability distributions with dependence trees. IEEE Transactions on Information Theory 14(3), 462–467 (1968)
[3] Dasgupta, S.: Learning polytrees. In: Uncertainty in Artificial Intelligence, pp. 134–141 (1999)
[4] Edmonds, J.: Optimum branchings. Journal of Research of the National Bureau of Standards B 71, 233–240 (1967)
[5] Gaspers, S., Koivisto, M., Liedloff, M., Ordyniak, S., Szeider, S.: On finding optimal polytrees. In: Twenty-Sixth AAAI Conference on Artificial Intelligence (2012)
[6] Tarjan, R.E.: Finding optimum branchings. Networks 7(1), 25–35 (1977)
[7] Verma, T.S., Pearl, J.: Equivalence and synthesis of causal models. In: Uncertainty in Artificial Intelligence (UAI). pp. 220–227 (1990)
[8] Wilf, H.S.: Generatingfunctionology. Academic Press (1990)

Enhanced N-Gram Extraction Using Relevance Feature Discovery

Mubarak Albathan[1,2], Yuefeng Li[1] and Abdulmohsen Algarni[3]

[1] School of Electrical Engineering and Computer Science, Queensland University of Technology, Brisbane, Australia
{m.albathan,y2.li}@qut.edu.au

[2] Al Imam Mohammad Ibn Saud Islamic University Saudi Arabia, P.O. Box 5701, Riyadh 11432

[3] College of Computer Science, King Khaled University Saudi Arabia, P.O. Box 394, Abha 61411
a.algarni@kku.edu.sa

Abstract. Guaranteeing the quality of extracted features that describe relevant knowledge to users or topics is a challenge because of the large number of extracted features. Most popular existing term-based feature selection methods suffer from noisy feature extraction, which is irrelevant to the user needs (noisy). One popular method is to extract phrases or n-grams to describe the relevant knowledge. However, extracted n-grams and phrases usually contain a lot of noise. This paper proposes a method for reducing the noise in n-grams. The method first extracts more specific features (terms) to remove noisy features. The method then uses an extended random set to accurately weight n-grams based on their distribution in the documents and their terms distribution in n-grams. The proposed approach not only reduces the number of extracted n-grams but also improves the performance. The experimental results on Reuters Corpus Volume 1 (RCV1) data collection and TREC topics show that the proposed method significantly outperforms the state-of-art methods underpinned by Okapi BM25, *tf*idf* and Rocchio.

Keywords: Feature selection, relevance feedback, terms weight, n-gram extraction.

1 Introduction

With the explosive growth of information sources available on the Web, search engines return large numbers of documents based on a term-matching approach, but most of the results are not relevant to what the user needs. It is becoming essential to provide users with tools that more effectively filter huge amounts of streamed text data in order to extract a set of features from feedback documents.

Various effective studies have been conducted on term-based and pattern-based approaches to solve this issue. Most of the studies involving term-based methods present an efficient method for improving the retrieval and performance of useful information needed by users. However, many terms or keywords can

S. Cranefield and A. Nayak (Eds.): AI 2013, LNAI 8272, pp. 453–465, 2013.

be extracted from the feedback documents. Phrases (n-gram) have also been studied in many information retrieval models since phrases carry more semantic information and are easy to obtain using pattern-mining algorithms. Phrases are also useful in building effective ranking functions [1].

Traditional information retrieval (IR) models usually represent documents with bags-of-words assuming that words occur independently [2]. Phrases and n-grams have also a special meaning; however, this meaning relies on the topic that user wants. For example, "Apple TV" has a special meaning beyond the appearance of its individual words that can be found in the documents that talk about the technology, not fruit. Thus, documents are represented based on the phrases or n-grams assumption, which are independent of what users want. Therefore, n-grams consist of specific features. Selecting theses features is essential in mining text and retrieving information. This process involves selecting the subset of features based on some criteria to remove the irrelevant, redundant, noisy features [3].

In this paper, we propose a new method for extracting n-grams, which uses different term-based models and features and compares the results using Relevance Feature Discovery (RFD) features to remove the noisy features. We, also try to enhance the n-gram extraction by extending the random set to calculate the n-gram weight accurately based on distribution in the documents and their terms distribution in the n-grams. The experimental results illustrate that the proposed method significantly outperforms the state-of-art methods.

2 Related Work

With the growing volume of published research and documents on the web, and therefore the underlying knowledge in these texts, data mining and IR assist researchers in extracting useful knowledge from a collection of texts and satisfy user needs [4].

Therefore, researchers have focused on extracting knowledge such as keywords from documents automatically to suggest keywords for researchers to use in their studies. Researchers focus on two types of statistics: corpus-oriented and documents-oriented [5]. Early studies focused on evaluating corpus-oriented statistics of individual words. For instance, in 1972, Jones [6], proposed fundamental study describing the positive results of selecting keywords as discriminating words over a collection.

In corpus-oriented statistics, however, a word that occurs in many documents in the corpus is not selected as a keyword. Therefore, documents-oriented statistics tries to avoid the limitation of corpus-oriented statistics by extracting the same keywords from a document in spite of the state of a corpus [5]. Recently, studies have compared the effectiveness of three approaches for selecting terms: noun-phrase (NP) chunks, n-grams, and POS tags, with four different features of these terms as inputs for automatic keyword extraction using a supervised machine-learning algorithm [7].

N-gram extraction has been have been used extensively in many areas related to data mining such as language modelling, information retrieval, information filtering, and information extraction. The n-gram is a sequence of characters or words generated from a document as a result of moving a window of n size [8].

Wei et al. [9] used n-grams with feature selection and extraction methods. Their study compared the use of different feature selection methods on Chinese text classification using n-grams. They performed two-step feature selection. First, they reduced the number of features in the created class (inter-class) which uses two methods: the relative text frequency method and the absolute frequency method. Second, they selected the best features among all the classes (cross-class) in the training set by assigning a weight to the feature based on the occurrence of the feature within the different document classes.

Evaluating the extracted terms or n-grams is an important task in the information extraction and retrieval system. Term weighting evaluates the terms by assigning a significant weight based on statistical information and indicates the importance of the term to a topic. The n-gram weights calculate the probability of the extracted features based on a probabilistic function [10].

Many different proposed weighting methods for estimating and evaluating the weight of the extracted features start with document frequency *idf* combine it with term frequency *tf* to be *tf*idf* [11]. In addition, a probabilistic weighting technique uses different probabilistic functions to estimate the probability of the extracted feature [8]. Some n-gram studies instead provide probabilities based on smoothed language models [12].

These studies showed useful and interesting results and conclusions; however, either a number of noisy features are extracted or an inaccurate weights are assigned to the features. These issues must be resolved. Compared with famous term-based methods, we aim to solve these issues to reduce the amount of noise by selecting suitable extracted features to extract the n-gram and give them an accurate weight.

3 Definitions

For a given topic, the objective of discovering relevant features in text documents is to find a set of useful features, including patterns, terms or keywords and their weights, in a training set D, which consists of a set of relevant documents, D^+, and a set of irrelevant documents, D^-. In this paper, we assume that all documents are split into paragraphs. Therefore, a given document d yields a set of paragraphs $PS(d)$. These definitions can also be found in [13].

3.1 Pattern Mining

Let $T_1 = \{t_1, t_2, ..., t_n\}$ be a set of terms or keywords extracted from positive documents D^+. Given a *termset* X, a set of terms, in document d, $coverset(X)$ is used to denote the covering set of X for d, which includes all paragraphs $dp \in PS(d)$ such that $X \subseteq dp$, and its absolute support ($supp_a$) is the number of occurrences of X in $PS(d)$, that is:

$$supp_a(x) = |coverset(X)| \tag{1}$$

Moreover, its relative support ($supp_r$) is the fraction of the paragraphs that contain the pattern, that is:

$$supp_r(x) = \frac{supp_a(x)}{|PS(d)|} \tag{2}$$

Therefore, *termset* X is called a *frequent pattern* if its $supp_a(x)$ or $supp_r(x)$ is greater than or equal to the minimum support (min_sup) [14]. However, given a set of paragraphs $Y \subseteq PS(d)$, we can define its *termset*, which satisfies:

$$termset(Y) = \{t | \forall dp \in Y \Longrightarrow t \in dp\} \tag{3}$$

and the closure of X is defined as

$$Cls(X) = termset(coverset(X)) \tag{4}$$

Therefore, the pattern X is called closed if and only if $X = Cls(X)$ [13].

In addition, The sequential pattern X is called a frequent pattern if its $supp_r \geqslant min\_sup$. A frequent sequential pattern X is called a closed sequential pattern if there exists no frequent sequential pattern Y, such that $X \sqsubset Y$ and $supp_a(X) = supp_a(Y)$ [14], where the relation $\sqsubset$ represents the strict part of subsequence relation $\sqsubseteq$.

Furthermore, For term-based approaches, weighting the usefulness of a given term or keywords is based on its appearance in documents. However, for pattern-based approaches, weighting the usefulness of a given term is based on its appearance in discovered patterns.

To improve the efficiency of the pattern taxonomy mining, an algorithm, *SP-Mining*(D^+, min_sup) [13], was proposed to find closed sequential patterns SP_i, for all documents $\in D^+$, based on a given min_sup. For example, let SP_1, SP_2, ..., $SP_{|D^+|}$ be the sets of discovered closed sequential patterns for all documents $d_i \in D^+ (i = 1, \cdots, n)$, where $n = |D^+|$. For a given term t, its *d_support* in discovered patterns can be described as follows:

$$d\_sup(t, D^+) = \sum_{i=1}^{n} sup_i(t) = \sum_{i=1}^{n} \frac{|\{p | p \in SP_i, t \in p\}|}{\sum_{p \in SP_i} |p|} \tag{5}$$

where $|p|$ is the number of terms in p.

3.2 Term Weighting

In this section, we introduce the RFD model as a term weighting technique for relevance feature discovery [13], which describes the relevant features in relation to three groups, namely: positive specific terms, general terms and negative specific terms based on their appearances in a training set.

In the RFD model, the *specificity* of a given term t in the training set $D = D^+ \cup D^-$ was defined as follows:

$$spe(t) = \frac{|coverage^+(t)| - |coverage^-(t)|}{n} \tag{6}$$

where $coverage^+(t) = \{d \in D^+ | t \in d\}$, $coverage^-(t) = \{d \in D^- | t \in d\}$, and $n = |D^+|$. $spe(t) > 0$ means that term t is used more frequently in relevant documents than in irrelevant documents.

Based on the *spe* function, the following are the classification rules for determining its general terms G, positive specific terms T^+, and negative specific terms T^-:

$$G = \{t \in T | \theta_1 \leq spe(t) \leq \theta_2\},$$
$$T^+ = \{t \in T | spe(t) > \theta_2\}, \text{ and}$$
$$T^- = \{t \in T | spe(t) < \theta_1\}.$$

where θ_2 is an experimental coefficient, the maximum boundary of the specificity for the general terms, and θ_1 is also an experimental coefficient, the minimum boundary of the specificity for the general terms. It is assumed that $\theta_2 > 0$ and $\theta_2 \geq \theta_1$.

To improve the effectiveness, the RFD used irrelevant documents in the training set to remove the noise. The first issue in using irrelevant documents is how to select a suitable set of irrelevant documents. Most models can rank documents using a set of extracted features. If an irrelevant document gets a high rank, the document is called an offender [4] because it is a false discovery. Offenders are normally defined as the top-K ranked irrelevant documents. The basic hypothesis is that the relevance features should be mainly discovered from the relevant documents. Therefore, RFD sets $K = \frac{n}{2}$, as half of the number of relevant documents.

Once the top-K irrelevant documents are selected, the set of irrelevant documents D^- is reduced to include only K offenders (irrelevant documents); therefore, we have $|D^+| \geq 2|D^-|$.

The *spe* function can get its maximum value, 1, if there is a term t such that $coverage^-(t) = \emptyset$, and its minimum value, $-\frac{1}{2}$, if there is a term t such that $coverage^+(t) = \emptyset$.

The RFD model uses the terms' support and the terms' specificity to define the terms' weights as follows:

$$w(t) = \begin{cases} d\_sup(t, D^+)(1 + spe(t)) & t \in T^+ \\ d\_sup(t, D^+) & t \in G \\ d\_sup(t, D^+)(1 - |spe(t)|) & t \in T_1 \\ -d\_sup(t, D^-)(1 + |spe(t)|) & \text{otherwise} \end{cases}$$

where the *d_sup* function is defined in Equation 5.

4 Feature Selection and N-Gram Extraction

Due to the increasing amount of data available today on the Web, user queries usually return with results that are irrelevant for the user needs. In information retrieval, the retrieval models use the documents index to retrieve the documents relevant to the user based on the keyword-matching approach. Each retrieved document is ranked based on the score of each document that will be presented to the user [15].

One of the objective of knowledge extraction is to find a set of features from feedback documents [4]. This issue has received attention from data mining and information retrieval researchers. Feature selection is the simplest way to solve this issue as it reduces the number of irrelevant terms. This method selects a set of terms as features in the training set to improve the efficiency and quality of n-gram extraction by decreasing the number of noisy features that cause errors in the new data [10]. As many n-grams are extracted from different datasets, n-gram extraction on a large corpus yields a large number of extracted n-grams. Only some will be interesting to the user; the others will be noisy or irrelevant to the user. Therefore, extracting n-grams from selected features will improve the quality of constructing the n-gram [16]. Therefore, we need a specific features for n-grams extraction which are interested for the topics.

This research requires two stages to extract and weight the n-gram in selected features. In the first step, we try to reduce the number of features and select the best specific ones to extract the n-gram. In the second step, we extend the random set of n-gram probability to include the terms' probability.

4.1 Extracting N-grams Using Feature Selection

The extracting stage includes two steps: first, selecting the best features and then extracting the n-gram based on the selected features. Selecting good features attempts to improve the quality of extracting the n-gram and reduce the computational complexity and noisy features. Feature selection is a crucial issue in data mining and has long been studied in data mining and machine learning [17].

To test the proposed method, different features have been tested in different feature selection methods, such as the n-gram, BM25, Rocchio, and RFD methods. As previously mentioned, in this experiment we attempted to select the best extraction method and features to enhance n-gram extraction. We found that the RFD model extracted good features for use in extracting n-grams[13].

Two reasons make the RFD model the best model. First, RFD features achieve the best performance compared with the different patterns (phrases) methods and term-based methods. The overall results for RFD compared with different pattern-based methods showed better performance with a maximum of 12.30% and a minimum of 6.92% in all five used measures. Furthermore, RFD also has good results compared with other term-based methods. It has achieved best results with a maximum of 17.50% and a minimum of 9.25% in all five measures. Second, the RFD results are closed to the real definition of the *spe* function as will be explained more in section 5.4. Therefore, we can conclude that, RFD shows a significant improvement in all five measures in both pattern-based and term-based methods [13] and extracted the best features that can be selected for extracting the n-gram in this experiment.

The next step in this experiment is to extract the n-grams. An n-gram is a sequence of n words over a sequence of given words. Fürnkranz [18] showed that using word sequences of 2 or 3 words usually improves the performance compared with using $n > 3$, which reduces the performance.

4.2 Weighting Extracted N-Grams

In this experiment, after the n-grams have been extracted from the documents, it is time to calculate the probability of the extracted n-grams. The n-grams are usually selected based on the sliding window technique, and the probability of a n-gram $= \{w_1w_2...w_n\}$ is calculated using the following equation [19]:

$$P(w_1w_2...w_n) = P(w_1)P(w_2|w_1w_2)...P(w_n|w_1w_2...w_{n-1})$$

It is hard to calculate this probability because of the noisy terms and the complex relationship between terms. As we described before, feature selection can largely reduce the number of noisy terms; however, it is still very difficult to understand the relationship between terms. The only information for the relationship is the term weighting function that uses to select the top features. We also observe in experiments, that the distribution of term weights in an n-gram could influence the probability of the n-gram. For example, let $n\text{-}gram_1 =< w_1, w_2, w_3 >$, if the w_2 in the n-gram has a very low weight (e.g., low frequency); that might lead to the decreased probability of the $n\text{-}gram_1$ and vice versa. Thus, if the search for "Apple TV" most of the retrieved documents will be about information technology, while if the search is for "Apple Fruit," the retrieved documents will be about food and fruit. Thus, we see that the words "TV" and "Fruit" affect the results, which affects the probability of the gram. Therefore, in this paper, we use extended random set (ERS) [20] to provide an alternative method for calculating the probability of n-grams.

Let the training set $D = D^+ \cup D^-$, G be the set of n-grams and T be the set of selected terms (or features). The relationship ξ between terms can be described based on their appearers in n-grams:

$$\xi : T \rightarrow 2^{G \times [0,1]}$$

where $\xi(t) = \{(g, tf(g)) | t \in g, tf(g) = \frac{tf(g,D^+)}{tf(g,D)}\}$.

The prior probability of terms can be described by the weighting function used for the phase of feature selection, which satisfies

$$p(t) = w(t) / \sum_{t_j \in T} w(t_j)$$

for all term $t \in T$.

Based on the above definitions, we then can calculate the probability of n-grams using the following equation:

$$pr : G \rightarrow [0,1]$$

such that,

$$pr(g) = \sum_{t \in T, (g, tf(g)) \in \xi(t)} (p(t) \times tf(g)) = tf(g) \times \sum_{t \in g} p(t) \tag{7}$$

for all n-grams $g \in G$.

5 Evaluation

The main objective of this research is to extract a high-quality n-gram from text documents by introducing a new method for weighting n-grams. The new method uses RFD features initially to extract the n-gram and then calculates the probability of the extracted n-grams. To support this idea, this section describes the experiment environment, including dataset, baseline models and the results and discussion of the experiment:

5.1 Data

To conduct the experiment, the Reuters Corpus Volume 1 (RCV1) have been used, which consists of all and only 100 topics; each topic contains different numbers of documents with relevance judgements in training and testing sets. These 100 topics from English language news stories produced by Reuters journalists between 1996 and 1997, comprising a total of 806,791 documents. The documents were structured in XML. The first 50 topics were developed by humans and the rest by the intersections of pairs of Reuters categories.

Before our method was applied, different operations were conducted on the data, such as preprocessing the documents and removing a given stop-words list. In addition, the terms were stemmed by applying the Porter stemmed algorithm for suffix stripping [13].

5.2 Baseline Models

In this experiment, we used well-known term-based methods: Rocchio and BM25, including the *tf*idf* terms .

The Rocchio model uses a centroid to describe a topic as follows:

$$\alpha\frac{1}{|D^+|}\sum_{\vec{d}\in D^+}\frac{\vec{d}}{||\vec{d}||}-\beta\frac{1}{|D^-|}\sum_{\vec{d}\in D^-}\frac{\vec{d}}{||\vec{d}||} \tag{8}$$

There are two recommendations for setting parameters α and β in the Rocchio model [21]: $\alpha = 16$ and $\beta = 4$; and $\alpha = \beta = 1.0$. Both recommendations were tested on RCV1, and $\alpha = \beta = 1.0$ gave the best result. Therefore, $\alpha = \beta = 1.0$ in Equation 8 [13].

Okapi BM25 [22] is a state-of-the-art term-based model. The term weights are estimated as follows:

$$W(t) = \frac{tf\cdot(k_1+1)}{k_1\cdot((1-b)+b\frac{DL}{AVDL})+tf}\cdot\log\frac{\frac{(r+0.5)}{(n-r+0.5)}}{\frac{(R-r+0.5)}{(N-n-R+r+0.5)}}$$

where N is the total number of documents in the training set; R is the number of relevant documents in the training set; n is the number of documents that contain the term t; r is the number of relevant documents that contain the term t; tf is the term frequency; DL and $AVDL$ are the document length and average document length, respectively; and k_1 and b are the experimental parameters (the values of k_1 and b are set at 1.2 and 0.75, respectively, in this paper).

In addition, in this experiment we used the Term Frequency Inverse Document Frequency (TF-IDF) weighting scheme [23], the most widely used measure for weighting terms. $TF - IDF$ is the combination of the exhaustive statistic (TF), which stands for term frequency, and the specificity statistic (IDF) is the inverted document frequency of a term.

$$TF - IDF(t) = TF(d,t) * IDF(t)$$

5.3 Evaluation Methods

To evaluate the effectiveness of this study, different means have been used, specifically precision p , the average precision of the *top 20* return documents, the $F_1 - score$ measure, and the *break-even point (b/p)*. Also, to evaluate the whole system, *interpolated Precision on 11-points* is used for comparison of the performance of different systems by averaging precisions at 11 standard recall levels which called Interpolated Average Precision (IAP). Moreover, Mean Average Precision (MAP) is used which is the average of precision of all experiment topics. These evaluation metrics are widely used in information retrieval research (for more information about these measures see [15]).

5.4 Results and Discussion

This experiment introduce a novel method of extracting the n-gram using RFD features and extending the random set by calculating the n-gram weight considering the n-gram content distribution. Thus, this experiment consists of two stages: extracting the n-gram using different features and extending the random set of n-grams to calculate their weight.

Extract n-Grams Based on Different Feature Select Methods: In this experiment, we test different term-based methods with *tf*idf* and RFD features to select the best features between these methods. Four different extracted features were tested: *tf*idf*, RFD, MB25 and Rocchio. We found that the RFD features show a significant improvement in performance compared with different baseline models. As mentioned in section 3.2, RFD extracted three different features(positive, general, negative), in this study, we excluded the negative features to focus only on positive and general features as *tf*idf* use positive features only.

Comparing the average number of extracted features, we observe that RFD extracted a small specific number of features, about 46 keywords in positive and general features, while the *tf*idf* extracted more than 600 features. However, if we compare the performance of these two methods, we found that using RFD features to extract the n-gram yielded an excellent improvement as shown in Table 2, which illustrates an excellent improvement in performance compared with *tf*idf* and other term-based methods in all five factors over the 50 topics. Thus, using RFD features has 23% maximum and 9% minimum percentage changes on average for all used measures. Even more, if we compare the use of *tf*idf* features with other used term-based methods in their performance compared with using RFD features, we found that RFD still beat them, which means that RFD features are more specific and relevant to the user while the *tf*idf* features contain noisy features.

In summary, the experiment results in this section show that RFD features are more suitable for extracting n-grams than using other features.

Extended Random Set (ERS)to Calculate n-Gram Probability: Estimating the probability of the extracted n-gram is based on calculating the n-gram frequency. However, the distribution of the extracted n-gram content could affect the results of calculating the probability. Thus, in addition to using the RFD features to extract the n-gram, we also revise the method for calculating the probability of n-grams by extending the random set based on their distribution in the documents and their terms distribution in n-grams as shown in Equation 7.

In this experiment, we extend the weighting function and run all the methods again to show the differences. As presented in Table 1, the performance of all methods has increased, especially the Rocchio and BM25 methods. RFD features is still the best method used in this experiment with 11% maximum and 5% minimum percentage changes on average over all five measures.

Table 1. Comparison of All Term-based Methods for Assessing Topics with an Extended Random Set (ERS)

Method	*top-20*	*b/p*	*MAP*	$F_{\beta=1}$	*IAP*
RFD+ERS	**0.539**	**0.467**	**0.484**	**0.460**	**0.506**
*tf*idf*+ERS	0.484	0.425	0.444	0.437	0.467
Rocchio+ERS	0.524	0.442	0.461	0.449	0.482
BM25+ERS	0.521	0.437	0.462	0.450	0.484
%change	**+11%**	**+10%**	**+9%**	**+5%**	**+8%**

In the overall results, as shown in Table 2, using the extended random set to accurately weight n-grams with RFD features improves the performance of the extracted n-gram significantly. Table 2 shows 29% maximum and 12% minimum percentage changes in the results. In addition, Figure 1 illustrates the 11 points, which indicates the improvement in performance between the proposed method and the other methods.

In summary, extending the random set to consider the n-gram's terms based on their distribution in the documents and their terms distribution in n-grams with RFD features would significantly improve the extraction performance of the n-gram compared with other term-based methods.

Table 2. Comparison of All Term-based Methods for Assessing Topics

Method	*top-20*	*b/p*	*MAP*	$F_{\beta=1}$	*IAP*
RFD+ERS	**0.539**	**0.467**	**0.484**	**0.460**	**0.506**
RFD	0.515	0.444	0.462	0.451	0.482
*tf*idf*	0.419	0.396	0.398	0.412	0.423
Rocchio	0.444	0.340	0.406	0.420	0.434
BM25	0.449	0.400	0.406	0.419	0.432
%change	**+29%**	**+18%**	**+22%**	**+12%**	**+19%**

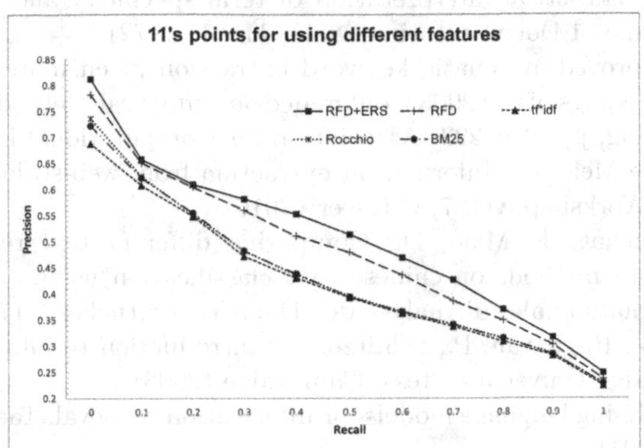

Fig. 1. 11's point for n-gram method on all assessing topics with features of other methods

6 Conclusion

This paper presents a new method for enhancing n-gram extraction in two stages. The first stage is to extract the n-gram using different features in various term-based methods. The second stage is to extend the random set to weight the extracted n-gram accurately.

The proposed method was also tested in a standard data collection (RCV1) for 50 TREC topics and compared with four up-to-date baseline models. The experimental results show that the proposed method can significantly enhance the n-gram extraction in the two stages. More than 14% percentage changes on average of of the five measures if RFD features are used. It also shows that the proposed method can significantly improve the performance when we extend the probability function for n-gram to consider both their distribution in the documents and their terms distribution in n-grams (the average percentage change is 20% for five measures).

References

1. Wang, X., Fang, H., Zhai, C.: A study of methods for negative relevance feedback. In: Proceedings of the 31st Annual International ACM SIGIR Conference on Research and Development in Information Retrieval, pp. 219–226. ACM (2008)

2. Wang, X., McCallum, A., Wei, X.: Topical n-grams: Phrase and topic discovery, with an application to information retrieval. In: Seventh IEEE International Conference on Data Mining, ICDM 2007, pp. 697–702. IEEE (2007)
3. Liu, H., Motoda, H., Setiono, R., Zhao, Z.: Feature selection: An ever evolving frontier in data mining. In: Proc. The Fourth Workshop on Feature Selection in Data Mining, vol. 4, pp. 4–13 (2010)
4. Li, Y., Zhong, N.: Mining ontology for automatically acquiring web user information needs. IEEE Transactions on Knowledge and Data Engineering 18(4), 554–568 (2006)
5. Berry, M.W., Kogan, J.: Text mining: applications and theory. Wiley (2010)
6. Jones, K.S.: A statistical interpretation of term specificity and its application in retrieval. Journal of Documentation 28(1), 11–21 (1972)
7. Hulth, A.: Improved automatic keyword extraction given more linguistic knowledge. In: Proceedings of the 2003 Conference on Empirical Methods in Natural Language Processing, pp. 216–223. Association for Computational Linguistics (2003)
8. Tandon, N., de Melo, G.: Information extraction from web-scale n-gram data. In: Web N-gram Workshop, vol. 7, Citeseer (2010)
9. Wei, Z., Chauchat, J., Miao, D.: Comparing different text representation and feature selection methods on chinese text classification using character n-grams. Journées Internationnales d'Analyse des Données Textuelles, 1175–1186 (2008)
10. Manning, C.D., Raghavan, P., Schütze, H.: Introduction to information retrieval, vol. 1. Cambridge University Press, Cambridge (2008)
11. Hiemstra, D.: Using language models for information retrieval. Taaluitgeverij Neslia Paniculata (2001)
12. Wang, K., Thrasher, C., Viegas, E., Li, X., Hsu, B.j.P.: An overview of microsoft web n-gram corpus and applications. In: Proceedings of the NAACL HLT 2010 Demonstration Session, pp. 45–48. Association for Computational Linguistics (2010)
13. Li, Y., Algarni, A., Zhong, N.: Mining positive and negative patterns for relevance feature discovery. In: Proceedings of the 16th ACM SIGKDD International Conference on Knowledge Discovery and Data Mining, KDD 2010, pp. 753–762. ACM, New York (2010)
14. Wu, S.T.: Knowledge discovery using pattern taxonomy model in text mining. PhD thesis, Queensland University of Technology (2007)
15. Liu, B.: Web data mining: exploring hyperlinks, contents, and usage data. Springer (2007)
16. Wei, Z., Miao, D., Chauchat, J.H., Zhao, R., Li, W.: N-grams based feature selection and text representation for chinese text classification. International Journal of Computational Intelligence Systems 2(4), 365–374 (2009)
17. Guyon, I., Elisseeff, A.: An introduction to variable and feature selection. The Journal of Machine Learning Research 3, 1157–1182 (2003)
18. Fürnkranz, J.: A study using n-gram features for text categorization. Austrian Research Institute for Artifical Intelligence 3(1998), 1–10 (1998)
19. Bertolami, R., Bunke, H.: Integration of n-gram language models in multiple classifier systems for offline handwritten text line recognition. International Journal of Pattern Recognition and Artificial Intelligence 22(07), 1301–1321 (2008)
20. Li, Y.: Extended random sets for knowledge discovery in information systems. In: Wang, G., Liu, Q., Yao, Y., Skowron, A. (eds.) RSFDGrC 2003. LNCS (LNAI), vol. 2639, pp. 524–532. Springer, Heidelberg (2003)

21. Joachims, T.: A probabilistic analysis of the rocchio algorithm with tfidf for text categorization. Technical report, DTIC Document (1996)
22. Robertson, S., Soboroff, I.: The trec 2002 filtering track report. In: Text REtrieval Conference (2002)
23. Salton, G., Buckley, C.: Term-weighting approaches in automatic text retrieval. Information processing & management 24(5), 513–523 (1988)

Generating Context Templates for Word Sense Disambiguation

Samuel W.K. Chan

The Chinese University of Hong Kong
Hong Kong SAR
swkchan@cuhk.edu.hk

Abstract. This paper presents a novel approach for generating context templates for the task of word sense disambiguation (WSD). Context information of an ambiguous word, in form of feature vectors, is first classified into coarse-grained semantic categories by topic features using the latent dirichlet allocation (LDA) algorithm. To further refine the sense tags, all feature vectors of the ambiguous word, under the same topic, are recast into a network. Various centrality measures are derived to figure out the features or context words in the context templates, which are highly influential in the disambiguation. The WSD is achieved by identifying the maximum pairwise similarities between the context encoded in the templates and the sentence. The correct sense of an ambiguous word is resolved by distinguishing the most activated template without being trapped in a subjective linguistic quagmire. The approach is assessed in a corpus of more than 1,000,000 words. Experimental result shows the best measures perform comparably to the state-of-the-art.

Keywords: Sense tagging; network-based approach; latent dirichlet allocation.

1 Introduction

Word sense disambiguation (WSD) tackles the problem of sense tagging and is perhaps one of the most challenging tasks in the area of natural language processing. It is also recognized as one of the foremost steps in sentence parsing [1,11]. Latest developments in WSD have been beneficial from the availability of large scale lexicons or sense-tagged corpus. A wide variety of techniques have been suggested, ranging from supervised methods in which a classifier is heavily trained for each distinct word in a sense-tagged corpus, to completely unsupervised methods that cluster occurrences of words in order to induce their senses. Although most current supervised approaches outperform their unsupervised counterparts, the importance of lexical semantic resources in WSD is well recognized. Even large lexical databases, such as WordNet [7], do not include all the words encountered in broad-coverage NLP applications. Meanwhile, the quality of these resources depends certainly, to a large degree, on the considerable efforts of lexicographers, who must keep pace with both language evolution and knowledge development. As a result, updating the resources, both lexicons and sense-tagged corpora, is an expensive and labor-intensive endeavor.

S. Cranefield and A. Nayak (Eds.): AI 2013, LNAI 8272, pp. 466–477, 2013.

On the other hand, words rarely work alone, but rather appear in tandem, forming various kinds of clusters or bundles. The recurrence of these bundles represents a meaningful chunk in a language, also called a *listeme*, that has to be memorized as part of a list for that particular context [6]. Based on collocation and colligation, words with similar meanings tend to be bundled with the same words to form similar listemes. In other words, given two words, the more comparable their collocations of the words in a listeme, the higher similarity the meanings of the two words in the context. In this paper, we propose a semi-supervised approach with less demand of lexical resources and address the WSD problem by clustering the context vectors of ambiguous words using a network-based model. Topic features are first constructed from an unlabeled corpus using the latent Dirichlet allocation (LDA) algorithm. Under the bag-of-word assumption, the algorithm produces coarse-grained clusters for different senses of the ambiguous words. Different network-based models are explored to refine the clusters and preserve the word order as well as the shallow semantic knowledge which are the prominent linguistic devices for sentence interpretation in all languages. Specifically, various measures of network centrality will be compared and contrasted. Our experiments try to attest these centrality measures are competent to discriminate the senses. We also attempt to unveil the relative contribution of each context word during the WSD. The contribution of this paper is three-fold. First, we demonstrate the LDA algorithm in topic modeling can significantly segregate different senses of an ambiguous word using our proposed context vectors. Second, we suggest a framework to fine-grain the sense tags using a network-based model. Third, we conduct an empirical experiment to compare a broad range of network centralities in the WSD problem. The organization of the paper is as follows. In Section 2, we first provide a review of the related work. The system architecture is also outlined. We then describe, in Section 3, the construction of context feature vectors as well as the topic modeling using the LDA which is an unsupervised technique. The technique produces a coarse-grained sense classification by imposing necessary, even not sufficient, constraints on their features. Section 4 presents in detail the network construction and different measures of centrality. Context templates for different senses of ambiguous words are generated. The templates unfold the fine-grained sense tags of ambiguous words by relying on the most prominent words with large centralities. In order to demonstrate the capability of the approach, the system is experimentally evaluated using an unlabeled corpus of more than 1,000,000 words. Every context template produced is calibrated with a sense tagged mini-corpus of 200,000 words. The detailed results are given in Section 5, followed by a conclusion.

2 Related Work and System Architecture

Primarily, all WSD methods can be sub-divided into two categories, namely supervised and unsupervised techniques. One of the state-of-the-art supervised WSD methods is the SenseLearner which uses a relatively small amount of training data [13]. Several semantic models are constructed for all predefined word categories. The word categories are defined as groups of words that share some common syntactic or semantic properties. First, the models are trained using the TIMBL memory based learning algorithm. It then makes generalizations of concepts learned from the training data. The best result reported is 66% on Senseval-2 data. GAMBL is another supervised approach using so-called word experts [5]. The word expert module consists of two cascaded memory-based classifiers: the sense predicted by the first classifier is used as a feature in the second classifier. The first classifier is trained on keywords selected according to a statistical criterion, and the second one is trained on the prediction of the first and on the local context of the ambiguous word-lemma-POS-tag combination. Genetic algorithm is deployed to optimize local context features and the output of a separate keyword classifier. The system incorporates both grammatical relations and chunk features into their learning. The best result reported is 65.2% on Senseval-3 data. Supervised WSD methods always outperform their unsupervised counterparts, in the expense of requiring a lexicon with high integrity as well as a reasonable size of sense tagged corpus, which may be difficult to come by in most other languages. Recently, unsupervised techniques have gained momentum, simply because the accuracy gap between the two major techniques gets closer. An example of unsupervised WSD model is the Structural Semantic Interconnections [14]. Their approach is to disambiguate words by identifying the sense with the highest similarity with its context, i.e., the senses of the words surrounding the current word. The context of a word can be the sentence that the word appears in, but also the paragraph or document in which the word is used. They introduce a graph construction method in which all candidate senses are connected and consequently ranked using network algorithms. Buoyed by the WordNet sense inventory [7], the graph is expanded by adding semantic edges and nodes from the thesaurus. Their approach is to maximize the degree of mutual interconnection among a set of senses defined under the WordNet. The WSD is determined by ranking each node in the network according to its importance. Tsatsaronis *et al.* [17] conduct an experiment in like manner using P-Rank which has an assumption that two nodes in an information network are similar if they are referenced by similar nodes.

In this paper, inspired by the work above, we propose and implement a mechanism to WSD based on one important concept: *distributionally similarity*. Distributional similarity suggests the more semantically similar two words are, the more distributionally similar they will be, and thus the more that they will tend to occur in similar linguistic contexts [4]. In other words, words that occur within similar neighbors are semantically similar. The architecture of the system is shown in Figure 1. We first extract all the context feature vectors of ambiguous words from a corpus after the text preprocessing. The vectors are subject to the topic classification under the LDA. The algorithm produces a coarse categorization and assigns a topic to each feature vector. In other words, words with the same bag of features will be delegated into the same topic. However, words in a piece of text are certainly not random. Their senses are highly influenced by their sequential order. At the same time, words with mono-sense always provide a good hint for the WSD. In addition, not all features in the vector play the same role or contribution in identifying the sense. It is mandatory to provide a mechanism to represent this sequential, well-connected and coherence characteristics among the features in the vector. We recast all feature vectors of the ambiguous word w, under the topic t, into networks and derive various centrality measures to figure out the words or features which are highly influential in the disambiguation using network analysis. As a result, features with large connectivity, or centrality, are distinguished and they are imperative for the sense tagging. The resulting network representation, or context template CT_w^t, are then used to disambiguate any unseen polysemous words. The context templates function as coalitions of concepts which can pinpoint the meaning of the words corporately through the saliency and coherence of its adjacent neighbors.

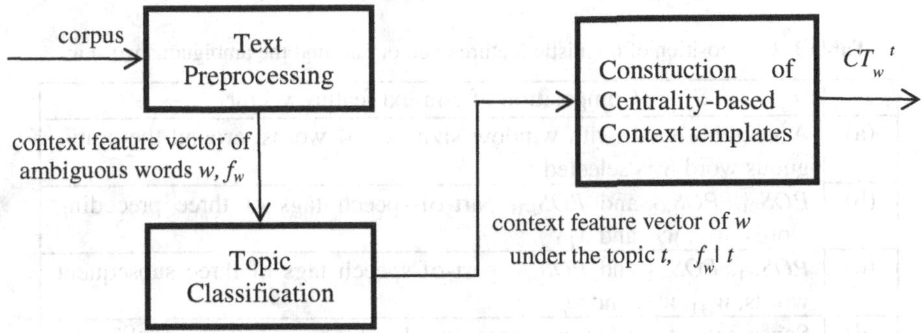

Fig. 1. System architecture for constructing context templates for WSD

3 Identifying Different Context of an Ambiguous Word

In this section, we first explain the linguistic features that are extracted for differentiating the context of the ambiguous words and then present a brief review of the latent dirichlet allocation (LDA) algorithm.

3.1 Text Preprocessing and Linguistic Feature Extraction

The preprocessing consists of the sequential application of two major components in a pipeline. The components include tokenization and part-of-speech (POS) tagging. The output of preprocessing is the words that contain the most important information in the sentences. In English, while punctuation marks, such as periods, may suggest the end of a sentence, they also signal the end of an abbreviation, or used in the specification of dates, times, initials, e-mail addresses or URLs. Similarly, spaces delimit English words and do not necessarily identify word boundaries, as in the case of many named entities such as *The Australian*. In other languages, such as in Chinese, the situation is even more taxing. Each Chinese morpheme or character carries meaning, new words can be simply constructed by the concatenation of morphemes, and there is no delimiter between words. As a result, the number of words in Chinese is huge. Both tokenization and POS tagging are the challenging tasks in Chinese NLP. In this research, we employ the segmenter and POS tagger developed by the Peking University [18]. At the same time, we eliminate all the rare and irrelevant terms in the sentences. While rare terms are commonly accepted as the words with occurrence frequency less than a threshold and irrelevant terms, in a stoplist, usually refer to the words with less indexing power, we adopt the *tf-idf* scheme to filter the rare terms and remove all words that do not fall under the categories of noun, verb, adjective and adverb. After the preprocessing, for each ambiguous word *w* in a corpus *C*, a set of features from its neighbor is elicited. The assumption is that the sense of any word in the sentence *S* in *C* is not independent but rather mutually related. One can model this dependence implicitly by including information about the preceding and subsequent words. Table 1 shows the composition of our context feature vector *f* extracted in the neighborhood of the ambiguous word *w* after the preprocessing.

Table 1. Composition of linguistic features vector *f* around the ambiguous word *w*

	Composition of context feature vector
(a)	A neighborhood, with window size of ± 4 words, around the ambiguous word *w* is selected.
(b)	POS_{i-1}, POS_{i-2} and POS_{i-3}, part-of-speech tags of three preceding words, w_{i-1}, w_{i-2} and w_{i-3}
(c)	POS_{i+1}, POS_{i+2} and POS_{i+3}, part-of-speech tags of three subsequent words, w_{i+1}, w_{i+2} and w_{i+3}
(d)	Sense tag of two mono-sense words, if any, which lie within the neighborhood of *w*
(e)	Pointwise mutual information in quantifying the collocation of the neighbor words of *w*

All these feature vectors are then subject to an unsupervised clustering as described in the following section.

3.2 Latent Dirichlet Allocation

Latent dirichlet allocation (LDA), first introduced by Blei *et al* [2], is a probabilistic generative model for discovering underlying topic structures of any discrete data. It has been applied extensively in text modeling as well as classification. LDA is usually represented as a hierarchical Bayesian model, in the format of a plate notation, as shown in Figure 2 [16,3]. The notation is used to represent repeated variables. The boxes in Figure 2 indicate the repeated variables. The number of iterations is shown at the right hand corner of the boxes. All the directed edges between the variables in the figure indicate the conditional dependencies between them. At the same time, the shaded and unshaded bubbles indicate observed and latent variables respectively. Given a corpus consisting of *M* documents, LDA models each document using a mixture over *K* topics, which are in turn characterized as distributions over words. Simply speaking, the outermost box represents all the variables related to a document *i*, over the topic distribution θ_i estimated from a Dirichlet prior with parameter α. The variables inside the outermost box are repeated iteratively for *M* times, as shown in the lower right hand corner of the box.

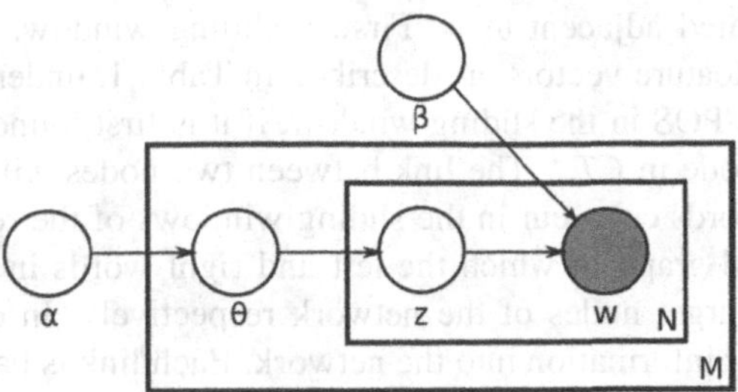

Fig. 2. Graphical model for LDA

The inner box represents the variables associated with each of the words in document i. A topic z_{ij} is first drawn from a multinomial distribution for w_{ij}, the j-th word in document i with the parameter θ_i. The word w_{ij} not only relies on the topic z_{ij}, but also a parameter β that is another uniform Dirichlet prior on the per-topic word distribution. During the inference, while it is intractable to solve the posterior distribution of the hidden variables, say the topic z_{ij} for all the word w_{ij}, Blei *et al* [2] has shown a lower bound on the log likelihood of the probability, $p(\theta,\mathbf{z}|\mathbf{w},\alpha,\beta)$, can be estimated. Interested readers can refer to the literature above for the detailed mathematical formulation and discussion. Instead of identifying the latent topic of a document as in other applications, all context feature vectors of the ambiguous word w are then subject to the classification using the LDA algorithm. LDA models each features vectors using a mixture over K topics, in the assumption that the maximum number of possible senses of any ambiguous word w is limited to K. The LDA algorithm also produces $p(f_i|z_j)$ which represents the probability of a feature vector f_i given a topic z_j. Basically, under each topic, the algorithm congregates all feature vectors that tend to occur in similar linguistic contexts and they are distributionally similar in meaning. The extraction of the context feature vectors from a large unlabeled corpus, as shown below, and the unsupervised clustering try to relieve the scarcity problem induced from any manually sense tagged corpus.

4 Construction of Context Templates Using Network Centrality

The LDA approach is akin to the standard bag of words model and has a strong assumption of exchangeability. That is, the words in a document are exchangeable. A document $\{w_1, w_2, ..., w_N\}$ is exchangeable if the joint distribution w_i is invariant to any possible permutations. In this research, we alleviate the strong assumption imposed by LDA into WSD since word order obviously has a pivotal role and is the most fundamental syntactic device in any language understanding. We propose a network approach which could incorporate both the word order information and shallow semantic knowledge into the WSD. The mechanism is described as follows.

4.1 Construction of the Network

Given an ambiguous word w, all context feature vectors under the topic t are used to construct a network, or called context template $CT_w^{\,t}$. The nodes in the network are the neighbor words located adjacent to *w*. First, a sliding window, size equal to three, slips through every feature vectors, as described in Table 1, under the topic. For each word tagged with its POS in the sliding window, if it is first found in the vectors, it is recorded as a new node in CT_w^t. The link between two nodes will only be reinforced whenever the two words co-occur in the sliding windows of the relevant vectors. The network is a directed graph in which the left and right words in the sliding window are the source and target nodes of the network respectively. In other words, we inscribe the word order information into the network. Each link is based on the proximity of the words in the feature vectors. The strength of the link between two nodes relies on how frequent the two words can be found in the sliding window across all feature vectors under the topic *t*. This approach translates the linear structure of the feature vectors into a network and allows us to visualize the proximity of the words by means of the intensity of the connections between the nodes in the network. In addition, shallow semantic knowledge could also be amalgamated into the network as described below.

4.2 Inclusion of Shallow Semantic Knowledge

While the collocations of words are being secured in above sliding window, the number of nodes in the network can be enormous. As a result, the main gist underlined in the network can be easily disregarded. Not every word in a sentence is ambiguous. As the statistics shown in Section 5, more than 90% of words in a sense tagged corpus have a unique sense. These mono-sense words certainly provide the anchor points in the WSD. The network will be restructured using the following criteria:

- Words with the same mono-sense are collapsed into a single sense node in the above network construction. The strength of the link between the sense node to other nodes is the sum of the links from all the individual words under the sense.
- All other mono-sense words with high semantic similarity will also be merged into a single sense node as above. However, the link will suffer from a depreciation of its strength.
- Jiang & Conrath [9] approach is adopted to measure semantic similarity/distance between words in the network. It combines a lexical taxonomy structure with corpus statistical information. The semantic similarity between nodes in the network is quantified with the computational evidence derived from a distributional analysis of corpus data as well as the taxonomy.

During the implementation, the network is not constructed in two different phases as perceived. The description in two separate sections only serves for clarification purpose. In sum, in lieu of using the LDA inference during the prediction for any unseen feature vectors, our aim is to generate, for each ambiguous word *w*, the context templates that anticipate the most appropriate sense of *w*. The network demonstrates all the possible collocations with some minimal senses, without a demand to manually tag the sense for each word in a large corpus. Even at this stage, the network may be too enormous to yield any conclusive predictions for the WSD. In this research, we take advantage of the ranking of each node in the CT_w^t according to its centrality, in the hope to unveil the relative contribution of the context words or sense nodes in the context template. In next section, we discuss in general several measures that operationalize centrality in graph-theoretic terms. We introduce some global measures which estimate the overall degree of connectivity of the network and explain how they can be exerted to identify the main gist of the context templates.

4.3 Measures of Network Centrality

We outline the general concepts of some measures of centrality of a network. Interested readers should refer more details in the literature [15].

Degree Centrality: Perhaps the simplest centrality measure in a network is just the degree of a node, the number of edges connected to it. Degree centrality of a node refers to the number of edges attached to the node. In directed networks, nodes have both an in-degree and an out-degree, and both may be useful as measures of centrality in the appropriate circumstances. In order to know the standardized score, we usually divide each score by n-1 where n is equal to the number of nodes.

Eigenvector Centrality and its Derivative: A natural extension of the simple degree centrality is eigenvector centrality which gives each node a score proportional to the sum of the scores of its neighbors. The centrality of the nodes can be represented in a matrix notation, $\mathbf{Ax} = k\mathbf{x}$ where $\mathbf{x}$ is the vector with element centrality x_i of each node i, $\mathbf{A}$ is the adjacency matrix of the network and k is the largest eigenvalue of $\mathbf{A}$. One of its most popular derivatives is the PageRank (PR) which is a link analysis algorithm, used by the Google web search engine. Different from the eigenvector centrality, PR includes an additive constant term in their definition and normalized by dividing by the out-degrees of its neighboring nodes. The PR centrality of the nodes can be represented in a matrix notation, $\mathbf{x} = \mathbf{D}(\mathbf{D} - \alpha\mathbf{A})^{-1}\mathbf{1}$, where $\mathbf{D}$ is a diagonal matrix with elements $D_{ii} = \max(K_i^{out}, 1)$, $\mathbf{1}$ being the vector (1,1,1, ..), and α is the damping factor.

HITS: There are two other different types of centrality for directed networks, the authority centrality and the hub centrality. The authority centrality indicates how often a node is being pointed by other hubs. Similarly, a node is a high hub centrality if it points to many nodes with high authority centrality. That is, a good hub is a node that points to many good authorities, whereas a good authority is a node that is pointed by many good hubs. The algorithm is also called as *hyperlink-induced topic search* or HITS. The centralities can be represented in matrix notations, $\mathbf{AA}^T\mathbf{x} = \lambda\mathbf{x}$, $\mathbf{A}^T\mathbf{Ay} = \lambda\mathbf{y}$, where $\mathbf{A}$ is the adjacency matrix of the network, $\mathbf{x}$, $\mathbf{y}$ being vectors which represent the authority and hub centrality of each node respectively. In other words, the centralities are given by eigenvectors of $\mathbf{AA}^T$ and $\mathbf{A}^T\mathbf{A}$ with the same eigenvalue λ.

Betweenness Centrality: An entirely different approach to centrality is betweenness centrality, which measures how often a node appears on shortest paths between nodes in the network. The idea of betweenness is usually attributed to Freeman [8]. Nodes with high betweenness centrality may have considerable influence within a network by virtue of their control over information passing between others. The nodes with highest betweenness are also the ones which will produce the most disaster effect in communication when they are removed from the network. It is a popular centrality measure used in the study of social networks. Mathematically, the betweenness centrality x_i of a node i are defined as:

$$x_i = \sum_{st} \frac{n_{st}^i}{g_{st}} \tag{1}$$

where n_{st}^i be the number of shortest paths from node s to t that pass through i and g_{st} be the total number of shortest paths that can be found from node s to t.

The context template of an ambiguous word w is a network in which the nodes of the network are the words or sense nodes at the adjacency of w under the same topic. Each node has its own centralities which are best described using various network-based centrality measures. For an unseen feature vector, its sense is resolved by the template which will return the maximum centrality score for all the words found in the feature vector.

5 Empirical Experimental Setup and Results

A large and accurate sense-tagged corpus provides a reliable resource for all kinds of computational linguistics research. Unfortunately, the construction of a large-scale Chinese sense tagged corpus is still underway [10]. In this empirical experiment, we employed a sense tagged Chinese mini-corpus that is originally from Harbin Institute of Technology. It contains more than 200,000 Chinese words in which POS and sense are manually tagged. The sense tags from a Chinese thesaurus called *Cilin* are adopted [12]. The number of senses of the words in the corpus is shown in Table 2.

Table 2. Percentage of ambiguous words in a sense-tagged corpus with 200,000 words

# of senses	% of words in the corpus
1	90.33%
2	6.94%
3	1.80%
4	0.60%
5	0.17%
≥6	0.16%

In our experiment, the words with more than one sense in the mini-corpus are all shortlisted. However, the number of feature vectors from the mini-corpus is far from sufficient to bring about the context templates. Instead, an unlabeled corpus of more than 1,000,000 words is used to produce the feature vectors and develop the context templates. The unlabeled corpus is then subject to the preprocessing as discussed in Section 3.1. During the preprocessing, we just point to the words with their POS tags that fall under the categories of noun, verb, adjective or adverb. A neighborhood, with window size of ± 4 words, around the ambiguous word is used to devise the feature vectors. As the result, the total number of context feature vectors produced from the unlabeled corpus and mini corpus are 93,500 and 19,000 respectively. All the feature vectors from the unlabeled corpus are used to yield the context templates as described in Sections 3.2 and 4. The feature vectors are first subject to the LDA algorithm. The choice of topic number K in the model can affect the interpretability of the results. A model with fewer topics may have a poor generalization while it may be over-fitting if the number of topics is too large. Another approach to estimation of the number is the measure of perplexity which is equivalent to the inverse of the geometric mean per-word likelihood [2]. A lower perplexity score indicates better generalization performance. Different from all previous topic model approaches that identify the hidden topics in a collection of documents, our intention is to differentiate the senses of an ambiguous word using their feature vectors. In this empirical study, we take a simple assumption that there are at most three different senses for each ambiguous word. All feature vectors of the word are then subject to train using 10,000 iterations of the Gibbs sampling. Hyper-parameter optimization, which allows the LDA model to better fit the vectors by allowing some topics to be more prominent than others, is also applied. The feature vectors are assigned to the topics with largest topic proportions under the iterations. All feature vectors under the LDA-assigned topics are used to construct the context templates under different centrality measures. As a result, more than 9,850 templates are produced from the unlabeled corpus for all 3,300 ambiguous words found in the mini-corpus. Feature vectors from the mini-corpus are reserved for the calibration and the test purpose. To evaluate the performance of the context templates, we reserve 85% of the feature vectors generated from the mini-corpus, with all known sense tags, for the calibration and the remaining 15% for the test purpose. The calibration is accomplished by feeding the feature vectors to the context templates. The known sense tag of the feature vector from the mini-corpus will be assigned to the template which is activated most. Similarly, during the test, the sense of an ambiguous word will be resolved by the context template with highest activation. Five centrality scores of the reserved 15% feature vectors for testing are recorded.

All network-based algorithms are compared against a naive baseline model that selects the most frequent sense of the ambiguous word in the mini-corpus. Our empirical results are summarized in Table 3 which reports the performance on ambiguous words only. The table demonstrates the accuracy of the six methods for four major POS tags in the feature vectors. As can be seen in the table, all the centrality approaches perform much better than the baseline model. The approaches have at least 10% increases in the overall performance and the differences are all significant. All the eigenvector approaches, described in Section 4.3, come by the overall accuracy more than 50%. The high similarity in their overall performances may be due to they all come from the same family. The HITS method, among all others, shows an impressive gain in accuracy close to 60%. While the performance between the HITS and PageRank is not significant difference, PageRank performs consistently well across all the POS tags. In fact, it is the best approach to disambiguate verbs which are notoriously difficult in WSD. Verbs usually have more polysemous than other POS. However, the betweenness centrality does not seem to perform equally well against the rest in sense tagging, even though it is a popular measure used in most social network analysis.

Table 3. Performance of different centralities in various POS tags during the testing of WSD

	Noun	Verb	Adj.	Adv.	Overall
Baseline model	25.8%	16.3%	27.7%	29.3%	22.80%
Degree	60.6%	36.2%	57.9%	62.2%	51.77%
Eigenvector	61.6%	37.2%	58.3%	66.4%	52.83%
HITS	**69.8%**	**42.2%**	**64.4%**	**69.7%**	**59.43%**
PageRank	66.5%	45.7%	61.5%	75.2%	58.90%
Betweenness	34.6%	28.1%	35.1%	37.7%	32.47%

6 Conclusion

In this paper, we have proposed a semi-supervised approach for the WSD with less demand of manually tagged lexical resources. The approach first makes use of a large, but unlabeled, corpus to generate the context templates and, subsequently, calibrate with a mini sense tagged corpus. This semi-supervised approach bridges the gap between the unsupervised and supervised paradigms in WSD. We have conducted an empirical experiment to attest the approach is competent to discriminate the senses. Certainly, further research should be investigated on how to uncover the ambiguous words that are present in the unlabeled corpus, but absent in the mini sense tagged corpus. Although we conduct the empirical experiment using a Chinese corpus, this does not mean the idea can only be applied in the language. The approach is applicable to all other languages.

Acknowledgement. The work described in this research was partially supported by the grants from the Research Grants Council of the Hong Kong Special Administrative Region, China (Project Nos. CUHK440609 and CUHK440913).

References

1. Agirre, E., Bengoetxea, K., Gojenola, K., Nivre, J.: Improving dependency parsing with semantic classes. In: Proceedings of the 49th Annual Meeting of the Association for Computational Linguistics: Human Language Technologies, ACL HLT 2011, Portland, pp. 699–703 (2011)
2. Blei, D., Ng, A., Jordan, M.: Latent Dirichlet allocation. Journal of Machine Learning Research 3, 993–1022 (2003)

3. Cai, J.F., Lee, W.S., Teh, Y.W.: Improving word sense disambiguation using topic features. In: Proceedings of the 2007 Joint Conference on Empirical Methods in Natural Languages Processing and Computational Natural Language Learning, pp. 1015–1023 (2007)
4. Dagan, I., Lee, L., Pereira, F.: Similarity-based models of word co-occurrence probabilities. Machine Learning Journal 3, 1–3, 43–69 (1999)
5. Decadt, B., Hoste, V., Daelemans, W., van den Bosch, A.: GAMBL, Genetic Algorithm Optimization of Memory-Based WSD. In: SENSEVAL-3: Third International Workshop on the Evaluation of Systems for the Semantic Analysis of Text (2004)
6. Di Sciullo, A.M., Williams, E.: On the Definition of Word. In: Linguistic Inquiry Monograph, vol. 14, MIT Press, Cambridge (1987)
7. Fellbaum, C.: WordNet: An Electronic Lexical Database. MIT Press, Cambridge (1998)
8. Freeman, L.C.: Centrality in social networks conceptual clarification. Social Networks 1, 215–239 (1977)
9. Jiang, J.J., Conrath, D.W.: Semantic Similarity Based on Corpus Statistics and Lexical Taxonomy. In: Proceedings of International Conference on Research in Computational Linguistics, pp. 19–33. International Committee on Computational Linguistics (1997)
10. Ker, S.-j., Huang, C.-R., Hong, J.-F., Liu, S.-Y., Jian, H.-L., Su, I.-L., Hsieh, S.-K.: Design and Prototype of a Large-scale and Fully Sense-tagged Corpus. In: Tokunaga, T., Ortega, A. (eds.) LKP 2008. LNCS (LNAI), vol. 4938, pp. 186–193. Springer, Heidelberg (2008)
11. Mackinlay, A., Dridan, R., Mccarthy, D., Baldwin, T.: The effects of semantic annotations on precision parse ranking. In: Proceedings of the First Joint Conference on Lexical and Computational Semantics (*SEM 2012), Montreal, pp. 228–236 (2012)
12. Mei, J., Zhu, Y., Gao, Y., Ying, H.: Tongyici Cilin. Commercial Press (1984) (in Chinese)
13. Mihalcea, R., Csomai, A.: SenseLearner: word sense disambiguation for all words in unrestricted text. In: Proceedings of the ACL 2005 on Interactive Poster and Demonstration Sessions, pp. 53–56 (2005)
14. Navigli, R., Lapata, M.: Graph connectivity measures for unsupervised word sense disambiguation. In: Proceedings of IJCAI, pp. 1683–1688 (2007)
15. Newman, M.: Networks: An Introduction. Oxford (2011)
16. Steyvers, M., Griffiths, T.: Probabilistic Topic Models. In: Landauer, T., Mcnamara, D., Dennis, S., Kintsch, W. (eds.) Handbook of Latent Semantic Analysis (2007)
17. Tsatsaronis, G., Varlamis, I., Nørvåg, K.: An experimental study on unsupervised graph-based word sense disambiguation. In: Gelbukh, A. (ed.) CICLing 2010. LNCS, vol. 6008, pp. 184–198. Springer, Heidelberg (2010)
18. Wu, Y., Jin, P., Guo, T., Yu, S.: Building Chinese sense annotated corpus with the help of software tools. In: Proceedings of the Linguistic Annotation Workshop. ACL, Prague (2007)

Evolving Stochastic Dispatching Rules for Order Acceptance and Scheduling via Genetic Programming

John Park, Su Nguyen, Mark Johnston, and Mengjie Zhang

Evolutionary Computation Research Group
Victoria University of Wellington, P.O. Box 600, Wellington, New Zealand
{parkjohn,Su.Nguyen,Mengjie.Zhang}@ecs.vuw.ac.nz,
Mark.Johnston@msor.vuw.ac.nz

Abstract. This paper focuses on Order Acceptance and Scheduling (OAS) problems in make-to-order manufacturing systems, which handle both acceptance and sequencing decisions simultaneously to maximise the total revenue. Since OAS is a NP-hard problem, several heuristics and meta-heuristics have been proposed to find near-optimal solutions in reasonable computational times. However, previous approaches still have trouble dealing with complex cases in OAS and they often need to be manually customised to handle specific OAS problems. Developing effective and efficient heuristics for OAS is a difficult task. In order to facilitate the development process, this paper proposes a new genetic programming (GP) method to automatically generate dispatching rules to solve OAS problems. To improve the effectiveness of evolved rules, the proposed GP method incorporates stochastic behaviours into dispatching rules to help explore multiple potential solutions effectively. The experimental results show that evolved stochastic dispatching rules (SDRs) can outperform the tabu search heuristic especially customized for OAS. In addition, the evolved SDRs also show better results as compared to rules evolved by the simple GP method.

1 Introduction

Order Acceptance and Scheduling (OAS) is an important planning activity within make-to-order manufacturing systems. OAS often occurs in manufacturing systems with limited capacities while the customer demand is high. In this situation, these manufacturing systems cannot accept all customer orders. The goal of OAS is to determine the set of accepted orders and decide how these accepted orders can be processed in order to effectively utilise the available capacity. This paper focuses on the OAS problem in a single machine environment with sequence dependent setup times [2,10,11]. In this problem, we need to determine which orders within the n customer orders are accepted (must be processed and delivered) and how the accepted orders are scheduled to maximise the total obtained revenue. Each order j is characterised by a release time r_j, a processing time p_j, a due date d_j, a weight/penalty w_j, a maximum revenue e_j,

S. Cranefield and A. Nayak (Eds.): AI 2013, LNAI 8272, pp. 478–489, 2013.

and a deadline $\bar{d}_j$. A specific setup time s_{ij} for order j is incurred if order j is processed immediately after order i ($i = 0$ if order j is processed first). When an order j is accepted, the manufacturing system will gain a maximum revenue e_j. However, if the order j is delivered after its due date d_j, a penalty $w_j T_j$ will occur, where $T_j = \max(0, C_j - d_j)$ is the tardiness and C_j is the completion time of order j. The actual total obtained revenue or profit from an order j is $rev_j = e_j I_j - w_j T_j$ in which I_j is 1 if order j is accepted; and 0 otherwise. If orders are finished after their deadline $\bar{d}_j$, no revenue is gained ($rev_j = 0$). In this paper, we try to maximise the total obtained revenue $\sum_{j \in A}(rev_j)$, where A is the set of accepted orders.

Due to its practical applications and computational challenges, OAS has been studied over the past two decades in the literature. Many optimisation methods have been proposed for OAS problems. Ghosh [6] showed that the OAS problem is NP-hard, and proposed pseudo-polynomial time and approximation methods for specific instances of the problem. Slotnick and Morton [16] developed a branch-and-bound algorithm to find an exact solution for OAS in a single machine environment with static arrival times, and also proposed two heuristics for this problem. The two heuristics were significantly faster than the branch-and-bound algorithm, although they performed poorly for certain instances. Rom and Slotnick [13] proposed a hybrid genetic algorithm with a local search heuristic to handle the same problem and showed very promising results. Some studies have also focused on OAS with sequence dependent setup times [2,10,11]. Oguz et al. [11] developed a simulated annealing based method (ISFAN) and showed that it can effectively solve large-scale OAS problem instances that cannot be solved by mixed-integer linear programming (MILP). Cesaret et al. [2] developed a tabu search (TS) heuristic that outperforms the ISFAN algorithm in many different scenarios. Nguyen et al. [10] developed a multi-objective GP (MOGP) method to discover Pareto efficient scheduling rules for multi-objective OAS problems. The GP method is employed in a two-stage learning/optimising system where GP is used to evolve rules that can be reused to initialise the population of an evolutionary multi-objective optimisation method (EMO). They showed that using GP in conjunction with EMO is superior to a pure EMO method. However, their work did not consider other representations in GP to enhance the performance of evolved rules. OAS problems in the job shop environment have also been investigated by Wester et al. [17] and Roundy et al. [14]. A comprehensive review of OAS is covered by Slotnick [15].

As compared to conventional scheduling problems, OAS is more complicated because we have to deal with acceptance and sequencing decisions simultaneously. Therefore, designing effective heuristics for OAS is difficult and time-consuming. Genetic programming based hyper-heuristics (GPHH) have been recently developed to automatically generate heuristics [1] for scheduling problems [3,8,7,4,5] and achieved very promising results. Nguyen et al. [10] made the first attempt to use GP for evolving scheduling rules for OAS. However, there are two major limitations with this GP method. First, the representation and the evaluation scheme of evolved rules are the same as those employed for conventional scheduling problems, and therefore, have not taken into account special characteristics of OAS (e.g. acceptance decisions) to enhance the quality of solutions generated by evolved rules. Second, evolved rules are restrictive in that they can only return a single solution for a problem instance, which make these rules less competitive compared to the state-of-the-art meta-heuristics such as tabu search (TS) [2].

1.1 Goals

The goal for this paper is to develop a new GP method to handle the two limitations discussed above. The novelty of the new GP method is the introduction of a new evaluation scheme which helps incorporate stochastic behaviours into evolved dispatching rules to generate effective *stochastic dispatching rules* (SDRs). The new GP method (GPSR) aims to evolve rules which can intelligently sample quality solutions by embedding some randomness into the order selection process. Three research objectives in this paper are:

(a) Developing a new GP method (GPSR) to evolve SDRs for OAS.
(b) Comparing GPSR and the simple GP method for OAS.
(c) Comparing the evolved SDRs evolved by GPSR and the TS heuristic.

1.2 Organisation

The organisation of the paper is as follows. Section 2 provides details about the proposed SDRs and the GPSR method to evolve SDRs. Section 3 describes the experimental setup used to evolve and evaluate the SDRs. Section 4 presents experimental results of GPSR and compares the performance of evolved SDRs to those of simple evolved rules and the TS heuristic. Finally, Section 5 provides conclusions and future research directions.

2 Genetic Programming for Evolving Stochastic Rules

This section describes how dispatching rules can be represented by GP programs. Then, details about the evaluation scheme for SDRs are provided. Finally, the fitness function used to evaluate the quality of evolved SDRs is presented.

2.1 GP Representation

To represent SDRs, we use a tree-based GP where the non-terminals represent the operators and terminals represent parameters and constants. An example of a GP program (dispatching rule) is shown in Fig. 1. Similar to the non-terminals that were previously used to evolve rules in OAS [12], we use basic arithmetic operators $+$, $-$, $\times$, and protected division $\div$ (returns one when the denominator is zero). We also use a ternary operator `If`, where `If` returns the value of the second term if the first term is greater than or equal to zero; otherwise it returns the value of the third term. The terminals that we use are the features of an

Table 1. Terminal and function sets for SDRs

Symbol	Description	Symbol	Description
R	release time r_j	P	processing time p_j
E	maximum revenue e_j	W	penalty w_j
S	setup time s_{ij}	d	due date d_j
D	deadline $\bar{d}_j$	t	current time
#	random number from 0 to 1		
Function set	+,−,×, % (protected division), If		

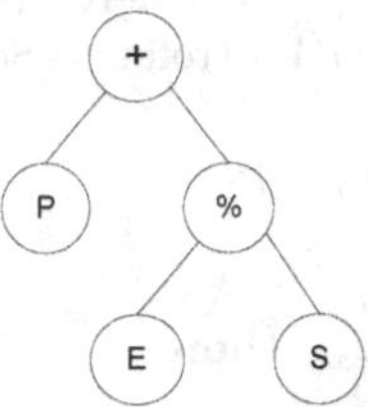

Fig. 1. An example dispatching rule P + E%S

order j such as processing time p_j or sequence dependent setup time s_{ij} (where i is the previous processed order). We also use current time t, which is the decision moment when the machine becomes idle. A complete list of terminals and functions are given in Table 1. Basically, a GP program is a priority function $f(\cdot)$. When we need to assign a priority for a particular order j, the GP program is evaluated with the terminal values extracted from order j. The output from the GP program $f(j)$ will be the priority assigned to order j.

2.2 Evaluation Scheme for Stochastic Dispatching Rule

The goal of SDRs is to generate multiple schedules instead of generating a single schedule (solutions) like simple dispatching rules. Algorithm 1 shows how SDRs can iteratively generate m schedules for a particular OAS problem instance. From each iteration (from step 3 to step 18), a new schedule S is generated. In an iteration, we start with an empty schedule S and a set of all available orders Ω, and incrementally add a new order into the schedule (from step 8 to step 14) until Ω is empty. Given a temporary set Ω, the procedure will first calculate the earliest completion time C'_j for each order in Ω. Any order j with $C'_j \geq \overline{d_j}$ will be removed from Ω because these orders are not able to help increase the total obtained revenue (step 8). Also, we are only interested in active orders in Ω (step 9) because non-active orders will result in a waste of manufacturing capacity (some orders can be completed before the release time of non-active orders). Then, the priority function $f(\cdot)$ evolved by GP is used to assign priorities for all orders in the set of active orders V (step 10). A stochastic selection scheme (described in Section 2.3) is used to select an order from V

Algorithm 1. Schedule construction procedure for a stochastic rule $f(\cdot)$

1: let S_{best} be the best schedule where initially $revenue(S_{best}) = 0$
2: **for** $count = 1$ **to** m **do**
3: let $\Omega \leftarrow \{1, 2, \ldots, n\}$ be the available orders
4: let S be the output schedule, initially empty
5: let $t \leftarrow 0$ be the current time
6: let $i \leftarrow 0$ be the previous processed order
7: **while** Ω is not empty **do**
8: remove any order with projected completion time $C'_j \geq \overline{d_j}$ from Ω
9: obtain the set of active orders $V = \{j \in \Omega : r_j < \min_{k \in \Omega} C'_k\}$
10: compute priority $f(j)$ for $\forall j \in V$ where $f(\cdot)$ is a priority function
11: let $j \leftarrow stochastic\_selection(V)$ (refer to Section 2.3)
12: append j to S
13: update $t \leftarrow \max\{r_j, t\} + s_{ij} + p_j$
14: update $i \leftarrow j$
15: **end while**
16: **if** $revenue(S) > revenue(S_{best})$ **then**
17: $S_{best} \leftarrow S$
18: **end if**
19: **end for**
20: **return** S_{best}

(step 11) to add into the schedule S (steps 12 to 14). These steps will be applied until Ω becomes empty. If the total revenue from the new schedule S is better than the best schedule S_{best}, it will replace S_{best}.

The stochastic selection scheme is the main difference between SDRs and simple dispatching rules [10,12]. In a simple dispatching rule, only the order with the highest priority will be selected to be processed next. Because of the stochastic selection scheme, SDRs can select orders with worse priorities; and therefore, they able to generate different potential schedules following the pattern governed by the priority function $f(\cdot)$.

2.3 Stochastic Selection Scheme

The stochastic selection is an important step in Algorithm 1. The key idea is that the probability of selecting an order will be a function of the priority assigned to that order. Orders with higher priorities are more likely to be selected. Assuming that the priority function can determine which orders are more suitable (higher priorities) to be processed next, this selection scheme helps explore potential schedules from combinations of potential orders selected at different decision moments.

Given a set of candidate orders V that can be selected to process next, a subset $W \subseteq V$ of orders with the highest priorities is obtained. Priority values of orders in W are then converted into their corresponding selection probabilities. Because the priority value $f(j)$ for each order j is an unbounded real number, a function g that uses the inverse tangent function, as shown by equation (1), is

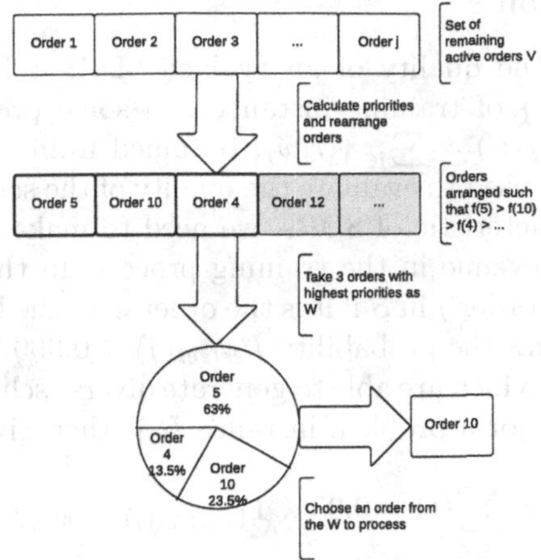

Fig. 2. An example of stochastic selection scheme with W of size 3

applied to transform the priorities into values in the fixed interval $(0, \pi)$. This ensures that higher priorities still give higher probabilities.

$$g : \mathbb{R} \rightarrow (0, \pi)$$
$$g \circ f(j) = \arctan f(j) + \pi/2 \tag{1}$$

After applying the transformation, the probability of selecting order j from W can be calculated by:

$$P_{select}(j) = \frac{\arctan f(j) + \pi/2}{\sum_{k \in W}(\arctan f(k) + \pi/2)} \tag{2}$$

The size of W is an important factor in the proposed selection scheme. If W only contains the order with the highest priority, SDRs are the same as conventional (deterministic) dispatching rules. If W contains more orders, SDRs will generate a more diverse set of schedules which depend less on the priority function $f(\cdot)$. However, if W is too large, SDRs are similar to random schedule generators because orders can be selected randomly at each decision moment. In order to achieve good results, the size of W should be chosen such that SDRs can take advantage of the priority function while exploring potential solutions.

Figure 2 gives an example to show how the stochastic selection scheme is used to select an order from W. In this example, W consists of the top three orders with the highest priorities from V. In this case, W contains orders 5, 10 and 4. Then, an order in W is randomly selected based on $P_{select}(j)$ of each order (similar to roulette wheel selection). Although order 5 has the highest probability of being selected, order 10 is chosen in the example because of the stochastic selection procedure.

2.4 Fitness Function

In order to measure the quality of an evolved SDR, it is applied to solve a set $\mathbb{I} = \{I_1, I_2, \ldots, I_K\}$ of training instances. In some previous works [10,12], the total revenue $TR_{I_k}(S) = \sum_{j \in A}(rev_j)$ obtained from a schedule S for each instance in $\mathbb{I}$ is directly used to evaluate the quality of the schedule. However, due to the stochastic characteristic of SDRs, we need to make a modification when calculating the total revenue in the training process. In this case, a penalty is applied to an accepted order j in S if it is the order with the highest priority in W (see Section 2.3) and has the probability $P_{select}(j) > 0.999$. This modification is made to promote rules which are able to generate diverse schedules. The modified total revenue $TR'_{I_k}(S)$ for a problem instance I_k is then given by:

$$TR'_{I_k}(S) = \sum_{j \in A}((1 - 0.05 \times \mathbf{1}_{\{P_{select}(j) > 0.999\}}) \times rev_j) \tag{3}$$

Because there are m schedules $S_1, S_2, \ldots, S_m$ generated by an evolved SDR, the average $\overline{TR'_{I_k}}$ of the modified total revenues obtained by all generated schedules will be used to calculate the fitness of an evolved SDR. The average $\overline{TR'_{I_k}}$ is used to assess the quality of a SDR here instead of $TR'_{I_k}(S_{best})$ from the best schedule S_{best} because we want the evolved SDR to robustly sample quality schedules rather than accidentally find a good (best) solution. After we obtain $\overline{TR'_{I_k}}$ for all training instances, the fitness value of an evolved SDR can be calculated as follows:

$$fitness = \frac{1}{|\mathbb{I}|} \times \sum_{I_k \in \mathbb{I}} dev'(I_k) \tag{4}$$

where $dev'(I_k) = (UB_{I_k} - \overline{TR'_{I_k}})/UB_{I_k}$ is the relative deviation between $\overline{TR'_{I_k}}$ and the upper bound UB_{I_k} (determined by MILP and relaxed LP [2]) of instance I_k. The fitness reflects the average performance of the evolved SDR across all training instances. Better SDRs will result in lower fitness values.

3 Experimental Design

This section describes the dataset used for training/testing and the parameter settings of GPSR and evolved SDRs.

3.1 Dataset

For training and testing, we use the dataset introduced by Oguz et al. [11]. This dataset has been used in the literature to evaluate the performance of optimisation heuristics for OAS with sequence dependent setup times [11,2,12]. The dataset is divided into different subsets by three parameters: the number of orders n, the tardiness factor τ and due date range R. With a fixed n, each subset $\langle \tau, R \rangle$ contains ten randomly generated problem instances. Release times and due dates of orders j are randomly generated such that $r_j \in [0, p_T]$ with $p_T = \sum_{i=1}^{n} p_i$ being the total processing time. Due dates are $d_j = r_j + \max_{i=0,1,\ldots,n}\{s_{ij}\} +$

Table 2. Parameters used for evolving rules

Parameter	Value
Population size	1024
Crossover rate	80%
Mutation rate	10%
Reproduction rate	10%
Generations	51
Max-depth	8
Selection method	tournament selection
Initialisation	ramped-half-and-half

$\max\{slack, p_j\}$, where *slack* is generated from $[p_T(1-\tau-R/2), p_T(1-\tau+R/2)]$. This means that instances generated with high τ and R values have very high volatility, and different decisions made early in the schedule can result in widely different total revenue values. On the other hand, early scheduling decisions do not affect the total revenue as much for instances generated with low τ and R values. We also use the maximum number of orders $n = 100$ for all experiments in this paper, as this is the most difficult category of problem instances to solve. The training sets are the first five instances of either the $\langle 0.1, 0.1\rangle$, $\langle 0.5, 0.5\rangle$ or $\langle 0.9, 0.9\rangle$ subsets. Each training instance only contains five instances in order to save the computational cost. All 250 instances (25 subsets from combinations of five values of τ and five values of R) in the dataset with $n = 100$ are used to test the performance of rules by GP.

3.2 Parameter Settings

The GP system for learning SDRs is developed based on the ECJ20 library [9]. This paper also compares the performance of GPSR and the simple GP method for OAS (GPOAS) [12]. In all experiments, both GPSR and GPOAS use the parameters in Table 2 and the terminal/function sets introduced in Section 2.1. For each GP method, 30 independent runs are performed and the best evolved rules are recorded for the comparison.

For SDRs, two important factors are the size of the set of orders W used in the stochastic selection scheme and the number of schedules m generated by SDRs, as shown in Algorithm 1. Preliminary experiments show that W of size 3 and $m = 30$ provide good results within reasonable computation times. Therefore, we will apply these two parameters for SDRs in our experiments.

4 Results

This section shows results of GPSR and compares GPSR with GPOAS. To show the effectiveness of evolved rules, we also compare the performance of a typical SDR evolved by GPSR with that from TS [2], an effective optimisation heuristic for OAS with sequence dependent setup times. The details of this TS heuristic was described in Cesaret et al. [2]. When solving an instance, the stopping condition of TS is when there is no improvement made in 50 iterations [2].

4.1 GPSR and GPOAS

Table 3 shows the performance of SDRs evolved by GPSR and dispatching rules evolved by GPOAS for each OAS subset. In order to evaluate the performance of a rule on a subset $\mathbb{D}$ of 10 instances, total revenue $TR_{I_k}(S_{best})$ obtained by the rule for each instance I_k is recorded to calculate the average relative deviation $\%dev_{avg} = 100 \times (\sum_{I_k \in \mathbb{D}} dev(I_k))/10$, where $dev(I_k) = (UB_{I_k} - TR_{I_k}(S_{best}))/UB_{I_k}$ is the relative deviation between $TR_{I_k}(S_{best})$ and the upper bound UB_{I_k}. This method is commonly used in the literature [2,10,12]. The average relative deviation is similar to the fitness used for training SDRs but the total revenue $TR_{I_k}(S_{best})$ is used in the calculation instead of the average modified total revenue $\overline{TR'_{I_k}}$.

Means and standard deviations of $\%dev_{avg}$ for each subset from evolved rules obtained by 30 independent runs of GPSR and GPOAS are presented in Table 3. In this table, column GPSR-x and GPOAS-x represents the results of SDRs and simple rules evolved by GPSR developed in this paper and GPOAS [12] with the training set $\langle x, x \rangle$. For each subset, the highlighted results indicate whether GPSR is significantly better than GPOAS. The results show that GPSR

Table 3. Performance of GPSR and GPOAS

τ	R	GPSR-0.1	GPOAS-0.1	GPSR-0.5	GPOAS-0.5	GPSR-0.9	GPOAS-0.9
0.1	0.1	1.7 ± 0.2	2.1 ± 0.5	3.2 ± 0.4	5.1 ± 0.6	3.1 ± 0.4	5.7 ± 0.6
	0.3	2.7 ± 0.4	4.1 ± 0.5	2.1 ± 0.4	2.9 ± 0.6	2.7 ± 0.5	4.3 ± 0.6
	0.5	2.9 ± 0.3	5.4 ± 0.6	2.2 ± 0.3	2.6 ± 0.4	2.0 ± 0.3	3.3 ± 0.4
	0.7	2.9 ± 0.6	6.8 ± 0.9	2.0 ± 0.5	2.2 ± 0.5	1.3 ± 0.3	2.2 ± 0.5
	0.9	3.0 ± 0.5	7.4 ± 0.8	1.9 ± 0.4	2.1 ± 0.3	0.9 ± 0.3	1.6 ± 0.4
0.3	0.1	2.9 ± 0.5	4.6 ± 0.7	4.7 ± 1.0	6.7 ± 1.0	4.1 ± 0.7	7.4 ± 0.7
	0.3	4.5 ± 0.9	6.9 ± 1.0	3.2 ± 0.9	4.4 ± 1.3	4.0 ± 1.0	6.1 ± 1.2
	0.5	5.4 ± 0.9	9.7 ± 0.9	3.4 ± 0.5	4.1 ± 0.8	3.5 ± 0.7	5.2 ± 0.9
	0.7	5.3 ± 0.8	10.8 ± 1.2	3.3 ± 0.7	3.6 ± 0.6	2.5 ± 0.6	3.9 ± 0.8
	0.9	5.3 ± 1.2	11.5 ± 1.7	2.9 ± 0.8	3.2 ± 0.8	1.9 ± 0.7	3.2 ± 1.1
0.5	0.1	6.1 ± 0.8	8.6 ± 0.9	8.8 ± 1.1	10.3 ± 1.4	6.8 ± 1.0	10.4 ± 1.0
	0.3	6.7 ± 0.6	10.4 ± 0.6	4.8 ± 0.7	7.0 ± 0.8	6.0 ± 0.9	8.7 ± 0.9
	0.5	8.0 ± 0.8	13.3 ± 0.9	4.9 ± 0.8	5.8 ± 0.8	5.7 ± 0.9	8.2 ± 1.3
	0.7	7.9 ± 1.0	15.4 ± 1.0	4.5 ± 0.8	5.3 ± 1.0	4.0 ± 1.0	5.9 ± 1.4
	0.9	8.6 ± 1.3	16.9 ± 1.5	4.9 ± 1.0	5.8 ± 1.1	4.0 ± 1.2	5.8 ± 1.7
0.7	0.1	8.4 ± 1.3	11.1 ± 1.3	10.8 ± 1.9	12.0 ± 1.7	7.9 ± 1.1	11.6 ± 1.1
	0.3	9.6 ± 2.3	13.3 ± 2.4	8.1 ± 2.9	10.6 ± 2.9	8.3 ± 2.4	11.2 ± 2.6
	0.5	11.9 ± 3.1	18.2 ± 2.9	8.3 ± 3.4	11.0 ± 3.7	9.1 ± 3.5	11.7 ± 3.6
	0.7	14.4 ± 2.5	22.4 ± 2.6	10.5 ± 2.3	13.4 ± 3.0	10.2 ± 2.3	13.0 ± 2.8
	0.9	15.4 ± 2.7	24.3 ± 3.1	11.4 ± 2.3	14.3 ± 2.9	10.5 ± 2.5	13.4 ± 3.1
0.9	0.1	14.5 ± 1.7	17.3 ± 1.9	16.0 ± 1.9	17.7 ± 2.2	12.5 ± 1.7	16.0 ± 2.0
	0.3	18.7 ± 2.0	24.2 ± 2.0	17.2 ± 1.9	21.0 ± 1.6	15.8 ± 2.1	19.3 ± 1.9
	0.5	22.0 ± 2.6	28.0 ± 2.7	19.2 ± 2.9	23.5 ± 3.3	18.3 ± 2.5	21.5 ± 3.0
	0.7	22.7 ± 2.1	30.3 ± 2.0	18.9 ± 2.3	23.4 ± 2.1	17.8 ± 2.5	21.1 ± 2.2
	0.9	23.0 ± 3.6	30.1 ± 3.8	19.0 ± 4.1	22.8 ± 4.4	17.5 ± 4.0	19.4 ± 4.6

* $\overline{x} \pm s$ represents the mean and standard deviation of $\%dev_{avg}$.

** Highlighted cell means the GPSR is better than GPOAS trained over particular training set is better with Z-test of 5% significance level, or vice versa.

Table 4. Performance of the best evolved rules and TS

τ	R	SDR-0.1	SDR-0.5	SDR-0.9	TS	GPOAS
0.1	0.1	1.7 ± 0.1	1.9 ± 0.1	2.9 ± 0.1	4.1 ± 0.2	2.7 ± 0.0
	0.3	1.7 ± 0.1	1.8 ± 0.0	2.2 ± 0.1	3.5 ± 0.3	2.5 ± 0.0
	0.5	1.4 ± 0.1	1.2 ± 0.1	1.7 ± 0.1	2.2 ± 0.3	1.8 ± 0.0
	0.7	0.8 ± 0.1	0.6 ± 0.0	1.1 ± 0.1	1.3 ± 0.2	1.1 ± 0.0
	0.9	0.4 ± 0.0	0.2 ± 0.0	0.8 ± 0.0	0.5 ± 0.1	0.7 ± 0.0
0.3	0.1	2.9 ± 0.1	2.9 ± 0.1	3.4 ± 0.1	4.9 ± 0.3	3.5 ± 0.0
	0.3	3.6 ± 0.1	3.3 ± 0.1	3.2 ± 0.1	4.8 ± 0.3	4.6 ± 0.0
	0.5	3.4 ± 0.1	3.0 ± 0.1	2.6 ± 0.1	4.2 ± 0.3	3.4 ± 0.0
	0.7	2.1 ± 0.1	1.8 ± 0.1	1.6 ± 0.1	3.1 ± 0.2	2.6 ± 0.0
	0.9	1.6 ± 0.1	1.1 ± 0.1	1.2 ± 0.1	2.2 ± 0.2	1.7 ± 0.0
0.5	0.1	4.5 ± 0.1	5.0 ± 0.1	6.2 ± 0.2	6.9 ± 0.3	6.3 ± 0.0
	0.3	5.3 ± 0.1	5.1 ± 0.1	5.0 ± 0.1	7.0 ± 0.4	6.2 ± 0.0
	0.5	5.5 ± 0.1	5.1 ± 0.1	4.8 ± 0.1	6.6 ± 0.4	6.4 ± 0.0
	0.7	3.8 ± 0.1	3.5 ± 0.1	3.1 ± 0.1	5.3 ± 0.3	4.5 ± 0.0
	0.9	3.7 ± 0.1	3.3 ± 0.1	2.8 ± 0.1	4.7 ± 0.3	4.3 ± 0.0
0.7	0.1	6.0 ± 0.1	6.6 ± 0.2	7.2 ± 0.2	8.2 ± 0.3	7.5 ± 0.0
	0.3	7.3 ± 0.1	7.4 ± 0.2	7.7 ± 0.2	9.0 ± 0.4	8.9 ± 0.0
	0.5	8.6 ± 0.2	8.5 ± 0.1	8.9 ± 0.2	10.1 ± 0.5	9.6 ± 0.0
	0.7	10.5 ± 0.2	10.0 ± 0.2	9.5 ± 0.2	11.9 ± 0.6	11.9 ± 0.0
	0.9	10.9 ± 0.2	10.3 ± 0.2	9.6 ± 0.1	12.7 ± 0.6	12.6 ± 0.0
0.9	0.1	12.4 ± 0.3	12.7 ± 0.2	11.3 ± 0.2	11.8 ± 0.5	13.3 ± 0.0
	0.3	16.0 ± 0.2	15.9 ± 0.3	15.4 ± 0.2	16.7 ± 0.6	17.6 ± 0.0
	0.5	19.2 ± 0.3	18.5 ± 0.3	17.7 ± 0.2	18.6 ± 0.6	21.2 ± 0.0
	0.7	19.5 ± 0.3	18.1 ± 0.2	17.6 ± 0.3	18.5 ± 0.7	20.7 ± 0.0
	0.9	19.6 ± 0.2	18.4 ± 0.2	16.8 ± 0.2	18.1 ± 0.5	18.9 ± 0.0

* $\overline{x} \pm s$ represents the mean and standard deviation of $\%dev_{avg}$.

** Highlighted cell means the SDR trained over particular training set is better than TS and GPOAS under Z-test with 5% significance level.

is significantly better than GPOAS in most subsets. When rules are evolved with $\langle 0.9, 0.9 \rangle$, GPSR is significantly better than GPOAS in all subsets. These results confirm the effectiveness of the proposed GPSR as compared to GPOAS. Similar to previous studies [10,12], the training set has a large impact on the performance of evolved SDRs. For example, GPSR evolved with $\langle 0.1, 0.1 \rangle$ provides very good results with the instances with low τ and R. However, the effectiveness of GPSR-0.1 reduces as τ and R increase as compared to GPSR-0.5 and GPSR-0.9. In general, GPSR-0.9 shows good performance in most subsets. The reason is that the subset $\langle 0.9, 0.9 \rangle$ is more complicated (especially for acceptance decisions), which help SDRs trained with instances from this subset deal with complicated situations better.

4.2 Evolved SDRs and Tabu Search Heuristic

In this section, we pick the best evolved SDRs evolved with the three training sets and compare them with TS [2]. The average relative deviation $\%dev_{avg}$

is used to measure the performance of these heuristics for each OAS subset. Since SDRs and TS are all stochastic heuristics, we will perform 30 independent runs of SDRs and TS to find $\%dev_{avg}$ for each subset. The mean and standard deviation values of $\%dev_{avg}$ from 30 runs for each subset will help us assess the effectiveness and the robustness of these heuristics. The results for these experiments are presented in Table 4. In this Table, SDR-x is the best SDR (based on its overall performance across all subsets) evolved with the training set $\langle x, x\rangle$. The last column in the table shows the best rule evolved by GPOAS. A highlighted result indicates that the corresponding rule/heuristic is significantly better than other rules/heuristics for the particular subset.

Within the group of the best evolved SDRs, SDR-0.9 shows superior performance in most subsets. SDR-0.1 and SDR-0.5 are able to dominate SDR-0.9 in a few subsets with low τ and low R. It is also easy to see that evolved SDRs are better than TS in most subsets. Only in the subsets with $R = 0.9$, SDR-0.1 and SDR-0.5 are slightly worse than TS. This is understandable because TS can handle acceptance decisions better in these subsets. Because SDR-0.9 is trained from these situations, it still shows its dominance here. It is also noted that the standard deviation values from the best evolved SDRs are lower than TS in all cases. This suggests that SDRs are more robust than TS. The best evolved SDRs also show their dominance against the best rule evolved by GPOAS. Although the evolved SDRs are m times slower than the rule evolved by GPOAS, we believe this is a good trade-off to significantly improve the performance of rules evolved by GP. Moreover, evolved SDRs are still much faster than TS (the average computational times of SDRs and TS are respectively 0.15 seconds and 6 seconds per instance).

5 Conclusions

Overall, this paper has shown that evolving stochastic dispatching rules using GP is an effective alternative to meta-heuristics in OAS. By introducing some randomness into the order selection process, evolved rules can explore more potential solutions. Such rules can provide superior performance as compared to the highly customised TS heuristic [11,2] and they are still very efficient. These results are very encouraging and show that automatic heuristic design methods such as GPSR in this paper are capable of generating very effective rules as compared to optimisation heuristics/meta-heuristics in the literature. These rules can be used either to quickly generate good solutions for OAS or to initialise solutions to reduce the computational effort of optimisation heuristics.

For future work, it would be interesting to extend the GP representation to take into account multiple conflicting objectives [10]. With the ability to explore multiple solutions, stochastic dispatching rules can be a good approach to explore non-dominated solutions in multi-objective problems. Moreover, future works could also focus on developing smarter stochastic selection schemes, which can help reduce the computational effort of stochastic dispatching rules and improve their effectiveness.

References

1. Burke, E., Hyde, M., Kendall, G., Ochoa, G., Ozcan, E., Qu, R.: Hyper-heuristics: A survey of the state of the art. Computer Science Technical Report No. NOTTCS-TR-SUB-0906241418-2747 (2010)
2. Cesaret, B., Oğuz, C., Salman, F.S.: A tabu search algorithm for order acceptance and scheduling. Computers & Operations Research 39(6), 1197–1205 (2012)
3. Dimopoulos, C., Zalzala, A.: Investigating the use of genetic programming for a classic one-machine scheduling problem. Advances in Engineering Software 32(6), 489–498 (2001)
4. Geiger, C.D., Uzsoy, R.: Learning effective dispatching rules for batch processor scheduling. International Journal of Production Research 46(6), 1431–1454 (2008)
5. Geiger, C.D., Uzsoy, R., Aytug, H.: Rapid modeling and discovery of priority dispatching rules: An autonomous learning approach. Journal of Scheduling 9(1), 7–34 (2006)
6. Ghosh, J.B.: Job selection in a heavily loaded shop. Computers & Operations Research 24(2), 141–145 (1997)
7. Hildebrandt, T., Heger, J., Scholz-Reiter, B.: Towards improved dispatching rules for complex shop floor scenarios: A genetic programming approach. In: Proceedings of the 12th Annual Conference on Genetic and Evolutionary Computation, pp. 257–264 (2010)
8. Jakobović, D., Budin, L.: Dynamic scheduling with genetic programming. In: Collet, P., Tomassini, M., Ebner, M., Gustafson, S., Ekárt, A. (eds.) EuroGP 2006. LNCS, vol. 3905, pp. 73–84. Springer, Heidelberg (2006)
9. Luke, S.: Essentials of Metaheuristics. Lulu (2009)
10. Nguyen, S., Zhang, M., Johnston, M., Tan, K.C.: Learning reusable initial solutions for multi-objective order acceptance and scheduling problems with genetic programming. In: Krawiec, K., Moraglio, A., Hu, T., Etaner-Uyar, A.Ş., Hu, B. (eds.) EuroGP 2013. LNCS, vol. 7831, pp. 157–168. Springer, Heidelberg (2013)
11. Oğuz, C., Salman, F.S., Yalin, Z.B.: Order acceptance and scheduling decisions in make-to-order systems. International Journal of Production Economics 125(1), 200–211 (2010)
12. Park, J., Nguyen, S., Zhang, M., Johnston, M.: Genetic programming for order acceptance and scheduling. In: IEEE Congress on Evolutionary Computation, pp. 1005–1012 (to appear, 2013)
13. Rom, W.O., Slotnick, S.A.: Order acceptance using genetic algorithms. Computers & Operations Research 36(6), 1758–1767 (2009)
14. Roundy, R., Chen, D., Chen, P., Cakanyildirim, M., Freimer, M.B., Melkonian, V.: Capacity-driven acceptance of customer orders for a multi-stage batch manufacturing system: models and algorithms. IIE Transactions 37(12), 1093–1105 (2005)
15. Slotnick, S.A.: Order acceptance and scheduling: A taxonomy and review. European Journal of Operational Research 212(1), 1–11 (2011)
16. Slotnick, S.A., Morton, T.E.: Order acceptance with weighted tardiness. Computers & Operations Research 34(10), 3029–3042 (2007)
17. Wester, F.A.W., Wijngaard, J., Zijm, W.R.M.: Order acceptance strategies in a production-to-order environment with setup times and due-dates. International Journal of Production Research 30(6), 1313–1326 (1992)

Detecting Mutex Pairs in State Spaces by Sampling

Mehdi Sadeqi[1], Robert C. Holte[2], and Sandra Zilles[1]

[1] Department of Computer Science, University of Regina
Regina, SK, Canada S4S 0A2
{sadeqi2m,zilles}@cs.uregina.ca

[2] Department of Computing Science, University of Alberta
Edmonton, AB, Canada T6G 2E8
holte@cs.ualberta.ca

Abstract. In the context of state space planning, a mutex pair is a pair of variable-value assignments that does not occur in any reachable state. Detecting mutex pairs is a problem that has been addressed frequently in the planning literature. In this paper, we present the Missing Mass Method (MMM)—a new efficient and domain-independent method for mutex pair detection, based on sampling reachable states. We exploit a recent result from statistical theory, proven by Berend and Kontorovich in [1], that bounds the probability mass of missing events in a sample of a given size. We tested MMM empirically on various sizes of four standard benchmark domains from the planning and heuristic search literature. In many cases, MMM works perfectly, i.e., finds all and only the mutex pairs. In the other cases, it is near-perfect: it correctly labels all mutex pairs and more than 99.99% of all non-mutex pairs.

1 Introduction

The aim of heuristic search and planning systems is to find a path (sometimes a least-cost path) from a given start state to a given goal using a given set of operators. The set of possible states is defined by specifying a set of variables and the possible values for each variable, and a particular state is specified by assigning a specific value to each variable. For example, in the familiar 8-puzzle (3×3-sliding-tile puzzle), there might be 9 variables, one for each position of the puzzle (e.g. variable UL might refer to the upper-left corner position), and the value of each variable indicates which tile (or "no tile") is in that position. We use the phrase "variable-value assignment" to refer to the assignment of a specific value to a specific variable, for example, UL="no tile" is a variable-value assignment. Some planning systems use propositional variables which we treat as variables that can take on one of two values (true and false).

The term "mutually exclusive pair" (of facts), or mutex pair for short, refers to a pair of variable-value assignments that do not co-occur in any valid state. By "valid state" one typically means a state that is reachable from a given start state. For instance, in the 8-puzzle, an example of a mutex pair would be

S. Cranefield and A. Nayak (Eds.): AI 2013, LNAI 8272, pp. 490–501, 2013.

(UL="no tile" and BR="no tile"), where BR is the variable saying what is in the bottom-right position. In the 8-puzzle, there is only one empty position, so there cannot simultaneously be two variables that have the value "no tile".

Mutex pair detection was first addressed by Dawson and Siklóssy [6], and has been in use for improving the performance of planning systems since the development of Graphplan [2]. In Graphplan, two actions or two facts at the same level of reasoning are mutex if there is no valid plan allowing both actions or both facts at the same time. The original success of mutex detection in Graphplan [2] led to the development of many efficient planners using mutex detection, such as temporal planners, different SAT-based planners and planners based on constraint programming. State-of-the-art techniques for detecting mutexes in binary domain representation often use invariants, i.e., properties that hold true in all reachable states of a state space. Using an invariant synthesis algorithm, HSP [3] is tailored to detect a certain type of mutex that Graphplan misses. Constrained abstraction [8] uses particular types of invariants to detect some mutexes in the description of an abstract state. For more background on domain analysis and examples of constrained abstractions using invariants the reader is referred to [7], where mutex pairs are discussed as a special case of "at-most-one" invariants consisting of only two atoms. In this work, Haslum also introduces the h^2 heuristic, which is a state-of-the-art method for finding mutex pairs.

One use of mutex detection is to translate propositionally encoded planning domains into representations with multi-valued state variables; e.g., FD [9] can thus reveal intuitive dependencies between variables which helps to explicate some of the implicit constraints of propositional planning tasks.

Another application of mutex detection is search space pruning. In regression planning, nodes in the search space are partial assignments of state variables, and edges are actions. Search proceeds backward from the goal until reaching a node consistent with the start state. To reduce search effort, one prunes nodes that contain the assignment of some previously encountered node. Further, one often prunes nodes that are "impossible" because they contain mutex pairs.

Mutex detection can also be applied for improving the quality of heuristic functions derived from abstractions. Heuristic functions estimate, for any state s, the distance from s to a goal state. Heuristic search algorithms like A* and IDA* are guaranteed to find optimal solutions when using *admissible heuristics*, i.e., heuristic functions that never overestimate the true distances. One popular method for obtaining admissible heuristics is to create an abstract version of the original state space and to use the true distances in the abstract state space as heuristic values. The key to the efficiency of A* and IDA* is the quality of the heuristic values: the closer the heuristic values are to the true distances, the more effective they will be in speeding up search. Unfortunately, standard efficient methods known for enumerating abstract state spaces, most notably pattern databases (PDBs) [5], may include abstract states to which no reachable original state is mapped by the abstraction. Such abstract states are called spurious; they may create short-cuts in the abstract space and thus lower heuristic values [15]. In many cases, spurious states contain mutex pairs. Hence, by removing some of the shortcuts created by spurious abstract states, mutex detection can help to improve the quality of heuristics, and thus to speed up search.

Unfortunately, there are no known efficient methods for detecting *all* mutex pairs. Existing algorithms usually make a compromise in the number of detected mutex constraints for the computational complexity of the algorithm. Various methods differ in the number and type of mutex constraints they detect.

In this paper, we propose the Missing Mass Method (MMM)—a new algorithm for detecting mutex pairs, based on sampling reachable states. We exploit a recent result from statistical theory, proven by Berend and Kontorovich in [1], that would allow us to bound the probability of missing a reachable pair (i.e., a non-mutex pair) in an i.i.d. sample of a given size. The main advantages of MMM over existing methods are the following.

- It is very simple to describe and to implement.
- As opposed to many state-of-the-art mutex detection techniques, MMM does not systematically restrict itself to detecting only a subclass of the mutex pairs.
- For several standard benchmark domains, MMM is *perfect* on reasonable domain sizes, i.e., it detects mutex pairs with 100% accuracy. For the same domains, it scales very well, mostly yielding perfect accuracy even for very large domain sizes, e.g., for Scanalyzer with 100 batches, Blocks World with 26 blocks, or the 10×10-Sliding-Tile Puzzle.
- While most of our experiments on MMM were on detecting mutex pairs, exactly the same method can be used to detect higher order mutexes. For instance, a mutex of order 3 in the Blocks World would be a triple of variable-value assignments that represents the facts *Block a is on top of Block b, Block b is on top of Block c*, and *Block c is on top of Block a*.
- MMM does not require backward reasoning, depending on the application for which it is used. Technically, MMM can be used with any kind of sampling method. (Theoretical guarantees only hold though for special sampling methods that will typically not be available in practice.)

All existing mutex detection methods err on one side: they might consider mutex pairs as reachable but they will never flag a reachable pair as mutex. As opposed to that, our method errs on the other side. It never considers a mutex pair as reachable, but it may consider reachable pairs as mutex. Depending on the application, this might be an advantage or disadvantage— we will discuss that below. Our experimental results will illustrate that MMM is very reliable in a large variety of state spaces, suggesting that it can be safely applied even in cases when it is crucial not to consider reachable pairs as mutex. In some such cases, we will demonstrate empirically that MMM can clearly outperform state-of-the-art mutex detection methods.

2 The Missing Mass Method for Mutex Detection

Assume that states are represented as variable-value pairs in m variables. We denote the variables with $x_1, \ldots, x_m$, so that the state vector $(a_1, \ldots, a_m)$ corresponds to the assignment vector $(x_1 = a_1, \ldots, x_m = a_m)$. Propositional logic variables, such as those commonly used in planning, are treated as variables that can take on one of two values (true and false). With this convention, we define the notions of reachable state and mutex pair formally.

Definition 1. *Let s^* be any fixed state.*

Since s^* is fixed, we will simply call a state reachable if it is reachable from s^*. For any i, j with $1 \le i < j \le m$ and any a_i, a_j, the partial original state $(x_i = a_i, x_j = a_j)$ is a reachable pair if there are a_k, for $k \in \{1, \dots, m\} \setminus \{i, j\}$ such that $(a_1, \dots, a_m)$ is a reachable state; otherwise $(x_i = a_i, x_j = a_j)$ is a mutex pair.

Our approach to detecting mutex pairs in large problem domains is based on sampling. The general scheme of our method, which we will call the *Missing Mass Method (MMM)* for reasons detailed below, is quite simple:

1. Determine a number N of pairs of variable-value assignments to be sampled.
2. Determine the smallest integer N_s such that $N_s \cdot \binom{m}{2} \ge N$. (Thus, sampling N_s states results in sampling at least N pairs of variable-value assignments.)
3. Sample N_s reachable states and extract all pairs of variable-value assignments from them.
4. Any pair of variable-value assignments not encountered this way is considered a mutex pair.

The two details that need to be defined are (i) how the sampling of reachable states is done, and (ii) how to fix the number N of pairs of variable-value assignments to be sampled.

Concerning (i), let us first assume we have fixed a method for sampling reachable states, which induces a probability distribution D over all possible pairs of variable-value assignments. (We experiment with a variety of sampling methods, as described in the following section.)

Statistical theory provides us with tools for addressing question (ii) under these circumstances. Berend and Kontorovich [1] give an upper bound on the expected probability mass of the elements not seen after taking N i.i.d. samples from any fixed distribution. This bound depends on the total number z of elements in the (finite) universe and is given by the following inequality.

$$\mathbb{E}_D[M_N] \le \frac{z}{eN} \text{ for } N > z\,. \tag{1}$$

Here M_N is the total probability mass of the elements not seen after sampling N times from the distribution D that results from the sampling method (called the *sampling distribution*), and $\mathbb{E}_D[M_N]$ is its expected value with respect to D.

We deploy this bound by choosing a number N of samples that is large enough for $\mathbb{E}_D[M_N]$ to be below a fixed threshold. The required number z may not be available, but an upper bound on z is obtained by computing the total number of possible pairs of variable-value assignments in the given representation of the state space. For example, if each of the m state variables can take one of k possible values, then there are $\binom{m}{2}$ many variable pairs, each with k^2 many possible value assignments, and thus a total of $\binom{m}{2}k^2$ possible pairs of variable-value assignments. We use this upper bound of $\binom{m}{2}k^2$ in lieu of z, since this never makes us sample less than when using the exact value of z.

Given this method for computing the sample size, we still need to fix a method for choosing the samples. The sampling process itself may be of crucial importance to the success of the scheme depicted above. Inequality (1) does not provide us with a sample size that bounds the probability of missing a pair, but with a sample size that bounds the probability mass of the non-seen pairs with respect to the sampling distribution. If some pairs of variable-value assignments have too small a probability of being sampled, then even with a sample resulting from a very low threshold for $\mathbb{E}_D[M_N]$ we might be missing these reachable pairs, because their cumulative probability mass with respect to the sampling distribution is too small. Consequently, if a poor sampling method is used, MMM might flag reachable pairs as mutex.

Designating reachable pairs as mutex could have positive or negative effects, depending on the problem that mutex detection is applied to. Hence, to minimize the risk of missing reachable pairs, it is desirable to use a near-uniform sampling process, so that no pairs of variable-value assignments have too small a probability of being sampled. Unfortunately, there is no known method for sampling states in a way that creates a near-uniform sample of the contained pairs of variable-value assignments, and further one does in general not sample pairs i.i.d. when sampling states. Because of the latter problem, Berend and Kontorovich's bound does not even yield theoretical guarantees in our case. We nevertheless use their bound to decide how many states to sample. Note that even with uniform sampling there would be no *guarantee* that our method finds all reachable pairs; however, we would have a guaranteed minimum probability of finding all reachable pairs. Our experiments suggest that for typical benchmark domains, even in large sizes, this probability is very high. In our experiments we tested a variety of sampling methods, which we will describe in Section 3.1.

2.1 Does MMM Err on the Wrong Side?

An important property distinguishing MMM from existing mutex detection methods is that it errs "on the other side". While existing methods never consider a reachable pair mutex, MMM never considers a mutex pair reachable, but might consider a reachable pair mutex.

In the case of state space pruning in regression planning, falsely considering a reachable pair as mutex might lead to the elimination of reachable states and thus to the elimination of paths from goal to start, in the worst case disconnecting the goal from the start state. The effect could be devastating, but a closer look at our experimental results will show that in many cases this would not be a major concern when applying MMM.

Considering the problem of improving heuristic values when a PDB contains spurious abstract states, the one-sided error of MMM may even have a positive effect. To address this problem with MMM, one (i) builds a PDB as usual, then (ii) uses MMM backwards from the goal in the original state space to find mutex pairs, (iii) builds an auxiliary PDB in the usual way with the exception that abstract states containing a mutex pair are not added to the open list, but are considered deadends. (This means one will find paths to abstract states that do not pass through abstract states suspected of being spurious.) Finally, (iv) one replaces entries in the PDB with those from the auxiliary PDB as long as the latter are not infinite. The resulting PDB would be used to guide the search. Since our method can flag reachable pairs as mutex, the resulting heuristic might be inadmissible and inconsistent, causing A* or IDA* to find only suboptimal solutions, but it may potentially find them much faster than with an admissible heuristic obtained before removing abstract states that contain mutexes.[1] The reader is referred to [14] for recent work on efficient suboptimal search with inadmissible heuristics.

We claim that MMM will be a very useful method for mutex detection, mainly because our experimental results demonstrate that for a large variety of state spaces, MMM is 100% accurate (it does not flag any reachable states mutex). In some cases, it outperforms all existing methods in terms of accuracy, while still being very efficient. The trade-off between efficiency and accuracy seems to be much less substantial for MMM than it is for existing mutex detection methods. (For a comparison of MMM with h^2 in terms of accuracy, see Section 4.)

3 Experimental Setup

Four planning and search benchmark problem domains, represented using production system vector notation (PSVN) [11], were selected for this study. All operators in all domain representations are invertible. While we describe the representations of the domains below, we omit a general description of the domains themselves, due to space constraints.

The particular representations were intentionally chosen so that many mutex pairs exist; hence they are not necessarily the most natural or the most compact.

In our experiments, we fix the choice of the state s^* with respect to which we consider states reachable, instead of running MMM on a variety of choices for s^*. In all the spaces we use for testing, operators are invertible, so that within any connected component, every state is reachable from every other state. Hence, the set of reachable pairs is the same no matter which s^* in a connected component we use. However, the distance from one choice of s^* to all the reachable pairs might be different than the distance from another s^*, and that could affect the success rate of some of the sampling methods. We believe that the standard goal state for each space is a good, representative choice for s^*, to measure the success of a sampling method. Hence, for any domain, we always chose s^* to be the standard goal state.

---

[1] We have initial empirical results supporting this claim for one representation of the Towers-of-Hanoi domain. Further, the described method can be implemented by flagging states considered spurious in the original PDB without actually building a separate PDB.

Domain 1: Towers of Hanoi. We encode a state of the n-Disks Towers of Hanoi with p pegs as a vector of length $p(n+1)$, where for every peg a sequence of $n+1$ components encodes the number of disks and the names of disks stacked on this peg (starting from the bottom of the peg); for a stack of k disks, the last $n-k$ components for this peg contain a 0.

This domain illustrates why we choose seemingly "unnatural" domain representations. A "natural" approach for representating the n-Disk Towers of Hanoi with p pegs would be to encode every state as a vector of n components. Each component corresponds to a disk; its value in $\{1, 2, ..., p\}$ represents the peg on which the disk is located. However, this representation does not yield any mutex pairs, because every one of the possible p^n vectors corresponds to a reachable original state. Hence this domain representation is not useful for our studies. (In addition, the number of operators in the "natural" representation described above is also exponential in the number of disks when $p > 3$, making this representation inconvenient for other reasons as well.)

Domain 2: Blocks World with Table Positions. We consider two PSVN representations of the n-Blocks World with p named table positions. In the first one, called the *top representation*, a state vector has $1+p+n$ components, each containing either the values 0 or one of n possible block names: (i) the value of the first component is the name of the block in the hand or 0 if the hand is free, (ii) the values of the next p components are the names of the blocks immediately on table positions 1 through p, (iii) the values of the last n components are the names of the blocks immediately on top of blocks $a, b, c, \ldots$.

In the *stack representation*, a state is encoded as a vector of length $p(n+1)+1$, where for every table position a sequence of $n+1$ components encodes the number of blocks and the names of blocks stacked on this position (starting from the bottom); for a stack of k blocks, the last $n-k$ components for this block contain a 0. The final component encodes the content of the hand. Note the similarity of this domain in this representation to our representation of the Towers of Hanoi.

The state s^* has all blocks stacked up in increasing lexicographical order, starting with block a, on table position 1.

Domain 3: Sliding-Tile Puzzle. In the *standard representation* of $n \times \ell$-Sliding-Tile Puzzle, states are represented as vectors of length $n \cdot \ell$, where each component corresponds to a grid position and contains a value in $\{1, 2, \ldots, n \cdot \ell - 1, B\}$, representing the number of the tile in this position (B, if the position is blank). In the *dual representation*, a vector component corresponds to either the blank or one of the tiles. The value of a vector component is an integer in $\{1, \ldots, n \cdot \ell\}$, representing the grid position at which the corresponding tile is located.

The state s^* contains the blank in the bottom right corner of the grid, while the remaining grid positions contain tiles with increasing numbers, row by row from top to bottom, each row being filled from left to right.

Domain 4: Scanalyzer. In the PSVN representation of the n-Belt Scanalyzer [10] (for even n), a state is encoded as a vector of length $2n$ in which each belt corresponds to two components: the name of the batch on that belt and a flag indicating whether that batch is analyzed. The state s^* corresponds to having all plant batches analyzed and placed on their original conveyor belts.

3.1 Sampling

In our experiments, we used a fixed threshold of 0.00001 by which to bound the estimated missing probability mass. For example, for the 4×5-puzzle in standard representation, the resulting number of pairs to be sampled was $2,795,883,753$. Since there are 190 pairs of variable-value assignments in every state, this corresponds to sampling $14,715,178$ states (note again that the pairs are then not sampled i.i.d., but our experiments will show that the missing mass bound is still effective). Similar calculations for various sizes of the domains we experimented with shows this approach to be scalable (e.g., the number of states to be sampled for the 10×10-puzzle is $367,879,441$). In other words, the sample size used in our approach is small enough for our sampling method to be feasible.

We tested a variety of sampling processes, in particular we report our experiments on single random walks (RW), and Frontier Sampling (FS) [12], always beginning at s^*. Furthermore, we tested uniform sampling of reachable states (USS). Note that USS does not necessarily mean uniform sampling of the reachable pairs of variable-value assignments. While USS is not a domain-independent method, RW and FS are.

For RW we conducted basic random walks without parent pruning and without restarts; thus the length of a random walk was the number of states we wanted to sample. FS conducts r dependent random walks in a search tree by keeping a list of r nodes [12]. Initially, some r nodes are sampled at random. From the joint list of all children of these r nodes, one child c is chosen uniformly at random as the next sampled node; c then replaces its parent in the list of r nodes. In our experiments, we set $r = 100$; the initial r nodes are chosen by 100 independent random walks conducted from s^*. The lengths of these random walks are chosen uniformly at random from $\{0, 1, \ldots, 1000\}$.

4 Experimental Results

We first tried MMM on small sizes of the domains described above. Here we enumerated the state space exhaustively and directly compared the true set of reachable pairs to those found by MMM, which, using any of USS, FS, and RW, found all reachable pairs and thus was perfect at mutex pair detection. For larger domains, we calculated the actual number of reachable pairs for every representation and compared it to the number of pairs MMM found.

For illustration, we show how to compute the actual number of reachable pairs for the 28-belt Scanalyzer, which is $331,184$. In the representation we use, any batch can occur in any location independent of where any other batch is located; any batch can be analyzed/not-analyzed independent of the status of any other batch; and the location of any batch is independent of the anlayzed status of any batch (including itself). Thus a pair of variables can take one of three possible forms: (i) Both variables represent batches; there are $28 \cdot 27/2$ such pairs of variables, each with $28 \cdot 27$ possible value assignments, resulting in $285,768 = (28 \cdot 27/2) \cdot 28 \cdot 27$ reachable pairs. (ii) Both variables represent an "analyzed" status; there are $28 \cdot 27/2$ such pairs of variables, each with $2 \cdot 2$ possible value assignments, resulting in $1,512$ reachable pairs. (iii) One variable represents a batch, the other represents an "analyzed" status. This results in $43,904 = (28 \cdot 28) \cdot 28 \cdot 2$ pairs. These numbers sum up to $331,184$. We do not show the calculations for the other domains, but report the resulting numbers below. Note that we do not use these numbers to calculate the bound in Inequality (1); instead in our experiments we assume that we only have the knowledge of how to compute the total number of pairs of variable-value assignments that can be expressed in the given domain representation language.

With any of the sampling methods USS, FS, and RW, MMM was perfect in the following testbeds: Blocks World with 12, 15, 18, 21, and 26 blocks, for 3 table positions in the top representation; 5×5-, 5×6-, and 10×10-Sliding-Tile Puzzle in the standard and 5×5-Sliding-Tile Puzzle in the dual representation; Scanalyzer with 12, 16, and 20 belts. For Scanalyzer with 28 and 100 belts, FS and USS were perfect, while RW missed a very small percentage of the reachable pairs. For Towers of Hanoi with 12 disks on 4 pegs, USS and RW were perfect, while FS missed a very small percentage of pairs. All methods missed a few pairs for the 26-Blocks World with 3 table positions in the stack representation.

Table 1 gives a representative sample of our results on larger domain versions. For each domain (in a particular representation and size), it shows the actual number of reachable pairs ("#Pairs"), the number of samples suggested by Inequality (1) ("Bound"), the percentage of reachable pairs that remained undetected after sampling as many states as suggested by the bound ("Missing Pairs"), and the actual number of samples after which all reachable pairs were found ("Minimum Sample Size"). The "Missing Pairs" number is averaged over 1000 repetitions of the whole sampling process. The "Minimum Sample Size" was obtained by repeating the whole sampling process 1000 times and recording the smallest multiple α of 100,000 such that all reachable pairs were always found when sampling α many states. The domains are the 12-disk Towers of Hanoi (ToH) with 4 pegs, the 26-Blocks World (BW) with 3 table positions, both in top and in stack (stk) representation, the 5×5-Sliding-Tile Puzzle (SP), both in standard (std) and in dual (du) representation, the 10×10-Sliding-Tile Puzzle in standard representation, and Scanalyzer (SCN) with 28 and 100 belts. Missing entries for RW on the Scanalyzer domain mean that even after sampling 100,000,000 (for 28 belts) and 1,000,000,000 (for 100 belts) reachable states, not all reachable pairs had been found.

MMM is very efficient in terms of running time. On a standard modern computer, MMM with FS for 1,000,000 sampled states takes on the order of one second for the 5×5-Sliding-Tile Puzzle in standard representation, less than one minute for the 10×10-Sliding-Tile Puzzle in standard representation or for the 28-belt Scanalyzer, and a few minutes for the 100-belt Scanalyzer. When taking

Table 1. MMM results for three different sampling methods, using a threshold of $0.00001 \geq \mathbb{E}_D[M_N]$. Minimum Sample Sizes are given as multiples of 1,000.

Domain	#Pairs	Bound	Missing Pairs USS	FS	RW	Minimum Sample Size USS	FS	RW
ToH (12,4)	51,642	6,217,162	0	0.005%	0	4,000	10,000	6,000
BW top (26,3)	285,551	26,818,411	0	0	0	30	2,200	2,700
BW stk (26,3)	1,547,049	26,818,411	0.0001%	0.006%	0.005%	35,000	70,000	70,000
SP std (5×5)	180,000	22,992,465	0	0	0	14	800	800
SP du (5×5)	180,000	22,992,465	0	0	0	14	900	1,100
SP std (10×10)	49,005,000	367,879,441	0	0	0	300	12,000	12,000
SCN (28)	331,184	33,261,751	0	0	0.004%	16	200	-
SCN (100)	51,024,800	382,890,408	0	0	0.0007%	300	7,000	-

as many samples as suggested by the bound, this would result in a time of less than half a minute for the 5×5-Sliding-Tile Puzzle (std), about half an hour for the 28-belt Scanalyzer, and a few hours for the 10×10-Sliding-Tile Puzzle (std). For the 100-belt Scanalyzer, sampling the full number suggested by the bound would take on the order of 1 day, but after less than half an hour actually all reachable pairs would have been found.[2] In general, the difference between the bound and the minimum sample sizes suggest that, in practical applications, one may set the threshold for $\mathbb{E}_D[M_N]$ substantially higher than 0.00001 and still obtain perfect results in many domains.

USS seems to be perfect whenever a feasible such sampling method exists, with the exception of Blocks World in the stack representation, where it misses on average 2 out of 1,547,049 reachable pairs. In many cases when a uniform method of sampling reachable states is not available, we can still expect that FS will work, though it probably needs to sample more before finding all reachable pairs and thus might not scale as well as USS. The simplest sampling method we tried, RW, works remarkably well. It is perfect on the top representation of Blocks World and on the Sliding-Tile Puzzle, and misses on average only 82 out of the 1,547,049 reachable pairs in the stack representation of Blocks World, 13 out of the 331,184 reachable pairs for the 28-belt Scanalyzer, and 350 out of the 51,024,800 reachable pairs for the 100-belt Scanalyzer.

When RW was missing a few pairs in the Scanalyzer domains, increasing the number of samples well beyond the bound (100,000,000 for 28 belts and 1,000,000,000 for 100 belts) still did not make the sampling perfect. The latter indicates that in these cases there were reachable pairs whose probability of being sampled by RW is so low that their cumulative weight under the probability distribution induced by the sampling procedure lies well below our threshold of 0.00001. Note that in some cases the percentages of missing pairs are slightly higher than $0.00001 = 0.001\%$, but the cumulative probability of the pairs under the distribution resulting from the sampling may still be lower than 0.00001.

[2] The time used for mutex detection could be amortized when solving a large number of search or planning problem instances in the same reachable component of the state space. Further, to the best of our knowledge, there are no methods that can solve an average instance of the 100-belt Scanalyzer in time on the order of a day.

4.1 Comparison with h^2

h^2 is a state-of-the-art method for mutex pair detection in planning [7] that errs on the opposite side when compared to MMM, i.e., it will never consider a reachable pair mutex. h^2 is very effective for a large number of domains, for example, it perfectly detects all mutex pairs for Scanalyzer, the Blocks World with Table Positions, and for almost all sizes of the Sliding-Tile Puzzle, in the representations we experimented with.

However, it systematically fails to detect mutexes of certain types in some domains in which MMM is perfect or almost perfect. Firstly, in the 2×2-Sliding-Tile Puzzle, h^2 misses a special kind of mutex pair, namely some of the pairs that state that two specific distinct tiles reside in two specific distinct locations. This amounts to 20% (12 out of 60) of the existing mutex pairs being missed by h^2. Such pairs are never mutex in larger versions of the puzzle [13]. Secondly, in the stack representation of the n-Blocks World with p table positions, h^2 will systematically miss all mutex pairs that state, for some $i < n$ and some $j > i$, that $n-i$ blocks are on position a while a specific block is at height j on position $b \neq a$. Such a pair of variable-value assignments is mutex as it would require the existence of more than n blocks. For $n = 26$ and $p = 3$, h^2 misses 81% (711,291 out of 873,960) of the existing mutex pairs. Similarly, in our representation of the n-Disk Towers of Hanoi with p pegs, h^2 will miss all mutex pairs stating that there are $n - i$ disks on peg a while a specific disk is at height j on peg $b \neq a$.

This demonstrates that in many domains MMM has an advantage due to sampling at random as opposed to being constructed in a way that systematically misses certain types of mutex pairs.

5 Conclusions

We presented MMM, a sampling-based approach for detecting mutually exclusive pairs of variable-value assignments that is applicable to any kind of domain representation. The method is easy to implement, very efficient, and, if a reasonably good sampling procedure is used, also very effective in detecting mutexes. It does not systematically restrict itself to detecting only a subclass of the mutex pairs, finds mutexes with perfect accuracy in almost all of the domains we tested, and is very near perfect in all other domains tested, when using either of the two domain-independent sampling method we tried (FS, RW). MMM scales very well to large domain sizes. Initial empirical results (not reported here) suggest that MMM may even be successful at detecting higher-order mutexes.

We have demonstrated that h^2, a state-of-the-art mutex detection method, systematically misses certain mutex pairs in some domains. In a small experiment on binary domain representations (not reported here), we showed that the same is true for LONDEX [4] , another state-of-the-art method. MMM outperforms both methods in the domains we experimented with, thanks to not being restricted to specific classes of mutex pairs a priori. In fact, all mutex detection methods we know of suffer from systematically missing certain pairs, and we are aware of only one method for which in some special kinds of domain completeness has been proven to be guaranteed (the CA method, see [13]).

Overall, we believe that MMM can improve the performance of planning systems and heuristic search without affecting runtime efficiency. MMM might further be of use when combined with a method that errs on the opposite side.

Acknowledgements. This work was supported by the Natural Sciences and Engineering Research Council of Canada (NSERC).

References

1. Berend, D., Kontorovich, A.: The missing mass problem. Stat. and Prob. Lett. 82, 1102–1110 (2012)
2. Blum, A.L., Furst, M.L.: Fast planning through planning graph analysis. Artif. Intell. 90(1), 1636–1642 (1995)
3. Bonet, B., Geffner, H.: Planning as heuristic search. Artif. Intell. 129(1-2), 5–33 (2001)
4. Chen, Y., Xing, Z., Zhang, W.: Long-distance mutual exclusion for propositional planning. In: IJCAI, pp. 1840–1845 (2007)
5. Culberson, J., Schaeffer, J.: Pattern databases. Comput. Intell. 14(3), 318–334 (1998)
6. Dawson, C., Siklóssy, L.: The role of preprocessing in problem solving systems. In: IJCAI, pp. 465–471 (1977)
7. Haslum, P.: Admissible Heuristics for Automated Planning. Linköping Studies in Science and Technology: Dissertations. Dept. of Computer and Information Science. Linköpings Univ. (2006)
8. Haslum, P., Bonet, B., Geffner, H.: New admissible heuristics for domain-independent planning. In: AAAI, pp. 1163–1168 (2005)
9. Helmert, M.: The Fast Downward planning system. J. Artif. Intell. Res. 26, 191–246 (2006)
10. Helmert, M., Lasinger, H.: The Scanalyzer domain: Greenhouse logistics as a planning problem. In: ICAPS, pp. 234–237 (2010)
11. Hernádvölgyi, I., Holte, R.: PSVN: A vector representation for production systems. Technical Report TR-99-04, Dept. of Computer Science, Univ. of Ottawa (1999)
12. Ribeiro, B.F., Towsley, D.F.: Estimating and sampling graphs with multidimensional random walks. CoRR abs/1002.1751 (2010)
13. Sadeqi, M., Holte, R.C., Zilles, S.: Using coarse state space abstractions to detect mutex pairs. In: SARA, pp. 104–111 (2013)
14. Thayer, J., Ruml, W.: Bounded suboptimal search: A direct approach using inadmissible estimates. In: IJCAI 2011, pp. 674–679 (2011)
15. Zilles, S., Holte, R.C.: The computational complexity of avoiding spurious states in state space abstraction. Artif. Intell. 174, 1072–1092 (2010)

Scheduling for Optimal Response Times in Queues of Stochastic Workflows

Michal Wosko, Irene Moser, and Khalid Mansour

Faculty of Information and Communication Technologies,
Swinburne University of Technology,
Melbourne, Australia
{mwosko,imoser,mwmansour}@swin.edu.au

Abstract. We investigate the problem of scheduling tasks of structured workflows, given a stochastic arrival of workflow instances, which gives rise to a queue. Each workflow conforms to a known structure expressed by a directed acyclic graph. However, within this model, the precise execution time of each atomic task and the delay of each communication edge are non-deterministic. Unlike in most scheduling approaches that minimize the schedule length, we additionally aim at minimizing the total time spent by a workflow instance in the system, as perceived by the end user on whose behalf the workflow is executed, i.e., the expected response time. Moreover, we do not make any restrictive assumptions on the nature of the involved distributions. We propose a novel risk-gain local trade-off mechanism to determine priorities at runtime that optionally can be made even more accurate by employing of conditional means for running activities instead of marginal mean execution times. Finally, the tasks that are unlikely to affect the makespan of an instance are delayed with a local look-ahead to allow incoming new instances to start earlier. We show that adding these features leads to a significant improvement in response time, particularly in situations of scarce processing resources.

1 Introduction

Workflows are collections of coordinated tasks designed to carry out a well-defined complex process [1]. Both in the business and scientific communities a range of workflow management systems, i.e., generic information systems that support modeling, execution and monitoring of workflows [2], have been devised, and languages and tools made available. Yet from the point of view of any possible end user, what matters most are not the internal workings or the provider-side efficiency of the system, but the accomplishment, in the best possible manner, of the complex task represented by the workflow specification when it is enacted on behalf of the user. To achieve this goal, the optimality of this execution must be targeted with respect to a set of objectives that describe measurable resources (time, money) needed or consumed during the execution.

In a dynamic environment, workflow instances can be executed following different actual workflows, even given a common workflow schema. Depending on

S. Cranefield and A. Nayak (Eds.): AI 2013, LNAI 8272, pp. 502–513, 2013.

the use case, the enacted workflow will be parametrized differently and the processing times of the tasks within it as well as the delays required for the communication between the subtasks will vary. The arrival of these workflow instances is governed by a stochastic process, which results in a queue of instances.

We investigate the problem under the assumption that the workflow schema be restricted in structure to an arbitrary directed acyclic graph (DAG). The DAG model is expressive enough to represent any combination of series-parallel groups of tasks and thus, even with this restriction, many real-life workflows can be represented, especially with non-deterministic activity execution times. DAG scheduling problems, with or without structural restrictions, have been the topic of a substantial body of work in the area of *deterministic scheduling in parallel processing* systems. Drozdowski [3] offers a complete monograph, discussing work such as the taxonomy developed by Graham et al. [4] and extended by Allahverdi et al. [5]. Kwok and Ahmad [6] provided an extensive survey for the *deterministic* DAG scheduling problem, including exact solutions for certain very restricted cases, heuristics with a performance guarantee under certain assumptions and, finally, state-of-the-art algorithms which are applicable to a wide range of arbitrary DAG structures with non-unitary communication delays. As it is well known, the last of these approaches do not provide a performance guarantee for any non-trivial instance, but have proved successful experimentally. The Critical Fast Path Duplication algorithm CPFD by Kwok and Ahmad [7] is an example of an algorithm that shortens the makespan by duplicating a preceding task to achieve the earliest possible start time for the succeeding task. For task priority determination, it uses what is termed a dominant sequence (DS). The DS is a total order of all task nodes, constructed around the critical path nodes (CPNs), with an addition of the in-branch nodes (IBNs), which are the parents of the CPNs not belonging to the CP, and of out-branch nodes (OBNs), which are neither CPNs nor IBNs and are added last. The authors later ported the algorithm (HCPFD, [8]) to a $Q|v_i, c_{ij}, prec|C_{max}$ (heterogeneous) setting with a bounded number of uniform processors. The most common benchmark in this kind of environments, however, is the much simpler Heterogeneous Earliest Finish Time heuristic by Topcuoglu et al. [9], which uses b- or t-level task priority lists. It inspired many other algorithms, such as the Critical Path on a Processor (CPOP, [10]), PETS [11] or Push-Pull [12]. None of these algorithms supports any of the distinctive features of the problem herein considered: workflows of stochastic nature and a queuing system. One of the very few algorithms for static scheduling of a *single* stochastic DAG, i.e., the $Q|v_i, c_{ij} \sim stoch, prec|E[C_{max}]$ problem, is SHEFT [13]. It is in practice an upward-rank HEFT where the activity times are expressed by means and standard deviations of the (exclusively exponential) probability distributions.

Research in the *stochastic scheduling* has yielded results in the form of optimal or near-optimal online policies for a set of objective functions, formulated as expectations on random variables such as makespan [14] or weighted completion time of a set of jobs [15]. It is argued by authors in this field (e.g., Pinedo [16], Moehring et al. [17] and Skutella and Uetz [18]) that in stochastic settings the problem naturally reduces to finding policies for online scheduling decisions, and should not be treated as a combinatorial optimization problem. For precedence-constrained job set scheduling problems stochastic scheduling algorithms provide even guarantees (e.g., [19]), however such solutions are far from optimal. Additionally, none of them support problem settings with communication delays.

All algorithms discussed so far schedule single sets of jobs, with or without precedence constraints, in an off-line or on-line fashion. In common workflow enactment scenarios, however, the processing resources form a continuously operated system that processes a periodic inflow of such jobs. Iverson and Ozguner [20] propose an algorithm for the minimization of the average schedule length in a heterogeneous environment that processes DAG-type workflows arriving according to a Poisson process. The DAGs are assumed to be deterministic and identical, and the solution is based on the low-complexity off-line heterogeneous Dynamic Level Scheduling heuristic [21]. Bender and Rabin [22] provide an online scheduler for multiple "competing" task graphs, under several structural restrictions, and optimize the utilization of the processor pool. In addition, their scheduler takes advantage of preemption. Grid-oriented approaches typically also target a system optimum (resource utilization or fairness, [23,24]), instead of a workflow-centric criterion, like makespan or throughput. This perspective has become predominant in the recent years in cloud-targeted approaches, with focus on resource utilization, energy efficiency or provisioning cost. While very challenging resource models are proposed, the application models considered there are oversimplified and rudimentary: mostly trivially parallel tasks (TPTs) that are hardly "workflows" [25,26]. Surprisingly few approaches share their objective function with the current work, while also confronting a problem setting similar with respect to the application model. Gallet et al. [27] provide a static (off-line) deployment algorithm that determines task allocations which are fixed for all workflow instances, while maximizing the throughput of the system, which strictly corresponds to minimizing the response time. Notably, their algorithm is based on the assumption of deterministic processing and communication times.

This paper is structured as follows. Section 2 presents the problem formalization. Section 3 formulates a general online scheduling algorithm structure by decomposing the problem into the sub-problems of choosing the highest priority task to be scheduled, choosing the machine assignment and determining when low-priority tasks can be scheduled. Novel approaches to these sub-problems are developed and combined into two alternative policies, evaluated experimentally against comparable algorithms in Section 4. For this purpose, in addition to the classical benchmarks, two novel parametrized ones are proposed. Section 5 presents conclusions and directions for future work.

2 Problem Formulation

As in all problems of the class $|prec|C_{max}$, let $\mathbb{G} = (\mathbb{V}, \mathbb{E})$, with $\mathbb{V}$ the set of nodes and $\mathbb{E}$ the set of edges connecting them, be a single directed acyclic graph (DAG) representing a process model, or workflow schema. The nodes represent atomic tasks and the directed edges represent precedence constraints between them. Each task has an associated execution time and each edge between two tasks has an associated communication delay. A delay is only caused when an actual transfer of data between two separate processing units is required, i.e., when two task nodes connected by an edge are scheduled to two different processors.

In addition, let all the execution times $w(v_i)$ and communication times $w(e_{ij})$ be described by random variables $w(\cdot)$, with arbitrary probability distributions, having a finite mean and a known cumulative distribution function (CDF). In particular, it is irrelevant whether the distributions are part of any specific family and whether they are continuous or discrete.

Let G be a stochastic process describing the arrival of new DAG instances into the system: it is not required to follow any specific distribution.

Let Q with $|Q| < \infty$ be a set of *uniform* parallel processors and $Q \times Q$ a set of *uniform* communication channels connecting them. As in the reported work on scheduling in parallel processing, we term the processors identical, if the processing times of the same tasks are equal across all processors, and uniform, if the processing times of the same tasks are equal across all processors when multiplied by a factor (speed), that is constant for any given processor. The atomic tasks of the instances in the system are allocated to idle time slots of these processors and executed without preemption. A task, once its execution has commenced on a processor, cannot be interrupted and later resumed.

We assume the existence of a monitoring system that reports the completion of a task and the arrival of a new job instance to the scheduler in real time. Similarly, the centralized scheduler can dispatch tasks instantaneously. The scheduler is not required to generate schedules for tasks that are not yet ready for execution. As there is no reward nor requirement for advance scheduling, the scheduling policies considered are fully online.

As outlined in the introduction, such a system can be described as a queue. In particular, in the Kendall notation [28], it is a $G/G/c$ queue, in which the jobs are the DAG instances, there are $c = |Q|$ homogeneous processors and both the job arrival and departure processes are described by arbitrary distributions.

Finally, let the objective function to minimize be the expected total response time of the system $E[W]$, where W is the sum of the in-queue waiting time W_q (queuing theory notation) and the service time, which is equivalent to the makespan of the instance C_{max} (scheduling notation), as Equation 1 shows.

$$E[W] = E[W_q + C_{max}] = E[W_q] + E[C_{max}] \tag{1}$$

We investigate the performance of different scheduling policies for the given objective as a whole, but design the policies considering the relationship in the above equation to take advantage of the objective's structure.

3 Scheduling Policies

In this Section, we propose two online scheduling policies, Loss-Gain Marginal Mean (LG/MM) and Loss-Gain Conditional Mean (LG/CM). They both provide at their core novel solutions for both the task priority list determination and machine assignment phases in a list scheduler. At the same time, they implement different levels of accuracy in dynamic priority list refinement, which result in different levels of complexity.

In an online scheduling algorithm, whenever a new job instance arrives or the monitor reports the completion of a task at an instant in time τ, for all processors that are idle, a decision has to be made which task to assign to that processor. Let S be the state of the system at τ. Algorithm 1 provides a generic (common) list scheduling algorithm called in a loop by an execution monitor, in accordance with the above approach to the problem. The variable π stands for any possible policy that determines the task priority list and machine assignment and thus shapes the behavior of the algorithm as a whole.

Algorithm 1. Schedule(π, S)

```
1  U ← UnscheduledTasks(S); t_max ← PrioTask(π, U)
2  q_max ← TopSpeed(Q)
3  if t_max is available and q_max is idle then
4      schedule t_max on q_max and remove it from U and S
5  while U not empty do
6      t ← PrioTask(π, U) ▷ consider the next task
7      M ← {} ▷ temporary set of potential task mappings
8      foreach processor q ∈ Q in non-decreasing order of speed do
9          (ST, FT) ← EstimateTime(π, t, q, S, M)
10         M ← M ∪ {(t, q, ST, FT)} ▷ cache a new potential mapping for t to M
11     m* ← ReMap(π, t, S, M) ▷ m* = (t*, q*, ST*, FT*)
12     if m* exists and ST* = CurrentTime(S) then
13         schedule t according to m* ▷ saves m* in the system state
14         remove t from U and S and all mappings for t from M
15     else
16         remove t from U ▷ consider another task
```

To initialize the priority list (*PrioTask*, Algorithm 1, Line 1) we employ two straightforward methods: the workflow instances currently present in the system are prioritized according to their arrival time (i.e., to the FIFO discipline), whereas, within the instances, the dominant sequence method is used, as the most accurate task graph analysis-based prioritizing strategy (refer to Sect. 1). This list is rearranged dynamically by the call (Line 11) to the subroutine *ReMap*, outlined in Algorithm 3. Lines 3-4 of Algorithm 1 implement a short-circuit strategy to immediately schedule the highest priority task t_{max}, if ready for execution, on the fastest processor q_{max}, if it is idle at the moment of the call. Starting with t_{max} if it could not be scheduled, otherwise with the next task in the list, in decreasing priority order, for all the processors, the estimated start and finish times (ST, FT) of that task on the given processor are computed (*EstimateTime*, Line 9) and cached (Line 10).

Unlike all static scheduling algorithms, that are unable to estimate distributed start/finish times and are not designed to take advantage of information available at runtime, the proposed scheduler with *ReMap* does not follow a strict priority list and is able to exploit runtime information, with different accuracy levels. To this end, we provide a mechanism to compute such an estimate that incorporates the actual task execution and inter-task communication times when available. In each subsequent call (Algorithm 1, Line 9; Algorithm 3, Lines 10-11, 18) to the corresponding function *EstimateTime* typically more actual values become available. Due to lack of space we omit the listing of this straightforward function, instead hinting at the fact that, to handle probabilistically distributed times of possibly ongoing activities, *EstimateTime* calls a subroutine *Weight*(π, *S*, $w(\cdot)$), where π is the policy selector, S is the system state, $(\cdot) \in \mathbb{V} \cup \mathbb{E}$ is a task or edge, $w(\cdot)$ is the corresponding stochastic variable. We propose and evaluate here two different specific implementations of *Weight*, that distinguish LG/MM and LG/CM. Note that the initial priority list (Algorithm 1, Line 1) and the way it is dynamically refined (Line 6) are also affected by the different values returned by the two implementations of *Weight*.

Algorithm 2. Weight(π, S, $w(\cdot)$)

1 $\tau \leftarrow CurrentTime(S)$
2 **if** $ST(\cdot) = null$ *or* π=*LG/MM* **then** ▷ $(\cdot)$ has not started yet or LG/MM
3   **return** $E[w(\cdot)]$ ▷ marginal mean of the distribution
4 **else**
5     $v_q \leftarrow Speed(Processor(S, (\cdot)))$ ▷ processor speed for normalization
6   **if** $FT(\cdot) < \tau$ **then** ▷ $(\cdot)$ is complete
7   **return** $\frac{FT(\cdot)-ST(\cdot)}{v_q}$ ▷ the actual execution time, normalized
8   **else** ▷ $(\cdot)$ has not started but not completed
9         $\tau_n \leftarrow \frac{\tau-ST(\cdot)}{v_q}$ ▷ normalized runtime until τ
10   **return** $E[w(\cdot)|w(\cdot) \geq \tau_n]$ ▷ conditional mean of the distribution

The implementation of *Weight* in LG/MM is based on the known marginal means of the involved probability distributions (Algorithm 2, Lines 2-3): it has thus the advantage of relatively low complexity. By contrast, in LG/CM, for running activities (Line 8), conditional means are derived from the known cumulative distribution functions, which mathematically is illustrated in Equation 2: $F(x)$ is the original known (continuous) CDF for $w(\cdot)$; the discrete equivalents are straightforward.

$$E[w|w \geq \tau_n] = \frac{\int_0^\infty F(\tau_n + r)dr}{1 - F(\tau_n)} \quad (2)$$

Both algorithms use either mean values in the absence of information about actual execution times, that is otherwise available through the execution monitor (Lines 6-7). As we compare means of *intrinsic* activity weights (processing/communication times) with actual running times on a set of processors with different speeds, a normalization step is necessary (Lines 5,7,9).

The idea of a trade-off-based mechanism is motivated by the properties of the makespan objective that are particularly relevant in a stochastic setting. As the minimization of the *global* makespan is an NP-complete problem, most heuristics minimize a *single local objective* instead: the earliest finish time (EFT) of the highest priority task considered for scheduling at a given step. In the general case, this approach results in leaving (wasted) idle time on the processors, as tasks of

Algorithm 3. ReMap(π, t, S, M)

```
1  FT_min ← ∞; m_min ← null
2  foreach mapping m = (t, q, ST, FT) ∈ M do
3      if FT < FT_min then
4          foreach mapping m' = (t', q, ST', FT') ∈ M \ {m} do
5              l ← Priority(t); l' ← Priority(t')
6              if l ≥ l' then
7                  FT_min ← FT; m_min ← m
8              else
9                  schedule t according to m in a copy of the state S*
10                 (ST*, FT*) ← EstimateTime(π, t, q, S*, M)
11                 (ST'*, FT'*) ← EstimateTime(π, t', q, S*, M)
12                 ΔFT ← (FT* − FT); ΔFT' ← (FT'* − FT')
13                 θ ← l'ΔFT' / lΔFT
14                 if θ < ε then
15                     FT_min ← FT; m_min ← m
16 if IsOBN(t, S) then
17     CP ← CriticalPath(Instance(t), S)
18     (ST*, FT*) ← max_{t_CP ∈ CP} EstimateTime(π, t_CP, q, S, M)
19     m_min ← (t, q(m_min), ST*, FT*)
20 return m_min
```

high priority that are not ready for execution may cause the scheduler to wait. On the other hand, schedules with less interleaved idle time are obviously closer to the optimum, which suggests to simultaneously minimize a *second local objective*: the amount of unused idle time. In the "classical" deterministic setting, assuming that the priority list is accurate, the only sensible strategy is to optionally insert a lower priority task into an idle time slot, if that slot is not smaller than the task's execution time on that processor: otherwise it is guaranteed that a higher priority task will be delayed. In the herein considered stochastic setting, however, the activity execution times are probabilistic. Therefore it is possible to compute a probability distribution of the delay potentially caused (i.e., of the risk to the EFT objective) and, conversely, the distribution of the gain in terms of the idle time recovered. The trade-off mechanism balances these two local objectives based on measures outlined below and, in more technical detail, works as follows.

In Algorithm 1, Lines 9-10, the estimate start and finish times for the currently considered task were cached, *without* direct *regard for* previous estimates for all the *higher priority tasks* that were not ready for execution and thus could not be scheduled. For each case where a task t is ready at the moment of the call (Algorithm 3, Lines 1-2), but scheduling it might cause a conflict with a future mapping for a task t' of higher priority (Line 8), a new estimate of the finish time of t' is computed under the assumption that t would be scheduled now (look-ahead for t': Line 11). $\Delta FT'$ is the estimated difference of the finish times for t', between the case when t is greedily scheduled now and the case when the given processor is left idle waiting for t' to be ready. Conversely, the potential gain is expressed by the improvement ΔFT in the finish time of t, should it be scheduled now. The original priorities l, l' act as a discount factors to compute the priority-discounted gain and loss risk, respectively, where the loss is a potential delay of the higher priority task t'. The trade-off is expressed by θ, which is compared to a fixed threshold value ϵ.

Unlike in makespan-centric online and offline algorithms, we also exploit the known decomposition of the expected response time objective into expected schedule length and waiting time (Equation 1). In a simplified formulation, the tasks that do not impact the critical path of a workflow instance, and thus are less relevant for the expected makespan of that instance, need not be scheduled to minimize their earliest finish time. This is reflected in Lines 16-19 of Algorithm 3. It is sufficient that these tasks (typically the out-branch nodes (OBNs) of the graph) be scheduled to finish at or slightly before (Algorithm 3, Line 19) the estimated finish time of the latest critical-path task (obtained in Line 18). Deferring these tasks does not harm the makespan sub-objective of the more privileged workflow instance (the instance that arrived into the system earlier) but, by allowing to schedule the highest priority task(s) of another (less privileged) instance sooner, reduces the waiting time of the less privileged instance.

The worst-case complexity of both policies is: $O(p(vO(E))^3)$, where p is the number of processors, v the number of unscheduled tasks in the system and, conventionally, $O(E)$ is the complexity of the mean determination (simply equal to one for LG/MM and, for LG/CM, linear in the number of the support points in the discrete representation of the CDF).

4 Experimental Results

For evaluation purposes, we created a set of 500 graphs and, for each single graph from the above set, 100 realizations, totaling 50000 different workflow instances. Each graph has an arbitrary structure, randomly generated by a procedure adapted from Topcuoglu et al. [9] and parametrized as follows:

- number of tasks $v \in \mathbb{N}$, chosen randomly from the interval [10, 25],
- maximum alpha α_{max} of 0.1, 0.25, 1, 4, 10; the actual values of α uniformly distributed in the interval $(0, \alpha_{max}]$, where $\alpha\sqrt{v}$ is the mean graph height,
- maximum out-degree of a node of 10 and actual out-degree uniformly distributed in the interval [1, max],
- mean activity execution times $E[w(\cdot)]$ uniformly random in [1, 10],
- computation-to-communication ratio (CCR) of 0.1, 0.2, 1, 5, 10.

The generated activity execution times are used to extrapolate arbitrary continuous four-parameter Beta distributions, which are capable of emulating many common continuous distribution families, symmetric and asymmetric. Each such distribution has a mean $\mu = E[w(\cdot)]$ and is extrapolated as follows: the support is defined to be in $[\frac{\mu}{2}, 10\mu]$ and a normalized standard deviation $\frac{\sigma}{\mu}$ is prescribed with different values, taken from the set $\{0.1, 0.25, 0.5, 0.75, 0.9\}$. Thus each instance has the same structure and precedence constraints, but different activity execution times. For each graph and number of processors, an estimate λ_{max} of the critical arrival rate is computed in the pre-simulation phase and the actual values of the average arrival rate λ are calculated as $\lambda = \lambda_{mul}\lambda_{max}$, where the factor λ_{mul} is chosen from {0.25, 0.5, 0.75, 1}. Similarly, the number of processors $|Q|$ is defined as a set of multipliers $p_{mul} \in \{0.25, 0.5, 0.75, 1\}$ of the nodes v of the task graphs as $|Q| = \lceil p_{mul} v \rceil$. The processor speeds are chosen as uniform random values in the interval (0,1], as are the speeds of the processor-to-processor physical channels.

The proposed scheduling policies LG/MM and LG/CM are evaluated by means of simplified competitive analysis (CA) [18]. We apply, for each policy π, a performance measure $E^N_{n_\pi}[W]$, expressed by the ratio of the total response time W (for all the N=100 instances) to the total processing time required for all the task realizations in each simulation run, i.e., we apply the following formula:

$$E^N_{n_\pi}[W] = \frac{\sum_{i=1}^{N} W_{\pi_i}}{\sum_{i=1}^{N} \sum_{v \in V} r_i(w(v))} \tag{3}$$

To evaluate the effectiveness of both LG/MM and LG/CM, we run the above defined set of simulations applying three other online scheduling algorithms. All of these, in the same way as LG/MM and LG/CM, employ the natural FIFO instance ordering in accordance with the arrival process. The first one is a multi-instance version of HEFT [9], which uses a static task priority list (identical across the instances). Given the stochastic setting, in this case, the priority list is computed by using the means of the distributions of the activities' durations, hence the designation mHEFT-M. The second one is mSHEFT, a multi-instance version of SHEFT [13]: as in the original, for the appropriate rank and priority calculations the sums of the means and standard deviations are used. For reference, a policy RND was added, which randomly assigns the highest priority to one of the tasks in the set of unscheduled tasks.

Overall, LG/CM performed best across almost all parameter ranges. The average measured total response time, normalized by the processing time required for all task realizations at the speed of the fastest processor, was approx. 12% shorter when scheduled by LG/CM compared to both mHEFT-M and mSHEFT, considering that mSHEFT only outperformed mHEFT-M by 0.3% in our simulation. Of the two proposed algorithms, LG/CM outperformed LG/MM by 4%. The improvement over a random strategy (RND) for LG/CM was around 22%.

As Figure 1 shows, LG/MM and LG/CM, perform clearly better than the competitors, notably in settings with scarce processing power (lower multiplier or at higher arrival rates). As the availability of processing power grows, the differences between all algorithms naturally disappear. In the range of arrival rates examined, LG/CM is almost insensitive to the change of the rate, whereas the response times grow for all other policies. However, with growing uncertainty as expressed by a larger standard deviation, all algorithms face similar problems, although LG/MM and LG/CM slightly outperform the reference algorithms, while RND falls well behind. LG/CM performs better than LG/MM at high standard deviations and small computation-to-communication ratios (CCR); growing CCR values give rise to longer response times for all algorithms.

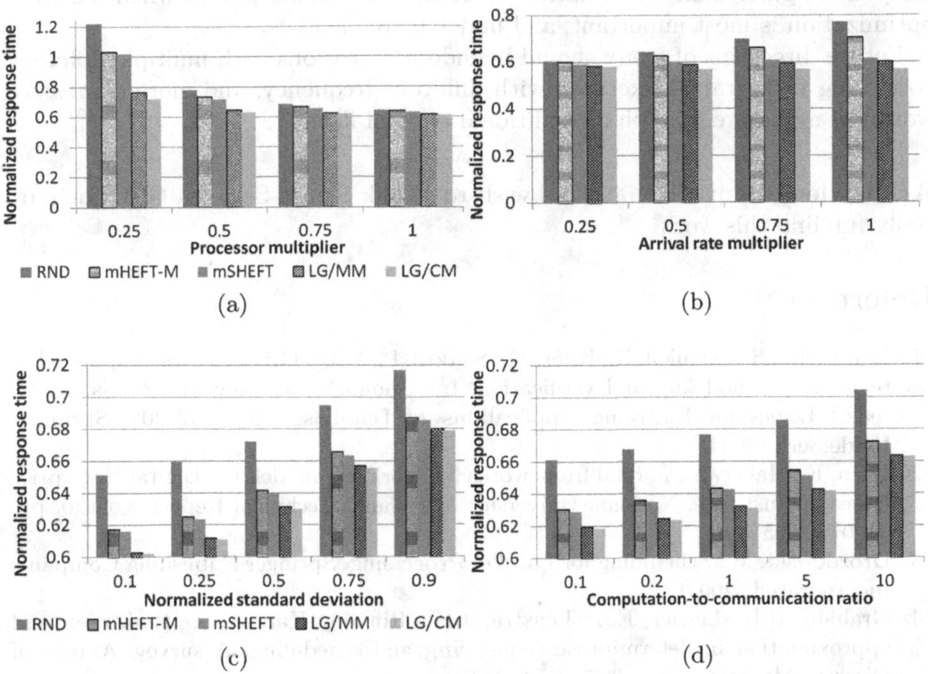

Fig. 1. Average total response times $E_n[W]$, normalized to the total processing time required for all the task realizations, as a function: (a) of the processor multiplier p_{mul}, given $|P| = \lceil vp_{mul} \rceil$; (b) of the multiplier λ_{mul} of the critical arrival rate, given $\lambda = \lambda_{mul}\lambda_{max}$; (c) of the normalized standard deviation of the task execution and inter-task communication times; (d) of the computation-to-communication ratio

5 Conclusions and Future Work

In this work, we investigated online policies for the scheduling of workflow task graphs to a heterogeneous set of processors connected by physical communication channels of different speeds, a problem that typically arises in a cloud computing environment. The task graphs are precedence-constrained and have uncertain communication delays as well as non-deterministic computation times. Additionally, we assumed the existence of a queue of workflows, with a prescribed FIFO discipline; the job instances are processed in the order they arrive. Minimizing the total response time is the objective of the tested policies, which, in summary, leads to a $G/G/|Q| : Q|v_i, c_{ij} \sim stoch, prec|E[W]$ problem formulation.

We have evaluated two new alternative policies that both extend static priority assignment methods with novel ways of handling information available at runtime. The algorithms combine a sophisticated trade-off-based mechanism which determines when tasks of lower priority should be scheduled greedily in the presence of higher priority tasks with a look-ahead mechanism designed to schedule new workflow instances in advance. The algorithms outperform the state-of-the art policies particularly in situations of scarce processing power, when schedule optimization is most important, and under heavy load.

Future directions of study should include applications with multiple different coexisting task graphs, executed with different frequency, and more expressive workflow models, e.g., with a conditional control flow.

Acknowledgment. The authors wish to thank Smart Services CRC for partially funding this work.

References

1. Mukherjee, S., Davulcu, H., Kifer, M., Senkul, P., Yang, G.: Logic-based approaches to workflow modeling and verification. In: Chomicki, J., Meyden, R., Saake, G. (eds.) Logics for Emerging Applications of Databases, pp. 167–202. Springer, Heidelberg (2004)
2. Oren, E., Haller, A.: Formal frameworks for workflow modelling. Digital Enterprise Research Institute, National University of Ireland, Technical Report, vol. 20, pp. 04–07 (2005)
3. Drozdowski, M.: Scheduling for Parallel Processing. Springer Publishing Company, Incorporated (2009)
4. Graham, R.L., Lawler, E.L., Lenstra, J.K., Rinnooy Kan, A.: Optimization and approximation in deterministic sequencing and scheduling: A survey. Annals of Discrete Mathematics 5, 287–326 (1979)
5. Allahverdi, A., Gupta, J.N., Aldowaisan, T.: A review of scheduling research involving setup considerations. Omega 27(2), 219–239 (1999)
6. Kwok, Y.-K., Ahmad, I.: Static scheduling algorithms for allocating directed task graphs to multiprocessors. ACM Computing Surveys (CSUR) 31(4), 406–471 (1999)
7. Ahmad, I., Kwok, Y.-K.: On exploiting task duplication in parallel program scheduling. IEEE Transactions on Parallel and Distributed Systems 9(9), 872–892 (1998)
8. Kwok, Y.-K., Ahmad, I.: Exploiting duplication to minimize the execution times of parallel programs on message-passing systems. In: Proceedings of the Sixth IEEE Symposium on Parallel and Distributed Processing, pp. 426–433. IEEE (1994)
9. Topcuoglu, H., Hariri, S., Wu, M.-Y.: Task scheduling algorithms for heterogeneous processors. In: Proceedings of the Eighth Heterogeneous Computing Workshop, pp. 3–14 (1999)
10. Topcuoglu, H., Hariri, S., Wu, M.-Y.: Performance-effective and low-complexity task scheduling for heterogeneous computing. IEEE Transactions on Parallel and Distributed Systems 13(3), 260–274 (2002)
11. Ilavarasan, E., Thambidurai, P.: Low complexity performance effective task scheduling algorithm for heterogeneous computing environments. Journal of Computer Sciences 3(2), 94–103 (2007)
12. Kim, S.C., Lee, S., Hahm, J.: Push-Pull: Deterministic search-based DAG scheduling for heterogeneous cluster systems. IEEE Transactions on Parallel and Distributed Systems 18(11), 1489–1502 (2007)
13. Tang, X., Li, K., Liao, G., Fang, K., Wu, F.: A stochastic scheduling algorithm for precedence constrained tasks on grid. Future Generation Computer Systems 27(8), 1083–1091 (2011)

14. Coffman Jr., E., Flatto, L., Garey, M., Weber, R.: Minimizing expected makespans on uniform processor systems. In: Advances in Applied Probability, pp. 177–201 (1987)
15. Rothkopf, M.H.: Scheduling with random service times. Management Science 12(9), 707–713 (1966)
16. Pinedo, M.: Offline deterministic scheduling, stochastic scheduling, and online deterministic scheduling: A comparative overview. In: Handbook of Scheduling — Algorithms, Models, and Performance Analysis (2004)
17. Möhring, R.H., Schulz, A.S., Uetz, M.: Approximation in stochastic scheduling: The power of LP-based priority policies. Fachbereich Mathematik, TU Berlin, Tech. Rep. 595-1998 (1998)
18. Skutella, M., Uetz, M.: Stochastic machine scheduling with precedence constraints. SIAM Journal on Computing 34(4), 788–802 (2005)
19. Chandy, K.M., Reynolds, P.F.: Scheduling partially ordered tasks with probabilistic execution times. ACM SIGOPS Operating Systems Review 9(5), 169–177 (1975)
20. Iverson, M., Ozguner, F.: Dynamic, competitive scheduling of multiple DAGs in a distributed heterogeneous environment. In: Proceedings of the Seventh Heterogeneous Computing Workshop, pp. 70–78. IEEE (1998)
21. Sih, G.C., Lee, E.A.: A compile-time scheduling heuristic for interconnection-constrained heterogeneous processor architectures. IEEE Transactions on Parallel and Distributed Systems 4(2), 175–187 (1993)
22. Bender, M.A., Rabin, M.O.: Online scheduling of parallel programs on heterogeneous systems with applications to cilk. Theory of Computing Systems 35(3), 289–304 (2002)
23. Zhao, H., Sakellariou, R.: Scheduling multiple DAGs onto heterogeneous systems. In: Proceedings of the 20th International Parallel and Distributed Processing Symposium, pp. 14–27. IEEE (2006)
24. Wieczorek, M., Siddiqui, M., Villazon, A., Prodan, R., Fahringer, T.: Applying advance reservation to increase predictability of workflow execution on the grid. In: Proceedings of the Second IEEE International Conference on e-Science and Grid Computing, pp. 82–89. IEEE (2006)
25. Calheiros, R.N., Ranjan, R., Buyya, R.: Virtual machine provisioning based on analytical performance and QoS in cloud computing environments. In: Proceedings of the International Conference on Parallel Processing, pp. 295–304. IEEE (2011)
26. Muthuvelu, N., Vecchiola, C., Chai, I., Chikkannan, E., Buyya, R.: Task granularity policies for deploying bag-of-task applications on global grids. Future Generation Computer Systems 29(1), 170–181 (2013)
27. Gallet, M., Marchal, L., Vivien, F.: Efficient scheduling of task graph collections on heterogeneous resources. In: Proceedings of the IEEE International Symposium on Parallel & Distributed Processing, pp. 1–11. IEEE (2009)
28. Kendall, D.G.: Stochastic processes occurring in the theory of queues and their analysis by the method of the imbedded markov chain. The Annals of Mathematical Statistics 24, 338–354 (1953)

Author Index